The Carpenter's Manifesto

Jeffrey Ehrlich and Marc Mannheimer

THE CARPENTER'S MANIFESTO

Drawings by Marc Mannheimer

Holt, Rinehart and Winston New York

To Bonnie and Lenore for their wisdom, love, and support

Published simultaneously in Canada by Holt, Rinehart
and Winston of Canada, Limited.
Library of Congress Cataloging in Publication Data
Ehrlich, Jeffrey.
The carpenter's manifesto.
Includes index.
1. Carpentry—Amateurs' manuals. I. Mannheimer,
Marc, joint author. II. Title.
TH5606.E38 1977 694 77–73865
ISBN Hardbound: 0–03–016756–6
ISBN Paperback: 0–03–016761–2

Designed, printed, and produced in association with
Chanticleer Press, Inc., New York, N.Y. 10017

Designer: Henry Altchek

Printed in USA

10 9 8 7 6

ISBN 0-03-016756-6 HARDBOUND
ISBN 0-03-016761-2 PAPERBACK

Acknowledgments

We especially want to thank David Frederickson, that good-natured gentleman of taste and precision, for his long hours of work and his magnificent job of editing this book. A finer editor is impossible to imagine.

Our gratitude goes also to Ellyn Polshek, our senior editor, for bringing David and us together, and for her tactful suggestions and enthusiastic support for our project from the very start and to Sally Smith, Ellyn's invaluable and imperturbable aide-de-camp.

Finally, the suggestion and general design for our potting table project come from Kathy Wenderoth and Lenore Stein; the name "finger chompers," for certain types of tools, comes courtesy of Howard Schnurnberger; and our title from Chuck Lindholm.

CONTENTS

INTRODUCTION

People seem to think carpentry is a big mystery. Smart as they are in their daily occupations, their brains brake when confronted with a carpentry problem. It doesn't need to be that way.

When one of your wise authors was four years old, he watched a carpenter named Schuster plane a board. He asked Schuster to let *him* try, but he couldn't get the blasted thing to cut at all. Schuster took the plane back and said, "No, no, you're doing it all wrong. You've got to wiggle your ears while you do it or it won't work." Schuster wiggled his ears and the plane started to take off big shavings. Your author unfortunately didn't know how to wiggle his ears and just couldn't get the plane to work. He was midway through college before he realized his ears probably had nothing to do with it.

We're both carpenters now, but one of us is also an artist, and the other is a writer and sometime musician. Therefore we feel we have a foot (or two) in both worlds, that of the confident professional carpenter and that of the puzzled amateur. So we believe we're in a good position to write the *Carpenter's Manifesto*—a book with which you can *learn* how to design and build your own projects, start to finish.

We've tried to avoid the usual jargon in both word and picture, while still explaining all the principles and techniques that the nonprofessional might ever need. We also explain hundreds of little tricks toward better work, tricks we've learned from daily experience.

We usually work in people's homes, without the benefit of a shop, without special vises or table tools, without a workbench. We use hand tools or portable power tools, and we work on the floor or, if we're lucky, on a couple of sawhorses. You don't have to have a shop to build anything in this book.

Most interesting carpentry projects may seem mysterious at first, their structure and manner of construction a big unknown even to professional carpenters. Our main goal in this book, therefore, is to communicate to you a *way of thinking, of seeing*—a method of approaching and solving carpentry problems.

Anyone can work with wood. If you can run a sewing machine, drive a car, or adjust the dials on your TV, you can do anything described in this book.

The notion that carpentry is something mysterious or beyond the nontechnical mind is wrong. We've found that the two types of carpentry books usually available haven't helped to dispel that prejudice. First there are the heavy technical tomes for the professional. Some are excellent, but they assume you have five years of fundamentals behind you, and they are therefore incomprehensible to most people. Then there are the light-reading crafts books, the how-to books, which may tell you how to build certain projects (a colonial-style magazine rack, a Danish-modern box), but little else. When you're done, you still have no idea what carpentry is *about,* how to design and build something on your own. The

Carpenter's Manifesto, we hope, will teach you not only how to make things but how to understand the principles and techniques behind the making.

You'll see in the Contents that we've divided this book into five parts, each being one step in our overall approach to any project. First is the Beginning, naturally enough, where you'll learn how easy it is to structure and design your own projects. Essential to any design is the Wherewithal (Part II), the materials and tools to work with. While the basic use of each tool is covered here, the next part, Know-How, contains the more general techniques and procedures of actual carpentry work, such as how to make joints, how to lay out complicated shapes, and so on.

Manifestations is Part IV. Here we show how all the knowledge and techniques of the previous three parts go together to complete actual projects—including specially designed furniture pieces and general construction jobs such as new walls, floors, and ceilings.

In Finishing we explain the various types of protective finishes for your work, and how to apply them so they'll last.

We think several of the chapters are particularly important to our overall approach.

"Structure" (Chapter 1) explains the simple principles on which all construction, no matter how complex, is based. Once you know these principles, you'll be able to design and build just about anything you want. You'll know how to make things strong, how to support heavy weights, what wood joints are all about, and much more. We feel a knowledge of structure is crucial for good carpentry, which is why we spend a whole chapter on it. Which, to our knowledge, no other carpentry book has done.

"Drawing" (Chapter 2) is another special feature, in which we show you quick, simple ways to draw the shapes and objects in your mind. You *can* learn to draw—it's no mystery. Good plans, no matter how primitively drawn, are essential. It's amazing how many ways a project can go wrong when you have no plan to work from.

The Tool chapters (Chapters 5 and 6) cover all the tools you might ever need and how to use them. We also try to tell you whether you *do* need them or not, based on what kind of work you plan to do; and if you do need one, what to spend for what you need.

The Furniture Projects we selected (Chapter 10) are the ones we are most often asked to build for people. We take you step by step through the entire life of a project, beginning from the moment you first notice piles of junk around the house and decide you need something to put them in. Our designs are simple and efficient—yet not without class. Since each project is organized to give you an understanding of its principles, you can change the design easily to suit your own needs. In any case, complete plans and construction steps are included for each basic project, either with detailed variations or suggested modifications.

Two questions that people constantly ask us are how they can attach something to their walls, and how they can soundproof their homes. "Walls and Ceilings" (Chapter 7) is our answer to the first question. The Soundproofing section in "Room Renewal" (Chapter 11) is our answer to the second one—it explains the different kinds of sound you have to deal with, how they travel, how to control them, and when, unfortunately, it makes more sense just to give up.

Don't neglect the Reference section, including the Glossary and Recommended Books, at the back of the book. The Reference tables are full of useful information on design, materials, and construction. Strange words are usually defined the first time they appear, but if you're skipping around and sneak up on one, the Glossary or Index is useful.

Consider the Index an essential part of the book. One thing we found while writing this book is that it's impossible to put *all* the information about one subject in just one place—nails, for instance, come up in "Hardware," "Structure," "Working," and many other spots throughout the book. The Index, along with the Contents and the cross-references we've included in the text, will help you find the information you want.

Good luck. Now let's begin.

1" = 1"

BEGINNING

1/STRUCTURE

Why Things Stand up and How to Make Them Strong

Structure is the guts of carpentry. It's the soul of your work, the built-in strength that withstands everyday stresses. The concepts of structure in this chapter are basic and relatively easy to grasp. Once you understand them, you'll be able to design and build more efficiently, more economically. And what you build will be strong and long-lasting.

You may ask, "How strong do I have to make it? How much trouble do I have to go to?" That's up to you. Think about the structural principles in this chapter. Look around at other things that are similar to what you want to build. See how weak or strong they feel to you, look at how they're put together. Check the Span charts in the Reference section, if relevant to your project. Look at our designs. Think.

Our own tendency is to overstructure, to make things stronger than they need be. That's our prejudice. We feel the small amount of extra work and materials is well worth it in the long run. You can build a bookshelf that will support the several hundred paperbacks you now own. But what happens if you inherit an encyclopedia? Will that mean you're going to have to build a stronger bookcase? What if six people sit down on your bed during a party—will it collapse? In that sense it's always better to overstructure. It certainly can't hurt.

On the other hand, don't get carried away. If you're planning to do a lot of rough carpentry and you need a temporary work table, just build a minimal structure. As long as it doesn't collapse, it's all you need. Figure your present requirements and whether they might change in the future.

There are any number of ways to build a loft bed, as we'll show later. If you're planning to move after your two-year lease is up, or if you're in a big hurry, you can build it one way. The loft bed might creak and even sway under certain conditions, but it will last the two years. But if this is your own home and you want a good piece that will last, or if you simply like doing good work, there are better ways to build it. We forget what good work is like sometimes and end up accepting automobiles and department-store furniture that self-destruct in a few years.

Fig. 1 *Fig. 2*

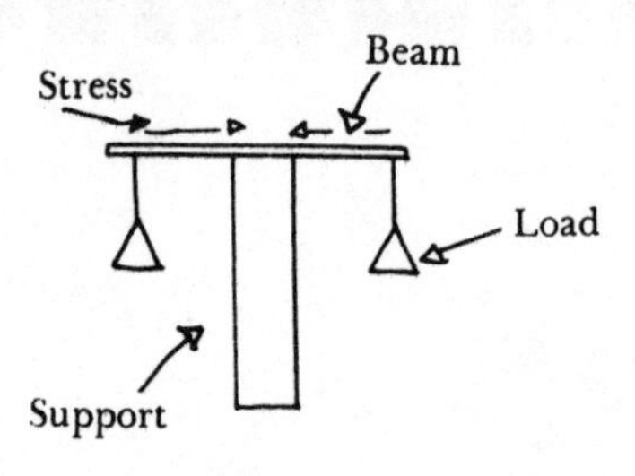

Fig. 3

Diagram of "Woman with Pails," structural system

Elements of Good Structure

Dealing with the Work as an Entire Structure

We see a woman walking on the road, carrying a pail of milk in each hand (*1*). Before long her arms start to get tired. She stops, thinks a moment. Then she picks up a good strong pole by the side of the road, hangs a pail on each end, and lays the pole across her shoulders (*2*). She smiles and walks on.

This woman has designed a structure that uses the strength of her whole body, rather than that of just her hands and arms. Actually we are all familiar with this type of "structure" solution. A backpack uses the same principle—we might say we have *transferred the weight.*

Similarly, with your own projects, find a way to make the *entire structure* work for you, not just some of its parts. This will enable you to use fewer materials, span longer distances, support more weight, and probably get a lot more enjoyment out of building and living with your work.

Transferring the Weight

We can clarify what it means to say "transfer the weight" in the pole-and-pails example, and also illustrate some basic structural principles, by looking at several other load examples. First, let's understand some basic terms.

When we look at the woman with the pole, we see in structural terms a *beam* balanced on a *support* while under *stress* from a *load* (*3*). That only *sounds* complicated.

The woman's back is the *support,* the *beam* is the pole. Look at the drawings; all of these are beams. The *load* that is being exerted on the pole is the weight of the pails of milk. Two more pails would double the load, and so on. The *stress* is that force that makes the pole bend, the force exerted on the pole by the load of the pails. Look at the other examples of beams in Figure 4; the same principles apply to them.

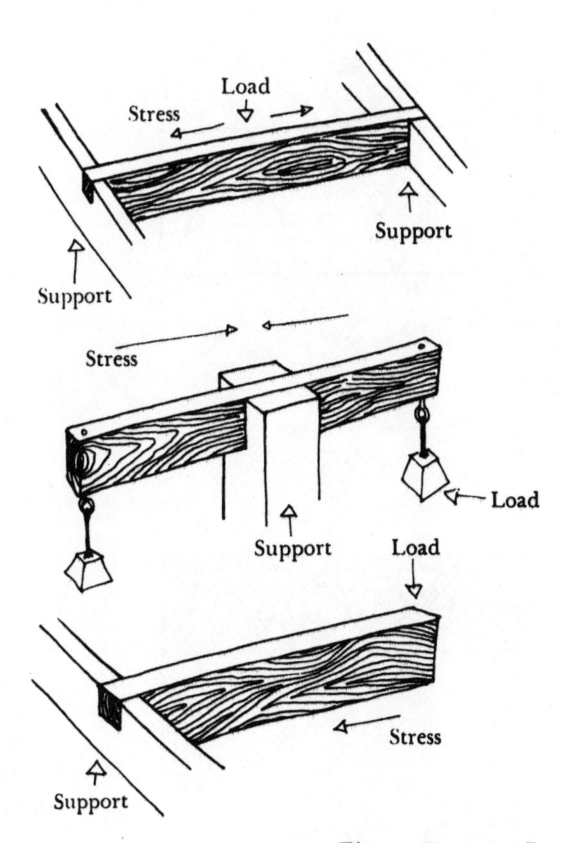

Fig. 4. Types of Beams

A ten-year-old who weighs 60 pounds runs across a frozen pond. His big brother, who weighs 150 pounds, runs after him. The ice starts to break under the big brother. His weight—the load he places on the ice—is more than twice that of his brother. The ice cannot withstand the increased stress. If the older brother lies down on the ice and slithers back to shore, he can probably manage not to break through the ice. His weight has remained the same, but he has increased the area of the ice on which he is directly placing his weight. He has *transferred his weight* to the surface of his whole body. He has thus decreased the stress on the ice directly beneath him by spreading his total load over a greater area.

Under your floorboards there are horizontal beams, or joists. They support the entire floor. Without them, your floor would sag and break; it would be as weak structurally as a sheet of ice (though ice is more brittle). The floor and beams together act as one total structure. The floor takes up the stress from your load, disperses it to all the beams and on out to the supporting structure.

If the maximum safe load for your floor is one elephant, that elephant could walk through your room and the floor would be fine. If you took away the support of the joists, the elephant, of course, would fall through the floorboards—the load could not be efficiently passed from the floorboards to the walls of the house. The thin floorboards, like a sheet of ice, would break.

Just as important, however, is the structural role of the floorboards. If you left the joists and removed the floorboards, and the elephant walked out on one joist, that joist would break. The floorboards are needed as a skin to tie all the joists together into one structure, to disperse the load equally to all of them.

The older brother can't do anything about the structure of

the ice; he has to rearrange his load so it is transferred to the ice along a larger area of contact. We can, however, build a floor structure that will take the load of the elephant transferred through the feet and disperse the stress from that load efficiently to the joists and out to the supporting walls.

This is what we always want to accomplish with good structure—either to transfer the load better or to disperse the stress efficiently over the whole structure. The distinction may seem vague at times, because it often is—most structures are a combination of the two principles.

Stress and How It Works upon a Beam of Wood

You will find it easier to work with the principle of dispersion through structure if you understand exactly *how* stress travels along a beam from one place to another. How *do* the supports on the *ends* of the cross beam manage to support the load in the *center* of the beam?

Basically, the stress from the load travels along the beam in a zigzag motion between the load and the supports (*5*). If you look at the lattice beam in a bridge (*6*) you can see the skeleton of what is actually happening within the wood fibers of a wooden beam. The stresses actually travel along these diagonal paths. The material that would normally be inside those triangular openings was eliminated because it wasn't needed. The supports at the ends of the beam exert forces equal—but opposed—to that of the load; such forces are transmitted along those same diagonal paths.

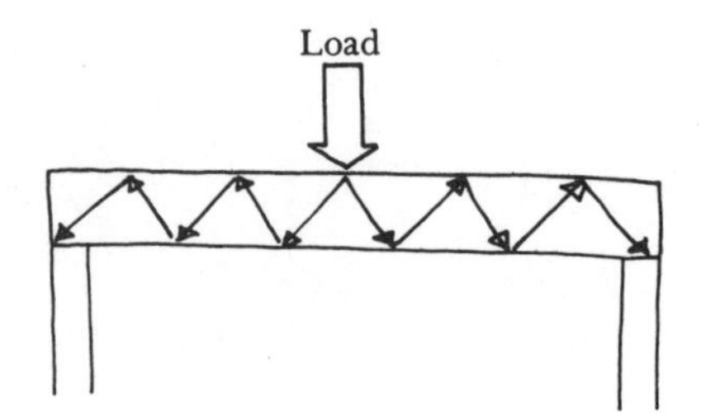

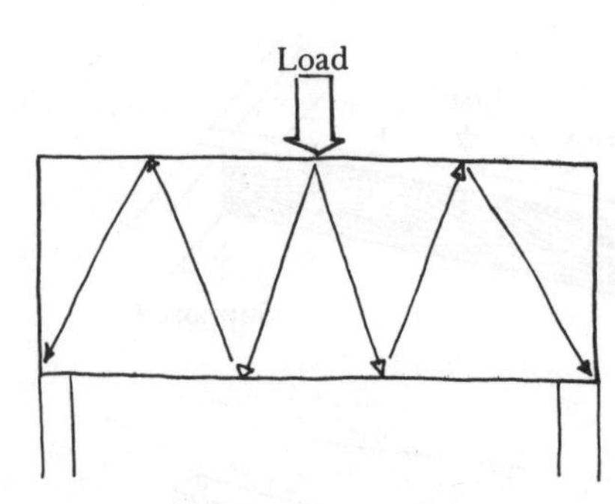

Fig. 5. Stress Paths

This concept of *equal resistance* is important. For every stationary load, there is a force of equal resistance pushing back. When you stand on the ground, your weight wants to force you downward, but the earth pushes back with a force equal to your weight. A good structure provides efficient paths for stresses to pass in both these directions.

There are also horizontal stress paths along the top and bottom of a beam, but they do not directly support the load. They tend mainly to keep the beam from bending. A large enough load will, of course, exert enough stress to bend or even break the beam. There are limits to the stress that can be dispersed along these paths. The Span Charts at the end of the book will show you how long and how thick a beam you need to use for different loads and spans.

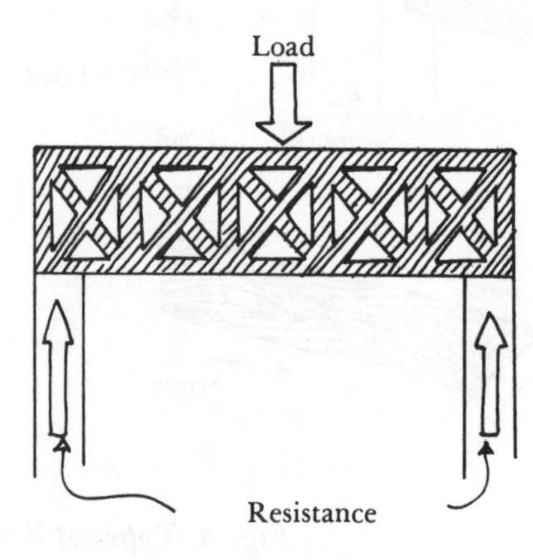

Fig. 6. Lattice Beam

Stress Dispersal in the Pole and Pail

If we return now to our friend with the two pails of milk and the pole, we can see more clearly what is happening structurally. The weight of the pails constitutes a load that creates stress on the pole at either end. This stress is transferred along the pole in zigzag fashion until it reaches the woman's shoulders, then along her body and down to the ground. The stress then returns, along the same lines, as resistance from the earth. What she has done structurally is to disperse the stress from the load to the strongest part of the supporting structure, in this case her whole body.

When she carries the pails in her hands, she is using only a small amount of the strength of her body. The load in her hands puts *tension stress,* a pulling force, primarily on her arm muscles and joints. You have felt this tension stress yourself whenever you've had to carry a suitcase for too long.

There is a reason that tension stress is not dispersed efficiently along the arms to the shoulders and rest of the body. The human body does not transmit tension forces nearly as efficiently as it does *compression,* or pushing, forces. This is why you hear of pulled muscles, but not pushed muscles. When the woman holds the pails in her hands, the stress is taken up mostly by her arm muscles before it reaches her shoulders. Her arms tire long before her body feels the strain. The pole, however, disperses the load directly onto her body through her shoulders as a compression load.

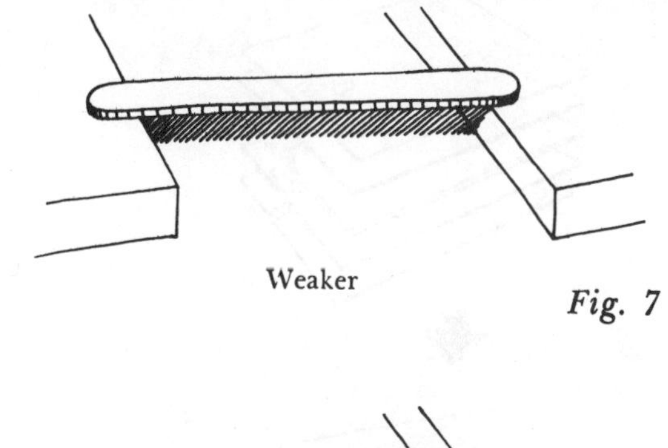

Fig. 7

Using Beams

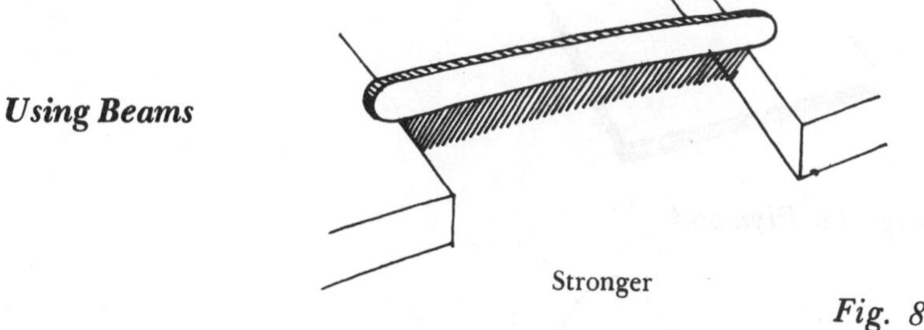

Fig. 8

Beams are always laid with the wider dimension as the vertical. That is, a 2x12 would be laid with the 12″ dimension vertical. You can see why with a quick experiment. Take a thin piece of wood, like an ice-cream stick or a ruler. Lay it flat across the gap between two books and press on it with your hand (*7*). Then do the same with the piece laid on edge (*8*). The wood will bend much more easily the first way.

The stress is dispersed outward to the supporting structure more easily in a beam with more vertical height. You might say (with something less than scientific rigor) that since the stress flows in a zigzag path, the more vertical height there is, the fewer zigzags the stress is forced to take.

A T beam is a beam made by fastening one beam laid flat to another beam laid on edge (*9*). When used as a spanning beam, a T beam made of 2x4s is much stronger than a simple beam structure of one 2x4. (It is also stronger than two 2x4s glued and nailed face to face to make a *laminated beam* as in Figure 10.)

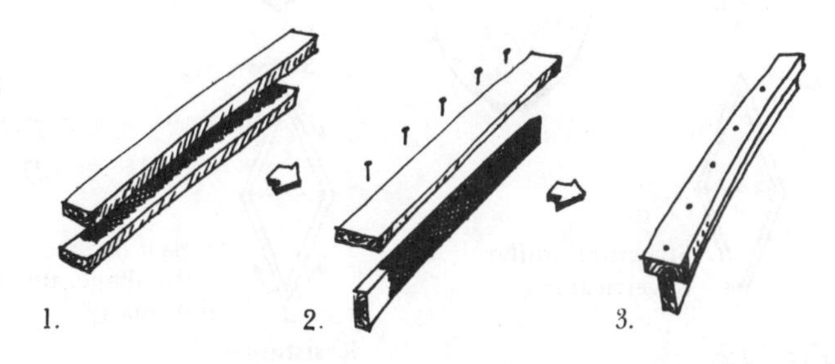

Fig. 9. T Beam Construction

The T beam has extra strength because the flat top member is extra wide and disperses more of the load directly inward to the vertical bottom member, which maintains resistance against the load and prevents the top piece from bending.

T beams can come in handy. Let's say you want to span an 11′ space with beams to support a platform, and you don't want to put any posts or vertical supports underneath within that 11′ space. The tables at the end of the book will tell you to use at least a 2x6 for your beams, but you have only 2x4s around. Rather than wasting time with your lumberyard and laying out more money, you can make T beams from the 2x4s you already have and use them in place of the 2x6s. They will be just as good, or better.

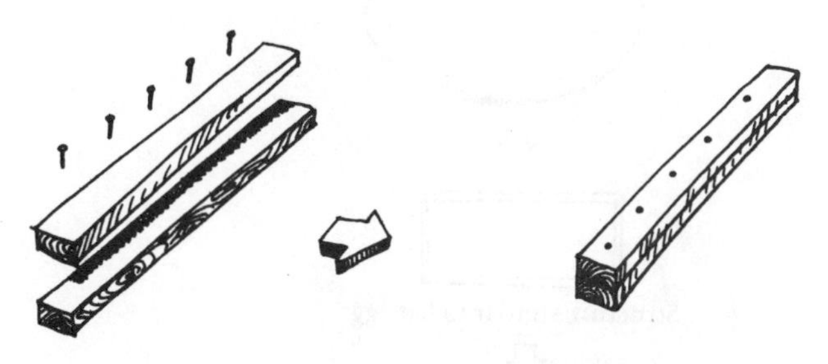

Fig. 10. Laminated Beam

Wood Grain

The grain running through wood gives extra strength for dispersing stress. The stress tends to follow the grain. Wood grain can be pictured as a bunch of thin tubes bound together very tightly. If these tubes are placed vertically, a load on top would be carried straight down. If these tubes are laid horizontally, they would tend to carry the stress from a load

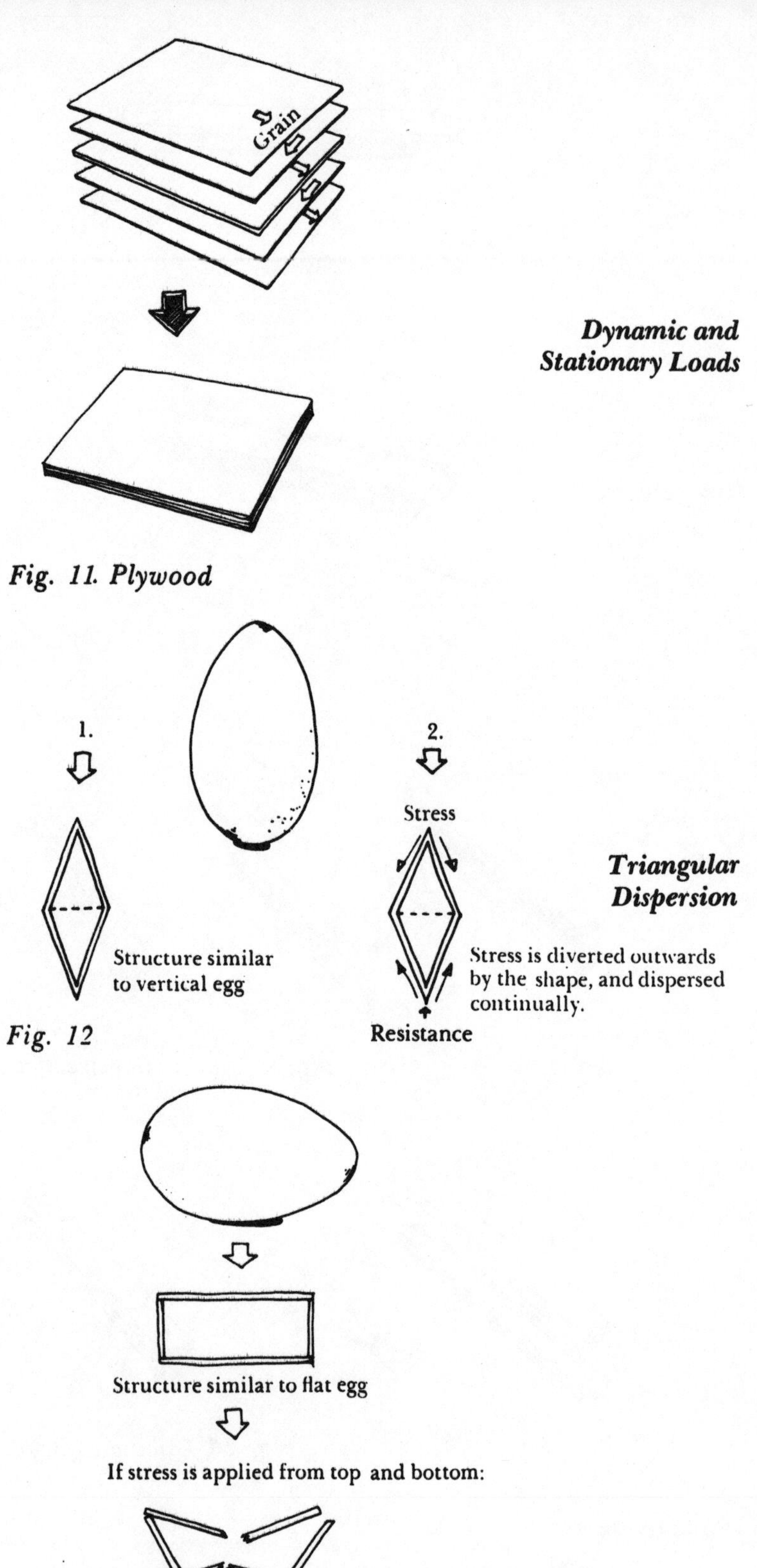

Fig. 11. Plywood

Fig. 12

Fig. 13

horizontally out to either end—but not to the sides.

Plywood is stronger than an equally thick piece of solid wood because it is designed to carry the stress to the sides, as well as to the ends. Plywood usually consists of five different layers of wood compressed and glued together, with the grain crossing in alternate layers (*11*). Therefore, as each layer takes up the stress, the alternate grains can carry the stress in all four directions.

Dynamic and Stationary Loads

A *stationary load* is exactly what it says—stationary. It doesn't move on its own. Books, records, and clothing are stationary loads on your shelves. The stress a stationary load exerts is constant.

A *dynamic load* is a load that moves. *You* are a dynamic load. Motors vibrate and are considered to be dynamic loads even if they stay in place. The stresses under a dynamic load are constantly changing, causing stresses to flow in every direction. Therefore a dynamic load requires a stronger structure than an equally heavy stationary load. To support you, a loft bed must be structurally stronger than a shelf system designed to hold your weight in books.

Before you design your structure, decide whether you are dealing with stationary or dynamic loads. This will help you to figure out how much trouble you have to go to, whether you need nails or bolts, 2x4s or 4x4s, bracing, and so on.

Triangular Dispersion

Hold an egg in your hand and try to break it by squeezing from the two ends. You'll find it takes a great deal of pressure. If you turn the egg around and squeeze along the flatter sides, it will break easily. In the first case the stress was diverted by the oval shape of the egg at the ends. In the second case you are pressing directly on the flatter part and there is no place for the stress to be diverted; the shell cracks under the stress. The egg illustrates the particular strength of a triangular shape in diverting stress. Figure 12 shows how, when stress is applied to the ends, the oval shape of the egg acts like two inverted triangles. Figure 13 shows the weaker structure, similar to a "flat" egg.

Any time you can adapt the principle of triangular structure to one of your own projects, you will improve its strength tremendously. The T beam shown earlier is a good example. In a cross section, you can see that the stress paths are basically the same as those in the triangle, even though part of the triangle is missing in the T beam (*14*).

Diagonal supports for shelves and counters also illustrate triangular dispersion (*15*). As long as all the joints at the corners are strong, any compression stress is dispersed along the diagonal to the wall.

Diagonal bracing (tightly joined, of course) is particularly useful to combat lateral stress in objects prone to sway, such as the open box in Figure 16. If pressure were applied to a corner—for instance, if you tried sitting on the edge—it could easily start collapsing to the side into the shape of a rhombus.

Once you put in the diagonal braces, the lateral stresses can be diverted down to the floor. The whole structure becomes rigid and much stronger.

A police lock with its long steel bar is based on the principle of triangular structure. As most big-city dwellers know, one end of the bar fits into the lock case on the door and the other end stays in a plate on the floor, several feet behind the door. The bar resists the lateral stress of someone trying to push your door in by transferring the stress to the floor and onto the entire house structure (*17*). A regular door lock could be ripped off the door jamb by enough pressure.

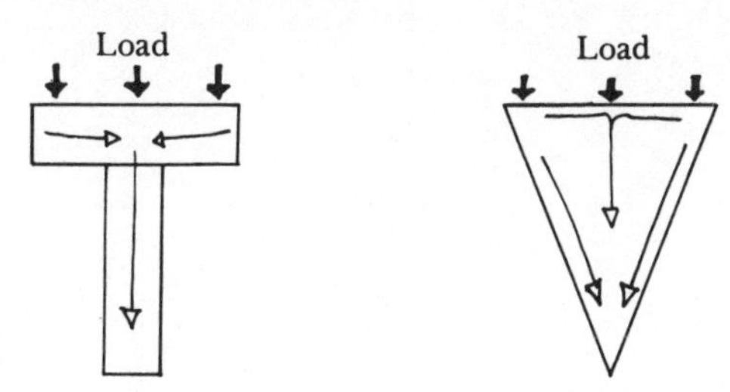

Fig. 14. Paths of stress end at same point in T beam as in triangle

Joints

A wood joint is the junction of two pieces of wood, held together by various means—nails, screws, glue, dowels, notches, string, rubber bands, spit, or magic. Obviously, certain joints are better than others. For structural strength, you want your joints to be as tight as possible. The tighter they are, the less movement there will be between the joined members when they are subjected to stress. The two pieces should act as much like one piece as possible.

A tight joint works better than a loose one because it enables the stress to pass through it and down to the ground, and return as resistance. A loose joint would break this chain, absorb much of the stress itself and not transmit the stress efficiently. This is what happens when the woman tries to carry the pails in her hands. Her muscles are loose joints as far as tension stress is concerned. Thus the stress is not efficiently passed down to the ground, and her shoulder joints quickly get tired from absorbing too much of the stress.

Loose joints can therefore cause a structure to fail. Loose joints are what make tables and chairs wobble. If you have a wobbling piece of furniture, take a look at it. Almost certainly you will find at least one weak or loose joint where a leg meets the rest of the structure.

In general, screwed joints are much stronger than nailed joints, and glue greatly strengthens either joint. Also, the more gluing surfaces you have, the better; that's why the fancier "interlocking" types of joints, such as the lap and dado joints, are stronger than the simple butt joint. In "Working Techniques" we'll show you how to make all these joints easily and when to use them.

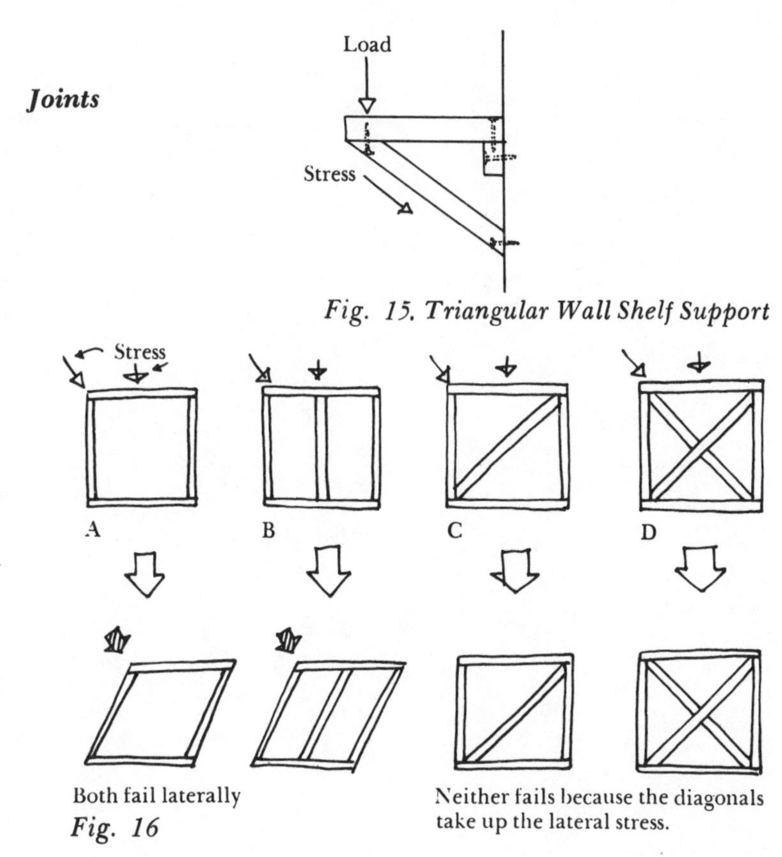

Fig. 15. Triangular Wall Shelf Support

Fig. 16

Stability

The stability of a structure is also important. Does it stand straight on its own? Will it fall over in the breeze? A chair can be built strongly, but if one leg is shorter than the others, the chair will wobble every time you sit in it. You will be putting its parts under unequal stress, since your load is transmitted to the floor along only three of its four corners. The joints will eventually weaken and come apart. More obviously, a bookcase that wobbles back and forth is a danger. If a cat jumps up, the whole piece can fall over.

Therefore it's important to build your piece with a firm foundation. If your piece is built evenly and squarely, but

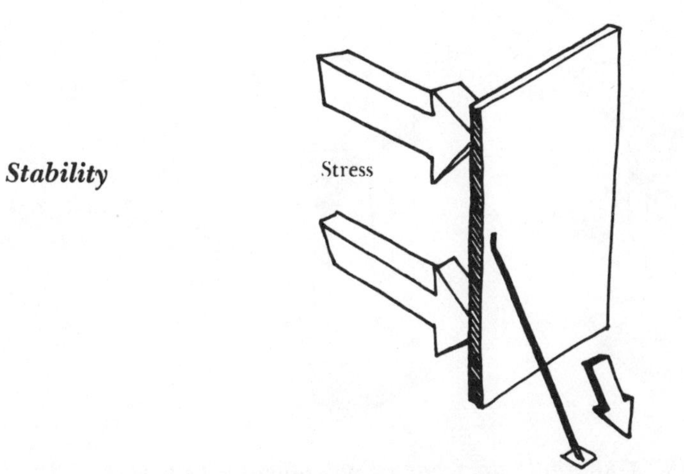

Fig. 17. Stress is taken up by bar along path of arrow

your floor is uneven so that the piece wobbles anyway, it's important to level and stabilize the piece by sticking some shims under it.

A standing piece like a bookcase can be easily stabilized another way, an even better way—by somehow fastening it to the wall. Of course, not every piece of furniture will be leaning against the wall, so this method is not always feasible. When possible, though, you could do it several ways. One is to simply put a couple of screws through the back piece into a solid part of the wall. If the piece is sitting an inch or so away from the wall (because of some floor molding, for instance) you could attach a cleat to the wall and then screw through the back into the cleat (see Chapter 7).

Another element in stability is the size of the base in relation to the upper portions of the object: the larger the base, the more stable the object. A 4′ cube on top of a 1′ cube can be easily tipped over. A 4′ cube on top of another 4′ cube is very stable. Similarly, a table is more stable the closer the legs are to the outside corners (assuming the table top is thick enough or structured so it won't sag in the middle). The amount of "overhang" in such cases is mainly a matter of common sense and usage. You don't want a table with a large overhang if most of the expected load will be on the edges. Experiment carefully, and try to keep the base of an object as large as possible. (See Cantilever in Chapter 9 for the limits of such overhang in certain cases.)

Lumber Imperfections and Structure

There are certain characteristics of the lumber you'll be using that you should know about. Unless you specifically order *clear* wood (as described in Chapter 3), most of your boards will have knots; some will even have cracks and splits running through them; and some pieces will be warped or bent, not at all as straight as you would like. Some of the wood may be impossible to use for any kind of structure—if so, you shouldn't even accept it from the lumberyard. You can, however, use a lot of this wood in certain situations if you understand the nature of the imperfections and the basic principles of structure we've discussed. This means that you may also be able to work with many of those seemingly useless pieces of lumber you may have lying around.

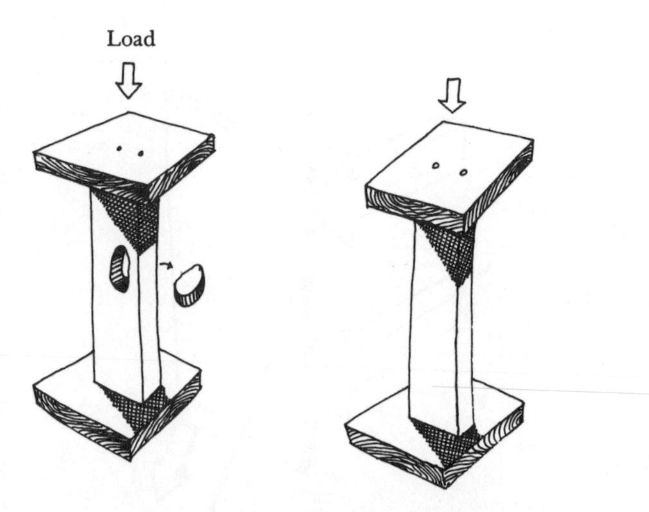

Fig. 18. A knothole does not considerably affect a post structure under compression

Splits Let's say you have an eight-foot 2x4 with what looks like a small crack near its center. One quick way to test the strength of a suspicious-looking piece is to try to flex it. You don't have to be particularly strong to do this. Let one end of the piece rest on the ground, hold up the other end, and then put pressure on the middle of the piece with your foot or hand. A weak piece will flex more than a strong one, of course. If there's a serious split, the wood might even begin to crack. Try this test first with a piece of lumber that looks fairly good, so that you can get the feel of it. After a few tests you'll be able to tell right away if you have a piece that has been terribly weakened by any imperfections.

Knots

The knots you see in wood are the result of the intersection of branches with the trunk of the tree. Most knots are of the *intergrown* type. There is no distinct physical separation between the knot and the surrounding wood. If you run your finger over the border of the knot, it will feel smooth. These knots affect the strength of the wood only to a negligible degree. You don't usually have to worry about them.

The *encased* knot, however, is the one you should look out for. You can feel the separation between this kind of knot and the surrounding wood with your fingers. If the knot is held in tightly, if there doesn't seem to be any way it's going to fall out, the wood will probably be all right for most uses. The tighter the knot, the more it becomes a part of the structure, and the more easily it passes on the stress. The looser the knot is, the weaker the piece of wood will be. A loose knot acts much like a loose joint in that it breaks the chain of flow of the stress. Additionally, all knots interrupt the flow of the wood grain, causing the grain to change direction. Any stress that might normally be carried along the grain will be detoured from its path.

Wood that has loose knots or knotholes in it can still be used as vertical posts, where the wood is under compression along its length *(18)*. For instance, you can put one or two pieces into a stud wall without weakening the wall very much. Or you can use one as a post helping to support a stationary load, such as storage space or shelving. Such lumber is *not* recommended for posts or legs of structures that support dynamic loads, such as chairs or loft beds. In a post, the downward stress can flow around the knothole and continue to the ground. The wood below and above the hole is too thick to snap. Only a strong lateral stress, such as one that might be set up by a dynamic load, could snap the wood.

Do not use wood with holes or loose knots as beams *(19)*. A large hole can diminish a piece's strength by half when the wood is used as a beam. A beam, as we said earlier, disperses the stress outward to the sides in a diagonal path. A hole in the beam will interrupt the direction of this flow. Unable to flow across the knothole, the stress will concentrate at and be absorbed by the thin wood remaining around the hole. This thin wood is forced to carry as much stress as a solid beam is supposed to. It can't do so without snapping.

If you find you *have* to use a piece of wood with a loose knot as a beam or joist, place the wood so the knot is in the upper half of the piece *(20)*. The load on the beam will compress the wood against the knot, so that a tight joint can be maintained between the knot and the piece. But a knot on the underside of a beam would be placed under tension stress. As the beam bends under a load, the wood on either side of the knot will pull away from it *(21)*. The flow of the stress will be interrupted by the loose "joint."

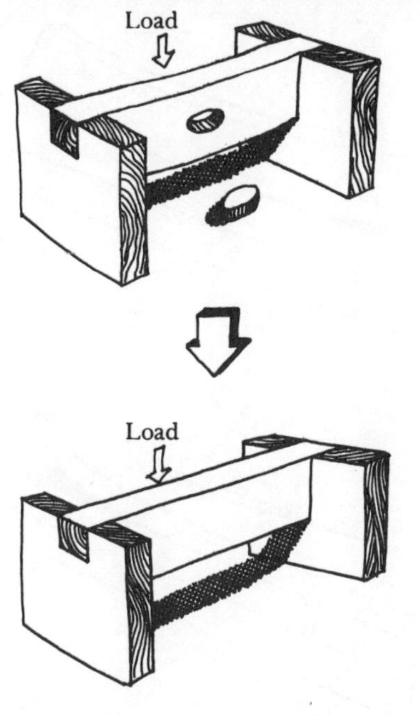

Fig. 19. A knothole greatly weakens a beam

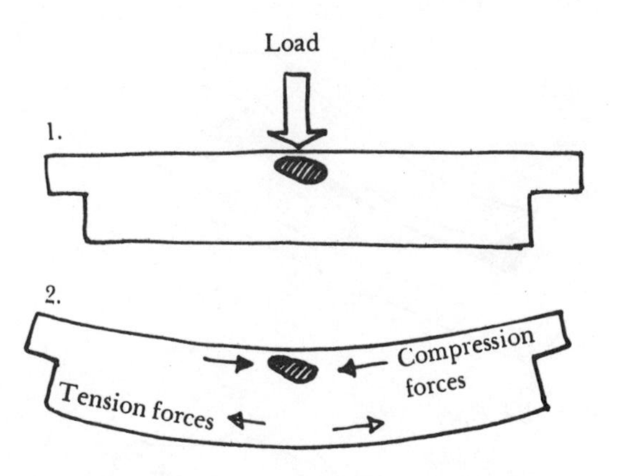

Fig. 20. Knot under compression does not fail

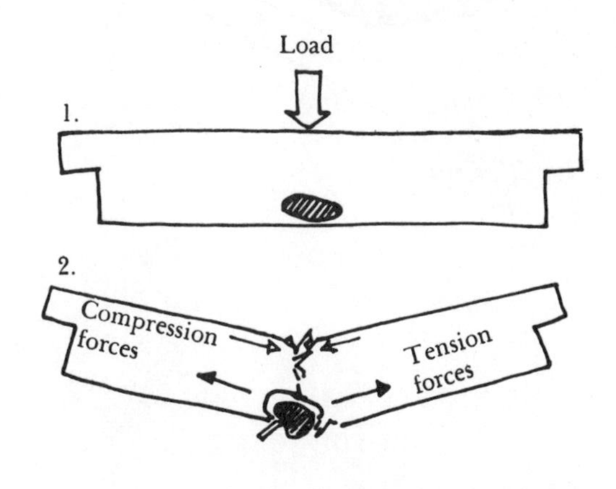

Fig. 21. Knot under tension does fail

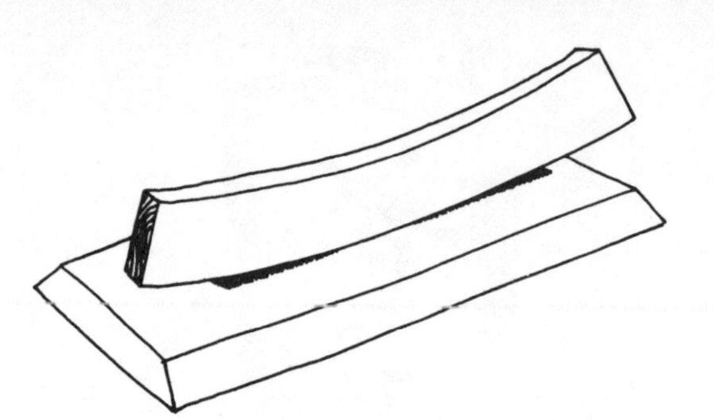

Fig. 22. Crook

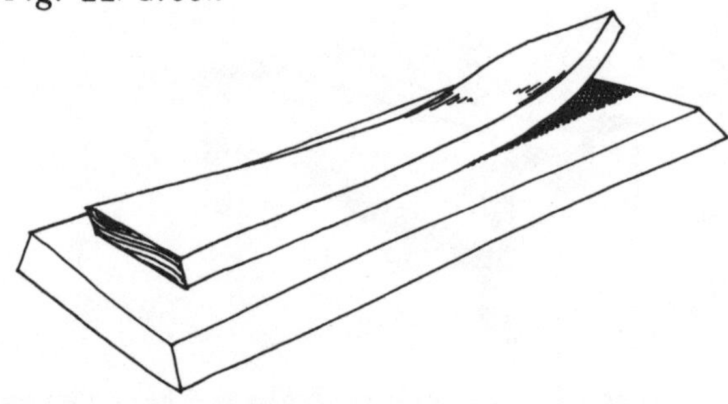

Fig. 23. Twist

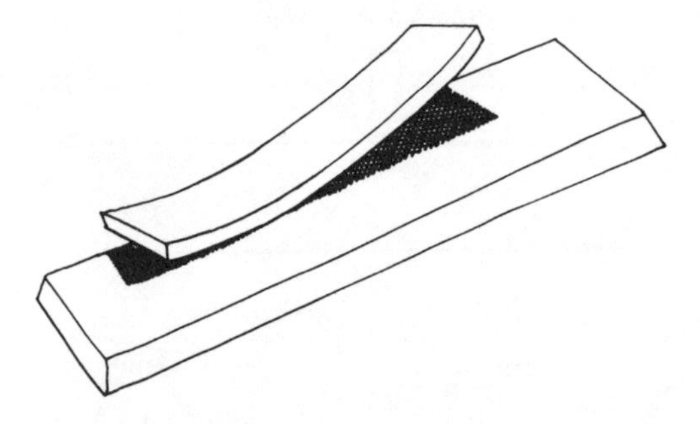

Fig. 24. Bow

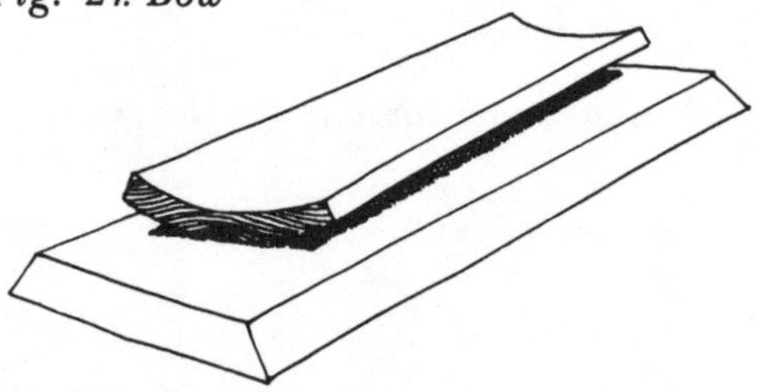

Fig. 25. Cup

Warps Some of your lumber may also be warped. There are four types of warps—a *crook,* a *twist,* a *bow,* and a *cup* (*22–25*). They can be caused by, among other things, improper storage or uneven drying and seasoning. Warps can weaken the structural strength of lumber, but not nearly as much as knots do. Use of warped wood depends more on common sense and planning than on structural considerations.

For example a cupped 1x12 will not make a very good shelf for knickknacks—they might fall or slide off, particularly if the hollowed side is down (*26*). But with the hollowed side up, records or books could sit comfortably on the shelf (*27*).

A bowed or crooked 2x4 is all right to use as an upright in a non-load-bearing partition, as far as structural strength goes. Each will create certain small problems, however, when the wallboard or paneling is applied. In the case of the bowed piece, you may have trouble finding the curve of the bowed wood with your line of nails. The crooked piece might make the wallboard bulge out too much for your taste, or it might even force some paneling to bend beyond its breaking point. A twisted piece can be used as a stud or beam, but it may also create a bulge and will not present a flat surface for the wall or floor surface to sit on.

If you must use a crooked piece as a beam, put it so the arch is up; then the load will push it back straight.

Common sense will tell you not to use warped pieces when you need straight edges—frames, tabletops, and so on. If you try to force warped pieces into forming a straight and square frame, you will be putting the joints under an extra inherent stress. The warped pieces will tend to return to their natural bend.

Remember that wood with a bad warp or large knotholes is not necessarily wasted wood. If you can't find a sound use for it in its full length, you can cut the wood into several pieces, leaving out the knots and lessening—if not eliminating—the warps. If you're building a wall, for instance, you will need small sections as crosspieces between the studs. Don't cut up the long straight lumber for that; use the warped lumber.

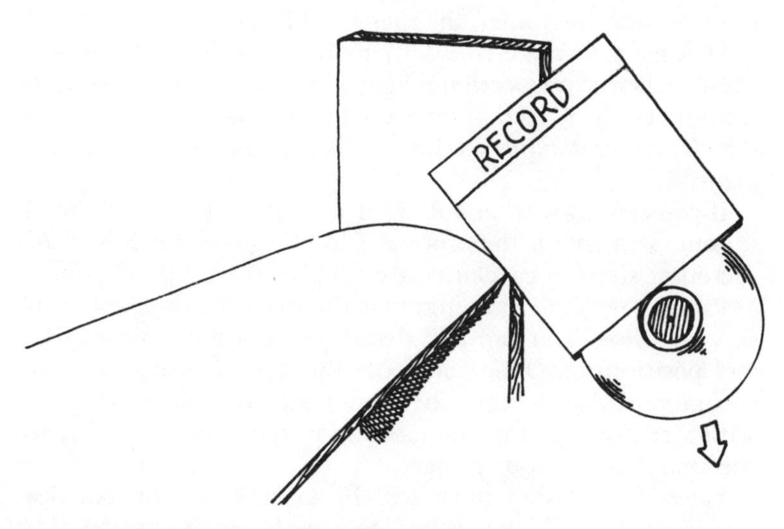

Fig. 26. Incorrect use of cupped board

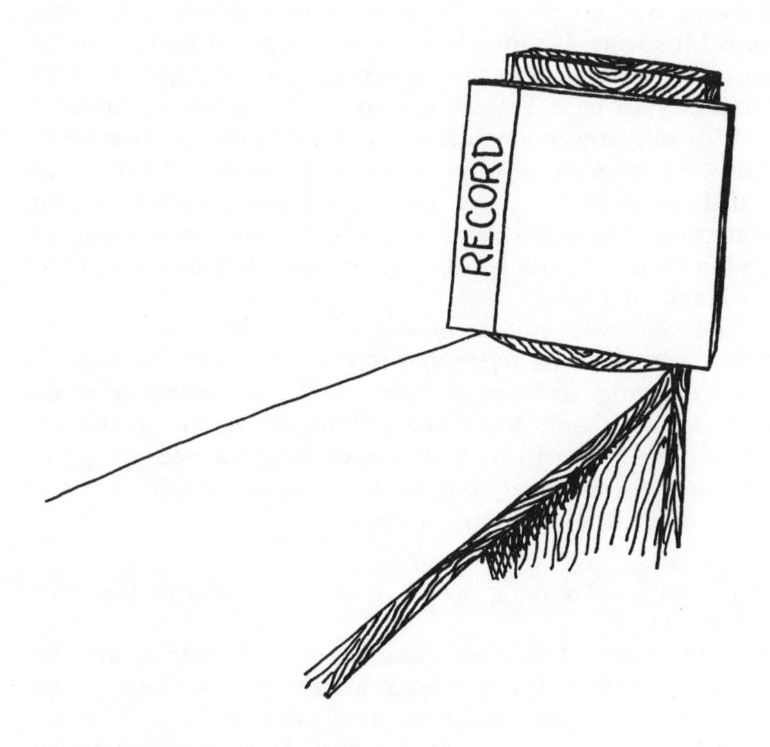

Fig. 27. Correct use of cupped board

2/DRAWING AND VISUALIZING

Putting Your Ideas on Paper

The first step in a carpentry project begins in your head. Maybe you visualize a bookcase for your boxes of books, or a partition for privacy—you detect a dim shimmer of a design, but you can't fully focus the shape and its parts.

Making a few sketches will help you solidify the shape. Most carpentry projects are right-angled—they're made up of combinations of rectangles like the one in Figure 1. *If you can draw a rectangle, you can probably draw most carpentry projects.*

If you can draw what you want to build, a lot of your work is done. Drawing is the process of working out the design. All the other steps in carpentry are simplified by a good plan.

Rough *freehand drawings* give the simplest, quickest shape to your ideas. You can add details or play with proportions and positioning. Once you focus the design, you can create accurate *scale drawings*. The extra time invested rewards you with a realistic picture in accurate proportions. You can see the final shape of your project.

Your ideal design must still be tailored to the practical considerations of carpentry. Do you have the space? The time? The money? The tools? Is it structured well? Does it lend itself to efficient construction?

You adjust your plans accordingly and make a final scale drawing of your project. This becomes your *building plan,* your blueprint. It shows how the piece should look, how the parts fit together. All the measurements are right there in front of your eyes. You see exactly what lumber you need.

We can't stress enough how much easier carpentry is when you work from a careful plan of your project. You can devise a method of construction, and collect and prepare all your materials. The actual woodworking becomes much simpler and more fun. You'll concentrate on your technique and create much finer work.

You may swear you are no artist. No matter. You *can* learn how to draw in the styles that follow. Practice will improve your drawing technique. Soon you'll be doodling three-dimensional shapes while you talk on the phone. You'll notice a piece of furniture you like, or maybe a photograph of one, and discover that you can draw a quick sketch of it and ultimately a work plan for its construction.

Terms Used in This Chapter

Right angle—An angle of 90°. The corner of this page is a right angle.

Perpendicular—A line or plane is perpendicular to another line or surface if they meet at a right angle. The top edge of this page is perpendicular to the side edge.

Parallel—A line or plane at the same distance from another line or surface at all points. The top edge of this page is parallel to the bottom edge.

Vertical—A vertical line or plane is perpendicular to the horizontal. A vertical line drawn on paper is perpendicular to the top and bottom edges of the paper.

Horizontal—A horizontal line or surface is parallel to level ground. A horizontal drawn on paper is a line parallel to the top and bottom edges.

Diagonal—Any line neither horizontal nor vertical.

Rectangle—A four-sided figure with four right angles.

Square—A rectangle with four equal sides.

Plane—A flat surface.

Freehand drawing—For our purposes, any drawing made without using a T square or triangle to arrange lines parallel or perpendicular to each other. A drawing made using a straightedge to make straight lines is still a freehand drawing, so long as the actual alignments of the various lines are done by eye.

Scale drawing—A drawing that has the same relative proportions and the same shape as the object it represents.

Two-dimensional drawing—One that shows only two dimensions, such as width and height. For instance, Figure 1; also the three views of the orthographic in Figure 55.

Three-dimensional drawing—One that shows three dimensions of an object—width, height, and depth. A "realistic" picture.

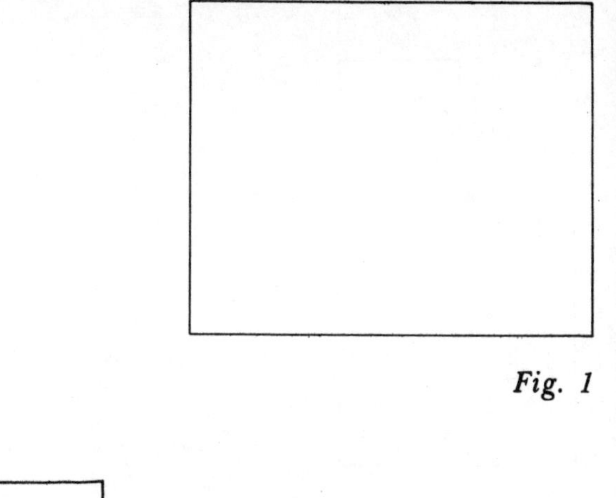

Fig. 1

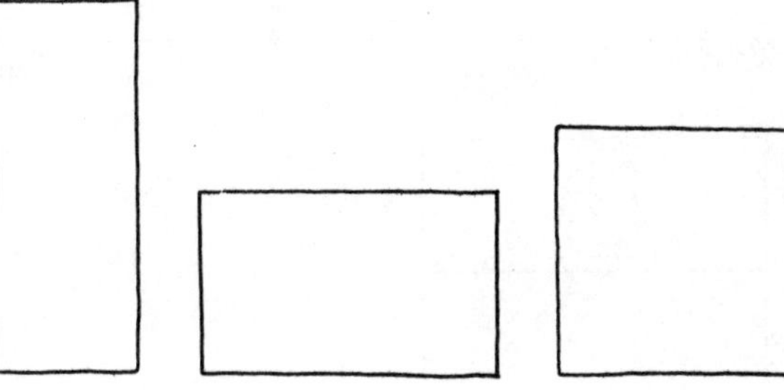

Fig. 2

Freehand Drawing—Nonscale

The tools you will need for a freehand drawing are a pencil and some paper. Any kind of paper will do. Graph paper is an important tool that will help you draw verticals and horizontals accurately, but it's not essential. You should also have a good eraser. It is not dishonorable to erase. Architects do it all the time.

Let's say you want to hang a bookcase on the wall for those boxes of books. What will the piece look like? What do you want it to look like?

The first step is to define the shape as much as possible in your mind's eye. Close your eyes and try to visualize the basic shape, the general outline of the bookcase—most likely some kind of rectangle. Take your pencil and draw a simple rectangle. It can be high, wide, or square *(2)*. Play around till you find one that seems both functional and pleasing to your eye.

Estimate how many shelves you need or can fit in. Represent them with horizontal lines. Later, your scale drawing will help you figure out exactly how many shelves you can fit in a certain space. If you like, add some vertical lines between the shelves as dividers and additional support *(3)*.

The sketch will look more like wood pieces joined together if you draw lines to show the wood thickness. Add them on the sketch you just made, and erase any wrong overlaps that result *(4)*.

This rough *front view* alone is often very helpful for simple projects. Side and top views help for more complex projects. They are drawn the same way. Just turn the object

Two-Dimensional

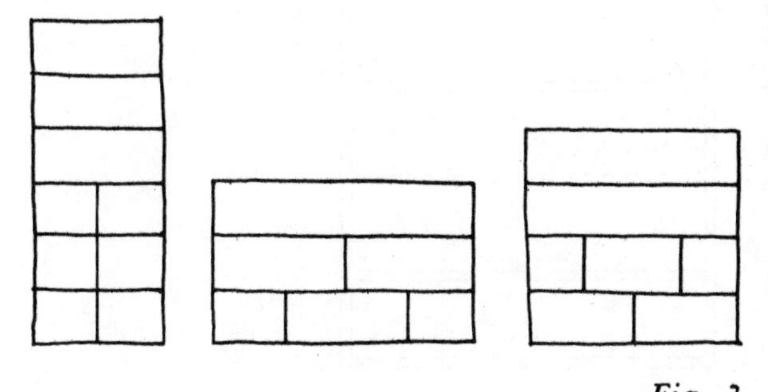

Fig. 3

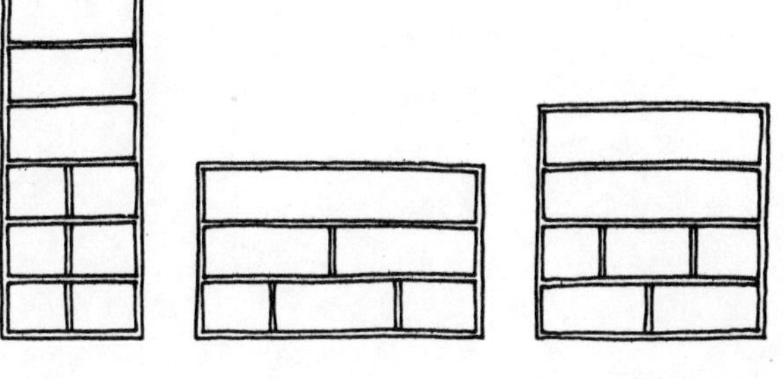

Fig. 4

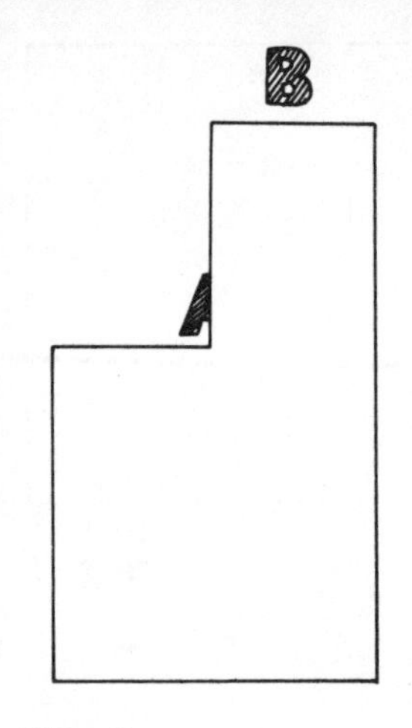

Fig. 5

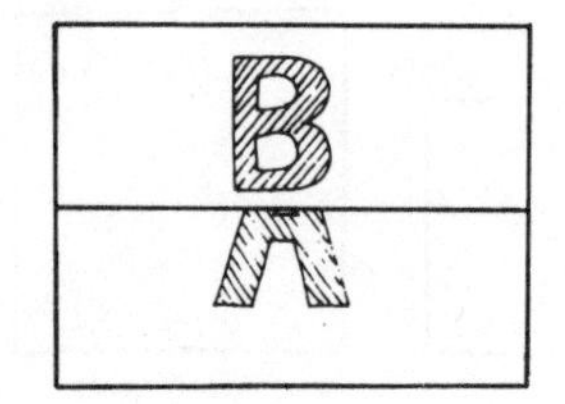

Fig. 6

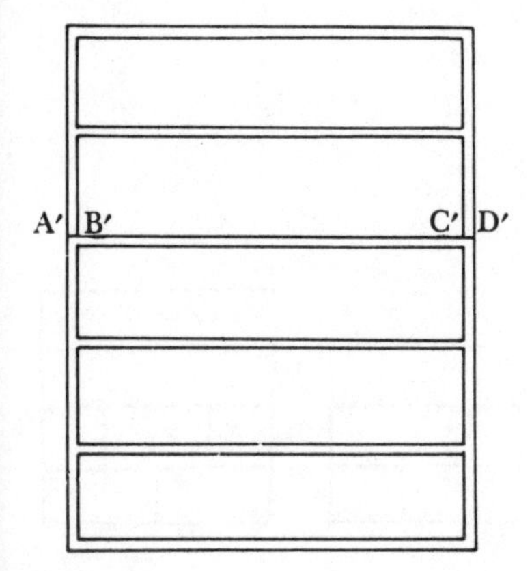

Fig. 7

around in your mind and think of the side view as being a front view of the object's side.

The *side view* of a bookcase with deep lower shelves and shallow high shelves might look like Figure 5. You might see it more easily as a smaller bookcase on top of a larger one, without the seams on the side showing.

The *top view* would be a little trickier (*6*). You see *B,* the entire top shelf, and *part* of *A,* the top shelf of the lower section.

The front view would be drawn exactly the same as that of the original straight bookcase, but adds the two little lines at *A′B′* and *C′D′* to show the top edge of the lower section (*7*). These lines, like all the lines in these sketches, would actually be seen if the real object were in front of you.

Three-Dimensional

It is a simple step to change your two-dimensional sketch to a view that shows three sides at once, giving an illusion of depth and distance on a flat piece of paper. Most people will find this three-dimensional sketch a much easier way to visualize the whole project.

For instance, imagine a solid box. Its front view would be a rectangle. Visualize what the whole box would look like if you moved your head a little to the right and a bit above the top of the box. You would still see the front, but you would also see the top and right side extending to the back. First, draw the square front view (*8*). The edges that run from front to back, which indicate the depth, are shown by parallel diagonal lines (*9*). The diagonals run up and to the right—the same direction you moved your head—at about 45° to the horizontal. The lines are equal and roughly parallel because the edges they represent are equal and parallel. The diagonal from the bottom left corner is not drawn because it is not seen from this angle. (You would see it if the front were open.)

Connect the ends of the diagonals with one horizontal and one vertical to represent the visible back edges of the box (*10*). *Voilà*—a three-dimensional box. Note that in a box, each edge on the back face is parallel to its corresponding edge on the front face.

The broken lines in Figure 11 are called *hidden lines.* They represent those edges, such as that fourth diagonal, that you cannot see in reality unless you have x-ray vision or the box is made of clear glass. Sometimes you won't be sure at first if a line is hidden or not. Just draw all your guidelines lightly and look. The hidden lines will be those lines behind the solid surfaces. You can use dots or dashes for the hidden lines if it helps you to envision the whole object, or just erase them.

This type of drawing, where a simple front view is extended back by diagonals, is called an *oblique view.* With a little practice you'll find it a quick and easy way to get a three-dimensional view of an object. Remember that all depth lines—that is, all the edges that go between the front and the back faces—are drawn as diagonals. If these edges are parallel in the object, they are parallel in the drawing. And of course, you should erase or make dotted any line not visible from that view.

Try drawing an oblique view of a square box with no front or back face. Draw the square front view *(12)*. Try to visualize which surfaces will show and which ones won't. The bottom left depth line is visible, because the front of the box has been removed. Draw all four diagonals *(13)*. Look some more before drawing in the back edges. The outer faces of the top and the right side will show *(14)*. Are any other parts going to be hidden even though the box is now open at the front? Yes. The two back edges extending from that bottom left diagonal toward the bottom right and top left corners will be partly hidden by the right side and top of the box. Draw these back edges just up to these sides *(15)*. You can extend these edges as dotted lines if you wish.

The bookcase we wanted to build earlier is just a variation on the open box. Draw an open rectangle and put in some shelves *(16)*. Draw the diagonals and back edges just as with an open box *(17)*. Note that *AB* is one straight line. Try to make it look straight, not like Figure 18. Again, don't be afraid to erase.

One hard shape to draw is the kind whose front view is actually composed of two different parts, one further back than the other. For instance, take a solid shape similar to the outline of our two-part bookcase. You must realize that, although in the front view *(19)* the two parts *look* like they're both up front, there is actually a distance between them. This is apparent in the side view *(20)*. Thus there must be a distance between them in the three-dimensional view of the oblique drawing. First, draw the part that is *most up front* for your front view *(21)*. Its horizontals and verticals remain horizontal and vertical. Then, run the diagonal depth lines on top back to where the second part of the front view begins *(22)*. This second part was horizontal and vertical in the two-dimensional front view, and it remains so in the oblique *(23)*. Now run the rest of the diagonals to the back of the piece *(24)*, so they can be connected by a horizontal top edge and a vertical side edge *(25)*.

These complex shapes are often clearer if you draw two-dimensional side views first, just to get an idea of what the side of the oblique will look like. You might even find it easier to use the side of the object as the front view of your oblique *(26)*—assuming the side view is not also composed of two parts in different planes. This eliminates the "double" front view. Note that the oblique diagonals can run to the left as well as to the right *(27, 28)*. It's all up to you.

These freehand styles are great for quick sketches, and for playing around with shapes and elements as shown in Figures 29–32. Only a scale drawing, however, will show the proper proportions of a project when you start to give it specific dimensions.

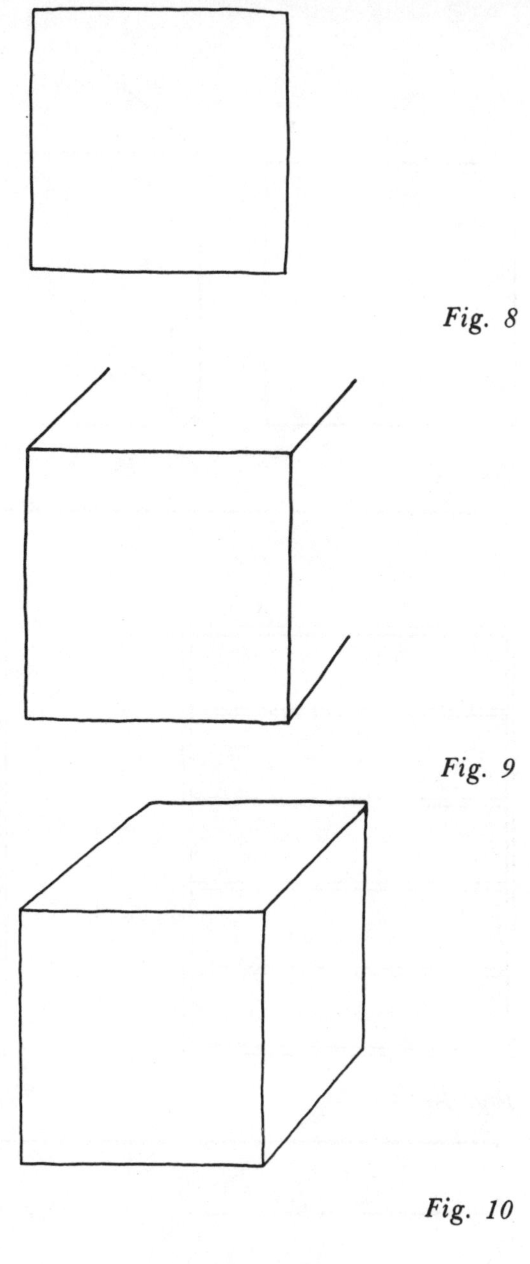

Fig. 8

Fig. 9

Fig. 10

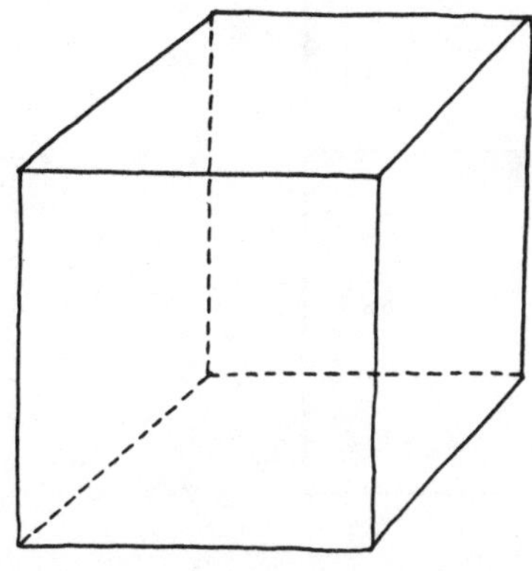

Fig. 11

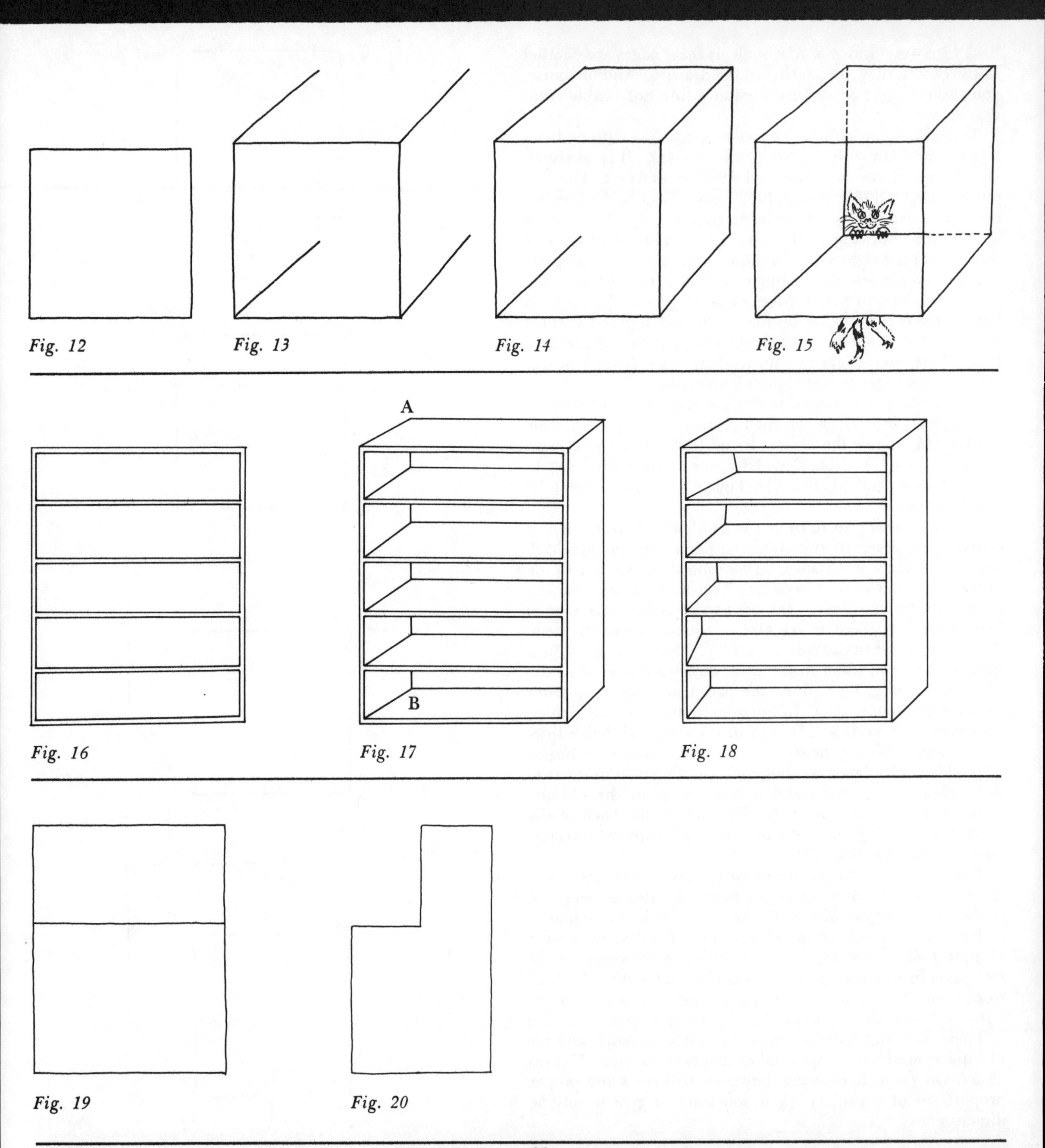

Fig. 12 Fig. 13 Fig. 14 Fig. 15

Fig. 16 Fig. 17 Fig. 18

Fig. 19 Fig. 20

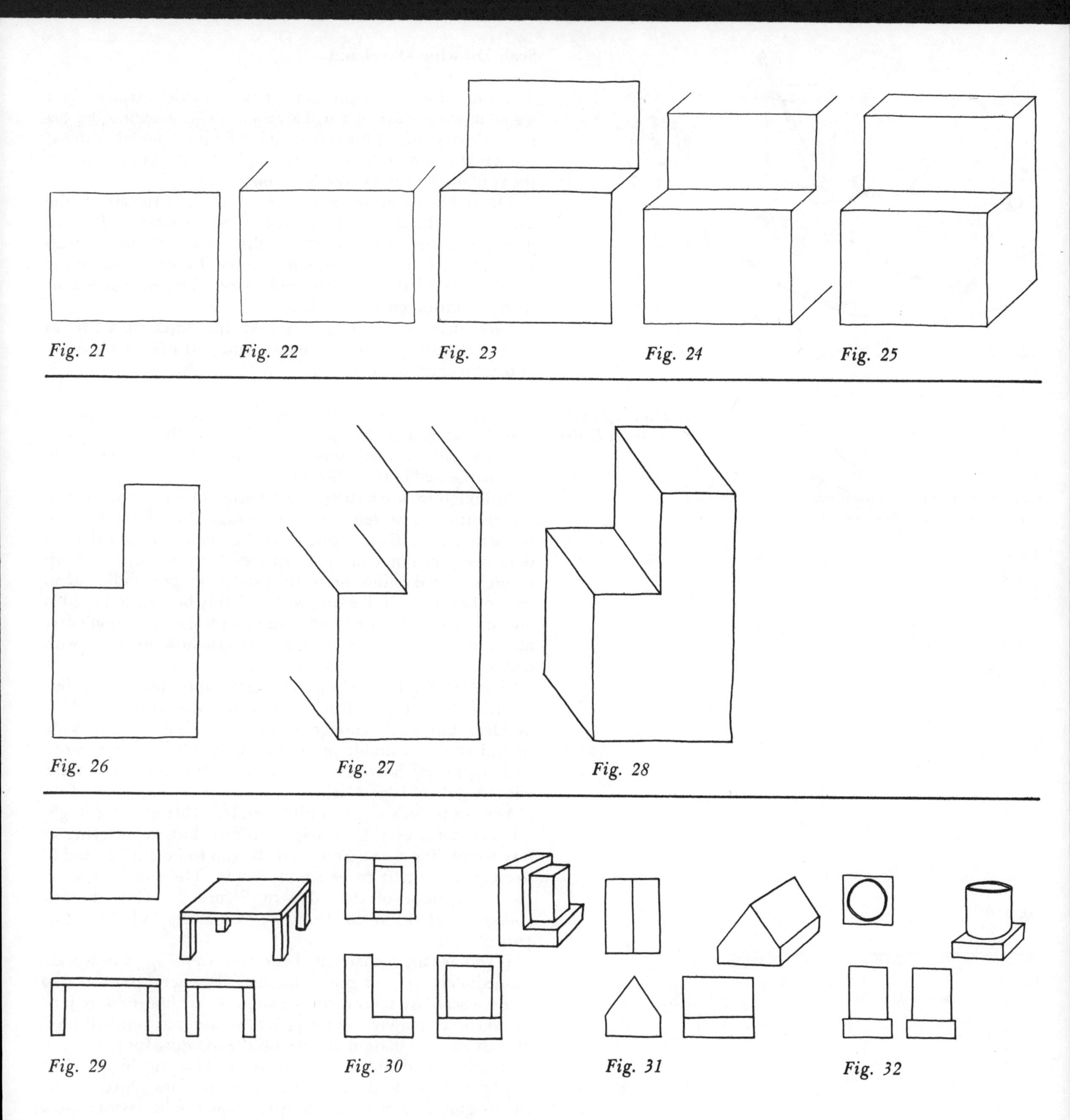

Fig. 21 Fig. 22 Fig. 23 Fig. 24 Fig. 25

Fig. 26 Fig. 27 Fig. 28

Fig. 29 Fig. 30 Fig. 31 Fig. 32

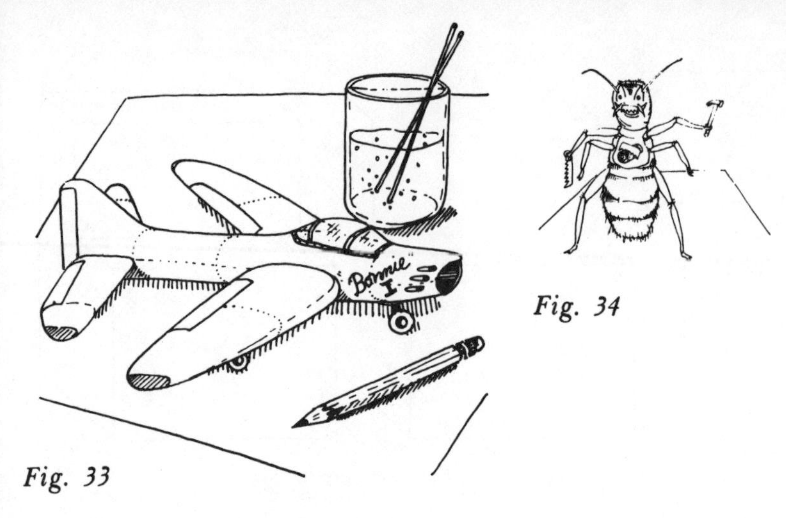

Fig. 33

Fig. 34

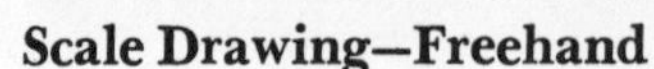

Scale Drawing—Freehand

A *scale drawing* is the next step. A model airplane is a scaled-down version of a real airplane (*33*). A map of Peoria is a scale drawing of the actual city. The scale model or drawing has the same relative proportions and the same shape as the real object. It looks the same, only smaller.

The *scale* of a drawing is the proportion of the size of the drawing to the size of the source object. A scale of $1'' = 10$ miles on a road map means just that—one inch on the map represents ten miles in ordinary reality. Three inches on the map represent thirty miles, and so on. The *proportion* of inches to miles remains the same.

Very small objects can be scaled up, larger than life, as with biologists' pictures of microscopic cell life, or blow-ups of termites and sawteeth (*34*).

Two-Dimensional—With Ruler

To make two-dimensional scale drawings, you need paper, pencil, and a ruler—also some very basic math sense. You can even do without math sense if you have an architect's scale rule, explained in the following section.

Although we use a ruler to draw straight measured lines in this technique, we refer to the drawings as freehand because the actual vertical and horizontal alignments of the ruler are done by eye, not with a T square. No one's eye is sharp enough to make lines perfectly parallel or perpendicular to each other, so your drawing will probably be a little lopsided and not a true square. Either use graph paper or don't fret about it. The concern here is to learn how to work with scales.

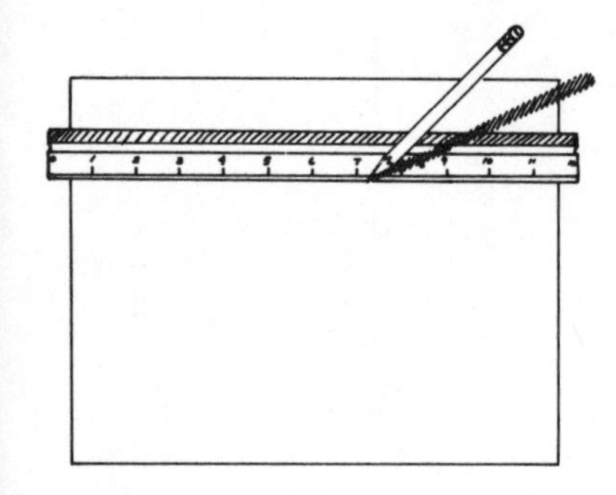

Fig. 35

The first step is somehow to pluck an appropriate scale out of the air—one that will make your drawing large enough to be clear but small enough to fit on the paper. The scale should also be a simple proportion that will be easy to work with mathematically. A scale of $1\frac{7}{32}'' = 1'$ is not easy to work with. A scale of $1'' = 1'$ is.

Fig. 36

You want, let's say, to build a coffee table about 2′ high. (That's not a very good height, in fact, but we're trying to stay simple for now.) You want the top to be 4′ wide and 3′ deep. You want it to be simply made. The rough sketches look like those of the table in Figure 29. How do you choose a scale to fit this front view on an $8\frac{1}{2}''$x$11''$ sheet of paper?

If you are mathematically bent, you would spy a neat scale immediately. If your genius inclines elsewhere, you can zero in on a scale by several simple steps. First, find exactly how much space you have on the paper. Figure that you will need at least an inch margin all around the drawing for writing in dimensions and notes. That leaves a space no bigger than $6\frac{1}{2}''$x$9''$. To fit 4′, the longest dimension of the object, in 9″, the longest dimension of the space, you would have to use a scale of $9'' = 4'$, or $1'' = \frac{4}{9}'$. Forget it. Ninths of a foot are too difficult to work with. Look for a neat, even multiple of 4

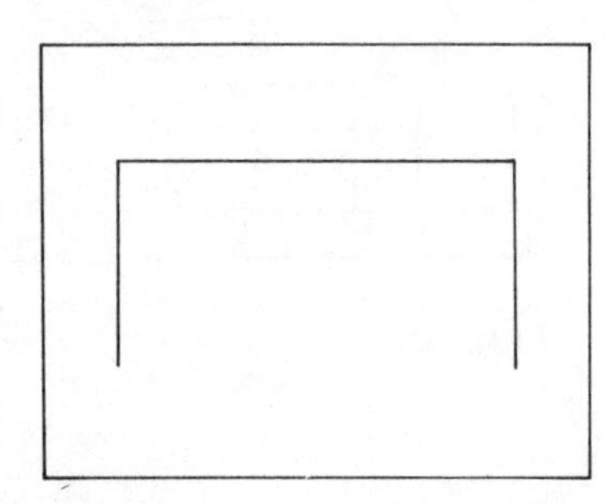

Fig. 37

(the length of the object) that is close to, but less than, 9. Try 8. Your scale is then $8'' = 4'$, or $1'' = \frac{1}{2}'$. Or $2'' = 1'$. They are all the same proportion, the same scale. They are just stated differently. You might find one easier to work with than another, but for consistency, we'll use $2'' = 1'$.

To be certain of this scale, we must also check that it works for the table height. A 2′ height means a 4″ line on the page, leaving a $2\frac{1}{4}''$ margin top and bottom—a good fit. If the table were 4′ high, the scale height would be 8″ in the $6\frac{1}{2}''$ space. We would need a smaller scale—or a bigger piece of paper. The same scale must, of course, be used for all the measurements in one drawing.

Now you can draw a scale front view of this table. Write the scale on the paper in one corner. It will be handy for future reference. Line the ruler up parallel to and about $2\frac{1}{4}''$ below the top of the paper *(35)*. Draw a light line 8″ long, roughly centered between the sides of the paper, to show the 4′ table top *(36)*.

Draw 4″ vertical lines down from each end of the tabletop, to represent the 2′ height *(37)*. This is the scale outline of your coffee table.

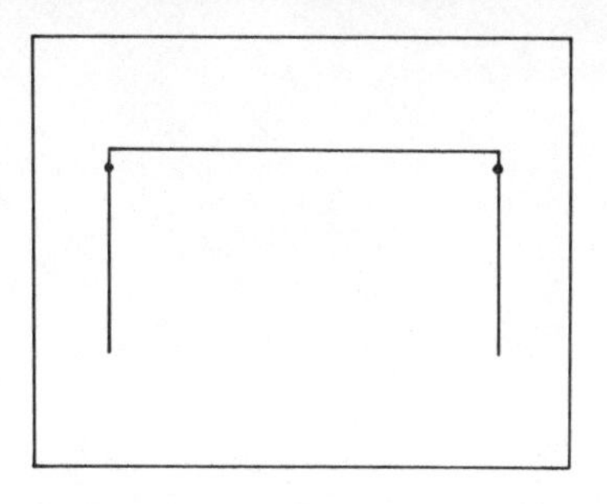

Fig. 38

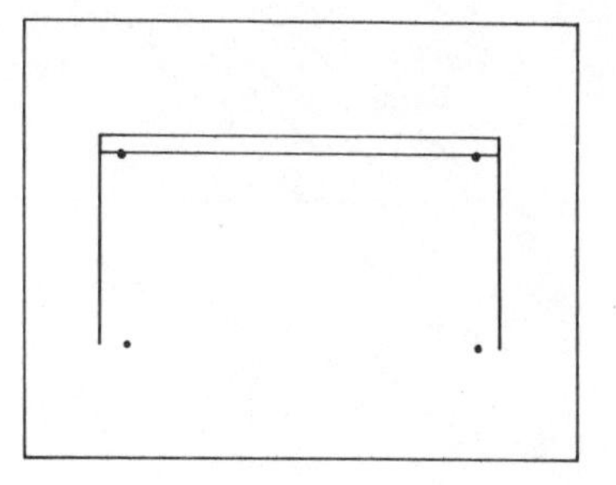

Fig. 39

Fill in the thickness of the wood. Say you want the top to be 2″ thick. In our scale of $2'' = 1'$, that thickness is $\frac{1}{3}''$. Mark $\frac{1}{3}''$ down from the top of the legs *(38)*. (You can approximate $\frac{1}{3}''$ on your ruler by making a mark at about $\frac{5}{16}''$.) Connect the marks with a horizontal.

Four-by-fours would make good, sturdy legs for this table. The actual size of what is called a 4x4 is $3\frac{1}{2}''$x$3\frac{1}{2}''$, as you will learn in "Woods and Other Building Materials." What is the $2'' = 1'$ scale equivalent of $3\frac{1}{2}''$? The exact answer, $\frac{7}{12}''$, is more trouble calculating than it's worth at this point. Approximate such troublesome fractions—$3\frac{1}{2}''$ is just a little bigger than 3″, which would be $\frac{1}{2}''$ in our scale. Mark off the leg thickness to be just a shade larger than $\frac{1}{2}''$ *(39)*—close enough for a freehand drawing. Complete the legs *(40)*.

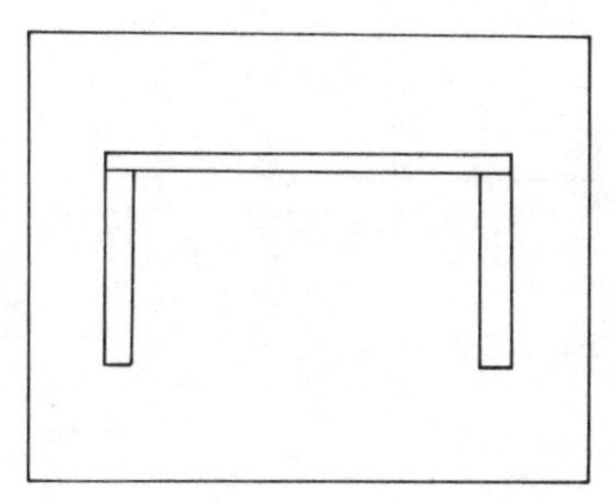

Fig. 40

When you are done, erase any wrong lines, such as the leg verticals running into the tabletop. Notice that the scale drawing looks slightly different in proportion than the rough sketch. If the scale view doesn't look like what you want, play around with the new measurements. Erase some lines or make a new drawing. Make the top thicker if it looks too thin, or maybe slide the legs in a bit from the corners. When you find what you like, darken in the lines and mark down the new dimensions in the margins *(41)*. You could also experiment with new elements—perhaps a shelf under the tabletop. Move it up and down till it looks good, and mark down its height *(42)*.

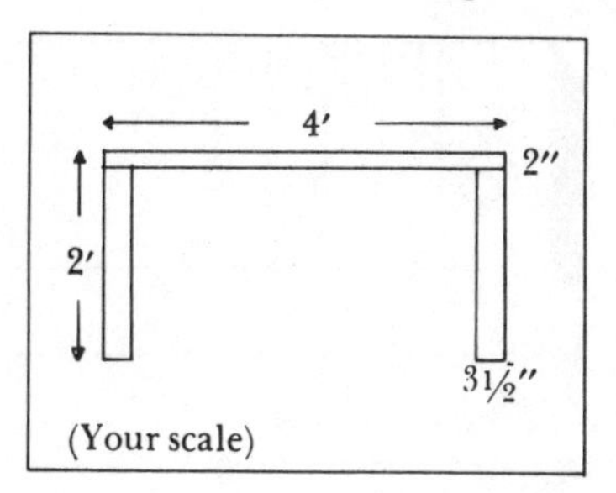

Fig. 41

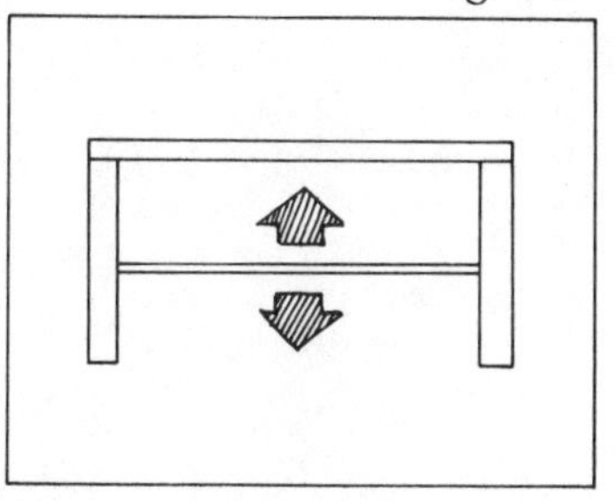

Fig. 42

Two-Dimensional— With Graph Paper

Any two-dimensional scale drawing can be made on graph paper without the use of a ruler for calculations. Just let a foot equal an appropriate number of little squares on the paper. If four squares equals a foot, for instance, you draw the top of the table 16 squares long. Many people find this easier than calculating with a ruler. Fractions of squares can just be approximated by eye. Unfortunately, graph paper is no help in a three-dimensional scale drawing, where you have to deal with diagonal lines. Therefore it's important to understand how to convert to scale with a ruler, or how to use the architect's scale discussed in the next section.

Two-Dimensional—With Architect's Scale

The beautifully simple *architect's scale* eliminates all the math. It costs only a dollar or two at stationery stores. It's a three-sided ruler, a bit more than 12″ long (*43*). It may seem difficult to orient yourself to it at first, but it's simple to use.

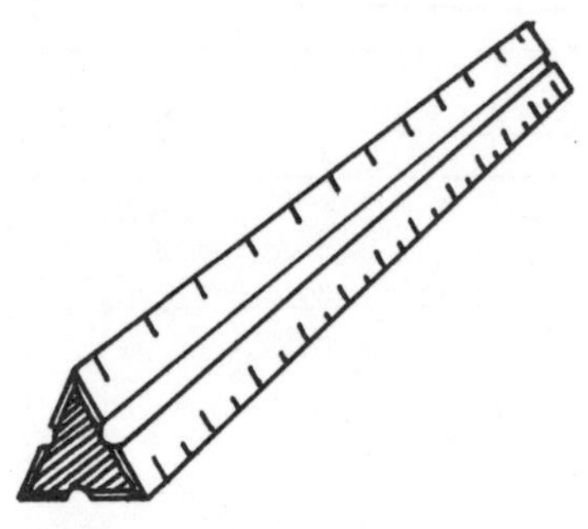

Fig. 43

Each side is divided into two measured lengths. One of the six sections has the normal inch markings, each inch divided into sixteenths. Each of the other five sections has on it two different scales, one running from right to left, and one running from left to right.

In Figure 44, the numbers closer to the top edge belong to the 1/2 scale. The numbers just below (but also in the top section of the rule) belong to the 1 scale, reading from the right. The fact that both scales share the same hash marks may present difficulties. This arrangement is meant to save room. Always treat each of the pair as a totally separate scale.

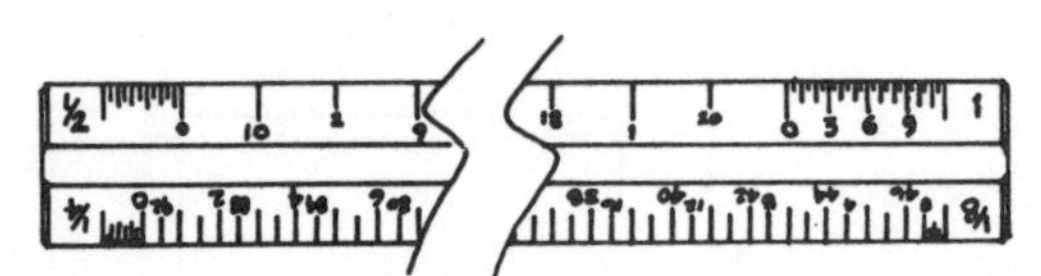

Fig. 44

Each scale is divided into a certain number of equal units, each representing a foot. The smallest scale, 3/32, contains 124 numbered units. The largest scale, the 3 scale, has just four units, representing four feet (actually, only two units are marked; you have to use the ends to get all four feet). A scale foot will obviously be drawn smaller on a smaller scale than on a larger scale.

The number-name of the scale is the number of inches (or fraction of an inch) that equals one foot—1/2 means 1/2″ = 1′; on the 3 scale, 3″ = 1′, and so on. Check it yourself with a regular ruler. Each unit on the 3 scale is 3″ long.

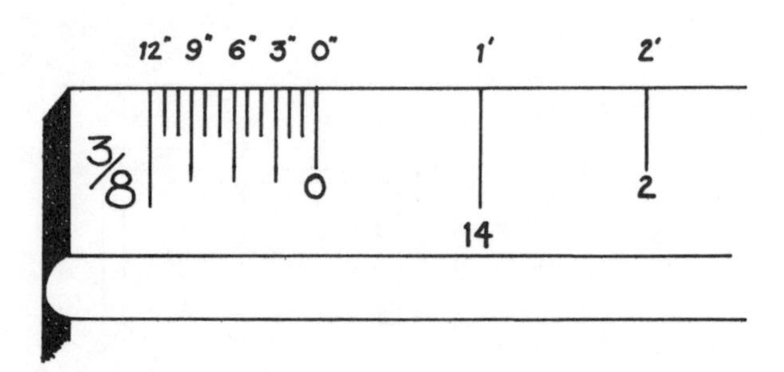

Fig. 45

To choose a scale for your drawing, first find the longest dimension of the object. On the rule, find a scale where that scale distance is a convenient size. If you are drawing a 100′-long house, choose a scale with that many units on it. The 3/32 scale is the only one. A line from 0 to 100 will represent 100 feet. If the house is 20 feet high, you draw the vertical line from 0 to 20 on that same *scale*.

Inches and fractions of feet are also easy to find by eye. There is one whole unit *before* 0 on each scale, divided into either six small parts or an even multiple of twelve parts, to correspond neatly with the twelve inches in a foot (*45*).

Fine draftsmen draw their guidelines with a straightedge or ordinary ruler, and use the architect's scale only to mark off distances (the excess lines are erased later). This protects the edges of the scale from being colored or chipped by the drawing point. Use your own judgment. At any rate, measure all your lines from the given number of feet *toward* the 0 mark, and then count off the inches, if any, from the 0 on that initial inch-marked unit.

Draw our table, using the architect's scale. The piece's long dimension is 4′. You will find that the 1 or the 1½ scale represents this distance admirably. Figure your margins and draw the top of the table, measuring off from 4 to 0 on, let's say, the 1½ scale. Draw the vertical legs from 2 to 0. Make sure you use the same scale for all these measurements, and that you don't get confused with the numbers on the 3 scale coming from the opposite direction. The top thickness is 2″. Mark off the scale equivalent of 2″ on that initial inch-marked unit. Mark off 3½ subdivisions for the 3½″ leg thickness. That's all there is to doing inches. The scale ruler does all the math. As long as you use the same scale for all the measurements, the drawing will be perfectly proportioned to its original source.

Write down the numerical name of the scale in one corner of the paper. You, or someone else, can return later, find that scale on the ruler, and read off the exact measurements with it right from the drawing. And if you know the meaning of the scale numbers, you can measure a drawing with only a normal ruler or a carpenter's tape and convert the inch measurements back to the original number of feet. This is how an architect communicates to his carpenter on the drawing plans. An architect can't be on the job all the time, so his plans answer layout questions when he isn't there.

Scale Drawing—With T Square and Triangle

The T square and triangle are as easy to use as a ruler. Made with these draftsmen's layout tools, your drawings will be square and true.

The *T square* (*46*) is a straightedge with a perpendicular crosspiece. The crosspiece, or head, is offset so that it can hook over the side edge of a drawing board or table, sliding up and down so the long straightedge of the T square can be used to draw parallel horizontals on your paper (*47*). Inexpensive ones made of wood and plastic can be bought in stationery or art stores for a few dollars. One about 1½′ to 2′ long will be fine. Check that the head is firmly attached to the straightedge.

The 45° and the 30°–60° right *triangles* (*48, 49*) are used for drawing verticals and angled lines. They are flat and made of plastic, wood, or metal. The clear plastic kind are great and cost a dollar or two each at stationery stores. They come in several sizes; pick a triangle whose longer leg is 8″ to 10″. The legs of a right triangle are those two sides whose intersection, or vertex, forms the 90° angle (*50*). The third side, which is always longer than either of the two legs, is called the hypotenuse.

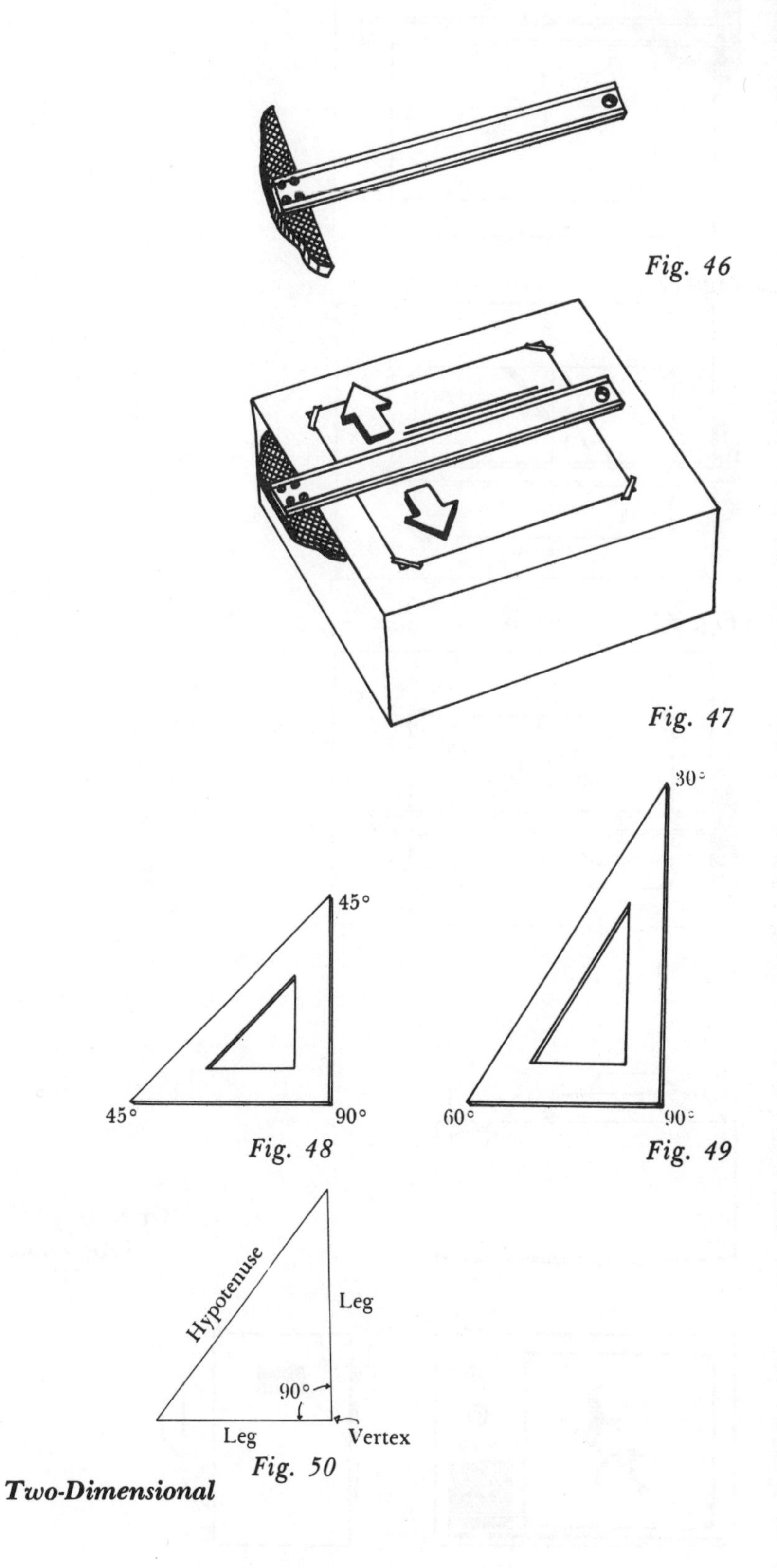

Fig. 46

Fig. 47

Fig. 48

Fig. 49

Fig. 50

Two-Dimensional

Find a table or other hard, flat surface with at least one straight edge on the side. Round tables don't make it. A piece of smooth plywood with one good edge will do, too; or buy a drawing board for a few dollars. Lay your paper on the table

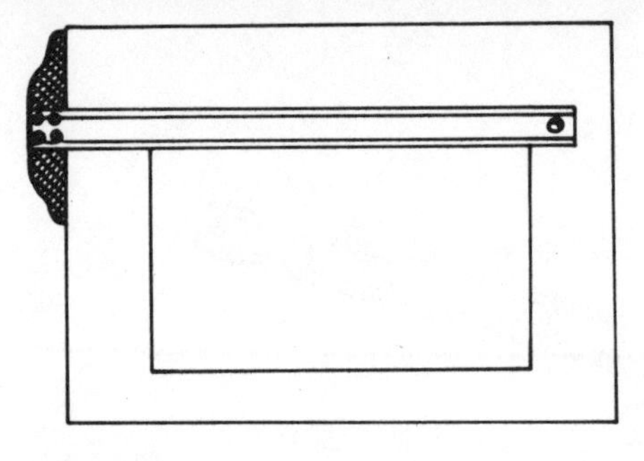

Fig. 51

Fig. 52

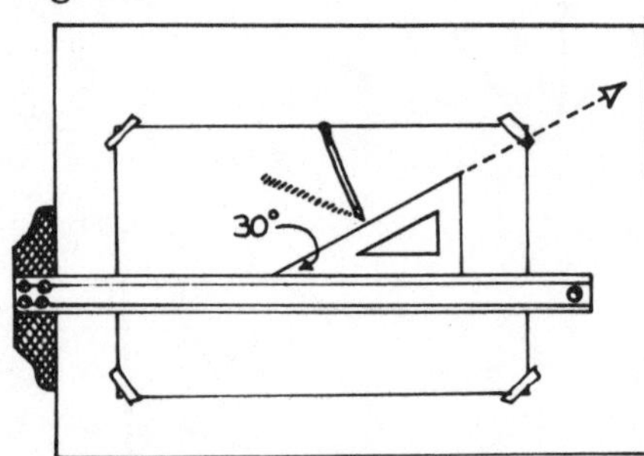

Fig. 53

and your T square across the paper, the head butted against the table side. The head goes on the left if you are right-handed, on the right if you are left-handed. Always keep the head tight against the edge of the table.

Line up the top or bottom edge of the paper parallel to and within the span of the T square (*51*). Tape the paper in place with some masking tape at the corners. Press the tape lightly so it won't rip the paper later when you remove it. Now all lines drawn along the T square will be perfectly horizontal and parallel to the top and bottom edges of the paper.

To draw a vertical, lay either triangle on the paper, with one leg against the top edge of the T square. Draw along the other leg (*52*). To draw a line at a 30°, 45°, or 60° to the horizontal, just place the leg that forms the side of the angle in question against the T square and draw along the hypotenuse (*53*).

Two-dimensional scale views are easy with these tools. Follow the same steps as with a freehand drawing, only make all your lines with the T square and triangle. Start from the bottom of the drawing and work up. This requires less shifting of the tools. Draw each line in lightly, and a little longer than you need. With your ruler or architect's scale, mark off the proper distance on the line. Continue the same process with each line before you go on to the next. Darken them in as you mark them off, or wait till all the lines are drawn and measured.

Begin the scale front view of our table by positioning the paper, taping it down, and drawing a light line along the T square for your bottom margin. This is also the base line for your table. Mark off the 4′ scale distance for the table width on this line. Move your T square a bit below the line and lay the leg of a triangle against it. Line up the other leg with one end of the 4′ line and draw a light vertical up. Repeat at the other end of the 4′ line, and mark off the proper 2′ scale distance on these verticals for the height of the table. Connect the top of these verticals. Continue in the same manner with all the thickness lines. When you're done, darken in the important lines.

Orthographic Projection

Your new layout tools also make it possible to construct a special set of scale views called an orthographic projection. Figure 54 is an orthographic projection of a TV, and shows the scale views of the top, front, and right side, at right angles to each other. (It would show more, only the screen is broken.) Once the front view is drawn, the layout tools eliminate a lot of the measuring that would normally have to be done for the top and side views.

This special set of views is one way of seeing three sides of an object at once. It provides a very full blueprint of your project. Details not visible on the outside can as usual be shown by broken lines, or by special inserts drawn elsewhere on the page. The bottom, left-side, and rear views can also be

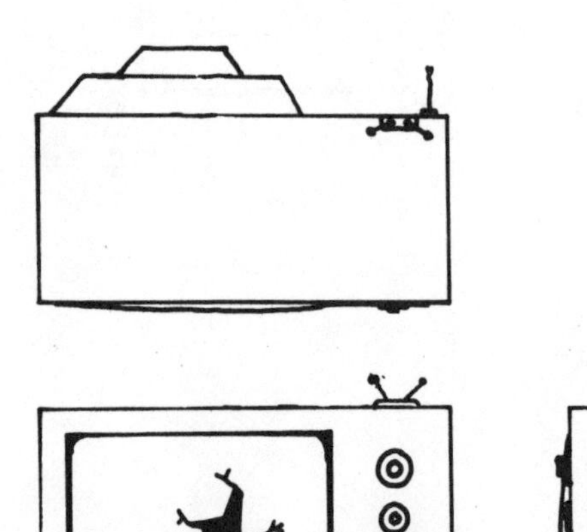

Fig. 54

drawn when necessary. Orthographic projections are not necessary for every project, but you will find them helpful on the more complicated ones. They can eliminate much grief from error in construction.

Use this style once you have a good idea of the size and shape of the object and want to see the exact form. You can still play around with the lines, but it will take more time than on a rough sketch.

Here comes that table again. Find a scale that will fit all three views on one page. You *must* use the same scale for all three views. Leave the minimum 1″ margin around the edges, and at least 1½″ between top and front views and between side and front views—the margins between each pair must be equal (you'll see why later). On an 8½″x11″ sheet you will be left with a 6½″x9″ drawing space, of which only 5″x7½″ is real drawing space after the 1½″ inner margins are subtracted. You know, or can see by a rough sketch, that you have a vertical total of 2′ (table height) + 3′ (table depth) = 5′; and a horizontal total of 4′ (front width) + 3′ (side width) = 7′. You have to fit this 5′x7′ total to scale within a space of 5″x7½″. A 1″ = 1′ scale would fit these views in very neatly.

Tape the paper in place on your drawing table, laying the longer dimension from left to right. Write your scale in one of the corners. Use your triangle and T square to draw all the lines. Draw a horizontal across the paper for your bottom margin of 1″, and a vertical for the left margin (*55*). The intersection of these lines will be the bottom left corner of the table. Draw your scale front view from this point just as you did earlier, only lightly extend all the verticals to the top margin and all the horizontals to the right margin (*56*). These will be guidelines for the other views. (In the drawings on these pages, we substitute broken lines for the light guidelines that you would draw.)

The top view will obviously be a view of the tabletop, as if you had leaned your head over the top of the front view. Mark point *C* on the left margin, 1½″ above the very top of the front view (your margin between views). Mark point *D* the scale equivalent of 3′ above *C*, representing the depth of the tabletop. Draw horizontals from each, continuing them to the right edge of the paper (*57*). The points where these horizontals intersect the extended verticals *A* and *B* of the front view define the corners of the top view. When you darken in the lines between these corners, you will have your top view.

These corners also represent the outside corners of the legs. These legs are hidden from sight by the table top. If you want to show their hidden lines, first mark point *E*, the scale equivalent of 3½″ above *C*, to show the thickness of the legs. Do the same with *F*, the same distance below *D*. Draw light horizontals from *E* and *F* to the right margin, and dot in the outline of each leg at the corners of the top view (*57*).

Because of the layout of these views and the accuracy of

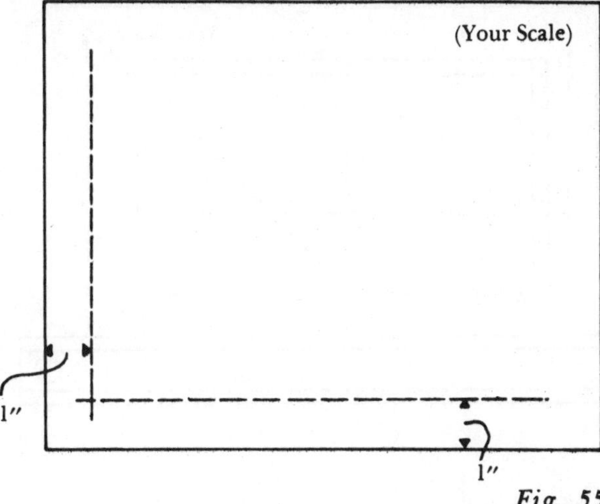

Fig. 55

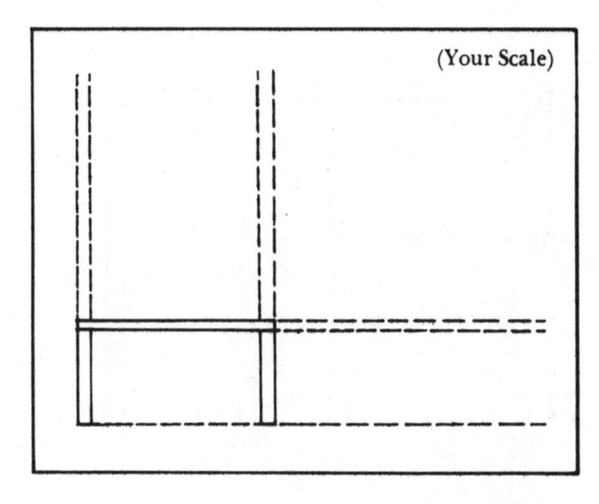

Fig. 56

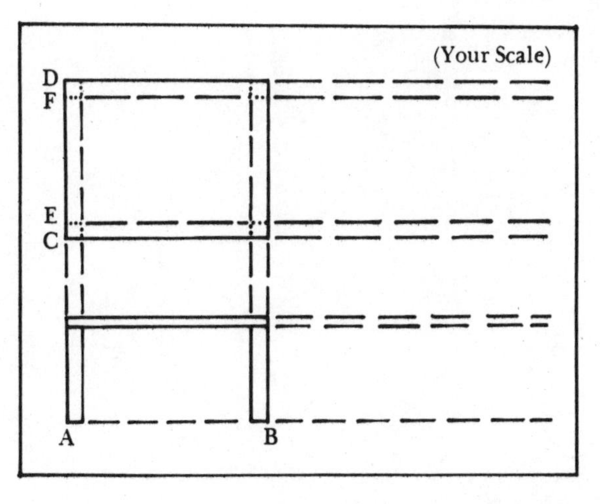

Fig. 57

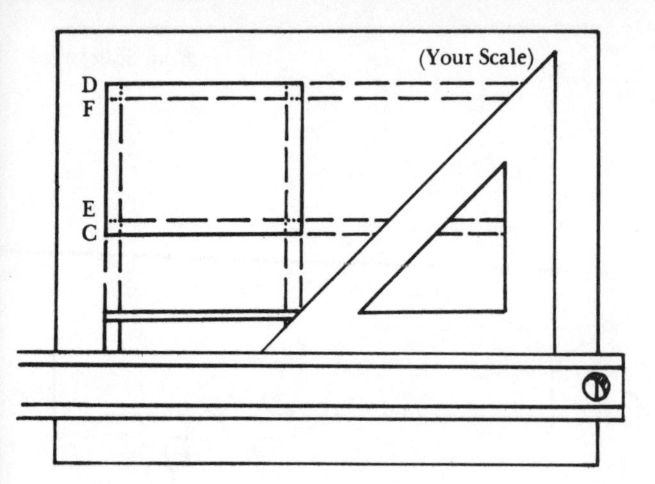

Fig. 58

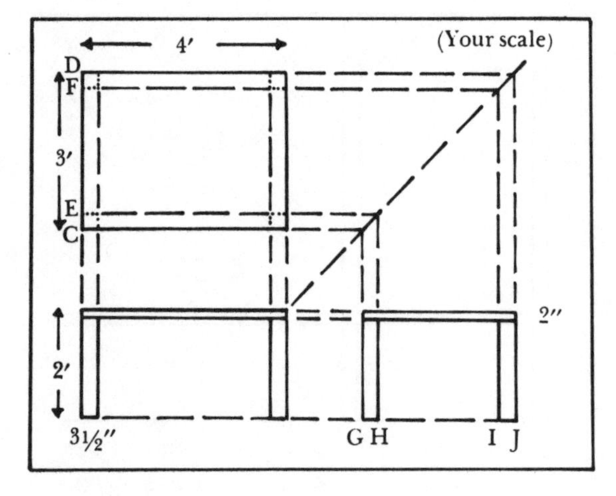

Fig. 59

your layout tools, you do not have to measure for every distance. The extended guidelines do it for you, saving trouble and eliminating careless mistakes. You can also *see* at a glance that certain dimensions in one view equal corresponding dimensions in another.

A neat trick with the 45° triangle helps define the side view without the need to make any measurements. (Now you'll know why you had to allow for the same 1½″ margin on the right, too.) Lay the triangle on your T square so that the hypotenuse intersects the upper right corner of the front view and continues on to the upper right of the paper *(58)*. Draw a light line along the hypotenuse. You'll notice that it intersects all the horizontals from the top view. At each intersection, draw a vertical down to the bottom margin *(59)*. As long as the diagonal is drawn at exactly 45°, or half of a right angle, the distance *between* any two horizontals will remain the same after they turn the corner and become verticals. $CE = GH$, $DF = IJ$, and so on.

The distance between the horizontals from *C* and *E* show the 3½″ depth of the front table legs in the top view. They turn the corner at our diagonals and become verticals intersecting *G* and *H;* but they still describe that same 3½″ depth on the front legs, only now in the side view. Similarly, all the other horizontals of the top view correspond exactly to the verticals they become.

The proper height of each side-view element is defined by its corresponding lines extended from the front view. Thus, no measuring at all is needed in this side view *(59)*. Its exact shape is defined by the intersections of the new verticals with the extended horizontals from the front view. If you have trouble choosing which lines to darken in, try to visualize the side of the table, or check one of your rough sketches. You can convince yourself of this method's undying accuracy by checking the side view with a ruler.

When the drawing is finished, write all the measurements in the margins. This now becomes your plan, your blueprint for construction.

Practice orthographic projections with other shapes. Start with simple ones, like rectangular solids, and work your way up to more complex shapes. Remember to draw the diagonal with the 45° triangle, not the 30°–60°. Remember to use the same scale for all the views on one page. The examples back in Figures 29–32 are actually orthographics; use them as reference if you have problems visualizing the views of difficult shapes.

Three-Dimensional—Isometric

The orthographic projection is a good building plan, but some people find it too fragmented to be able to grasp the full picture of the object. The oblique drawing, unfortunately, cannot be drawn to scale realistically. If each line in an oblique were drawn to scale, the object would look deeper than it should. The 5′ cube in Figure 60 is drawn to scale, and yet *BC looks* longer than *AB*. The effects of perspective make the depth line *BC* look smaller than the front line *AB*. Place a book or a box of some sort in front of you and you can see this effect, called *foreshortening*.

You can draw obliques to scale and just accept the visual distortion; or you can shrink all your diagonal lines by some constant proportion—say ⅔—to approximate a more realistic drawing.

A much simpler and more realistic three-dimensional view is the *isometric*. Take your box and turn it so that one vertical edge faces front and its two sides recede at equal angles to the front horizontal (*61*). The top of the box can then be tilted toward you till the vertical edge recedes and is foreshortened at the same angle as the horizontal side edges. In this position, all the lines of our 5′ cube can be drawn exactly to scale, since they all recede equally and all look the same length.

Four rules simplify the drawing of isometrics. One: Always align the object with a vertical edge facing front. Any edge will do, but have the more important sides facing front. Two: All horizontals in a rectangular object are drawn at a 30° angle to the horizontal (*62*)—running in the corresponding direction, of course. Three: All verticals in the object remain vertical in the isometric. Four: All horizontals and verticals in a rectangular object are drawn exactly to scale in a scale drawing. Don't be confused by the slanted lines of the isometric. You could, of course, draw a freehand isometric sketch without worrying about scale at all.

Nonrectangular objects are trickier. We'll return to them later. But first, let's draw an isometric view of our table, a rectangular object.

Tape your paper down as usual and choose a proper scale. The final size of an isometric is a little uncertain at first, so begin with a small scale which you're sure will fit the whole object on the page. Draw the bottom margin with the T square. Choose one vertical edge of the table to be the front edge of the isometric. Since the 3′ side of the table to the right of this edge is shorter than the 4′ side to the left, position this vertical edge to the right of page center (*63*). This will help center the whole picture. Draw this vertical the scale equivalent of 2′ high, measuring from the bottom margin. Draw the 30° lines for the horizontals with your 30°–60° triangle and T square. Keep the long leg of the triangle against the T square, with the hypotenuse rising to the left or the right, depending on which way the line is rising. Draw your lines lightly as always. Mark off the scale measurements.

Note that the fourth leg in the back is not visible. If the legs had been longer, it might be visible (*64*). This is something to watch for. The visible part of the back leg can be located either by drawing its guidelines down from the top back corner, or by extending 30° guidelines from the bottoms of the outer legs.

With a little practice, the isometric style will become almost automatic. You will even be able to make creditable

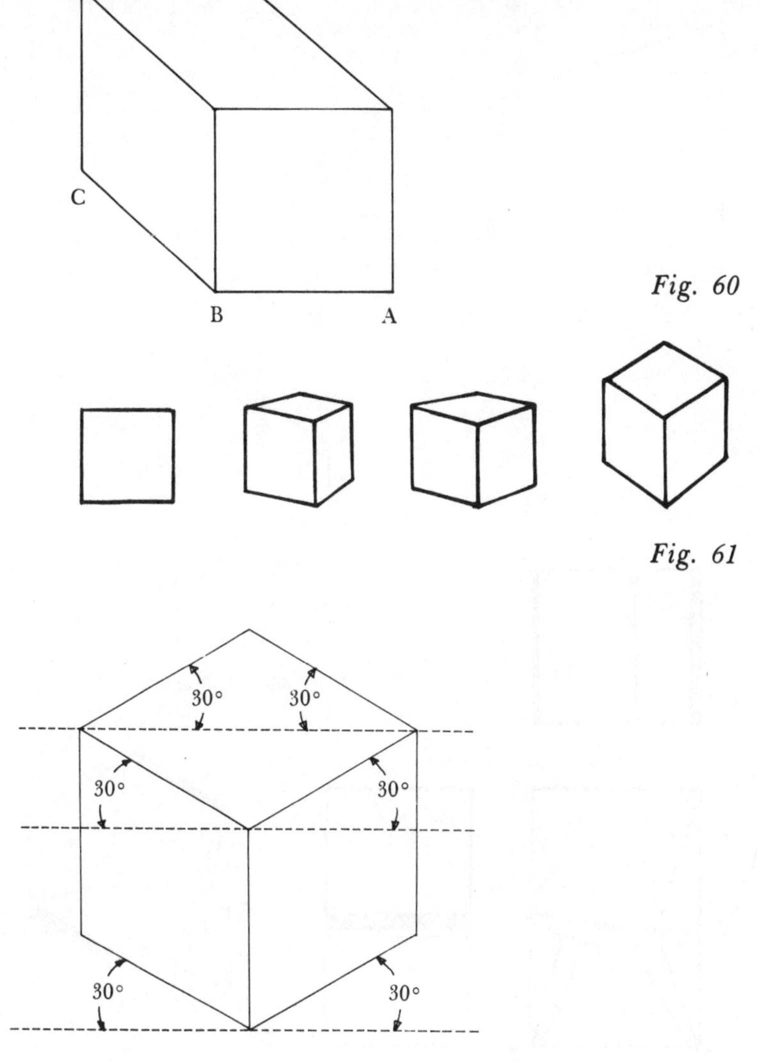

Fig. 60

Fig. 61

Fig. 62

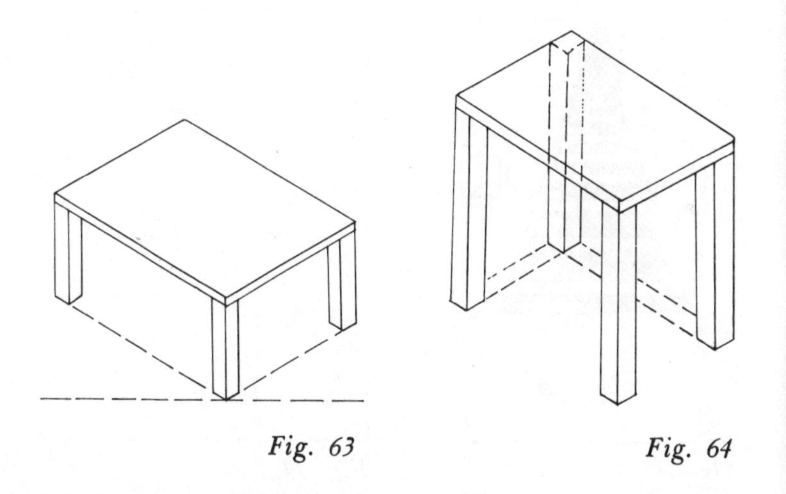

Fig. 63

Fig. 64

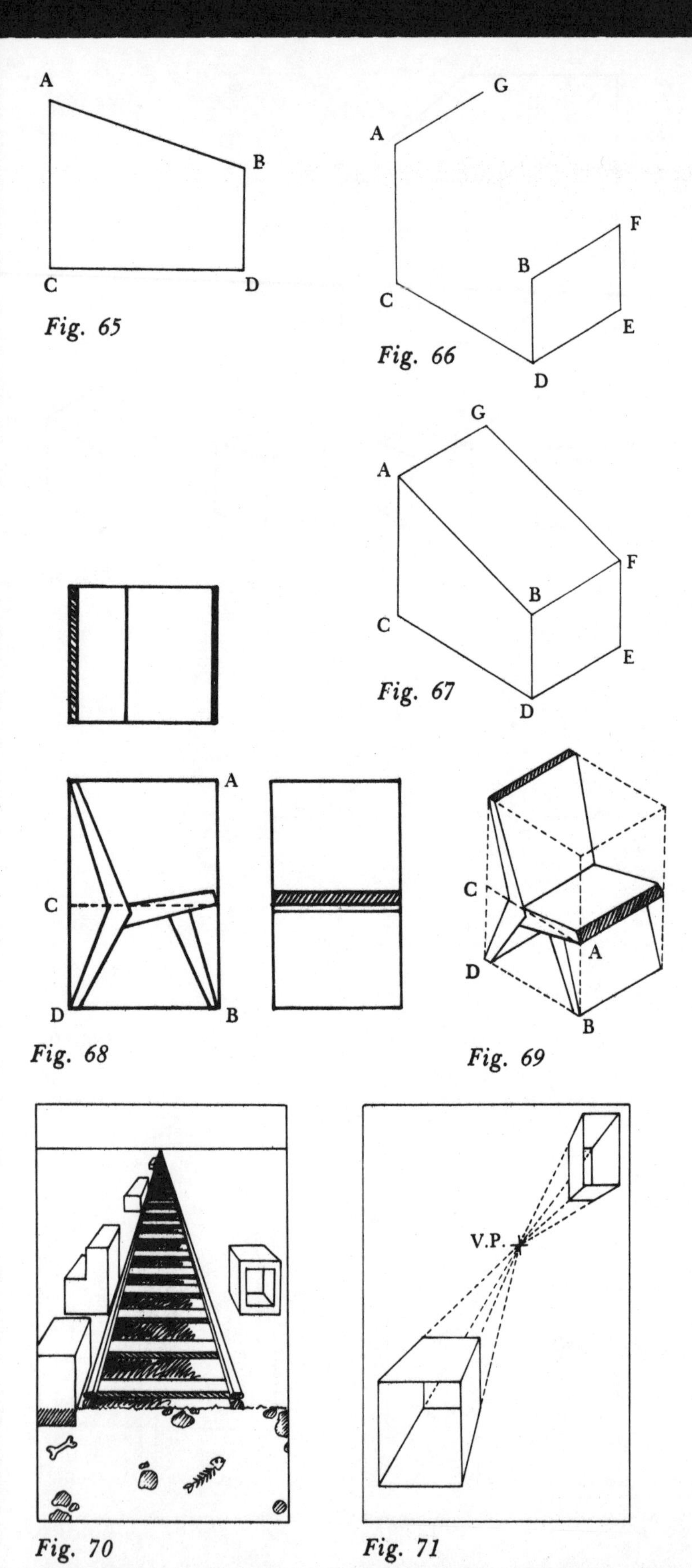

Fig. 65

Fig. 66

Fig. 67

Fig. 68

Fig. 69

Fig. 70

Fig. 71

freehand sketches, estimating the 30° angle with the eye of an eagle. It is probably the most realistic-looking of all the styles.

Complications arise in nonrectangular objects. Simpler shapes of this sort can be finessed within the previous techniques. Because line AB in the front view of Figure 65 is neither horizontal nor vertical, AB in an isometric view would recede at a different angle than the horizontals and verticals of the object. Line AB would be foreshortened to a different degree, and therefore could not be drawn directly to scale. You finesse the problem by drawing all the verticals and horizontals as usual *(66)*, leaving AB and GF till last. At this point, AC and BD are their proper scale heights. Their tops, A and B, define the length of the line AB. Draw a line from A to B. Do the same at GF. The figure is completed *(67)*. It would not have turned out the same shape if you had drawn AC, CD, and AB to scale, and then connected BD.

More complicated nonrectangular shapes are beyond most amateur draftsmen's patience. They are better drawn in orthographic projection or a very rough freehand isometric. For the brave ones among you, however, we'll show the basic technique for drawing such a scale isometrically. The trick is to draw the three views of the orthographic projection first, and box each view in with the smallest possible rectangle *(68)*. Draw an isometric scale view of the box defined by these three rectangles. Imagine the box as made of clear glass. Mark off on the isometric box the points of intersection with the troublesome shape, as measured in the orthographic views. $AB = A'B'$, $CD = C'D'$, and so on. Since these marks are measured along the horizontals and verticals of the two-dimensional boxes, their corresponding scale distances from each other along the edges of the isometric box will remain the same. After all the points of intersection are located, connect the dots with solid lines *(69)*.

It's really a multiple finesse, and not all that difficult, once you are familiar with orthographic projections and isometrics. It does take time, though.

Perspective

If you stand on a railroad track and look along its length, you will see the rails converge in the distance *(70)*. The ties grow thinner and closer together. This is an illusion, of course. The tracks are not really smaller up ahead; but they do look as if they were converging toward a single point.

One-point perspective drawing creates this same illusion by running all the depth lines of an object to a single point in the distance called the *vanishing point* (*V.P.* in the drawings). The front of the object is drawn as a simple front view, as in the oblique. The back edges are also drawn as in the oblique, only smaller. They are farther away, and are foreshortened by the perspective *(71)*.

Two-point perspective uses two vanishing points, one to

each side (*72*). The object is oriented with a vertical edge in the foreground, as in the isometric, and the sides recede toward their respective vanishing points. All horizontals receding to the left would, if extended, intersect at the left vanishing point. All horizontals receding to the right run to the right vanishing point. The two vanishing points are located anywhere along the *same* horizontal, at about eye level. If the points are too close to the object, however, an unrealistic foreshortening of the object appears.

Nonscale perspective sketches are easy to make. They give another type of realistic three-dimensional view (*73*). Scale perspective drawings are great for seeing an object in its planned environment. You can draw the walls of a room, the existing furniture, and then draw in your new project. Such drawings are not particularly difficult to construct, but the technique involved is extremely tedious and time-consuming. Fortunately, you can get small interior perspective charts that will simplify scale construction immensely. They are available at art stores for about four to five dollars. A chart consists of the outline of a room and a grid pattern, radiating from the vanishing point(s). The squares of the grid represent scale distances. They can be thought of as tiles covering the room, every tile the same size (*74*). Lay a piece of tracing paper on the chart, trace the wall outlines, and draw your project in place, using the squares as your guide for measuring distances.

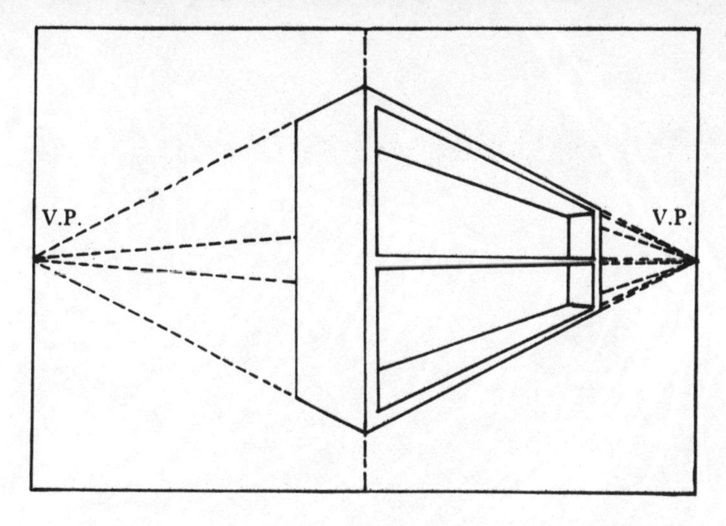

Fig. 72

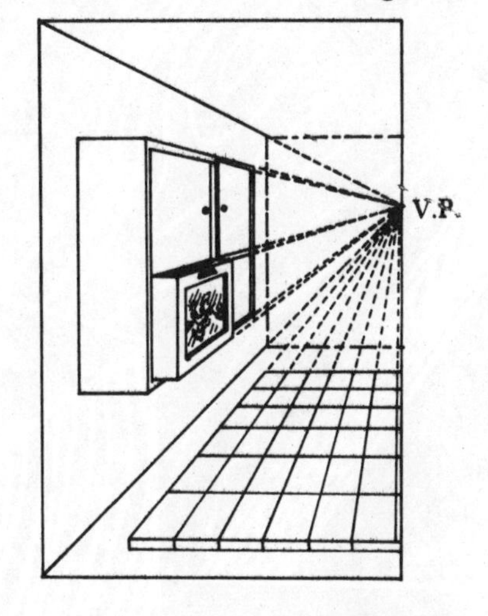

Fig. 73

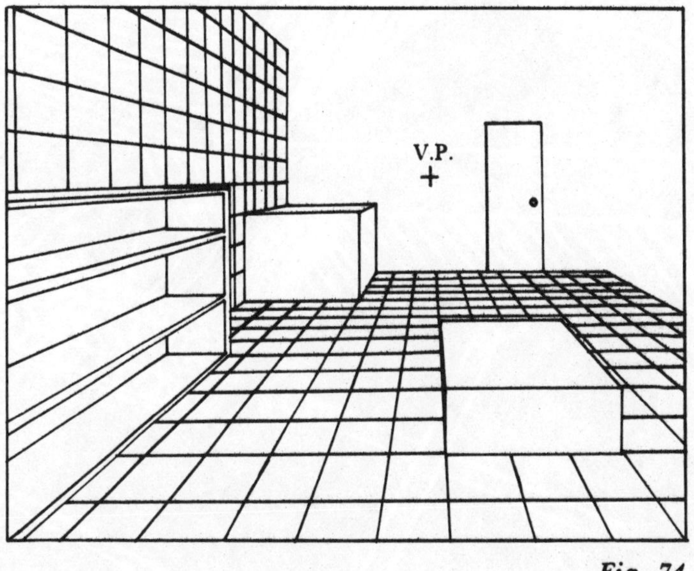

Fig. 74

WHEREWITHAL

3/WOOD AND MATERIALS

Descriptions and Uses of Common and Uncommon Wood and Materials

Wood is the main ingredient in carpentry, so it pays to get good wood and reliable service. Therefore we'll first discuss how to find and deal with a lumberyard, then the variety of materials you can buy there. Of these materials, we'll cover the different types and sizes of lumber, their names, and their uses; we'll then do the same for the most common types of other building materials that you might be interested in, such as paneling, fiberboard, plastic laminates, and plasterboard; next we'll describe a number of important "soft" building materials—adhesives, caulks, wood putties, plaster, and joint compound; and finally some miscellaneous materials that are midway between lumber and hardware.

We also want to mention here another invaluable source of wood and materials: your local streets. Many of the materials you need may be lying abandoned and homeless around the block. Find out the days of the week for garbage pickups of large objects. Check out places where renovations are being done. Look for the large "dumpsters" out in front of buildings; usually anything in them is fair game. Also, look or ask at glass suppliers and other businesses for crates they don't need. Some of the finest materials can be found free in the streets.

Finding and Dealing with a Lumberyard

Wood and other building materials are bought in a lumberyard. Get to know a good reliable yard. Ask friends if they know any. You may end up going to a few different ones for different things. One may be great for 2x4s and plywood but weak in their molding selection. The important thing is to take your time and look around. Find out about delivery and cutting costs. Will they let you pick out the wood yourself? Compare prices—you'll be surprised how much they vary. Talk to the people at the yard. Be friendly and they will usually be friendly in return.

If you order a load of materials and some of it turns out to be bad, don't hesitate to refuse delivery of the bad or damaged pieces. If you accept it, they may keep sending you bad wood each time. With common-grade lumber, the wood will not be too great anyway—don't expect perfectly straight pieces. A lot depends on what you are building. Certain deformities will not interfere with some applications (see "Structure").

Plasterboard should come in one piece, with no damaged corners. Plywood "good one side" should be just that, with no gouges in the veneer. The standards vary with each type of material, of course.

When buying unboxed nails, try to watch the weighing or weigh them yourself (not that all yards or even some are crooked, but when people rush, mistakes can happen).

If your materials are to be delivered, try to get an arrival time from the yard. Usually delivery is the day after you give the order. Your order will be one of several on the truck, so that "We'll be there before noon" may mean any time from 7:30 to 12:00. This may also be the time the yard is giving out that day to all who call. Plan on waiting, because you can be pretty sure you will. If it's late, don't be afraid to call and ask where your order is, and don't be surprised to hear the quaint, "It's on the truck!"

Most yards will take payment when the material is delivered; some might ask you for a deposit. Some take checks. One good reason for getting to know the people at a particular yard is that they will take your check and will send you better wood. Most yards will give you a free nail apron (with their name on it) and pencils just for the asking. You may also find racks with free literature on new products and how-to-do-it information from manufacturers; some sort of hardware section, large or small; and even a tool-rental section. Don't hesitate to ask questions—lumber people respond to interest in their work.

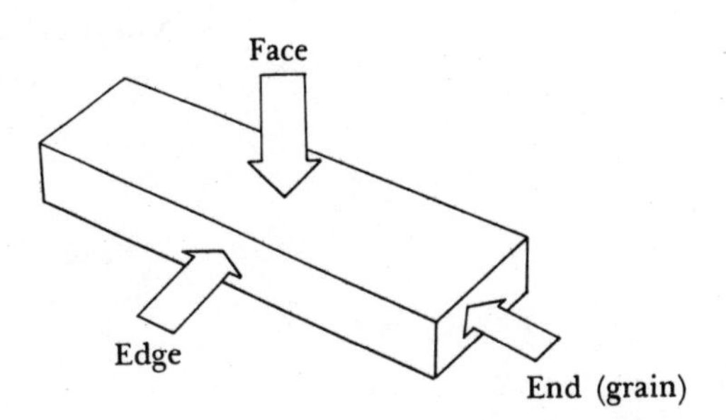

Fig. 1. Wood Surface Names

Order your materials to get the most efficient use out of your wood. Get common or construction grade lumber a little longer than needed. The ends may be split or have a knot on them that you will have to cut off, so if you need a 10′ piece, get a 12-footer. Remember also that you have to allow for the saw cut. If you need two 4′ pieces, an 8′ one will not be sufficient; order a 10-footer. Also if you need two 5½-footers, get a 12′ piece, not two 8-footers, to minimize waste.

Your lumberyard can probably cut things to size and rip special widths. (A rip is any cut done lengthwise.) For instance, if you need a piece 2″x6½″, they can cut a 2x8 down for you. Cuts at the yard may also make it easier to fit materials into your elevator so you don't have to carry them up stairs.

Lumberyards are the source for all types of building materials. Certain materials—like Acoustiglass (soundproofing), special doors, etc.—can be bought straight from the manufacturer at a cheaper price. Most lumberyards don't stock specialized items, so they get a high markup for them. If they say it will take a week or so to get something, you can assume they're ordering it and might want to check into getting it yourself. For large orders of materials, $200 or more, check around and compare prices. Don't hesitate to bargain a little.

What They Stock

Lumber for Construction

Most wood for general carpentry is *softwood*—probably fir or pine (sugar or ponderosa pine). Most is also kiln dried. This means that it should not shrink or check extensively (checks are cracks that appear along the grain due to uneven seasoning), but it's not always true. The wood comes in three main grades—construction, common, and clear. Some yards use common and construction synonymously. (*Hardwoods,* such as oak, maple, teak, walnut, etc., are used almost exclusively for making fine furniture.)

Construction-grade lumber may be a little rougher than common. Common will have knots and surface mars. Clear, as its name implies, is supposed to be free of knots. This is not always true, but clear *is* the best. Rough surface appearance, knots, etc., may not matter with some work and may enhance others. Clear will usually be about twice the price of the common or construction grades.

Nominal size and actual size

Lumber is identified by its *nominal* size—for instance, a 2x4 (read: "two-by-four") is nominally 2″ thick (measured on the *edge*) and 4″ wide (measured on the *face;* see Figure 1). The *actual* sizes are less than the nominal sizes—a 2x6 is actually 1½″x5½″. When the exact size to 1/16″ is important for your project, measure the actual piece of wood; you may find a further variation in individual pieces, up to ⅛″ either way. The nominal size is larger than the actual size because wood is cut at the mill to the full measurement, and then planed down for surface uniformity; it also shrinks somewhat in drying. (The nominal and actual sizes for commonly available lumber are shown in the Softwood Lumber Sizes table in the Reference section at the end of the book.)

Another trick of the names is that all sizes with the same thickness—2x4s, 2x6s, 2x8s, etc.—can be lumped together as *2-bys.* You'll find many references to 1-bys and 2-bys throughout the book.

1-Bys

1x14s are probably the widest boards you can easily find, though some boards are cut up to 24″ wide. Widths start at 1″, 2″, 3″, and 4″, then increase in 2″ increments. 1-bys are used for such things as shelves, trim, baseboards, door jambs, and boxes. Common-grade 1x2s are used for furring strips.

The lengths will start with 6′ or 8′, then move upward in increments of 2′. 1-bys may be hard to find any longer than 14′ or 16′.

2-Bys

These are standard for structural framing, general construction and so on. They come in standard lengths of 8′ to 16′, in 2′ intervals; available, but hard to get, are lengths up to 24′.

2-bys are usually fir and are available in three grades. Construction grade is very rough and used for on-site temporary work. Common has knots but is in much better shape; it's the most widely used. Clear is sometimes hard to find and isn't needed that often, since 2-bys are seldom used for finished work.

3-Bys

These are special load-bearing beams used for floor joists, etc. They will usually be fir, sometimes cedar. Lengths are the same as 2-bys.

Square Stock

4x4s are by far the most common. They are used as legs and posts in all sorts of furniture and large constructions. Stock 5″, 6″, and 8″ square is also available. They are all usually fir or cedar, in the same lengths as 2x4s.

While pine and fir are the most common woods for 1-bys and 2-bys, you can also find cedar, redwood, oak, mahogany, teak, and many others. Redwood and cedar are mainly used for outdoor applications because of their weathering characteristics (and for saunas for the same reason). The other types are usually much more expensive, but vary greatly by geographic location.

Some yards offer treated wood, usually fir, that has been dipped or saturated with a preservative for outdoor use (posts, etc.). It will usually be cheaper than redwood or cedar. Don't build an outdoor structure such as a deck out of plain fir or pine, because in a few years the wood will start to rot. This is why preservatives and paints are used on bare wood for houses.

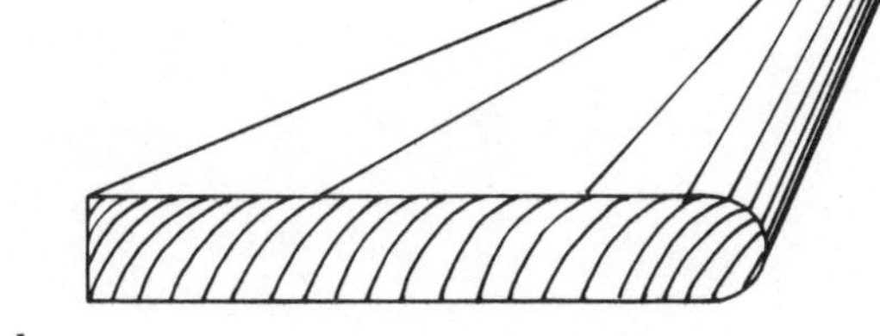

Fig. 2. Bull Nose Board

Odd Sizes

The following sizes and shapes come in clear only:

Five-quarter wood

"Five-quarter" wood (1¼″ thick) is used mainly for steps and decks (*2*). It comes bullnosed for steps or with edges squared, in pine, redwood, and oak, in lengths 6′ and up (in 2′ intervals). The full width of a bullnose step is measured from the back edge to the very front of the curved edge (*3*).

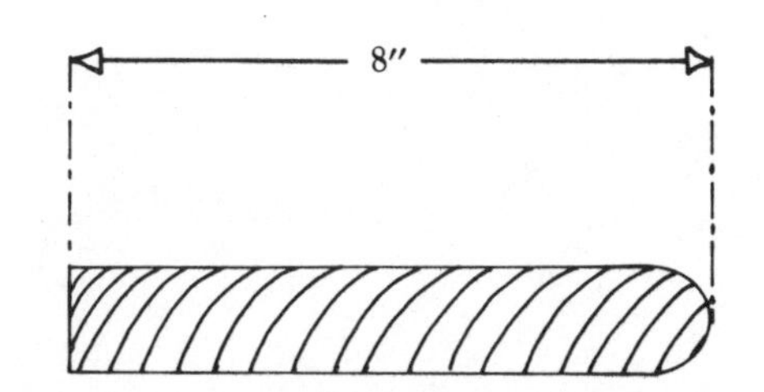

Fig. 3

Moldings

Many other sizes and shapes are available. Moldings, of course, come in a variety of cross-sectional shapes. Half- and quarter-round are among the most common, being half or quarter cross-sections of the round dowel. Dowels (full round pieces) come in various thicknesses from ⅛″ up to 2½″. Lattice is flat pieces ¼″ thick, and in widths from ½″ to 2″. Look around at your lumberyard to get an idea of the many varieties. They come in handy.

Shims

Shims or shingles are invaluable for a lot of carpentry jobs and projects (*4*). They are sold in three grades at most lumberyards. Number 3 is the roughest and used for shims, leveling, wedges or wherever you need small pieces of wood. They are wedge-shaped and come in lengths of about 18″ with random widths. The top grade is used for actual shingling on houses, the middle grade straddles both uses. All types are made of cedar and are sold by the bundle, prestrapped.

Fig. 4. Shingles or Shims

Veneers

Some yards will offer veneers. These are very thin sheets of wood, usually 1/64″ thick, sold by the square foot. They are of rare woods like ebony and rosewood that would be expensive if bought as solid pieces. These veneers are meant to be glued onto a cheaper wood to give the effect of a solid piece.

Veneer tapes

Veneer tapes are paper-backed rolls 1″ or 2″ wide meant to be glued onto the *edges* of plywood for the same solid-wood effect. They come in many common types of wood—mahogany, birch, oak, fir, etc. If your lumberyard doesn't carry veneers, look for a specialized woodcraft store in your area. There are also several that advertise widely and ship to anywhere in the country.

Panels

All of the materials that we mention are available at lumberyards, but they may be called by different brand names in various places. As long as you know what you want and what it looks like, you should be all right.

Plywood A very useful member of the lumberyard battery is plywood. The most common kind is made up of 3 or 5 "plys" or layers of fir veneer; the outside layers are usually thinner than those in the core. Each layer's grain crosses the next. (See "Structure" for a detailed explanation.) These sheets commonly come in 4′x8′ sizes, with others available. They are made in thicknesses of 1/4″, 3/8″, 1/2″, 5/8″, 3/4″ and 1″.

Besides the ply core described above, there are also *lumber-core* or *composition-core* panels, often with exterior veneers of finer wood. Lumber core is the strongest and most expensive, ply core the standard, and composition the weakest and the heaviest. The edges of plywood show the layers, hence veneer tape or an edge piece of 1/4″ wood should be applied for looks.

Plywood comes in many different types. Construction grade is rough, unsanded on both sides, and used for forms and fences at construction sites. Underlayment board, which has knotholes on its surface, is used to "underlay" other coatings such as roofs and floors. The best grades are "good one side" and "good both sides." (See Softwood Plywood Chart in the Reference section.) There are plywood panels covered with a fancy veneer, birch and oak being the most common. These are always good on at least one side, and these are used in conjunction with the veneer tape to give the appearance of solid wood, for a fraction of the cost. There are also specially treated types for outdoor use called *exterior plywood,* and a type called *marine plywood,* to be used where water resistance is required.

Composition board *Nova ply, particle board,* or *chip board,* is a panel made up of compressed wood flakes and chips. It also has a tendency to include tiny pieces of metal which throw off sparks and eventually dull every type of circular saw blade used on it. It comes in the usual 4′x8′ size (other sizes available), and in thicknesses from 1/4″ to 2″. It has also been known to crush a finger or two due to its excessive weight. Be careful.

The main use of composition board is for underlayment or for countertops that will be covered with plastic laminate, or where looks are not important. It has no grain, therefore is not as strong as wood or plywood. For the same reason, it is very brittle. Its main advantage is that it's cheaper than plywood. It usually comes sanded on both sides, and looks surprisingly good stained or polyurethaned.

Hardboard This is another very useful type of panel. You're probably familiar with it as Masonite (a brand name). It comes in two types—standard and tempered. The tempered is darker, stronger, and more resistant to weather and impact. Both types come either smooth with a gloss on both sides, or smooth one side and textured on the other. They come in thicknesses of 1/8,″ 1/4″, 3/16″, and 5/16″; and in widths of 4′ or 5′ and lengths of 4′, 6′, 8′, 10′, 12′, and 16′. They are made of refined and highly compressed wood fibers much smaller than those used in composition board (undiscernible without close scrutiny). Hardboard comes in a variety of patterns and styles, such as pegboard, and with enameled or vinyl surfaces for around showers. Most decorative paneling is made from this type of board. It has many uses, all nonstructural.

Paneling

Paneling is a type of board used for covering up old surfaces to make them look new. These boards come in 4′x8′ pieces in thicknesses of 1/4″, 3/8″, 1/2″, and sometimes 3/4″ or 1″, the thicker two being mainly for outdoor use such as siding on a house. They are made of either hardboard or plywood, and come in many different variations, patterns, and styles, as well as prices. If you decide that you want to panel a room, check the papers for sales and shop around.

Fiberboard

Known commonly by brand names like Homasote or Celotex, this board is made up of condensed wood, cane, or vegetable fibers, not as dense as the hardboard panels, usually grayish or yellowish, and almost always 1/2″ thick. It is used as an insulation panel, and is sometimes as much as 2″ thick. Usually used in 4′x8′ sheets, fiberboard lengths vary from 4′ to 12′, in 1′ increments. Fiberboard is good for absorbing sound due to the pockets of air trapped in the condensing process. It has virtually no structural strength and can be broken off by your hands.

Plastic laminates

Plastic laminates are sometimes called by brand names such as Formica. Most yards probably don't have a large stock and you have to order what you want. These are layered sheets of resin-impregnated board used for things like kitchen counters and bathroom-vanity surfaces. These laminates come in a standard thickness of 1/16″, with thicknesses of 1/20″ for curved surfaces and 1/10″ for surfaces that have defects under them. They are available in many sizes, common ones being 4′x10′ or 4′x12′. Many different colors, patterns, and textures are available. Your lumberman will have sample chips for you to look at.

More than most materials, these laminates vary greatly in price from place to place. Their surfaces are extremely tough and resistant to water, and some are burn-resistant. They are not structurally strong, however. This type of board is applied over a smooth clean surface, and glued down with contact cement. (See the Built-in Kitchen Cabinet project, in "Furniture Projects".)

Plasterboard

This is probably covering the walls in your house. It's called wallboard, plasterboard, Sheetrock, or gypsum board. This type of panel is made up of a sandwich of a smooth, paper-thin cardboard, a layer of plaster, and another one of cardboard a little thicker than the first. It is available in thicknesses of 1/4″, 3/8″, 1/2″ and 5/8″, and in sheet sizes of 4′ by 4′, 8′, 10′, and 16′; 4′x8′ is the most common.

It is designed for covering walls and ceilings; 5/8″ thickness is required for fire walls. Its fire-resistant rating is judged by how many hours it will take for the fire to burn through. A waterproof kind is also made, with a green-blue surface. It's meant to be covered with tile or the equivalent.

The most common of all plasterboard used is the 1/2″. The thinner, lighter ones are for patching or for ceilings. Most are tapered along the vertical edge for the joint compound and tape. When they are applied correctly you cannot see the seams (see Plasterboard in "Room Renewal"). Plasterboards are shipped in pairs, the good faces against each other, with a strip of paper covering the edges of both, but they can be bought separately.

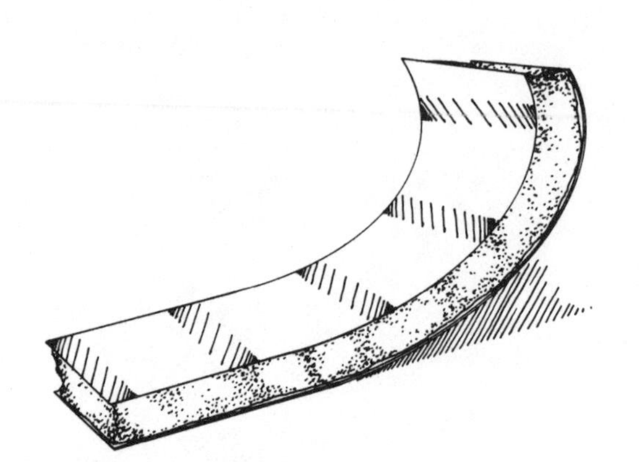

Fig. 5. Fiberglass Insulation

Insulation The material that keeps us all snug and warm in the winter and cool in the summer is insulation. Fiberboard has been described above. Your local yard may not carry a wide selection of insulation, but it should have some of the many different kinds available.

Fiberglass The most common is fiberglass (5). This is usually pink like cotton candy. It comes in various thicknesses, 3″ being the most common, with paper on one side and nothing on the other; or with foil for heat reflection on one side and either paper or nothing on the other; or with paper on both sides. It's designed to go between the studs of your walls or the beams of your floor or ceiling. It comes 15″, 19″, and 23″ wide, and in long rolls. It's also available in batts or sheets which may be 24″ and 48″ wide, mainly for attics.

Bags of loose insulation are also available, meant to be poured into attic spaces between rafters.

Fiberglass is very dangerous to handle without protection. *Always* wear long sleeves and long pants, gloves, goggles, and a breathing mask for maximum safety. This cannot be emphasized enough. It may not feel harmful, but it is. To cut the sheets, use a mat knife; attach them with a staple gun.

Styrofoam sheets Another type of insulation is rigid styrofoam sheets, usually 2′x4′. They are very light and easy to apply with glue or nails, and can even be left exposed. Styrofoam is not dangerous to breathe the way fiberglass is.

There are also sound blankets, and sound-deadening fiberboards for sound insulation (see Soundproofing in "Room Renewal").

Asbestos Asbestos sheeting is used for heat protection. You can separate heat sources from damageable surfaces, such as stoves from walls or the floor.

There are two types of asbestos, flexible and utility. Flexible is made in thicknesses of 1/8″, 1/4″, 3/16″ and 3/8″, with the utility coming in 3/16″, 1/4″ and 3/8″. Both come in sheet sizes of 32″x48″, 48″x48″, and 48″x96″. The color of both is light stone gray. The thicker types must be drilled before nailing. The sheets are very brittle and also dangerous if a piece enters an open cut or is swallowed.

Adhesives No matter what you want to stick together, there is most likely an adhesive that will do it. There are literally hundreds of adhesives. If you have a specific problem not covered here, ask your dealer.

White glue White glue is the one most commonly used for woodworking. It is cheap, particularly if you buy it by the gallon. Buy a brand name (like Elmer's or U.S. Plywood). A good bond with white glue is as strong as the wood itself. Using it is simple. Spread glue well on both surfaces and join them together. The pieces can still be moved and adjusted for a minute or so. For the bond to set well, it is necessary that the pieces be held together tightly while the glue dries, which may be twenty minutes to an hour. If the pieces are not also screwed together or otherwise held tightly, clamp them together.

Gluc that squeezes out on the finished surface should be wiped off immediately with a cloth soaked in warm water. Glue that dries on the surface will turn clear, but it will not take a stain and is very difficult to sand.

Hide or animal glues

Hide or animal glues are similar to white glues in texture; they are a tannish color and produce an even stronger and quicker bond. They dry a semiwood color and can be sanded. They require clamping just as white glue does.

Contact cement

Contact cement is useful for lighter bonds, like gluing thin paneling, veneers, and plastic laminate to wood and other materials. It is not good for heavy wood structures, or for wood joints under any stress. It is applied to both surfaces with a brush, stick, or almost any kind of spreader. If one surface is horizontal, you can pour it on and spread it around. Let both surfaces air-dry for ten to fifteen minutes until the glue surface becomes dull and dry to the touch. (You can wait up to an hour.) Then stick the pieces carefully together—*very* carefully, because they cannot be moved about once the surfaces touch. One trick is to put a slipsheet—a sheet of paper or plastic—over one surface. Lay the other piece in place, and slowly pull the slipsheet out. The glue will not stick to it—the glue sticks only to itself at this point. When pieces are in place, press hard on them, or run a roller over to make sure the bond is tight. Press from the center outward, to avoid trapping air bubbles. With something like plastic laminate, you might even tap lightly on a block of wood with a hammer over the entire surface.

No clamping is needed with contact cement, since it achieves up to 75 percent full holding power immediately upon contact. It can be cleaned off with a scraper when still tacky, and later best with a commercial solvent or thinner. Follow directions on the can.

Mastic

Mastic is a heavy glue or paste, usually black or dark brown. It's the only adhesive that works well with tile and linoleum. It is also used for wall paneling and other sheathing applications. It comes in one- or five-gallon cans or in caulking tubes. With paneling you have to tack each sheet in three or four spots to hold it up while the mastic sets. You can adjust the panels for several minutes after application.

Mastic should be applied from cans with a tooth-edged trowel, or with something improvised like a piece of wood that will leave a rough surface and uniform thickness.

Epoxy

Epoxy is a super glue, the kind you see in advertisements boasting that one drop will support an elephant from a crane. It comes in two tubes that have to be mixed for application.

It's good for small areas. Be careful not to get it on you. There are reports of people who got some on their fingers, rubbed their eyes, and glued their eyelids shut.

Silicon Silicon glues are waterproof and flexible when dry. They are used mainly in bathrooms, and come in caulking tubes. They are good for improvised solutions—for instance, if you have a car ashtray that is too big for its holder, set it in a bed of silicon glue.

With all adhesives, follow directions on the container. Ask your dealer for suggestions if you're not sure what to use. Expect that whatever applicator you use will be rendered useless for practically anything else.

There are many other glues, such as hot glues and casein glues. The ones we've listed, however, should take care of any woodworking need you have.

Caulk Caulk is a gooky substance, slightly thicker and stickier than toothpaste, used to plug holes and seal cracks in walls and around the house. It never fully hardens; it remains flexible enough to allow for slight expansion or contraction between wood or other materials over the changing seasons. It's essentially a weatherstripping, a better seal than felt strips. You don't want to use it, however, in places where it might be stepped on or where something like a door or window will move against it. It can break easily under any stress and may stick to whatever touches it.

Caulk comes in a cylindrical tube with a plastic nozzle. You apply it with the aid of a caulking gun. (See "Hand Tools.")

For simple caulking jobs, cheap caulks are fine. They won't last as many years as the more expensive kind, that's all. The better ones are a little easier to apply and last longer. Use the better kind in spots that are a hassle to caulk, that you don't want to redo every year or so. Or where neatness counts.

Caulks can also be used for waterproofing, as in bathrooms. Here it pays to use a good caulk. Buy silicon caulk, which dries like a single band of plasticky rubber. The bond is so good the caulk won't come off even when you pick at it. Silicon caulk is also the best thing to use in soundproofing situations, particularly around set-in windows and panes.

Wood Putty Wood putty is used to fill cracks and holes in the wood in your projects. It's particularly useful for filling in the holes over countersunk nails and the hairline gaps between joined pieces. One kind is a ready-mixed *plastic putty* that you can apply right from the can. It comes in a natural color, and in several different tones like oak, dark oak, and walnut to match the wood you use. Another type is a *water putty,* which comes in a powder form. Mix it with water and apply. This type dries very fast. You can add powdered pigments to it to get the tone you want. We prefer this powdered putty for any wood that stays natural or just gets stained.

Both types shrink a little as they dry. Apply more than you

need, then sand down the surface when dry. For deeper holes, the putty is best built up in several thin layers, letting each one dry before applying the next.

Plaster Materials

Plaster

Loose powdered plaster is sold in 1-lb. lots from a large barrel at some yards. Most, however, sell it in bags of 5, 10, 25, 50 and 100 pounds. Note there are different types of plaster for different jobs. If you want to patch small holes, use patching plaster or spackling plaster—both are sold in various sizes. *Plaster of Paris,* mixed with lime, is used for finished wall surfaces; a thicker, more cohesive type of aggregate plaster, such as Structolite, is for basecoats and rough filling of large surfaces, and can also be used for imitation stucco interior surfaces.

Joint compound and paper tape

Used together, these articles cover the seams between plasterboard sheets. (See Plasterboard in "Room Renewal.") Compound comes in powder or as a ready-mixed paste in cans of a pint, a quart, one gallon, and five gallons. It is a plaster derivative with an acrylic (plastic) base. The larger amount you get the cheaper it is; you can save even more by getting the powder form and mixing your own, but it's a lot of trouble and not really worth it.

The tape is a paper tape made just for wallboard seams, with a crease down the center for corner applications. It is specially made to allow the compound to permeate; some kinds have prepunched holes.

Miscellaneous Materials

J-molding

Metal corner bead

Also available for your work with plasterboard are metal corner bead and J-molding. The corner bead is used instead of the paper tape on outside corners; the metal protects the vulnerable corner against damage. The bead is nailed on both sides from floor to ceiling and the compound is applied directly over it. J-molding is often used on exposed single edges, such as around openings, again for protection from damage. The J-molding is slipped over the edge before the sheet is fully nailed down; then you just put a few nails through the plasterboard into the wider back lip to hold the strip in place. Some people then tape over the J-molding; others leave it exposed or paint it.

RC-1 sound clips

RC-1 sound clips are used in special soundproof wall constructions. (See Soundproofing in "Room Renewal.") They are long metal channels that are attached horizontally to a stud wall; the plasterboard is then attached directly to them rather than to the studs. The isolation the clips create between the plasterboard and the studs helps cut down the transmission of sound vibrations.

Nails and general construction hardware

All lumberyards also carry a wide selection of nails and general-construction hardware like joist hangers and post and framing anchors, some of which are listed in "Hardware." In many parts of the country, in fact, lumberyards have full hardware sections, including all types of fasteners, furniture hardware, and tools.

4/HARDWARE

Threading Your Way Through Screws, Bolts, Nails, Brackets, Catches, and Latches

This and the next two chapters deal with hardware, hand tools, and power tools. Your lumberyard may carry all these items; more likely you'll be buying them at a hardware store.

Therefore we begin this chapter with a section on how to find and deal with a good hardware store, in terms of both hardware and tools. We then describe the fasteners you will need for carpentry—the many types of nails, screws, and bolts; then the more important types of cabinet and door hardware, and general fastening and support hardware.

Hardware stores have endless amounts of gadgets and doodads, with more arriving every day. Some of them are quite useless. Some may be just what you need. That's one reason why you should get to know a good dealer, one who can listen to your specific problems, reach into the dark recesses under the counter, and pull out exactly what you're looking for.

The hardware we deal with should cover most of your normal carpentry needs. Just to mention a few of the many other items you might look for in the store—there are many styles of ornamental cabinet hardware, door and window locks, door closers, see-through door-viewers, drawer rollers, bin swings, folding support hinges for table or desk leaves, casters, springs, turnbuckles, etc.

How to Deal with Your Hardware Store

Find a *good* hardware store. Good means a store where there's at least one person who knows what he or she is talking about and can make you understand it. Good means a store where the prices are reasonable. The store does not have to be big or fancy, with all the tools laid out in neat lines. The people there do not have to wear well-pressed smocks.

A word about hardware-store dealers. Contrary to popular opinion, *they* are people too—not petty tyrants put there to make you feel stupid. For years they have had to deal with people asking the same elementary questions over and over again, people who would know the answers if they would just think a little. It's all tied up with the big mystique and fear surrounding carpentry. Treat your dealer with respect—don't expect him to lead you by the hand and show you every little thing you have to do. Try to understand these things yourself first. Read the applicable section in this book, and then go talk to your dealer. You'll find him much more helpful. Most good dealers love to talk about their wares. Once a dealer knows you, you'll find him a wealth of information. He may know of a new product or esoteric tool that will be the perfect solution for your problem.

Finding a good store may take time. If you're lucky, you may have a friend who knows one. If not, you'll just have to shop around yourself. You may well find one store with good prices on quality hand tools, but exorbitant prices on power tools; another store may have high tool prices but a wonderful stock of hardware.

Comparison shop. Check out handsaws, since there are several brands carried by most stores. Hammers. A box of flathead wood screws, say 1½″ No. 10s. Do they carry nails? Price circular saws, both expensive and cheap ones. Some stores may have special prices on just one or two brands. Tell the dealer you are thinking of making a reasonable investment in tools, and you're looking around for a good store. Ask about service on power tools—will they ship your tool for repairs or is it up to you? What are their hours? A store with late hours is a godsend for many people.

Fasteners

Nails

Nails are the most common fasteners. They are quick and easy to use, but not as strong as screws or bolts. Nails are sold by the pound; the individual size of common, finishing, box, and casing nails is referred to as 2d, 3d, 4d and so on, the "d" standing for "penny." A 10d common, for instance, is 3″ long. See the Reference section for a full table of nail sizes. All other nails are just numbered by their length. Here are some of the most commonly used nails (*1*) .

Common nails

Common nails are used for any kind of wood construction where you don't mind the head showing. They are not used for finished work.

Common nails can be bought coated with *cement* or *resin* for extra holding strength. These are also good for outdoor work, since the coating makes them water-repellent. *Galvanized* common nails are also waterproof and rustproof.

Spikes

Spikes are very long common nails. One use might be fastening the top plate of a wall to the ceiling, when there's a big space between the ceiling surface and the joists.

Box nails

The box nail is thinner than a common nail, and has a flatter head. Other than for making boxes, its original use, it is used mainly for toenailing (at an angle) , when a thicker common nail might split the wood. Not easy to find, and not really necessary.

Casing nails

The casing nail is the same thickness as the box nail, but it has a small head, almost like a finishing nail. It's used in finishing work where strength is required, particularly for door jambs and molding.

Finishing nails

Finishing nails are very thin, with a small head, and used for finishing pieces where you want the head to disappear. The head is countersunk—sunk below the wood surface with a nail set of corresponding size. They are not strong enough for rough work, like a 2x4 framing.

Brads

Brads are finishing nails less than an inch long, sold by the box. They are so small that it's hard to hold one as you start hammering it. Try holding it with a pair of needle-nosed pliers (hammer the nail, not the pliers, of course) , or stick the brad through a matchbook cover and hold the cover. When the nail is almost in you can slip the cover off.

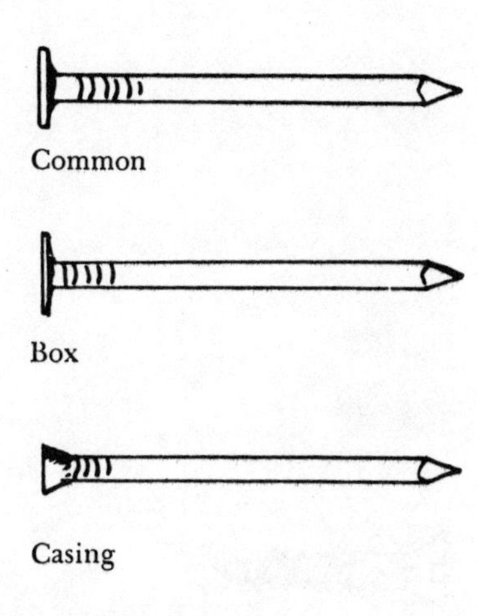

Common

Box

Casing

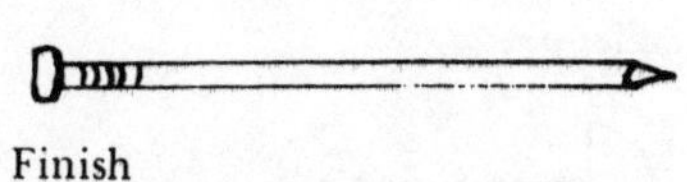

Finish

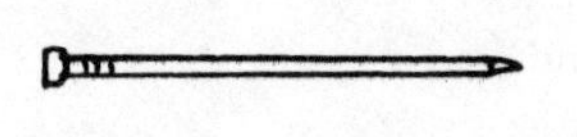

Brad

Flooring

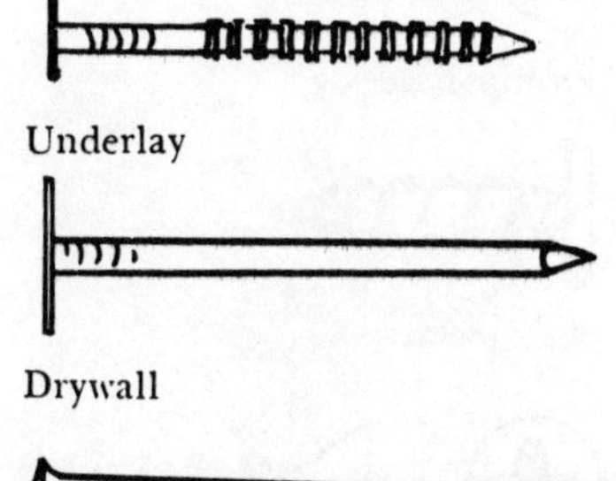

Underlay

Drywall

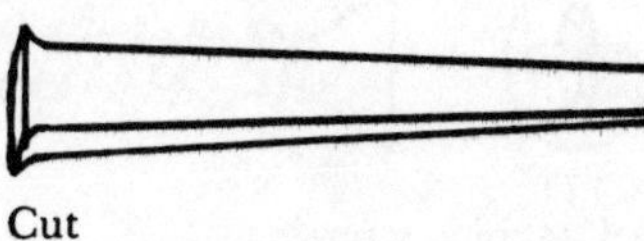

Cut

Fig. 1

Flooring and underlay nails

Flooring and underlay nails come with two types of ridge design, spiral and annular, designed to grip the wood extra tight, somewhat like a screw thread. Flooring flexes so much that regular nails just pull up. Flooring nails grip so well that you can't even pull them up with your claw hammer without damaging the wood. You have to hammer these nails sharply. The underlay annular (ringed) type will bend easily if you don't nail straight on; and the spiral will break off. If you must remove one, it's usually best to break it off above the wood and hammer the remaining part below the surface. Wear goggles for all nailing, but particularly for these nails, because they tend to break and fly off.

Plasterboard or drywall nails

Plasterboard or drywall nails come in two styles—annular, like flooring nails, and smooth galvanized with wide heads. The annular ones require a few minutes of practice before you can drive them straight without bending them, but they hold much better than the galvanized ones once they're in.

Cut nails

Cut nails are four-sided and slightly tapered along their length. They are meant for driving into masonry. *Definitely* wear goggles—they break because they're so brittle, and they also bounce off the masonry from a wild hammer blow.

A smaller type of cut nail can be used for tongue-and-groove hardwood flooring.

Nail hints

For large general-construction jobs, buy nails in twenty-five- or fifty-pound boxes, usually available from lumberyards. It's much cheaper in the long run.

Don't bother trying to straighten out old nails, or looking for the ones you drop—it's not worth it unless you're really low on funds. Any good nails left over, however, should be stored and labeled as to size and type.

Carry nails in a nail apron. When you're finished with one type, empty those nails into the bag or container before putting other kinds in your apron.

Don't leave old nails sticking out of wood. They can only hurt you. Remove them instantly or, if you're throwing the wood out, hammer the nails flat. (See also Hammers in "Hand Tools" and Fasteners in "Working.")

Screws

Screws have much more holding power than nails because their threads grip the wood (*2*). Never hammer a screw in—this defeats the whole purpose of the threads. Screws are also useful because they can be removed then reinserted, without hurting the wood. Thus a screwed structure can be taken apart easily.

Wood screw

The wood screw is the one you will use most in carpentry. About two-thirds of the length is threaded, while the shank just below the head is smooth. Optimally, the entire threaded part should enter the second piece of wood; then as you tighten the screw, the smooth part allows the top piece to be drawn tighter to the second piece.

Wood screws come with heads of many different shapes. Choose the kind that best suits your purpose. The *flat head* is the most common; it is easily countersunk so the top of the

Flat head

Round head

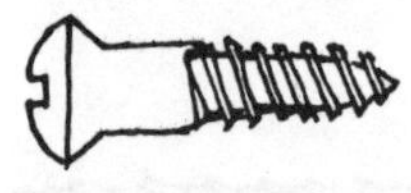

Oval head

Phillips head

Sheet metal

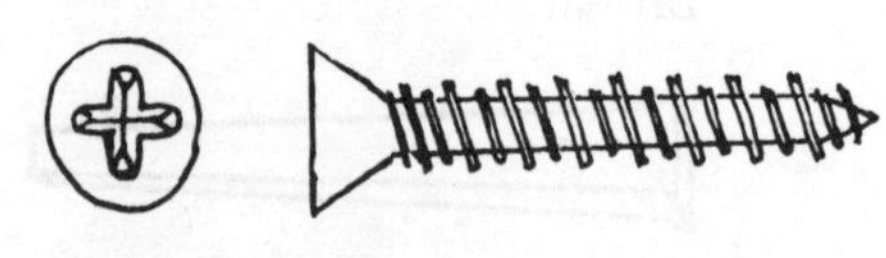

Self-tapping screw

Fig. 2

head is flush to the surface, or counterbored below the surface to be covered later with putty or a plug. A *round head* shows above the surface. The *oval head* is a little of each—it can be partly countersunk, but part of the head will show above the surface. There are other, more esoteric heads, including a *headless* head with a slot for the screwdriver right in the top of the shank; but the flat, round, and oval heads are satisfactory for most situations.

There are two major slot shapes for wood screws. The most common is the *straight* or *standard slot,* which takes a standard screwdriver blade. The other is the *Phillips head,* a special cross-shaped slot that requires a Phillips-head screwdriver. (A standard screwdriver usually rips the slot.) The Phillips slot gives a screwdriver a better grip, which makes it easier to drive the screw in. It is also easier for a screw gun to hold onto. The slot tends to be damaged easily, however. Use the standard slot unless you have a good reason to prefer the Phillips.

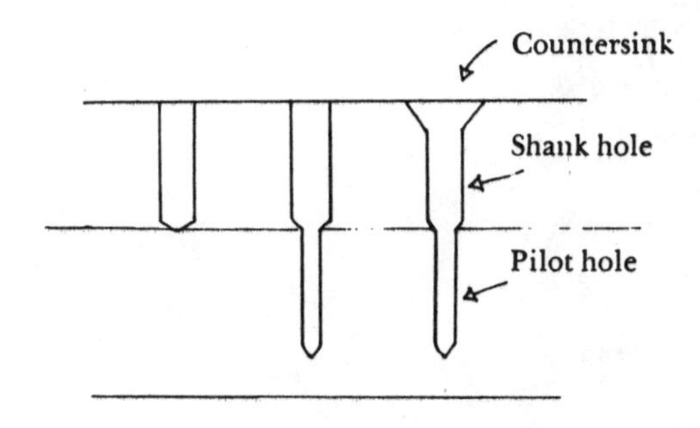

Fig. 3. Drilling Steps for a Screw

You need to drill three different size holes for a wood screw (*3*). The first, the *shank hole,* is the same diameter and depth as the smooth shank. Then drill a *pilot hole* almost as long as the screw. Finally *countersink* the top of the hole to the size of the screw head (assuming you want to countersink it). You want the screw to fit as tightly as possible, without being too difficult to turn. This may call for some trial-and-error drilling in a scrap piece till you find the proper-size bits. In hardwoods, the pilot hole is drilled almost as thick as the thread; in softwood, it should be smaller. One trick, if a hole is too tight, is to soap the threads before driving the screw. (With a power drill, you can use special screw bits that drill all three holes at once. See under Drill Bits in "Power Tools.")

Screws come in lengths up to 5″, and in varying diameters identified by gauge number. Every gauge comes in several different lengths, as shown in the Screws and Bolts table in the Reference section. Screws of 8 through 14 gauge are the most common for general carpentry. Lower gauges are used for fine, small work; when much larger ones are called for, lag bolts are usually easier to deal with.

Wood screws also come in different finishes. The common kind are plain steel, called "bright." If you want your screws to show, you can add some class to the work with stainless steel or brass screws, or zinc-, nickel-, or chrome-plated ones. Brass ones are particularly beautiful, but be careful since they are weaker than steel and tend to bend if forced.

Sheet-metal screws

Sheet-metal screws are threaded the whole length of the shank. The threads are farther apart than on the wood screw. Drill a hole in the metal slightly smaller than the thread diameter and turn the screw in. Hold the screw tight and straight for the first couple of turns, till it "taps in"—that is, makes grooves to match its threads. Besides the usual heads, sheet-metal screws also come with a hex head which may or may not include a standard slot in it. These can be driven with a wrench, nut driver, or, if slotted, a screwdriver.

Self-tapping screws

Self-tapping screws are usually used with a screw gun. They have a double thread, one a little wider than the other. This screw drills its own hole as it goes. The main use is in attaching plasterboard and other surface coverings to wood or metal framing.

For more information on screws, see under Fasteners in "Working."

Bolts

Bolts are heavy-duty fasteners, even stronger than screws. They are very important tools to understand. Try to be clear on the use of each kind.

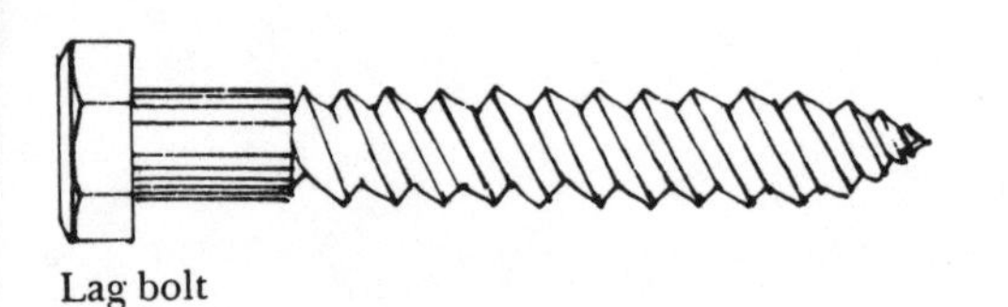

Lag bolt

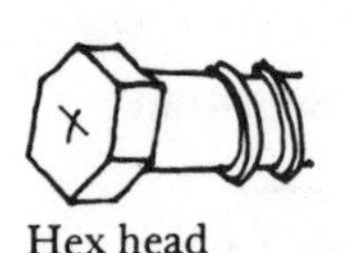

Hex head

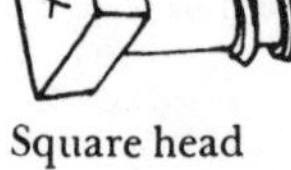

Square head

Fig. 4

Lag Bolt

The *lag bolt* is really a large screw with a hex or square bolt head (*4*). In fact, it is also called a *lag screw*. It would be torture to drive such a large screw with a screwdriver; the bolt head lets you use a wrench to get much more torque (turning power). Lags come in lengths from an inch to more than a foot, and in thicknesses from 1/4″ to 1″. (See the Screws and Bolts table in the Reference section.) Always use a washer with a lag to prevent the head from digging into the wood. Usually you have to drill only one hole for the lag, slightly thinner than the thread diameter. The wrench gives you enough torque to drive the thicker, smooth part of the shank. In some cases, however, particularly with harder woods, you should drill a thicker hole for that smooth part, just as with a screw. If you want to countersink the head, use a spade bit to drill a hole wider than the washer. Be sure to buy a bolt whose head matches the wrench you have—hexagonal head for hex socket wrench, for instance.

All other bolts, except for the lag bolt, have cylindrical shanks, either partially or wholly threaded, with a special attachment on the end of the threads—such as a nut, a shield, or a wing nut. As the bolt is tightened, the pieces it is holding together are sandwiched between the bolt head and the threaded end piece. The tighter the bolt, the more the pieces are compressed and the better the joint.

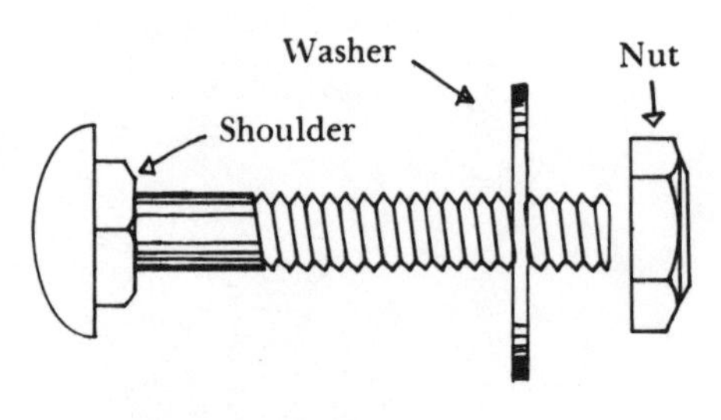

Fig. 5 Carriage Bolt

Carriage bolt

The carriage bolt has an unslotted round head at one end, and is threaded either for an inch or so at the other end or for the whole length of the shank (*5*). A *washer* and *nut* slip onto the threaded end and are tightened with a wrench. A small square section of the shank just below the head, called the *shoulder,* bites into the wood as the nut is tightened and prevents the bolt from turning with it. The tighter the nut, the stronger the joint. The washer is important because it spreads the stress so the nut doesn't sink into the wood.

A carriage bolt tightened well is actually stronger than a lag bolt of the same thickness, because the compression tension of the joint is greater. At the same time, carriage bolts can be removed more easily than lags, so they are good for structures designed for possible disassembling. Obviously, however, a carriage bolt can be used only when you have access to both sides of the joint; you can't use it to fasten something to a wall, for instance.

Get a carriage bolt at least ¼″ longer than the thickness of the two pieces you're joining, which leaves enough room to fit on the nut and washer. Make sure, however, it's not so long that the thread stops before it reaches the back of the joint. Carriage bolts come as short as ¾″ and as long as 20″; the thickness ranges from 3⁄16″ to ¾″.

Drill the hole for the carriage bolt the *same* thickness as the bolt—a ¼″ hole for a ¼″ bolt. If the bolt doesn't slide through the hole by hand, tap it gently with a hammer. The threads aren't meant to catch on the wood like those of a screw or a lag bolt. The threads are solely to engage the nut.

Lag and carriage bolts come in galvanized steel, standard black steel, or shiny chrome- or nickel-plated steel. They are all equally strong.

Machine bolt

The machine bolt is a carriage bolt with a hex or square head to accept a wrench. The *stove bolt* is a smaller carriage bolt with a slotted screwhead (round or flat). Both these bolts can be tightened from either end.

Molly bolt

The molly bolt is designed to fasten things to a plaster or hollow wall, when you can't get at a stud or joist *(6)*. It's really a stove bolt with a threaded anchor device over it. In the piece that is to be fastened to the wall, drill a hole equal to the thickness of the bolt. In the wall, drill a hole just large enough for the back part of the anchor device to pass through; the front of the anchor is slightly wider and has teeth that will set into the front surface of the wall. Usually the easiest way to attach a molly with a thick piece of wood is to first insert the anchor and bolt into the wall, then tighten the bolt to secure the anchor, remove the bolt, put it through the piece of wood, set the piece in place, and rethread the bolt into the anchor. Tightening the anchor up first insures your being able to reach the nut at the back of the anchor after the bolt is in the wood.

As the bolt is tightened, the anchor crimps up against the inside of the plaster and creates a compression tension similar to the way a carriage bolt works. The crimp and the face of the shield, by increasing the surface area of the fastener, enable you to get maximum holding strength out of the plaster. The plaster is limited in strength, however, so that mollies should be used only for light- and medium-weight objects like picture frames, mirrors, maybe a small cabinet. Naturally the use of several mollies disperses the stress further. Do not use them for heavy objects like record cabinets and loft beds, however. Expect even less strength from a molly when used in the ceiling; there, gravity is working directly against the thickness of the plaster and it is more likely to crack. (See also "Walls and Ceilings.")

Toggle or butterfly bolt

The toggle or butterfly bolt is a long thin stove bolt with two spring-loaded metal wings attached to a nut on the end *(7)*. It works similarly to the molly bolt in attaching a piece to a hollow wall. The wings tighten against the inside of the plaster as you tighten the bolt *(8)*.

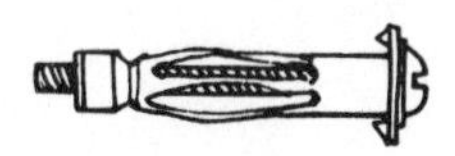
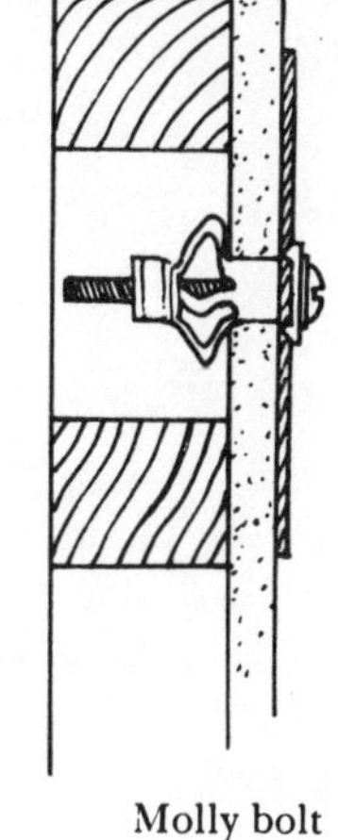

Molly bolt

Fig. 6

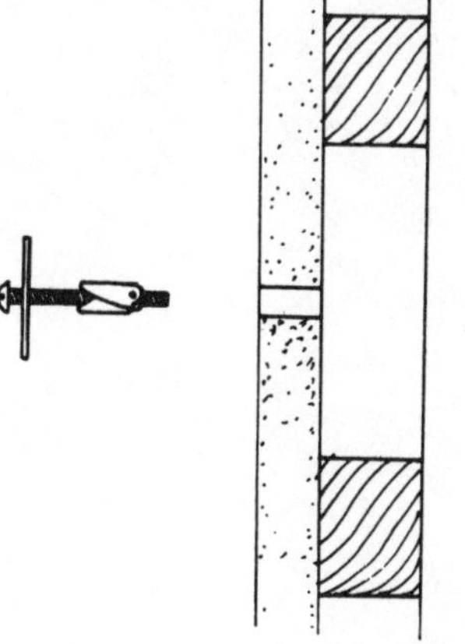

Toggle bolt

Fig. 7

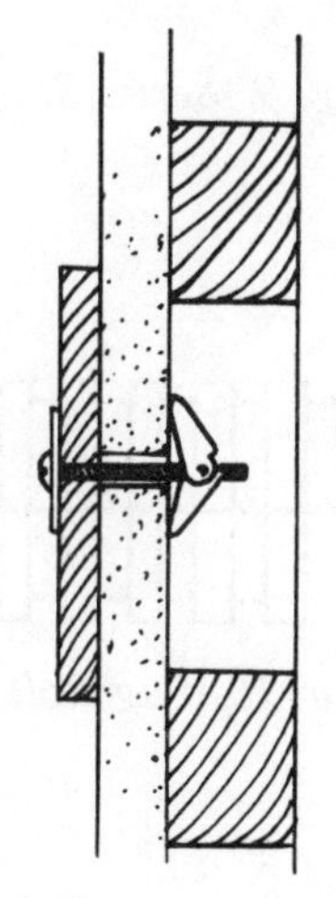

Toggle bolt expanded

Fig. 8

The hole in the piece should be as wide as the thread; the hole in the wall should be wide enough to let the closed wings pass. That latter size is usually printed on the wings. Insert the bolt with washer through the wood, thread the wings just on the end. While pinching the wings shut, set the piece in place and the wings through the hole in the wall. When they get behind the plaster they will spring open.

Toggles are also meant only for light- and medium-weight objects. Toggles can be tricky. Make sure there's enough room in the hollow for the wings to get all the way in before they open. Stick a nail into the wall to see how far back it goes. Pick a bolt long enough to carry the wings back far enough, but not so long that it will hit the surface behind the hollow. And be sure you thread the wings facing the right direction.

Once a wing opens inside the wall, it's almost impossible to take it out. If you make a mistake and have to remove the piece, you may have to unscrew the bolt till the wings fall off into the hollow. Pull the bolt out as you unscrew it, so the wings press against the wall and don't turn with the bolt.

If just one wing is opening, you're probably not getting it into the wall far enough. Wedge it out, if possible, and run the wing back on the thread as far as possible. If you still have no luck, remove the bolt and countersink a hole into the wood which will allow the bolt and washer to move forward enough.

If a bolt is too long you can cut it down with a hacksaw. But first, insert it in the wood and thread the wings on. The saw blade may damage the end of the thread so much that you won't be able to put the wings on after the cut.

Shields

Lead shields

Use lead shields in masonry and plaster walls to give wood screws something to grip (*9*). With a carbide bit, drill a hole the thickness of the shield, then insert the shield and hammer it gently flush to the surface. When the screw is threaded in, the shield expands and presses hard against the masonry (or plaster). The expansion tightens everything so neither the shield nor screw will come out.

Fig. 9. Screw Shield

Shields for screws also come in plastic or fiber. Avoid the plastic completely; fiber ones are often useful for very small screws.

Your hardware dealer should be able to give you the proper size shield for the screw you're using. The most common ones, however, are the 1/4″ shield for Nos. 6–8 screws; and the 5/16″ shield for Nos. 9–11 screws. The lengths vary from 1/2″ to 1 1/2″. You want the screw to reach almost to the end of the shield.

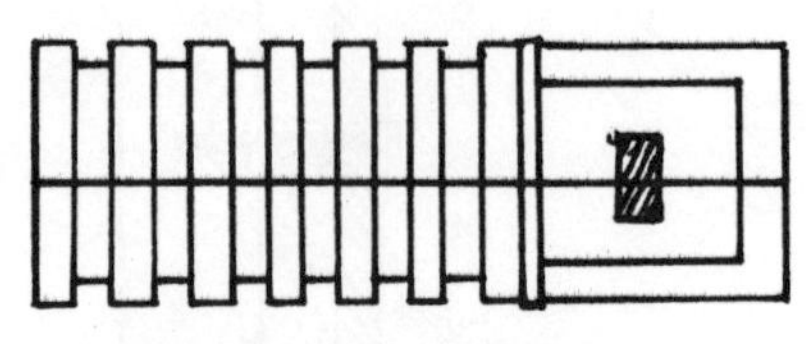

Fig. 10. Lag Bolt Shield

Lag bolt expansion shields

Lag bolt expansion shields (or anchors) are used with lag bolts in the same way that lead shields are used with wood screws (*10*). The lag shield is thicker, tougher, and in general meant for stronger loads.

Use a 1/2″ diameter shield for 1/4″ and 5/16″ bolts; a 5/8″ shield for 3/8″ bolts; and a 7/8″ shield for 5/8″ bolts (which you will probably never need). Each shield may be in two lengths, short or long. The 5/8″ shield, for instance, comes either 1 3/4″ or 2 1/2″ long. Unlike the lead shields for screws, lag shields allow the screw tip to go through the bottom of the shield (assuming the hole has been drilled that far). This gives you some extra leeway in matching the shield length to bolt length.

See "Walls and Ceilings" for advice on how to choose the proper length and thickness for your situation. Also, see the Loft Bed (in "Furniture Projects") for a practical example of how to fasten a piece of wood to the wall with lag shields and bolts.

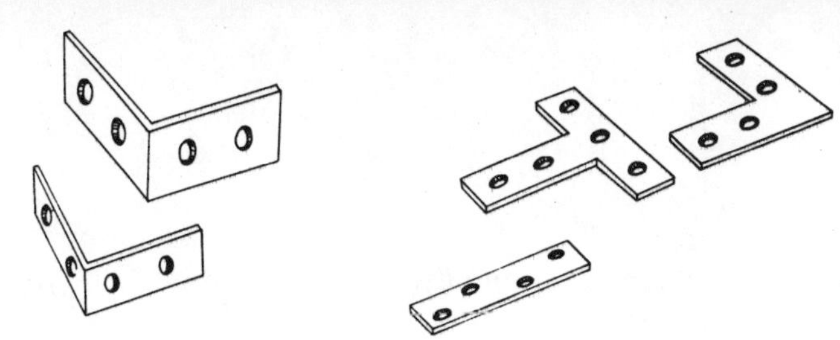

Fig. 11. Angle Irons *Mending Plates*

Fastening and Support Hardware

Angle iron

An angle iron is a piece of steel bent at a right angle, with predrilled screw holes (*11*). Use it to reinforce corner joints, support shelves, steady table legs and so on. *Corner braces* and *mending plates* are flat pieces of steel used to reinforce or hold together flush pieces.

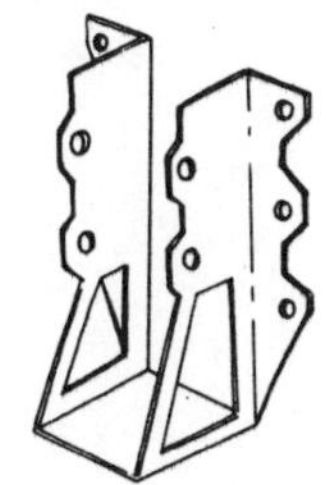

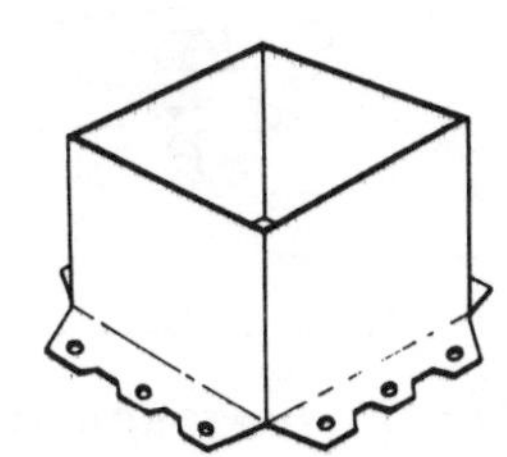

Fig. 12. Joist Hanger *Post Anchor*

Joist hangers, framing anchors, post anchors

These are all special metal pieces designed for specific joining problems (*12*). They make very strong joints. The joist hanger is particularly useful, allowing you to hang a joist without notches or cleats.

You can find this hardware at lumberyards, and sometimes in hardware stores. Be sure to get the special nails that go with the hangers—they are included in the price. (See "Working Techniques" for how to attach joist hangers.)

Standards and brackets

Standards and brackets provide a quick, easy way to support shelves. *Wall standards* are screwed firmly to the wall (use screws and lead shields or molly bolts, if you can't reach the studs) and the brackets slip into the standards at any height you want (*13*). The shelves, of course, sit on the brackets.

Fig. 13. Wall Standard and Bracket

Interior shelf brackets

Interior shelf brackets can be mounted inside cabinets, flush or mortised (*14*). They are mounted on the sides, and hold small shelf clips.

Shelf support

Another way to support a shelf inside a cabinet is to simply drill holes into the sides and insert shelf-support pins (*15*). Get the metal pins only—the plastic ones are very weak (see Cabinet project).

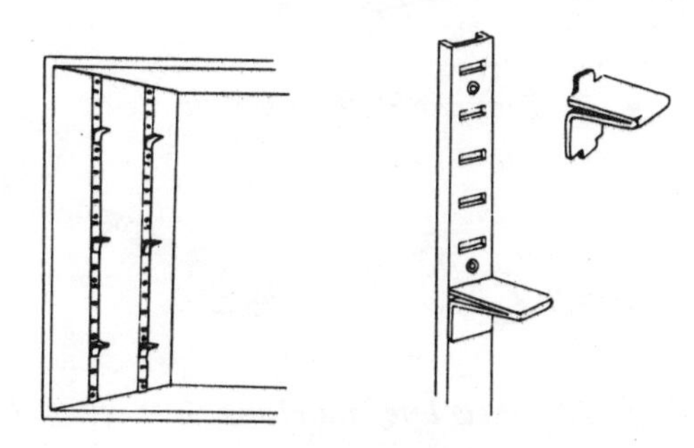

Fig. 14. Interior Shelf Brackets

Mirror brackets

Mirror hangers are specifically designed to hold unframed mirrors (*16*).

Ceiling hooks and screw eyes

Ceiling hooks and screw eyes are useful for hanging all sorts of objects, particularly from the ceiling—plants, pictures, fluorescent lights, a high trapeze (*17*). The ends are threaded like a screw. Drill a hole slightly smaller than the thread and turn the hook in by hand. If it gets hard to turn, slip a screwdriver (preferably an old one) through the eye for leverage, or use a wrench (*18*).

Hooks and eyes come in many sizes. If you're threading one into the wall or ceiling, make sure it's long enough to reach well into the supporting structure. If you're threading

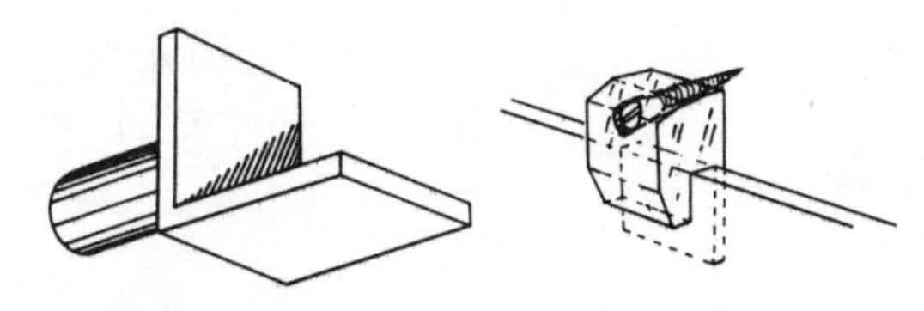

Fig. 15. Shelf Pin *Fig. 16. Mirror Hanger*

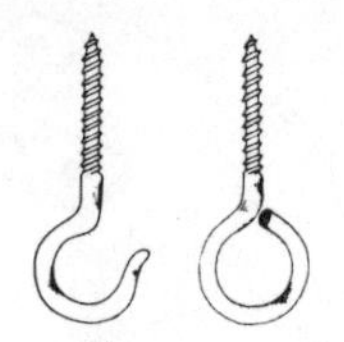

Fig. 17. Screw Hook and Eye

it into a picture frame, however, be sure that the tip doesn't break through the other side of the frame, and that the shank is not so thick it will split the wood.

The heavier the weight, the thicker the hook should be, of course. When you're uncertain, use two or more hooks for extra strength.

Picture wire

Picture wire comes in different weights, according to how much weight it will support. You can get by with a lighter wire by doubling or tripling the strands. It's very useful when hanging things from hooks or eyes, or for tying things together for temporary repairs (supporting pipes, joints, etc.) .

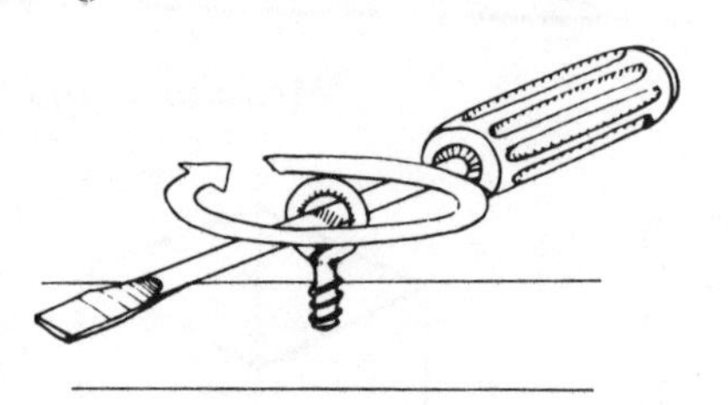

Fig. 18. Inserting a screw eye

Cabinet and Door Hardware

Cabinet catches

You're probably familiar with most of the catches shown in Figures 19–21. The *pressure-release* or *touch-latch* is a special spring-loaded kind that opens if you just push lightly on the door. It catches automatically when the door shuts. The only problem with this catch is that it sometimes opens when you don't want it to—when you accidentally kick it or your beagle leans against it.

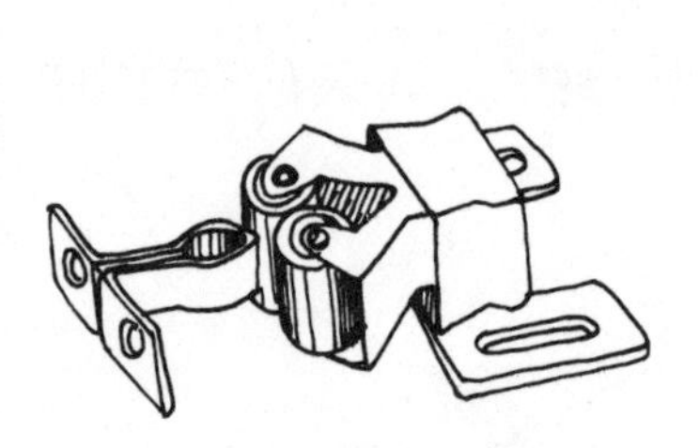

Fig. 19. Roller Catch

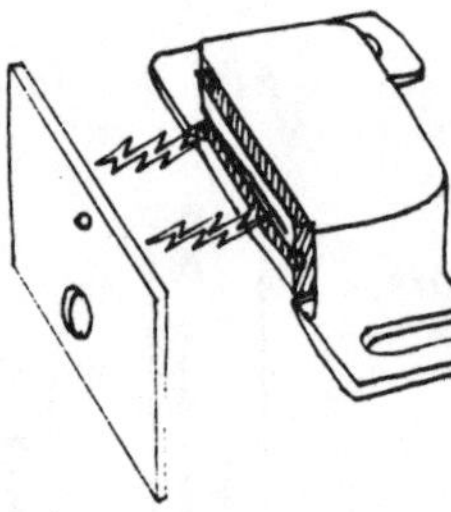

Fig. 20. Magnetic Catch

Installation of any of these catches is very simple. Specific directions are printed on most packages. In general, though, you first attach the larger part, called the catch, to the inside of the cabinet frame, close up against the closed door. You then position the smaller part, called the strike, on the back of the door so the parts will meet when the door is closed. Attach the strike; then loosen the screws in the catch and adjust it back or forth along the slotted holes till the door closes exactly the way you want it to.

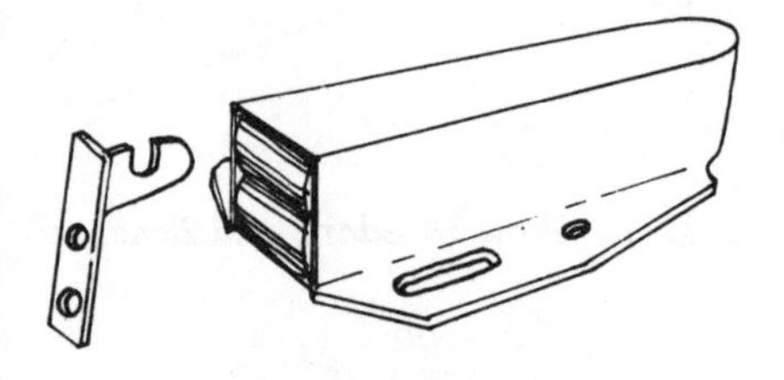

Fig. 21. Pressure-Release Catch

The *screw eye* and *hook* arrangement *(22)* and the *sliding bolt latch (23)* can also be used on cabinet doors. They're not as fine work, but they're easier to install and they do the job.

Handles and knobs

Besides standard doorknobs and locks, there are endless styles of handles you can put on a door or drawer. Some stores sell nothing but handles. Some of the more useful kind are the flush handles. One kind is mortised into the surface; another lies flat but swings out when you pull on it. All of them are simple to install—usually requiring nothing more than a couple of screws.

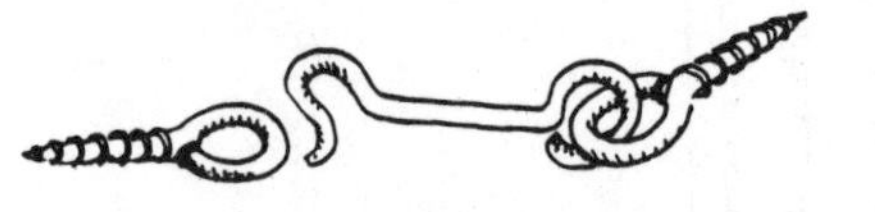

Fig. 22. Screw Eye and Hook

Tracks for sliding doors

Tracks for sliding doors are simple devices likely to cause you trouble. The working principle *(24)* is clear: attach a track to the top of the door opening, and usually some sort of guide along the floor; attach rollers to the doors; and hang the doors by the rollers into the track and guides. The problems occur when the door opening is not plumb or level, or when the floor is badly off level; then the doors get jammed or slip out of their tracks as they move. Therefore, check your door frame before deciding on a sliding-door system. If you're building it yourself, build carefully. When they work, sliding doors are delightful.

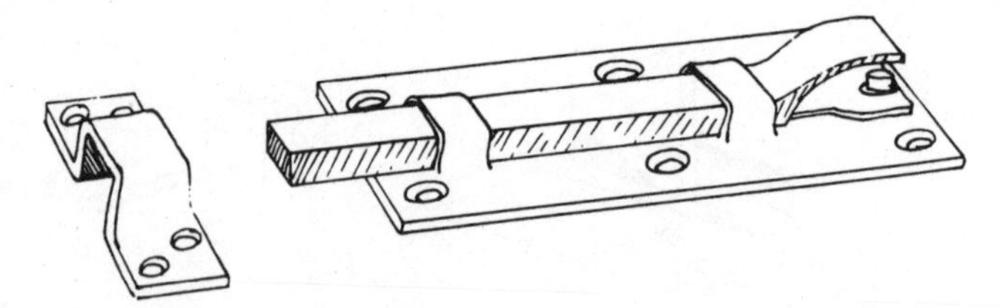

Fig. 23. Sliding Bolt Latch

Tracks come with complete instructions. Here are a few extra tips. Be careful not to bend the track. Don't use roundhead screws that stick out into the roller path. Don't force the doors if they get stuck. You can shim the track if the door opening is warped a lot.

The common hinge you see on a door is the *butt hinge* (*25*). It comes in various sizes. A 3½″x4″ hinge, for example, is 3½″ wide when open, or 1¾″ wide when closed, and 4″ high. The hinge leaf can be wider than the door thickness, so long as the screwholes all fit on the door edge. (See Doors and Locks in "Room Renewal" for mortising techniques.) The *rounded butt hinge* is used when the mortise is cut with a router bit. Use brass-plated steel hinges for inside doors, and solid brass for outside doors. Solid brass won't rust, but brass-plated steel is stronger. Butt hinges also come chrome-plated and in plain steel and other materials. Other hinges available are shown here.

The *pivot hinge* is useful on overlay doors when you don't want the door to go beyond the side of the cabinet when it's open (*26*). (See under Cabinet-door Construction in "The Ubiquitous Box" for types of doors.)

The lipped cabinet door needs a special, semiconcealed *offset lip door hinge* like the one shown in Figure 27. You can use a narrow butt hinge or a piano hinge for a flush door, but the *offset flush door hinge* shown is preferred in finer work. Its big advantage is that you can screw into the back of the door, rather than the edge; since most cabinet doors are plywood, this gives you much more holding strength.

The *strap hinge* is attached to the outside surface of a flush door and frame (*28*). It's easy to attach, and gives a rougher "garage-door"–type look.

The *piano* or *continuous hinge* is just a narrow, very long, butt hinge (*29*). It comes in various lengths up to 6′, which you can easily cut to fit with a hacksaw. It is stronger over a long span than several butt hinges, and looks nicer. It's usually mounted between the door and frame, so that just the barrel of the hinge shows when the door is closed; but it can also be mounted on the outside face.

Hinges

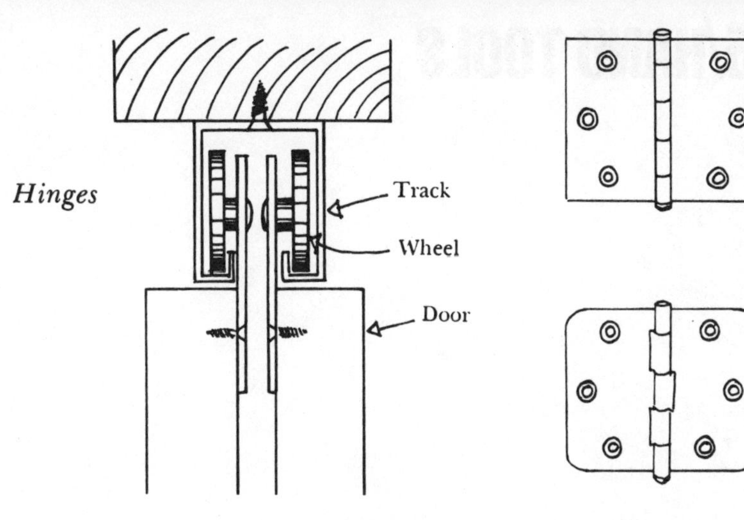

Fig. 24. Sliding Door Hardware

Fig. 25. Butt Hinges

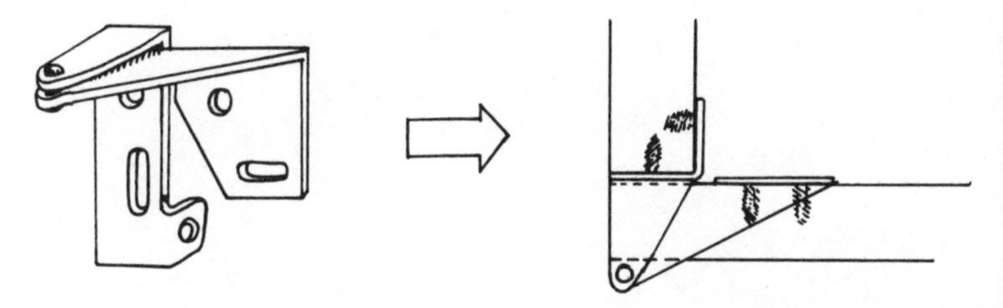

Fig. 26. Upper Pivot Hinge

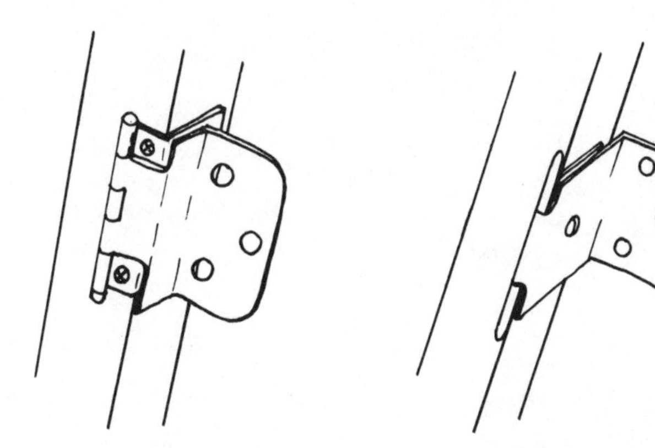

Fig. 27. Lip Door Hinge

Flush Door Hinge

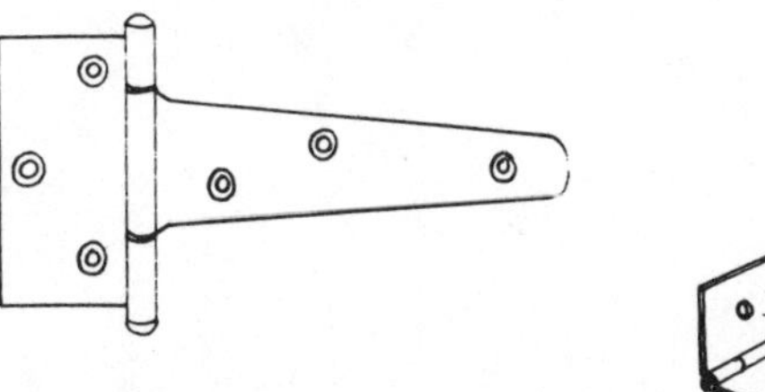

Fig. 28. Strap Hinge

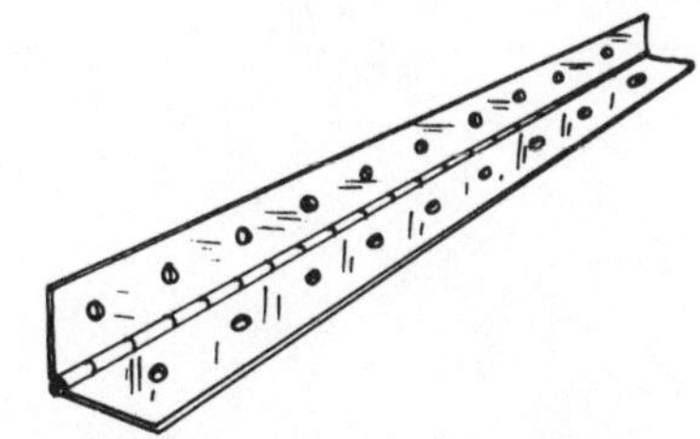

Fig. 29. Piano Hinge

5/HAND TOOLS

What They're Good for, How to Use Them, Which Ones You Need, How Good a One to Buy

Tools are not mysterious or terrifying. They are logical extensions of the human body, just like implements used by prehistoric man. A hammer is a rock. A saw is a more efficient cutting tool than your teeth. Early man made holes with his fingers or with a stick. We do it with a drill and drill bit.

With the right tool you can do any job. The following pages contain descriptions of all the tools you'll need for all but the most specialized carpentry jobs; this chapter deals with hand tools, the next with power tools. There are explanations of how each works, how to use it, and how good a tool you need. You certainly don't need all the tools, but it's good to know about them. That way you can choose and shop for the most efficient tool for your needs and not be at the mercy of a salesman.

While there are a few basic tools you probably can't do without—a hammer, screwdriver, a saw, and possibly some kind of drill—we suggest that in general you buy tools as you need them. What is basic for one person may be a luxury for another, depending on the type of work planned.

Buying a Tool

How do you decide which tool is necessary to solve a specific work problem? Either you know the general kind of tool you need, but not how good a one; or you don't know at all what tool to use. Think about the job. What is the basic principle? Are you cutting, for instance? The following list names the types of tools dealt with in this chapter. Check under the category you need. A tool for—

Measuring, layout, and marking
Pounding and fastening
Making holes
Twisting, grabbing, and holding
Cutting
Shaping and smoothing
Applying gook
Prying and pulling
Safety and convenience
Climbing

You should be able to narrow the search to just a few tools, if not to a single one. Then think about whether you need a very accurate job done, whether you will need such a tool for future jobs, and how much money you can afford. We usually give an idea of just how good a tool you need for your work. Also, consider whether you already have a tool that could be adapted for the job, though it might be slower and rougher.

For instance, you are laying linoleum with mastic over a very small area. You need a glue spreader of some sort. If you ask the hardware dealer, he'll probably show you a special trowel with teeth on the edge. The teeth, he'll explain, are to rough up the surface of the mastic so it'll stick better and help get a uniform thickness. The price may be six dollars. Do you need it? How often will you use it? Just this once? Can you improvise a tool to do the same thing? All you really need is a piece of scrap wood as a spreader. You can score lines in the mastic with the wood edge to rough it up. The scrap wood may take longer than a trowel, but it'll do a decent job and save you six bucks.

It's always best to buy as good a tool as you can afford if you'll use it more than a few times. Except where we tell you otherwise, the more expensive tool is almost always a better buy in the long run. It works more easily, more accurately, and it lasts much longer. There is a point of diminishing return, of course; you don't need the very best tools (though they are nice) unless you plan to become a full-time professional. Just stay away from the real cheapies—they're more trouble than they are worth.

If you've found a hardware store you can trust, you might ask the people there for advice on what tool to get. (See How to Deal with Your Hardware Store in "Hardware" and Buying the Tool in "Power Tools.") Examine the tool carefully in the store when you buy it; check that it has all its parts and is in perfect shape—not nicked, cracked, bent, rusted, etc.

When you get the tool home, read any instructions that come with it and also the corresponding entry in this book. Understand the tool's principle. Practice with it on some scraps to get the feel of it. Think of it as an extension of your body. Some tools you'll be able to use well with just a little practice. Others, like chisels, you'll be able to use well enough at first, but it will probably take you years to become really expert with them.

Be sure you practice on the kind of wood you'll be using for your finished project. Tools have different feels, require different pressures on different woods. Try to duplicate the final work situation as closely as possible. You don't want to run into too many surprises on the work piece.

General Tips and Care for Hand Tools

Clean tools after each use. Oil them, if need be, with a good, light machine oil, like 3-in-1, and a rag. Wiping oil on your saw or chisels, for instance, will prevent them from getting rusty. Clean your files with a file card to keep them workable. Clean your scrapers, putty knives, and trowels immediately after use—don't wait till the gook dries.

Store cutting tools so their cutting edges are protected. If you don't get protective cases with your saw and chisels, improvise some. Fold some cardboard over the saw blade; plastic slipcases that some drill bits and saber-saw blades come in are perfect cases for chisels. Store your plane on its side.

Have your saws sharpened periodically. It's much cheaper than buying new ones.

As you work, keep your tools organized. If you carry them in a tool belt or nail apron, use them and return them immediately to the proper pocket—particularly your pencil, which can easily get lost in the sawdust. Put the other tools down in the same spot each time. Keep small tools, or flat ones like squares, on a raised surface such as a table or elsewhere away from the work so they don't get covered with sawdust.

Tools for Measuring, Layout, and Marking

Measuring tools give you proper dimensions, layout tools give the proper shape. Good marking tools keep your measurements and layout lines accurate, which is essential for good carpentry work. Since the tools for all these operations are extremely simple, there's no reason why you can't be accurate the very first time you use them.

For measuring either small or large dimensions, the steel tape measure is the most useful tool. The carpenter's rule is also very useful. The square—particularly the rafter square—is probably the most useful layout tool, enabling you to get perpendicular lines and to build things square; every time you cut a 2x4 to length, you need a squared cutting line. In general construction, the level is important for keeping things horizontal and vertical; and also the plumb bob, for testing verticals and for locating points vertical to other points. A common pencil and a chalk line take care of most marking operations.

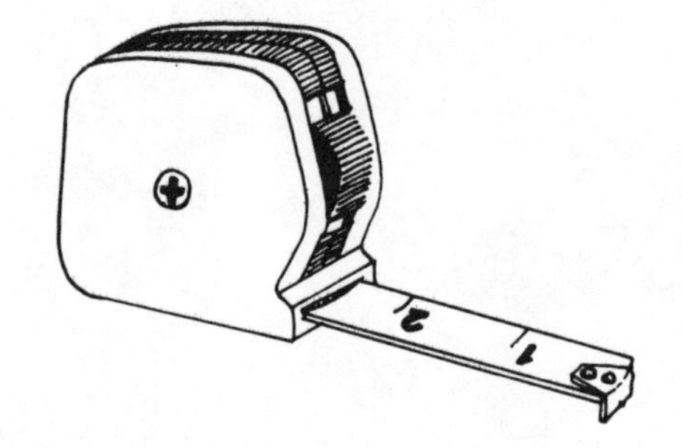

Fig. 1. Tape Measure

Measuring Tools

You must have a good measuring device to carpent.

Tape measure

A long thin strip of flexible steel tape with measured markings wound inside a small case (*1*). A little hook at the zero mark catches onto the end of a board, lets you pull the tape along without needing a second person at the end. Don't think the tape is broken if you notice the hook is loose; usually it can move in and out about $\frac{1}{16}''$ to compensate for the thickness of the hook whether you're taking inside or outside measurements.

Buy at least a 12′-long tape, with a width of at least $\frac{3}{4}''$ (most lumber you'll get is 12′ or less). The greater the width, the stronger the tape, and the more you can run it into mid-air before it bends and drops down. For a few dollars more you can get a 24′ tape, useful for general construction—partitions, rooms, etc. Tapes up to 24′ rewind automatically when you let them go, which saves incredible amounts of time. (For longer layouts there are 50′ and 100′ tapes, but you have to rewind these by hand with a key handle.) Most tapes can be locked at any measurement with a sliding switch; make sure yours has such a switch. Try to buy one that takes refills, since the tape can get bent or ripped through carelessness.

Some tapes have special markings at 16″ intervals, the standard interval for wall studs, floor joists, and other construction modules. This is another great time-saver. Some tape measures have useful guides printed on the back of the tape—metric conversions, decimal and fractional equivalents, nail and screw sizes, and so on. Other tapes are fully or partially marked off in metric measurements. This is not particularly useful now in the United States, but it will be eventually (we hope).

Care and use: Take good care of the tape. Don't let it lie around with the tape out and locked. It can get stepped on easily, creasing the tape or breaking the sliding hook. Once the tape is creased, it will eventually rip at that point. Don't let the tape snap back too fast—if the hook hits the case at full speed, it will soon break off. Run the tape through your fingers or stop it the last few inches. Watch out for curious cats.

Many tapes are designed for inside measurements, though they're not as accurate for this as a carpenter's rule. Besides the hook being movable, the case will often be squared and identified as, for example, 2″ wide. You push the hook end against one side of a window opening, say, and the case against the other side. Add the two inches to the measurement shown where the tape meets the case.

Always run the tape as straight or as level as possible between any two points. A diagonal or drooping path is longer than the true straight path.

Carpenter's rule: folding or zigzag

This is essentially a wooden ruler (usually 6′ or 8′) that folds up (*2*). It is more cumbersome and time-consuming than a tape, but it is more rigid and more accurate for inside measurements. Better rules have a thinner brass rule that slides out of one end for inside measurements. Unfold the rule to the length just short of the opening to be measured, and then slide the brass extension out till it touches the inside of the opening. Whatever the brass rule reads is added to the measurement at the other end of the rule.

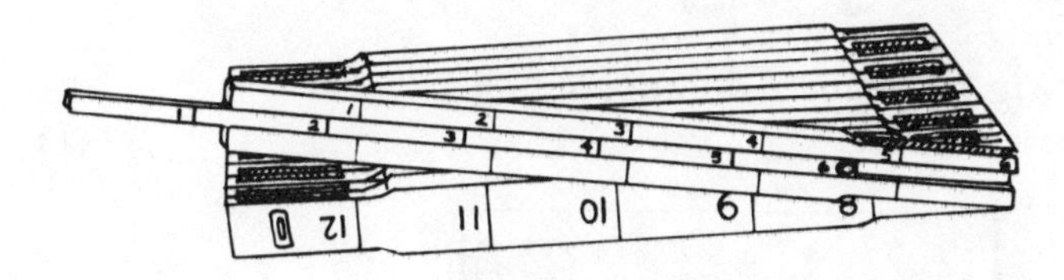

Fig. 2. Carpenter's Rule

The rule is good in cases where the tape would collapse—measuring from the ceiling down along the wall, or horizontally in mid-air, or even from the floor upward, where you might have trouble keeping a tape straight.

In the past, the rule was the rule. Now most carpenters prefer the tape measure. If you get a rule, get a good one. The cheap ones break. Don't leave it lying around open. A carpenter's rule is in that class of tools called Finger Chompers. As you fold the rule, the joints can pinch your skin if you're not looking. Be careful.

Metal rule

Aluminum or steel, lengths from 1′ to 8′. This is a good tool to have, not just for measuring, but because it gives you a good straightedge to use as a guide for power tools or for drawing straight lines. It's not as precisely measured off as a tape or folding rule, so it's not that accurate and certainly not as handy. It is handy for giving you quick *rough* measurements.

You should have one 8′ long. This is standard length for plasterboard and plywood, for which it is a handy cutting guide. If you need one more portable, get a 5′ one. Get one as thick as possible, for strength as a guide. The stainless

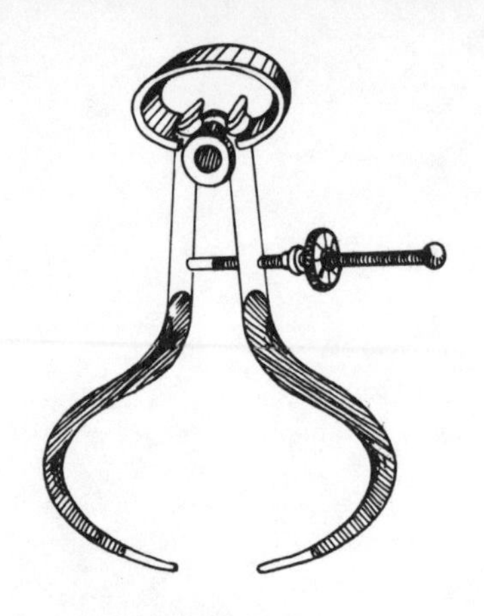

Fig. 3. Outside Calipers

steel ones are better, but save money and get an aluminum one.

Calipers

Calipers are used for small accurate measurements (*3*). Straight-legged calipers are sometimes called *dividers;* legs may be curved for either inside or outside measurements (*4*). They are mainly for measuring round objects, such as dowels or pipe, and not really necessary for most work.

General tips on measuring devices

If you have several measuring tools, check them against one another for relative accuracy. Surprisingly they won't always match. Try to use one device for all the most important tolerances. If you're working with someone else, make sure your rules match. Or use just one. Be sure the measurements of the space you are fitting something into were taken with the same measure you are cutting the wood with. Check the measures on your square and level. These will tend to be less accurate, meant just for handy reference.

A good tape is an absolutely necessary tool. It makes the best foundation you can have for your total structure. If you begin with incorrect measurements, no amount of fine craftsmanship can make the pieces fit together.

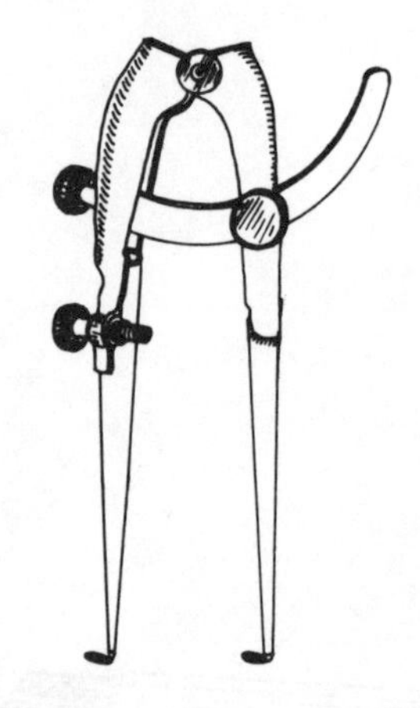

Fig. 4. Dividers or Calipers

Squares

It's as essential to get things square (that is, at right angles, or whatever angle you intend) as it is to get things the right size. Pieces just don't fit right unless the angles are right. Pieces that don't fit right weaken your structure.

Squares are used to measure and check angles, mainly right angles. You lay the square against a joint you're making to see if it is a true right angle. Or you lay one leg of the square against an edge and the other leg of the square gives you the perpendicular to that edge, for marking square cuts, etc.

Try square

A steel blade 6″ to 12″ long attached at a right angle to a thick, squared-off wood or metal handle (*5*). Use it for testing the square of lumber edges and other surfaces, and for drawing lines perpendicular to the edge of a piece of wood for cutting. You can hold it on a 2-by as a saw guide for a straight cut. Use it to place one thing perpendicular to another, like a pipe coming out of the wall. It measures the square on the inside or outside of a joint.

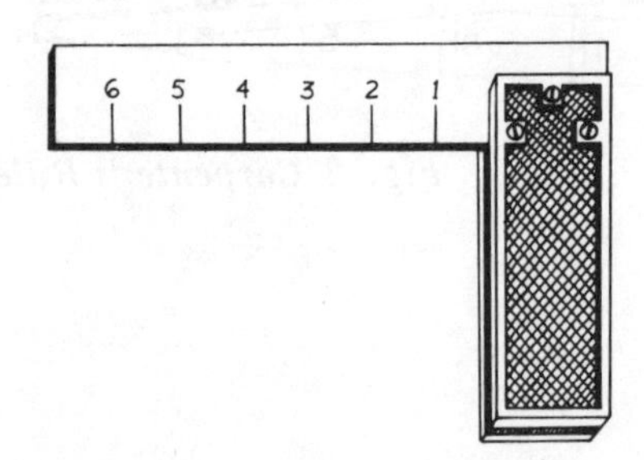

Fig. 5. Try Square

Buy at least a medium-priced one. Test its accuracy in the store by placing it within another. If they don't match up square, try two others. Don't knock it around, or it will go out of square.

Combination or adjustable square

A try square with a 12″ sliding blade that marks right angles on one side of the handle and 45° angles on the other (*6*). If you have a good one with accurate markings, you can use it as a depth gauge. Most have a scribe (used like an awl) inside a little pocket and a single leveling bubble—not as accurate as a long level, but handy.

Rafter square

A flat steel square, one leg (the tongue) sixteen inches, the other (the blade) twenty-four; some are larger (7). Measures inside or outside squares. To draw a perpendicular to the edge of a board, lay the inside of the blade flush to the edge, with the blade sloping down toward its end, and mark a line

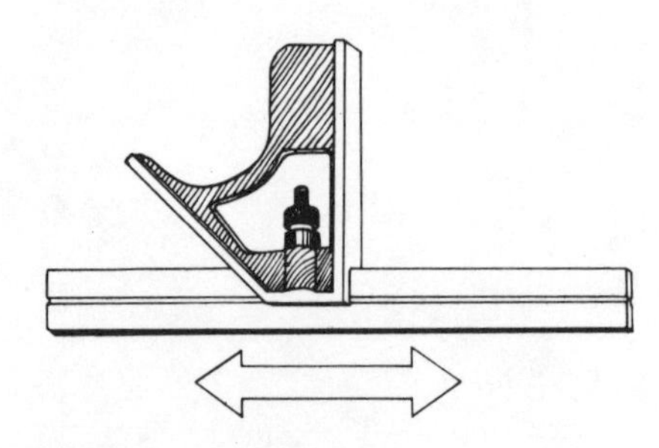

Fig. 6. Adjustable Square

along the inside of the tongue; or line up the outside edge of the blade with the edge of the board and mark on the outside of the tongue. Also very handy as a saw guide.

Better squares have tables on them—board-feet tables, rafter tables, angle tables. Complete instruction booklets come with these squares. One example: To get a line at any specific angle to a given line, consult the table for the inch markings on blade and tongue that correspond to that angle. Line up those markings with the given line. The angle between the given line and the blade will be the desired angle. To transfer this angle to another board easily, clamp a thin piece of wood to the square at those points. This piece acts as a guide to lay up against any other edge.

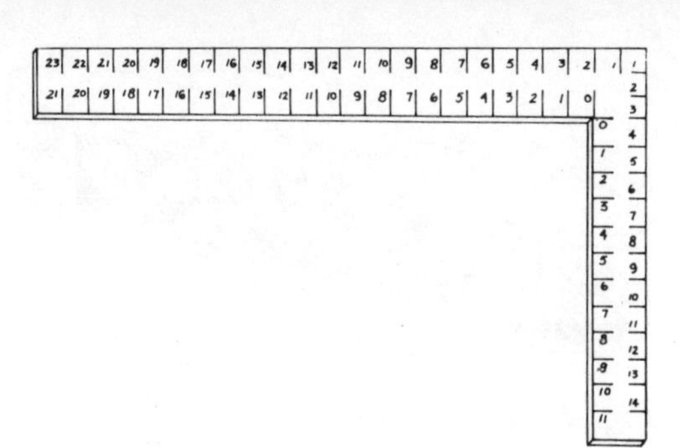

Fig. 7. Rafter Square

Buy a good rafter square. Test it in the store by placing one within another. When using it, don't leave it lying around on top of things. It can get bent if you step on it, and a bent square is as useless as a stopped clock.

T bevel

An infinitely adjustable angle (*8*). The blade pivots on the handle and can be locked in any position. Use it to transfer any angle from one place to another. Good for adjusting the angle of a power saw.

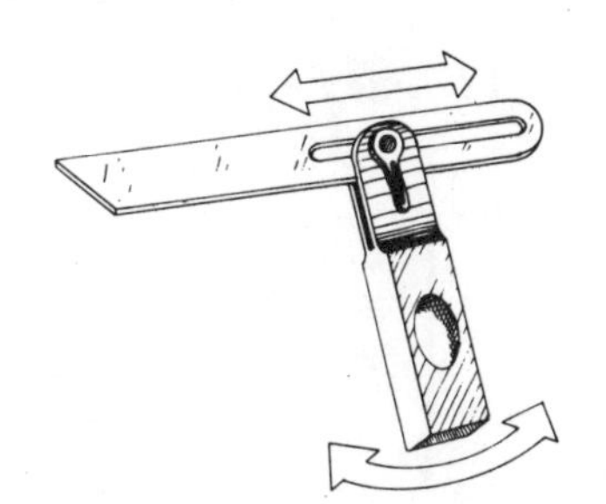

Fig. 8. T Bevel

Levels

The working part of all levels is the leveling vial, a thin glass tube slightly curved upward in the middle, filled *almost* completely with liquid (*9*). Air being lighter than water, the bit of air remaining inside forms a bubble that stays on the top of the curve. Two hash marks are painted on the tube to mark the center. When the bubble is exactly centered between them, the tube is level, parallel to the earth. If the bubble is to the left of center, the level and the surface it is sitting on are too high on the left (*9*).

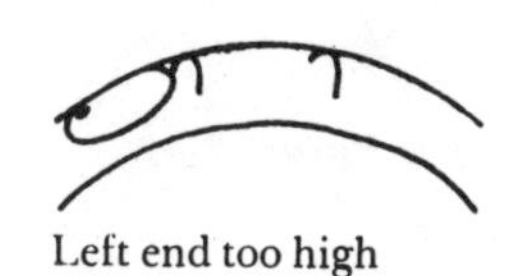

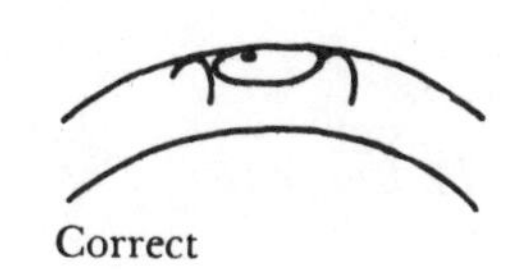

Fig. 9. Reading a Level

Carpenter's level

The most common level, used for checking any surface, horizontal or vertical, that it can be laid against (*10*). It contains from one to six bubbles encased in a long wood or metal frame, 2′ to 6′ long. Lay the edge flat against the work surface. The correct bubble to read is the one now running parallel to the earth with its center curved upward.

Buy a fairly good medium-priced level with bubbles for both horizontals and verticals. (Some also have tubes for 45° angles.) A 2′ one is fine, and will fit in some spots that a 4′ one can't. The only advantage of a longer level is that it can span a longer distance—a more representative sample of the floor, for instance. If you have any long straightedge, however, you can use it with a 2′ level and get the same effect. Good levels have replaceable bubble assemblies—a good idea.

In the store, sight down the length of the level for straightness. Place one edge of the level against a wall, then the other. It should read the same both ways, whether the surface is true or not.

Don't leave your level lying where it can be stepped on, or where something might drop on it. Don't leave it leaning against or sitting on something you might be hammering. Use it, then set it away from your work. Don't hammer it or

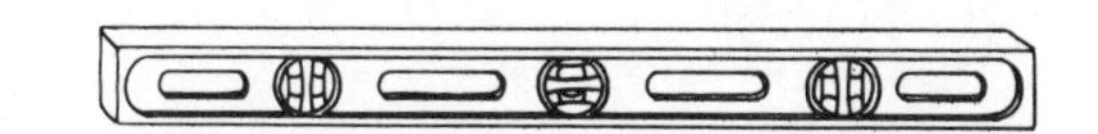

Fig. 10. Carpenter's Level

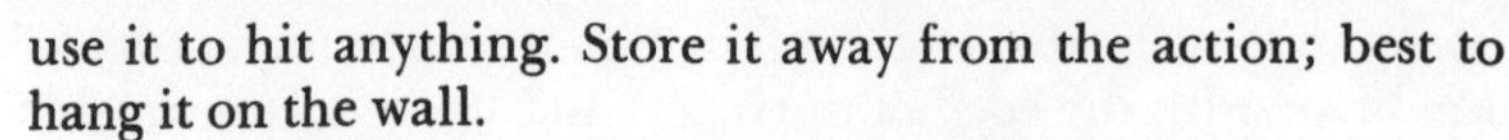

use it to hit anything. Store it away from the action; best to hang it on the wall.

Torpedo level A small level, 4″ to 12″ long, for tight spots (*11*).

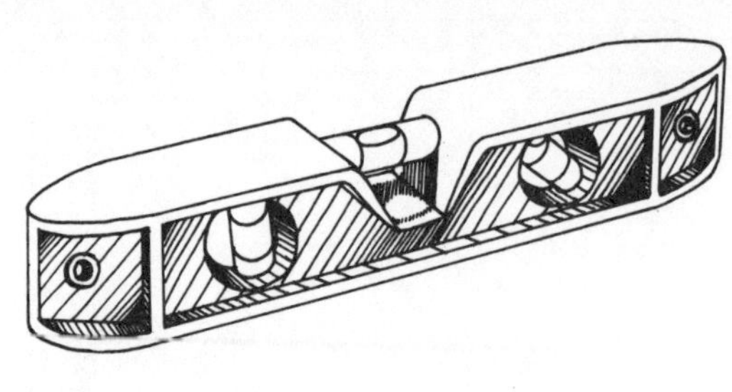

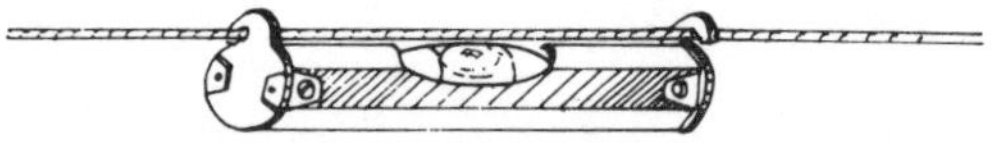

Fig. 11. Torpedo Level

Line level Used to find levels over long distances (*12*). A single vial in a small metal casing that hooks onto a string. The string is stretched taut between two points. Since the string will sag slightly no matter how taut, hang the level on the *midpoint* of the string and read it.

Fig. 12. Line Level

Plumb bob The plumb bob is used to test the vertical, or to find a point directly above or below another point (*13*). It's a weight, tapered to a point at the bottom, attached to a cord. Once the weight stops swinging, the cord will always hang exactly vertically.

The plumb bob is most often used to plumb a point between floor and ceiling. This requires two people, one up on a ladder and one on the floor. If the mark is already on the ceiling and to be located on the floor, the person above holds the cord on the mark so that the weight is hanging just above floor level. The person below marks the floor just below the point of the bob. The closer the bob is to the floor (without touching it), the easier it will be to get an accurate mark.

To plumb a point on the ceiling from a previous mark on the floor, the person below directs the person above to move the cord gradually until the bob is above the mark. Then the person above marks the ceiling where the cord meets it.

Fig. 13. Plumb Bob

Marking The pencil is the most common marker. The sharper your pencil is, the more exact your work will be.

Carpenter's pencil The carpenter's pencil is a thick flat pencil, with a thicker core of graphite (*14*). It will not make very fine lines, but it's great for rough lines on wood because it doesn't wear out quickly. For instance, use it to mark long layout lines or identifying marks on pieces, such as "Top right."

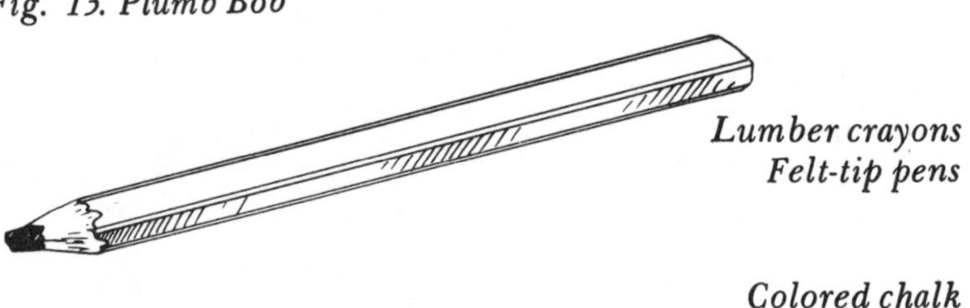

Fig. 14. Carpenter's Pencil

Lumber crayons
Felt-tip pens Lumber crayons and felt-tip pens are other familiar marking tools, but be aware that felt-tip lines are difficult to remove and will bleed through any number of layers of paint. Don't use them for surfaces that will show later.

Colored chalk Colored chalk is good for layout lines on the floor and other places where pencil lines don't show up. The chalk rubs away when you want.

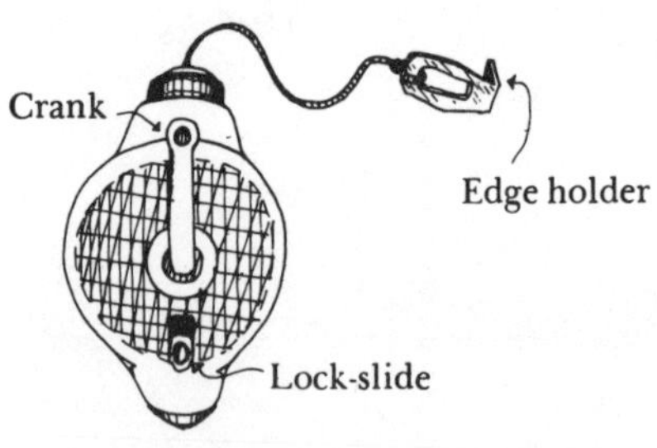

Chalk line A chalk line is a useful layout tool for marking straight lines over long distances quickly and without assistance (*15*). It's also used to mark cutting lines on plasterboard and plywood when you don't have a straightedge long enough to reach between two distant points.

It is a length of string, usually about a hundred feet, wound on a reel enclosed and covered with colored chalk dust inside a case. The end of the string is tied to some sort of hook. With most reels, the line can be locked in place with a sliding switch. To draw a line between two points, hook the end over one point or, if there's nothing to hook it onto, hammer a finishing nail partway in at one point and loop the hook over it. Pull the reel to the other point, unwinding the

Crank
Edge holder
Lock-slide

Fig. 15. Chalk Line

string as you go. Lay the string directly on the second point, pulling the line taut with one hand. With the other hand, reach as far over toward the middle of the line as possible, pull it *straight* up several inches, then let go sharply. The line will snap back down and leave a straight chalk imprint between the two points.

Before each use, shake up the reel to chalk the line. You can snap a reel no more than two or three times before the chalk wears off and you have to rewind it. Usually you'll be winding it up immediately anyway. Never leave the line out when you're not using it.

Chalk lines come without chalk. Buy a small plastic container of chalk that you can reclose. It will last a long time. To fill the reel, unscrew the knob at the top and simply pour the chalk in. Many chalk lines are also sold as plumb bobs, but they rarely work that way very accurately. Either the line lock won't hold against the hanging weight of the reel case, or the case keeps spinning about. Use them as plumb bobs only in a pinch. Get an inexpensive chalk line—all you need it for is the string and the reel.

Awl

The awl is a pointed metal shaft attached to a handle (*16*). With the point you can make starter or pilot holes for drill bits and nails by placing the point on the mark and lightly tapping it with a hammer. It will also punch holes in softer materials, and scratch or scribe very fine lines on your work.

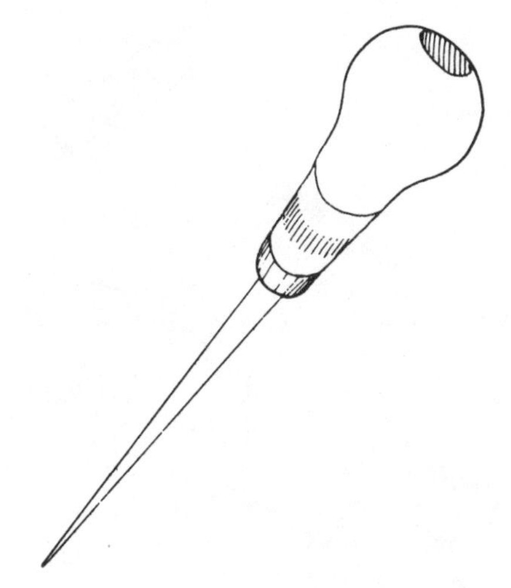

Fig. 16. Awl

The scribe is similar to the awl; it's a small awl shaft, no more than a couple of inches, with a small metal head. Many combination squares hold scribes in them.

The awl and scribe make finer lines than a pencil. If the line is on a surface to be finished, you will have to sand it away, so don't make the line too deep.

Dowel centers

Dowel centers are used for lining up dowel holes on pieces of wood to be joined with a blind dowel joint (*17*). (See Joints and Joining in "Working Techniques.") Dowel centers come in several sizes; the bottom part of the guide is the same size as a standard dowel, above it is a wider lip, and on the top is a point. After you drill holes for dowels in one piece, place the dowel centers in the holes, with the lip resting on the surface and the point pointing up. Lay the second piece on top in the exact position in which it will be joined. Press hard or tap lightly on the top piece. The points leave exact center marks on the second piece for all the other holes.

Doweling jig

The doweling jig is also very useful if you want to do a lot of blind doweling. It can be adjusted to clamp onto the piece of wood, with a sleeve automatically centered over the dowel hole. It usually contains several sleeves, to correspond to several standard dowel sizes. You drill through the sleeve, which guides the bit and keeps it straight.

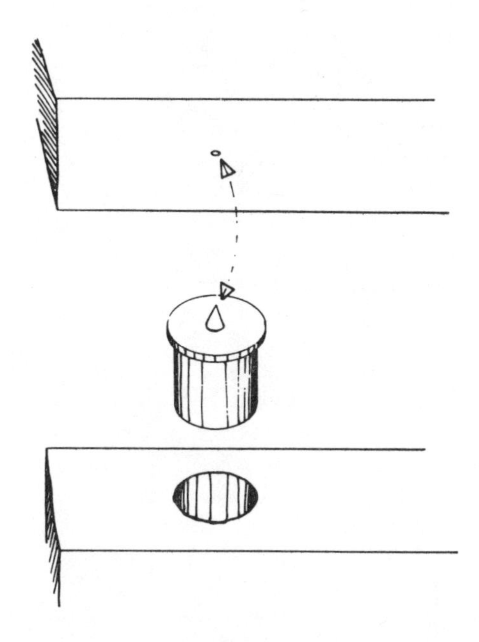

Fig. 17. Dowel Center

Tools for Pounding and Fastening

Pounding tools are needed to drive nails, sink dowels, and sometimes to knock joints together. The claw or nail hammer is the most basic and useful of these tools, but there are other, more specialized hammers and mallets. The staple gun is a fastening tool, useful for lighter materials in carpentry and many other jobs around the house. Some pounding aids such as nail sets are included here also.

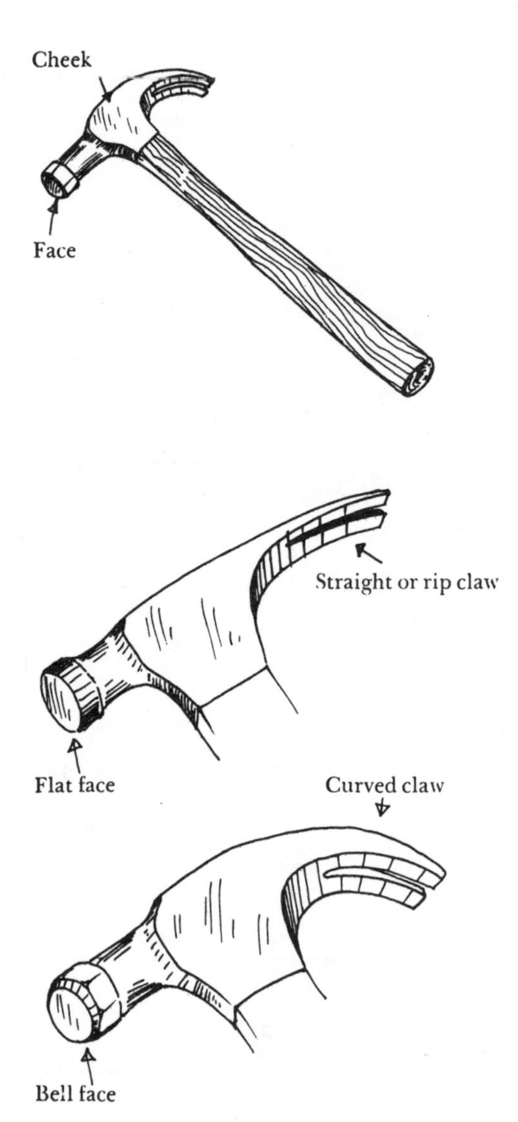

Fig. 18. Claw Hammer

Hammers and Mallets (and the Nail Set)

Claw hammer or nail hammer

The claw hammer or nail hammer is probably the most common carpentry tool (*18*). It is used, of course, to drive nails. You should definitely have a good hammer. The two major kinds are the *curved claw* and the *rip* or *straight claw,* referring to the part on the back of the head. The curved claw is designed for pulling nails, the rip claw for prying things apart, much like a crowbar. The curved claw will be far more useful to you.

The face of the hammer head, the nail-hitting part, can be either flat or "belled." With the flat face, it is slightly easier to hit the nail flush to the surface. Use it for rough work where you don't mind marks showing. The slightly convex shape of the more common bell-faced head drives the nail flush without leaving marks. Of course, it can also be used for rough work.

Cheap hammers have cast-iron heads. These heads can shatter and are very dangerous. You can usually identify them by the visible casting seam. Good hammer heads are drop-forged. The heads are smooth and will not shatter.

The other big differences between claw hammers are in the weight of the head and the kind of handle. Hammers weigh 7, 10, 13, 16, or 20 ounces, with some ripping hammers as much as 22 ounces. The heavier a hammer you can comfortably swing, the faster you'll drive the nail. Try out a few in the store to get the feel. The 20-ounce one is for muscular types with hammering experience. A 16-ounce one is a good weight for most people, men and women. With 13 ounces or less you'll be barely disturbing the nails.

Wood makes the weakest handles. Wood handles can crack under rough use, or the head can work loose. This is not likely on the very good ones, but definitely possible. If you prefer the feel of a wood handle (we do), get a good one and be careful. If the head loosens, you can sometimes fix it by one of three methods. First, soak the head and handle top in linseed oil for several hours to expand the wood. If this doesn't help, tighten by driving small metal wedges down into the top of the handle. Or drive a finishing nail into a thinner predrilled hole in the top of the handle.

Steel and fiberglass handles are both very strong and well-bonded to the head. A good model will last forever.

Pay the extra few dollars to get yourself a good hammer. We suggest a curved claw, bell-faced model, 16 ounces if comfortable, and whatever handle you like.

Using the hammer

Hammering nails is easy with a little practice. The closer you grip to the end of the handle, the more power and leverage you get, just as with a tennis racquet or golf club. The further toward the head you hold it, the less power but the more accuracy you'll have. The hammering motion should be smooth, involving the elbow and shoulder as much as the wrist. The more you swing, the more power you have; the less you swing, the more accuracy.

To start a nail, hold it between the thumb and second finger of one hand, down near the bottom so you can keep it in place, or up near the top with your little finger braced on the wood. Keep your fingers below the nailhead. Tap the nail lightly once or twice with the hammer to get it started, then swing harder. Once it's well in the wood, take your fingers away, and hammer as hard as you want, with no fear of smashing your thumb. Try to hit the nail squarely on the head. On impact, the hammer handle should be parallel to the surface. It's all a matter of feel—after a while it seems natural.

Choke up on the hammer if you're having problems hitting the nail. As you get more accustomed to the motion, lower your grip. For rough work, grip near the end and use a full-bodied swing (once your fingers are off the nail). On finishing nails, choke way up and practically tap them in—and use a nail set for the last fraction of an inch.

Hammering tips

Never hit anything with the side, or cheek, of the hammer head. This can ruin your hammer.

Some people always wear goggles when they use a hammer—a piece of wood or a sliver of metal from the head or nail could fly off. Certainly you should use goggles with masonry or case-hardened nails. They are very brittle and break if you don't hit them squarely. Wear goggles when hammering above your head or close to your eyes.

Don't hit plaster, brick, concrete, stone, cast iron, or other similar materials with your claw hammer. Just wood and nails, or you'll damage the hammer face.

The nail hammer can be used in place of a soft-faced mallet to knock pieces of wood together if you hold a scrap piece of wood against the work pieces. Only hit the scrap.

When pulling out nails, save the wood surface from being marred by slipping something between the hammer head and the wood, like a thin piece of wood or hardboard. With longer nails, use a thicker block of wood to get leverage at the end of the pull.

At ends of 2x4s, or other places where wood might split, drill a hole slightly thinner than the nail, then hammer the nail in. Or blunt the nail point by hitting it once or twice with your hammer. A blunt tip breaks the grain as it goes through, rather than spreading and splitting it.

To hold a small finishing nail or brad while starting, stick it into a piece of cardboard and hold the cardboard. When the nail is almost sunk, the cardboard will pull off over the head. Use nail sets to drive nails flush or below the surface.

Nail Set

The nail set, used with the hammer, sinks the heads of nails without marring the surrounding wood (*19*). Used mainly on finishing nails; the tip is usually cupped to fit just over the head.

Fig. 19. The nail set

Hammer as usual till nailhead is about 1/8" from surface. Place the point of the set on the nailhead and hit it with a hammer till the nail is below the surface about 1/16" to 1/8". Later you can fill the hole with wood putty. The only trick is to be sure the set doesn't slide off the head and damage the wood itself. Support the set with your little finger resting on the wood next to it.

Nail sets come in different sizes to fit different size nails. Buy good ones—cheap sets won't hold their point long. Wear goggles when using.

Tack Hammer

A small light hammer for close, fine work with finishing nails and tacks (*20*). Handy, for instance, if you're nailing molding around a window—it's less likely to hit the glass or mar a finished surface. The back of the head is often magnetized for picking up and starting tacks too small to hold easily. An inexpensive tack hammer is fine, since you won't use it much for general carpentry, and it doesn't undergo much stress.

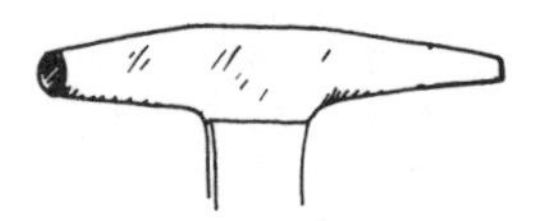

Fig. 20. Tack hammer

Mallet

A mallet is a soft-faced hammer (*21*). The handle is made of wood and the head of rubber, plastic, rawhide, or wood. The mallet is used when the surface you're hitting might be damaged by a metal-faced hammer. Use the mallet for hitting wood-handled chisels; for driving dowels; for knocking wood joints or pieces into place.

The weight of the head can vary anywhere from about 1½ ounces to 2 pounds. One of around 6 to 10 ounces should be fine for most work.

Fig. 21. Mallet

Hatchet

The hatchet has a hammer face on one side of the head and an ax blade on the other (*22*). The ax blade is handy for rough work like splitting shingles or knocking things apart. Obviously, it can be dangerous and requires care and concentration in use. Definitely wear goggles. Store it carefully; protect the blade and keep it sharp.

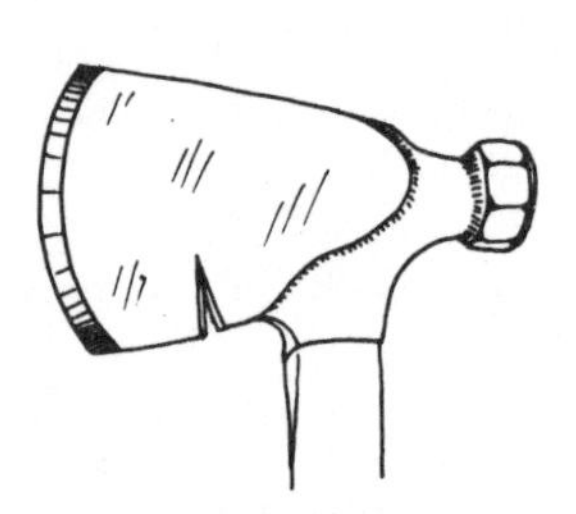

Fig. 22. Hatchet

Other Fastening Tools

Various *drive tools,* known by brand names such as *Ham-R-Tool* and *Shure Set,* can be used with a hammer or small sledge hammer to drive special fasteners into tough masonry. The tool is basically just a holder that keeps the fastener straight and perpendicular to the surface and prevents the fastener from bouncing back. Some fasteners have threaded shanks that remain above the surface, others flatten against the surface. Each tool comes with a list of what size fasteners to use for different materials.

Power-actuated drive tools

Power-actuated drive tools, powered not by electricity, but by special industrial power charges, are wonderful tools for fastening into masonry; but they should be used only by someone with a great deal of experience and skill with tools. They are perfectly safe when used correctly, but because of the powder charges they can be as dangerous as guns when misused.

One type is similar to the hammer-driven tools above. You load the fastener and charge into the holder, set the tool in place, then hit the plunger sharply with a sledge. Another type, called a *Ramset* (*23*), looks much like a long-barreled pistol, and it works by pulling the trigger. You must be careful with either model to hold the fastener point and front of the tool firmly and flush on the surface you're fastening into, so there is no chance of the fastener shooting or rebounding into the air. Wear goggles. Also use earplugs.

Both types have different size and style fasteners, and also different strength charges. You *must* use the proper charge for the material you're driving into; the manufacturers provide tables of this information. One of the dangers of these tools is that if the wall you're fastening into is too thin, you might shoot a fastener right into the next room.

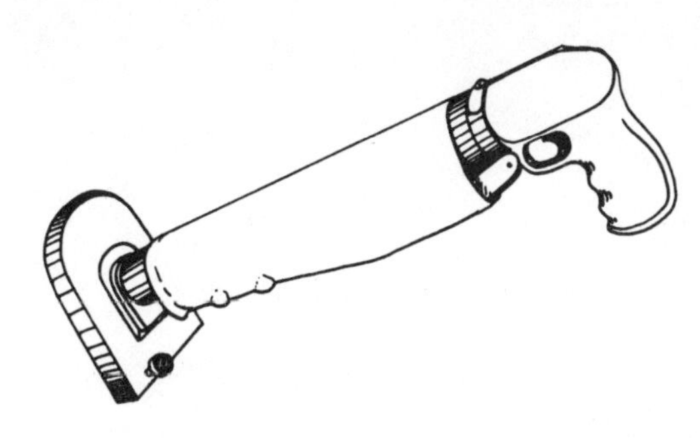

Fig. 23. Ramset

Again we stress that power-driven tools can be dangerous if misused. We strongly warn against using them till you've had a great deal of experience and confidence with tools, and then only with full and proper instructions from the manufacturer or representative. Until then, if you're faced with a concrete-aggregate wall impenetrable to a hammer drive tool or carbide drill bit, either hire someone else to do the fastening, or redesign your structure, if possible, so it need not be fastened into that surface. When absolutely needed, these expensive tools can be easily rented. Be sure to tell them what type of surface you want to fasten into, so they can provide you with the proper equipment; and be sure to get full instructions and, if possible, a demonstration, of the tool.

Staple gun

The staple gun is a wonderfully versatile tool (*24*). Essentially a powerful desk stapler, it drives the staples in straight rather than crimping the ends flat. It can eliminate the time and trouble of having to hammer in lots of tacks or smaller nails. It attaches insulation, thin paneling, vapor barriers, and plastic to 2x4s; weatherstripping to doors and windows; screening and fabrics over frames; and many other things. It can be used for anything thin that is not under great stress. Get a good stapler, one that takes different sizes of staples, and buy the staples designed for your stapler.

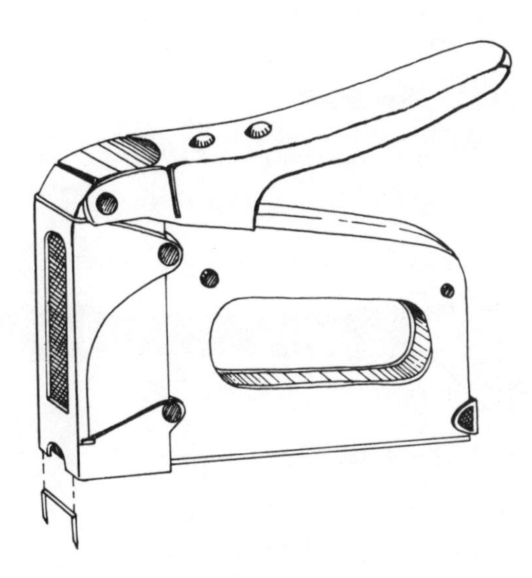

Fig. 24. Staple Gun

Tools for Making Holes

Drills and the Brace

You use a drill and drill bit to make a hole. Holes are needed in carpentry for fasteners like screws, bolts, dowels, even nails, sometimes; and also for larger objects like pipes or lock cylinders. Hand-operated drills will make holes in wood, metal, plastics (like plexiglass), and some other construction materials. While electric drills are much faster and so cheap that many carpenters don't bother much with anything else, hand-operated drills can be very satisfying, particularly for small jobs.

Hand drill

The hand drill is the one that looks something like an eggbeater (*25*). Turning the handle turns gears that rotate the drill bit, the cutting element of the drill. Clockwise rotation of the handle drills the hole; counterclockwise backs the bit out (though usually the bit will just slip out of the hole).

Hand drills usually take standard twist drill bits, the same

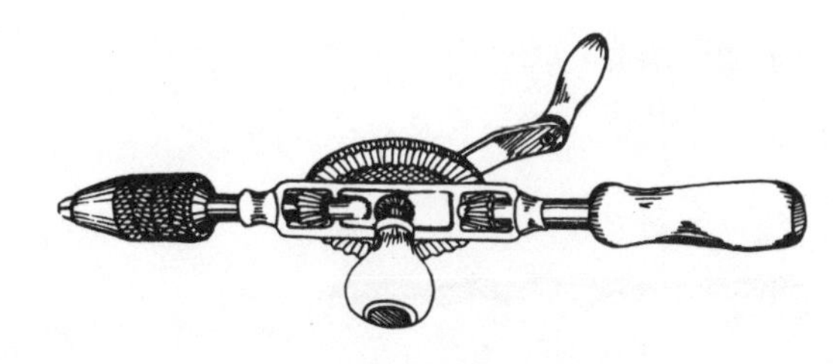

Fig. 25. Hand Drill

kind used in electric drills. You can get an inexpensive set of bits 1/16″ to 1/4″ in diameter, graduated in 1/64″ steps. The bit is inserted and tightened in the jaws of the drill chuck by holding the chuck with one hand and turning the handle with the other. Some hand drills will take special bits called drill points, with fluted shanks, ranging up to about 3/16″ diameter.

To drill an accurate hole, first mark the point to be drilled, then place a center punch or awl on the mark and tap lightly with your hammer. This makes a depression in the surface for the drill bit tip to sit in—on a flat surface, the bit would otherwise tend to slide away from the starting point as you begin to drill. Hold the drill vertical as you work. Go slowly at first, till the hole is well started and you're sure it's straight; then you can speed up, but keep the pace steady.

Push drill

The push drill is similar in design and handling to the Yankee screwdriver (*26*). Push in on the handle and the bit turns. Releasing the pressure returns the handle without rotating the bit. A ratchet device can be switched so that pushing reverses the rotation of the bit, allowing you to remove it from tight holes more easily.

Fig. 26. Push Drill

Drill points are used for bits (*27*). On most push drills you just pull the chuck back a little and slip the drill point in till it catches, then let the chuck spring back.

Fig. 27. Drill Point for Push Drill

This drill is used mainly for small holes, such as those for cabinet hinge screws. It's harder to keep straight, but very handy because it can be used with only one hand. Most models store the drill points inside the handle.

Pin vise

The pin vise is really nothing more than a handle for small bits (*28*). To drill, just turn the handle like a screwdriver. Its main use is in miniature work like models; but, without the knob, it can also be used as an adapter in a large drill whose chuck is too large to grip a very small bit.

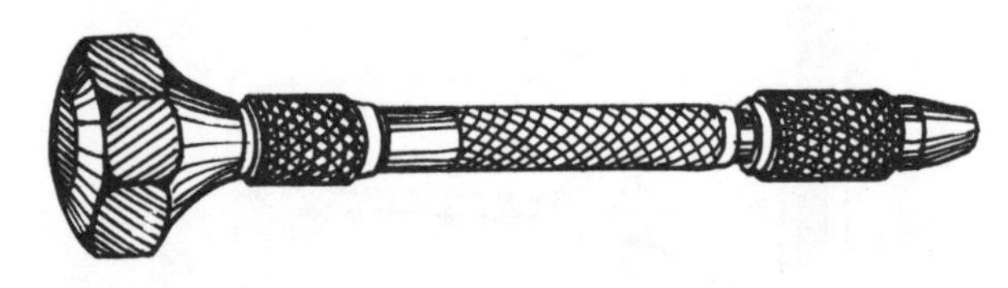

Fig. 28. Pin Vise

Brace

The brace is a tool that's been used for ages to make larger holes (*29*). It has a large crank in the middle of the handle. You hold the knob at the top, and turn the crank to turn the bit.

Auger Bits

Special auger bits are used with the brace (*30*), ranging in diameter from 1/4″ to 2″. Also used is the *expansive bit,* which can be adjusted to a diameter of anywhere from less than an inch to 3″. Each bit has a small screw centered at the tip that helps draw the bit through the wood. Unfortunately, auger bits are very expensive. A set of thirteen standard bits, from 1/4″ to 1″, can cost anywhere from $20 to $100.

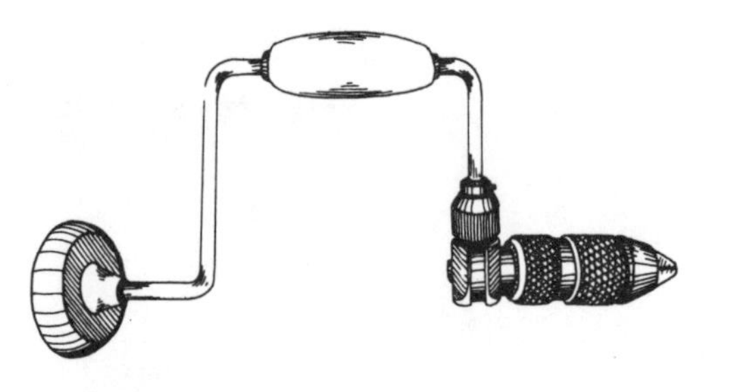

Fig. 29. Brace

To drill straight, push straight down on the knob at top, either with your hand or body. The idea is to keep that knob steady in the center so that the crank revolves around it.

Some braces are equipped with a ratchet device that allows you to drill by just turning the crank back and forth. This helps in tight spots where you have no room to completely rotate the crank.

Fig. 30. Auger Bit for Brace

General drilling tips

Always keep the drill straight as you make the hole. Bending or angling the bit after it is started in the hole may snap the bit in two. If you have trouble keeping a drill straight, place a try square next to it and sight the bit along the square's edge.

Any bit tends to splinter the bottom face of the wood as it comes through. Prevent this by clamping or holding another piece of wood tightly under the work piece as backing.

If you have a lot of holes to drill at the same depth, improvise a depth guide by wrapping a piece of tape around the bit at that distance from the tip. When the tape reaches the wood surface, the hole is the proper depth. You can also use manufactured *drill stops* that slip over the bit.

The brace and the hand drill can take a special cone-shaped bit called a *countersink,* which is designed to enlarge the top of a previously drilled hole. This allows the screw head to be set flush with the surface of the wood. Also adaptable are special screwsink or screw bits, which are really meant for electric drills. These bits perform all three drilling operations for a screw hole at once—the pilot hole, the shank hole, and the head hole. (See Screws in "Hardware.")

Tools for Twisting, Grabbing, and Holding

In some way, all of these tools grip or grab hold of objects much more strongly than you can with your hands. The main uses of these tools are to turn and drive fasteners—as with screwdrivers and wrenches, and to hold things together tightly—as with clamps.

Most of these tools are very useful. The ones you buy will depend on the work you are doing, and on how broad a selection you can afford. Probably the most all-around useful tools for the price (in this section) are the standard screwdriver, the adjustable wrench, and a pair of C clamps.

Screwdrivers

The screwdriver, one of the most misused tools, is a beautifully simple tool designed for one thing—to drive screws (*31*). This tool is rudely violated by using it as a paint-can opener, prying bar, chisel, or scraper. When you find yourself cursing at your old screwdriver, it is usually because misuse has ruined the tip or you've chosen the wrong size for that screw.

Standard screwdriver

This is the most common type of screwdriver for the most common screw. The flat blade is narrower at the end than near the shank.

The longer and thicker the screwdriver is, the easier it will be for you to drive a screw. The tip should fit the slot, or kerf, of the screw as exactly as possible, both in thickness of the tip and in width. Too wide a tip will rip the wood around the screw as the head is sunk. Too thin a tip makes it hard to turn the screw—you lose leverage. Also a thin tip keeps slipping and sliding out, eventually damaging the sides of the slot so that you never get the screw in.

Hold the screwdriver straight as you turn it. Grasp the end of the handle firmly with one hand, and with the other keep the blade tip in the screw slot. Push in hard, particularly when the screw is tight; this helps keep the blade in the slot.

You should have several standard screwdrivers. First you need a big one, say with a six-inch shank and a thick handle. Rubber-coated handles seem to be easier on the hands, but plastic handles (or wood if you can find them) are almost as good. A big screwdriver should have a square shank. This enables you to grab it with a wrench for extra turning leverage on stubborn screws. Some screwdrivers have a square or hex section on the handle for the same reason. Buy a good one, be sure the handle is well attached to the shank, and that the blade is straight. The harder, better steel of a good screwdriver will not get bent or chewed up easily.

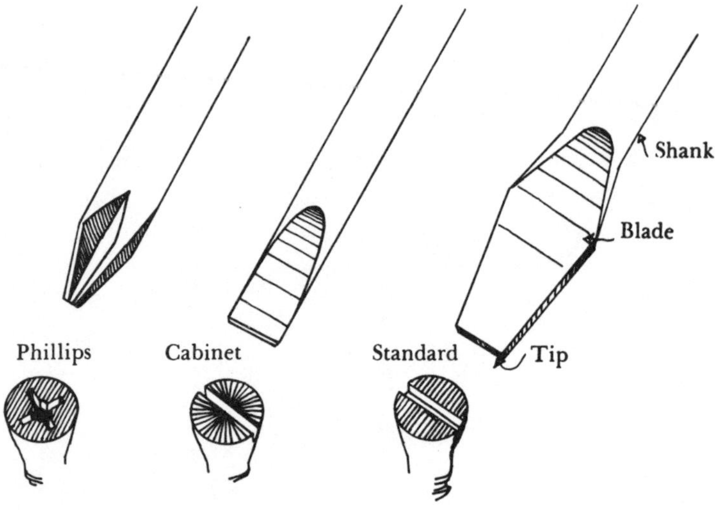

Then you will need a small one and a medium-sized-one. Here you can get by with less expensive models, particularly for the small size, since most smaller screws won't be hard to drive. A stubby-handled screwdriver is also useful, for tight spots where you can't fit in a long one.

Treat your screwdrivers with respect. Any kind of prying or chiseling will wear away the edge of the tip, and maybe even bend the blade. It's tempting, admittedly, to use your screwdriver in such ways; if you can't restrain yourself, put away an old one or a cheapie for those purposes and save your good ones for screws. An old, worn blade can sometimes be restored by filing it.

Fig. 31. Screwdrivers

Cabinet screwdriver

This is similar to a standard bit, but the blade is wider at the *tip,* or completely straight. Useful for countersinking in fine cabinet work. A standard blade might scrape the inside of the hole. This fineness comes at the loss of some strength in drive and leverage.

Phillips screwdriver

This has a tip like a cross to fit Phillips-head screws. You may never buy these screws yourself (unless you use a screw gun), but you'll find them in appliances such as TVs, stereos, and power tools. So it's a good idea to have a good Phillips screwdriver around. In an emergency, if you're careful and the screw isn't in too tight, you can sometimes remove a Phillips-head screw with a small standard screwdriver.

There are five sizes, 0 to 4, 0 being the smallest. Any size from 1 to 3 should be fine for most uses.

Nests

These are small sets of different sizes of standard and Phillips blades that fit into one universal handle. They're not as strong as a solid screwdriver, but much handier.

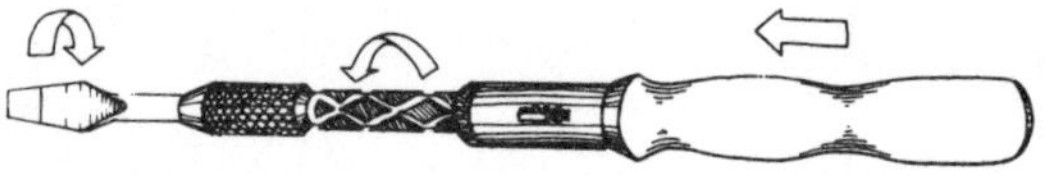

Spiral ratchet screwdriver

Spiral ratchet or Yankee screwdriver has a ratchet gear, which means you can push in the handle and the blade will rotate (*32*). When you release, the handle pushes back out, but the blade stays in place—a beautiful tool that can save you a lot of time. It has a chuck that accepts different size and type tips, eliminating the need for a variety of screwdrivers. A switch can reverse the action of the screwdriver to remove screws, or lock the shank so it can be turned like an ordinary screwdriver. Even the best ones have difficulty driving really stubborn screws at the last few turns, except when locked.

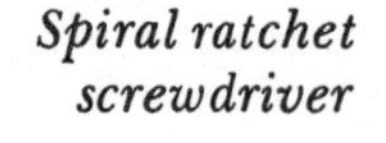

Fig. 32. Spiral Ratchet Screwdriver

Offset screwdriver

The offset screwdriver, with a blade at each end of a flattened S, is indispensable for tight spots (*33*). The tips come in all sizes and shapes. Often they are the same at both ends, but at a 90° angle to each other; you use them alternately on the screw when you don't have the room to swing the screwdriver more than half a turn.

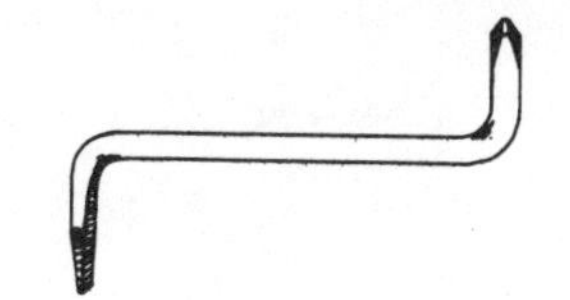

Fig. 33. Offset Screwdriver

Spline screwdriver

The spline screwdriver has a slightly flexible double tip that wedges into the screw kerf and holds the screw on, freeing your hand.

Pliers

Pliers are very powerful fingers. They grab and hold things tight, hold nails in tight areas, grab and turn your screwdriver. Your hand provides the tension that makes them grip tight, making them handier and quicker to use on small jobs than a wrench. Where a lot of power is required pliers might slip off; therefore avoid using pliers on bolts and nuts.

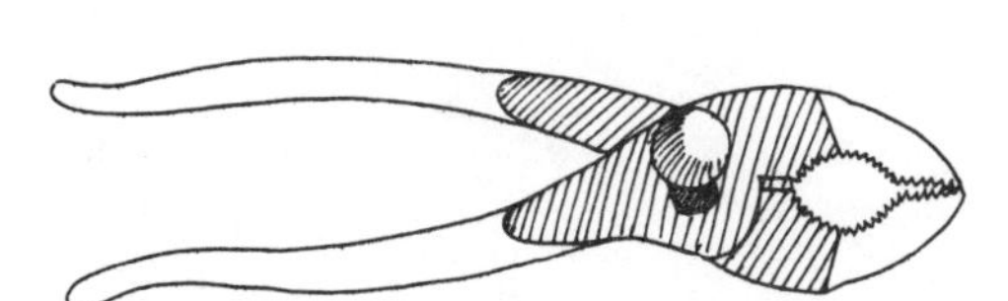

Fig. 34. Slip-Joint Pliers

The legs of the *slip-joint pliers* (*34*) can be adjusted in two positions—one with the jaws parallel to each other, the other with the jaws touching just at the ends, for irregular shapes and a wider grip. An inexpensive pair of these ordinary pliers will last a long time if you don't use them under a lot of pressure.

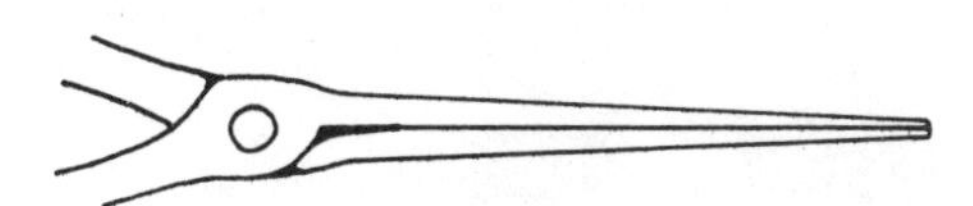

Fig. 35. Needle-Nose Pliers

Needle-nose pliers

Needle-nose pliers have very thin jaws for getting into small spaces and for holding small objects. Wait till you absolutely need them, then buy a decent pair.

Finger chompers or Lineman's pliers

Another useful pair of pliers is the one with a wire cutter at the bottom of the jaws. These are affectionately known as Finger Chompers to some people (*lineman's pliers* to others), because of their predilection for pinching your skin when you leave a finger dangling near the jaws. Handle them carefully (*35*).

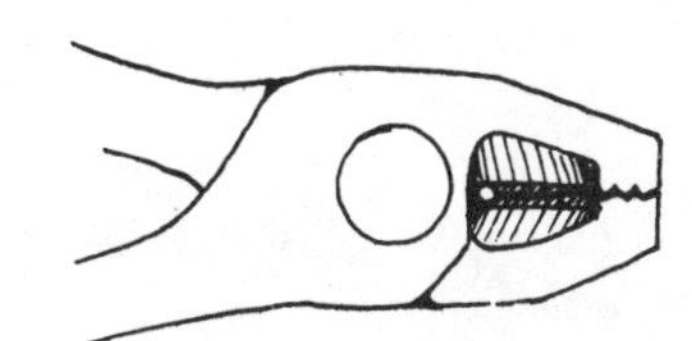

Fig. 35. Lineman's Pliers

Canvas pliers

Canvas pliers have wide jaws designed to hold canvas, screen, or other fabric while stretching it over a frame (*36*). They are also useful for breaking off very narrow strips of plasterboard after scoring.

Groove-joint pliers or channel locks

Groove-joint pliers or channel locks are a cross between pliers and a wrench (*37*). The two jaws are set at an angle to the handles; their opening can be adjusted to several different depths by means of a tongue-and-groove arrangement of arched channels. Although your grip is what holds the tool tight on the work, as with pliers, the channels help "lock" the pliers and prevent them from slipping off. Get a good pair.

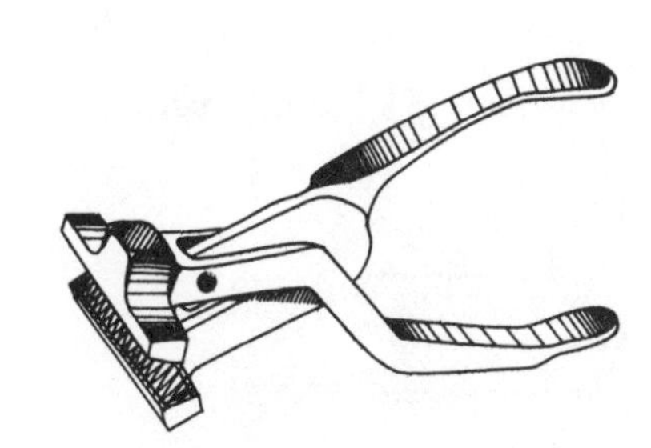

Fig. 36. Canvas Pliers

Wrenches

Wrenches are used to tighten or loosen such things as bolts, nuts, pipes. They can also be used as pliers to hold objects firmly. Their advantage over pliers is that either they fit or can be adjusted to fit exactly on the work, and hold tightly without the pressure of your own grip. Therefore you can apply much more pressure *turning* the wrench. The longer the handle, the more torque you can apply.

Box wrench

The box wrench (*38*) has a nonadjustable hexagonal or twelve-point head that fits over the corresponding size nut or

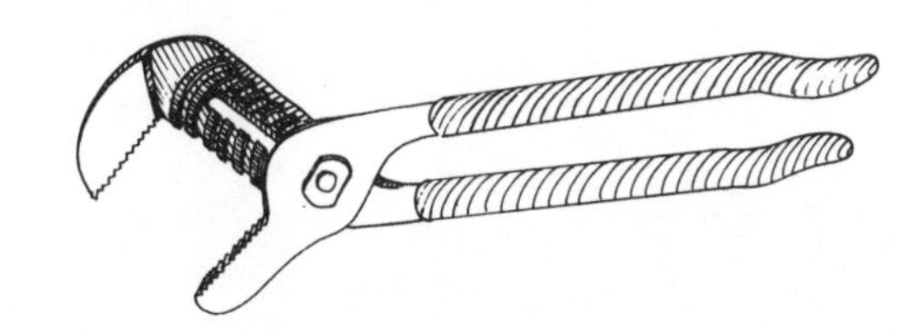

Fig. 37. Groove-Joint Pliers

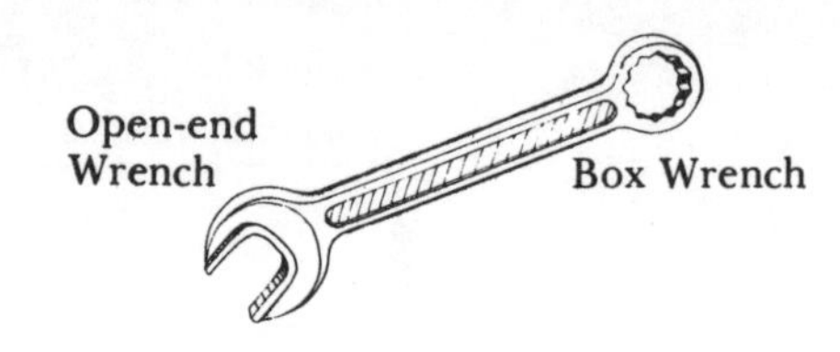

Combination Wrench

bolt. The exact size is needed. They come in American and metric sizes.

Open-end wrench

The open-end wrench (*38*) looks like the box wrench, but with one side of the head open. Where one side of a bolt is against an obstruction, you can slip this wrench on the bolt, make a half-turn, take it off, and repeat. The *combination wrench* has a box head at one end and the same size open head at the other.

Adjustable wrench

The adjustable wrench (also called *Crescent wrench*) has two parallel jaws that can be adjusted to any size, within a certain range, by a nut in the handle (*38*). It will also work on any shape of bolt or nut, thus eliminating the need for different kinds of nonadjustable wrenches, or for having to buy the proper bolts for your wrenches. The only disadvantage is that it is not as strong as the nonadjustable wrench. A cheap one will slip off or strip its gears under a lot of pressure, but a good one should last for just about any kind of normal work. However, it is not meant to be used on plumbing pipe. Always push this wrench in the direction shown to avoid breaking it. A good standard one is about 8″ long. The adjustable wrench is probably the best for you to buy first.

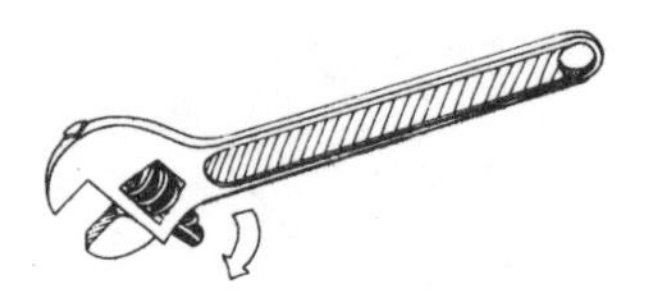

Fig. 38. Adjustable Wrench

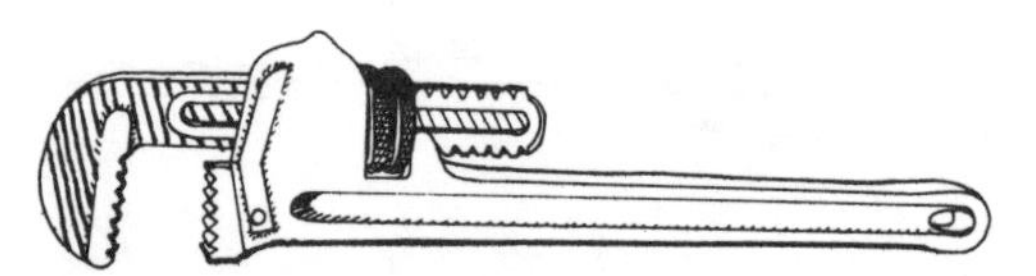

Fig. 39. Stilson Wrench

Stilson or pipe wrench

A Stilson or pipe wrench is a strong adjustable wrench designed to be used on plumbing pipe (*39*). These are the red wrenches you see with very rough-toothed jaws. They come in many sizes, much longer than other wrenches because pipes usually need more torque than bolts do. There are small ones also, of course, for tight spots. Don't buy cheap ones—loosening plumbing pipes requires all the strength you can get. You need two of these wrenches—one to hold the pipe firm, the other to turn the connecting pipe.

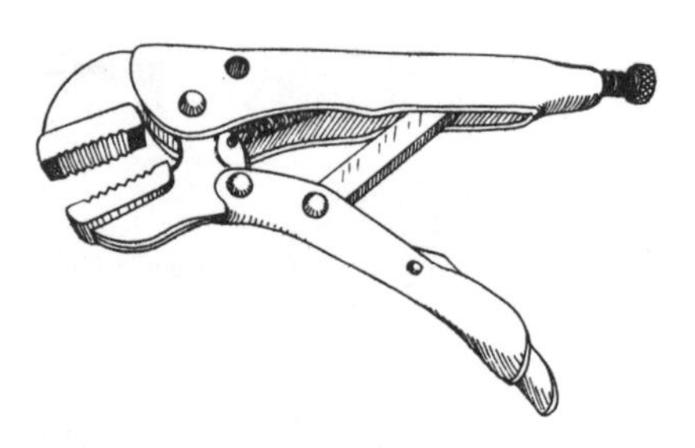

Fig. 40. Locking Plier Wrench

Locking plier wrench or vise grip

A locking plier wrench or Vise grip is a wonderfully adaptable tool—another cross between pliers and a wrench (*40*). The jaws are like those on pliers but they can be adjusted and set to any opening. You can actually lock them on the work by means of a spring-loaded clamp in the handle, and then turn like a wrench. They are easily released by opening the handles. They can be used as a small vise, for holding something small when hacksawing, and for grabbing and pulling all sorts of objects. Buy a good pair.

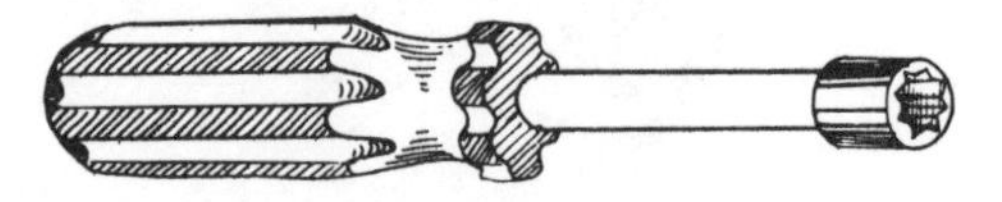

Fig. 41. Nut Driver

Nut driver

A nut driver is a screwdriver handle and shank ending in a box-head sleeve that cups over the bolt or nut (*41*). The sleeves come in 6, 8, or 12 point heads. It is designed for quickness and ease in driving nuts and bolts that do not require the extra torque of a wrench—smaller things, like the bolts in TVs and stereos. It's also good for getting at bolts set deep behind things, in a place where you can't maneuver a wrench. Some handles can take different-size sleeves; others are all one piece and fit only one kind of bolt.

Ratchet wrench

A ratchet wrench is a wrench handle with a ratchet device at the end that accepts different nut-driver sleeves and/or other attachments and extensions (*42*). The ratchet engages the sleeve in one direction and disengages it in the other. This means that you can tighten a bolt with a quick back-and-forth motion of the handle, never taking the sleeve off the bolt. This saves a great deal of time and labor over a normal wrench, particularly when you have a lot of bolts to tighten. You can also reverse the action of the ratchet to remove bolts.

Standard ratchets come in what is called ¼″, ⅜″, and ½″ drive. The size of the drive is simply the measurement across the square male protuberance at the end of the ratchet head that accepts the corresponding female part in the sleeve. A larger drive gives more torque and turning leverage.

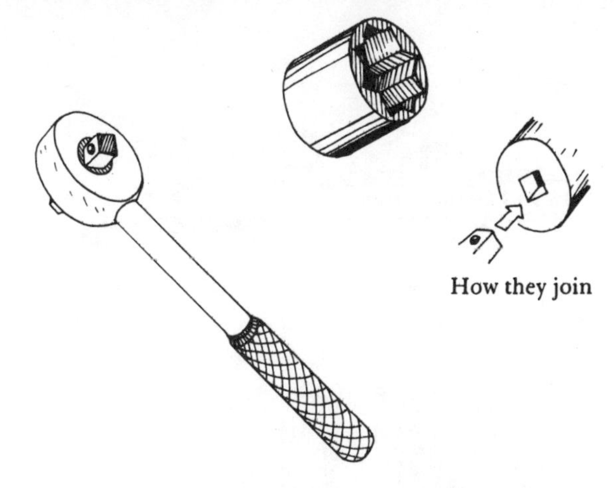

Fig. 42. Ratchet Wrench

If you want a ratchet, it's definitely wise to consider a ratchet set. You'll get an assortment of sleeves and attachments, rather than just a ratchet handle with the one or two sleeves you might need at the moment. If you do much general construction, you will need a lot of different-size sleeves. Sets are much cheaper than the prices of the individual components, and there are several good brand-name sets on the market that are almost always sold way below list price.

Sets come with varying combinations of components. All have a ratchet handle and a certain number of sleeves. Other components might be extra-deep sleeves, adapters for changing the drive size to accept very small or very large sleeves, extensions for the male plug so you can reach into deep areas, right-angle extensions, and nut-driver handles.

Also make sure whether the set is in metric or inch measurements. The *drive* is always marked in inches; don't be misled by that—examine the sleeve-size markings. Metric sleeves will not work on standard American bolts and nuts; they slip and strip the bolt.

Allen or hex wrenches

Allen or hex wrenches are L-shaped rods of metal, hexagonal in cross section, designed to fit *into* hex-shaped openings in certain kinds of bolts and screw heads (*43*). You will find such fasteners mainly in machines and tools, such as saber saws. The advantage of these fasteners is that they can be set flush to the surface, out of the way of moving parts. When you need an Allen wrench, get the cheapest you can find. They come as sets in little bags, or in a pen-knife type nest that is very handy. They shouldn't cost more than a dollar or two for a dozen different sizes.

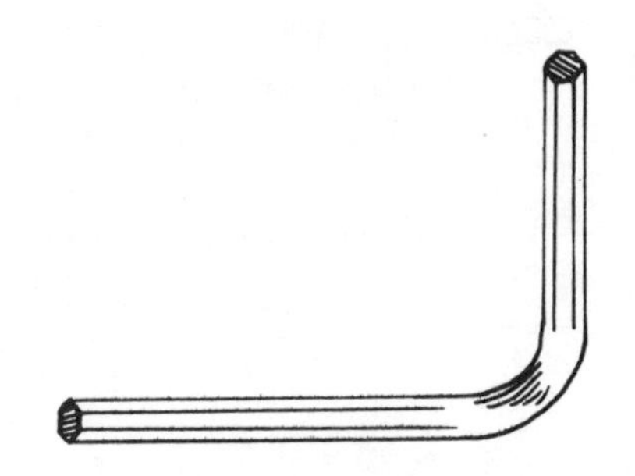

Fig. 43. Allen Wrench

Clamps

Clamps are extra-powerful mechanical hands. They will hold something together tightly, freeing your own hands. Clamps hold pieces exactly in place while you hammer, chisel, shape, or whatever. Clamps hold a guide on the work while you saw or rout. They steady the work when you sand it. They hold pieces tightly together while the glue sets.

C clamps or carriage clamps

C clamps or carriage clamps are shaped like a C (*44*). One end of the C comes to a small cupped shape; the other end has a similar cup on a swivel ball attached to an adjustable turn bolt. The work pieces are placed between the ends, and the bolt is tightened up to the work. Clamps range in size from ones with maximum opening of about 1″ to large ones of about 18″. Within the maximum opening, the jaws can be adjusted to any dimension at all.

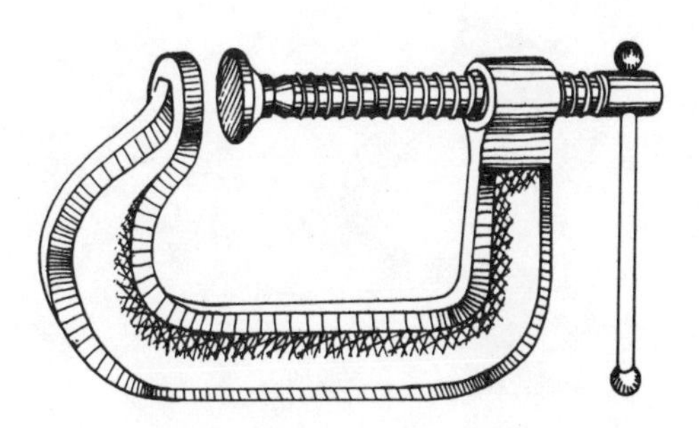

Fig. 44. C Clamp

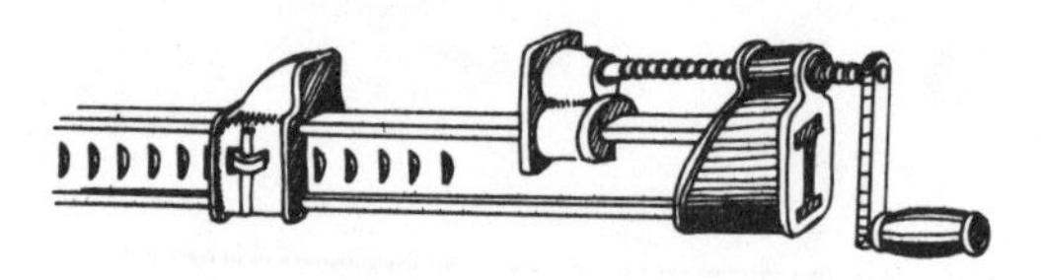

Fig. 45. *Bar Clamp*

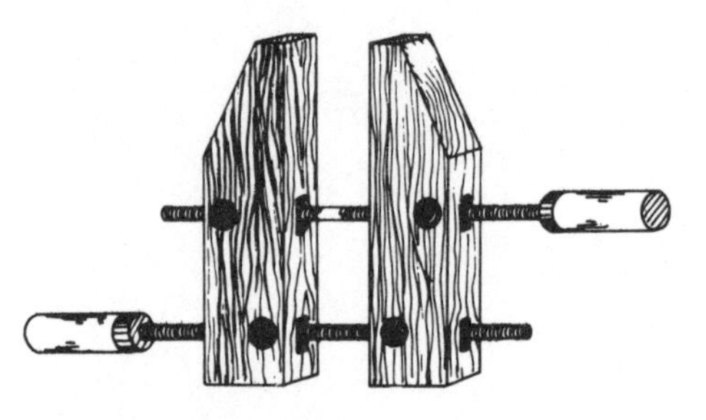

Fig. 46. *Wood Screw*

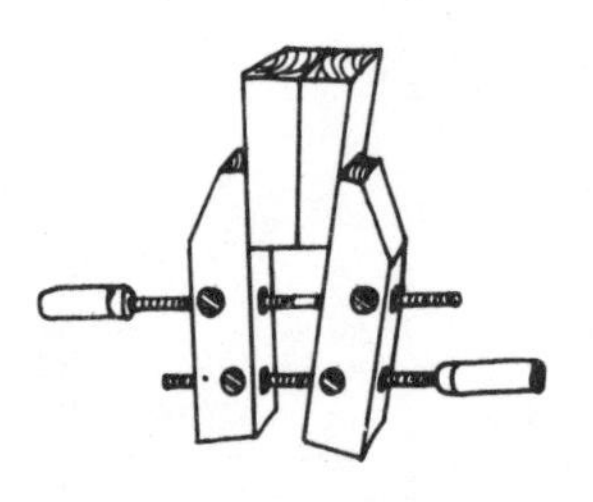

Fig. 47. *Clamping an irregular shape*

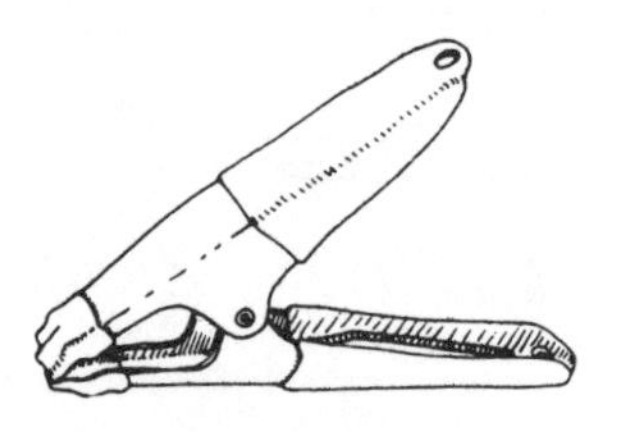

Fig. 48. *Spring Clamp*

Fig. 49. *Web Clamp*

On fine work, it's a good idea to slip a thin piece of wood between the jaws and the actual work to protect the surface from cup marks. These clamps are handy to use on a work guide, and hold as tight as you need. Inexpensive ones are fine to start with—just make sure they turn smoothly all the way along the bolt.

Edging clamps

Edging clamps are shaped like C clamps, but with an extra bolt coming out of the middle of the C at a right angle to the other bolt. They are used to clamp edging strips down.

Miter clamps

Miter clamps are used to hold together mitered pieces, such as the sides of a picture frame. Each side of the L-shape has a runner into which a mitered piece is placed and a clamp that holds the piece in place at right angles to the other. Buy a fairly good one, since accuracy is essential in mitering.

Bar clamps

A bar clamp is used in the same way as a C clamp (it clamps pieces between two opposing jaws), but it can have a much larger jaw opening (*45*). It's used for things like table tops, butcher blocks.

It has three parts. First is a bar or rod. Second is a cap that attaches to one end of the bar, with a flat surface to lie against the work. The third part is a sleeve that slides over the other end of the bar and can be tightened at any spot along the bar. Attached to the sleeve is a bolt device similar to a C clamp that tightens up against the other end of the work.

The sleeve and cap can be bought without the bar, and you can use standard plumbing pipe instead of the bar (check the size hole on the sleeve; the larger the pipe it takes, the stronger the clamp). The advantage of this clamp is that it can be extended to the length of the longest pipe you can dig up.

Wood screws

Wood screws are two parallel pieces of wood joined by two adjustable bolts (*46*). The bolts are tightened evenly to keep the jaws parallel to each other. Because these wood jaws are much larger than the C clamp jaws, they apply pressure much more evenly over the surface of the work. Thus wood screws are useful for clamping wider and more irregular pieces together (*47*). Buy a decent set.

Spring clamps

A spring clamp works like a spring clothespin—you squeeze at the wide end and the jaws open at the other end to accept the work (*48*). The spring holds the jaws tight when you release your hand pressure. The cheaper ones are simply bare steel; the more expensive ones are covered with a soft plastic to protect the clamped wood. Spring clamps are quick and easy to use and very handy for work that does not require a large jaw opening, and especially for clamping saw guides. They are not as strong as C clamps, however. Cheap ones are fine for small work; the plastic-covered ones are well worth the extra money if you plan to use them a lot.

Web clamp

A length of thick webbed tape that wraps around a large work, and adjusts and tightens by means of a bucklelike device with a ratchet action (*49*). Put small blocks between the tape and the work, so the tape won't bite into the work edges. Buy one only when you absolutely need it.

Bench vise

A bench vise is basically a clamp fixed in place (*50*). The work clamp is made of two large flat jaws—one steady, the other adjustable on a threaded track; the jaws should be lined with pieces of wood to avoid marring the work piece. Mainly for holding wood while you work on it. Not necessary unless you're setting up a shop.

Improvising clamps

Anything that will hold pieces together tightly is a good clamp. Rope, wire, even twisted-up cloth can serve. You can tighten it with a small piece of wood like a tourniquet. Strong rubber bands or tape can work on small objects; special thick and wide rubber bands are sold by specialty stores as clamps for things like chair legs.

For a temporary bench vise you can use a C clamp to hold the work against a stationary object. To clamp a guide down for a saw, spring clamps are quicker and easier than C clamps. They are also closer to the surface of the wood, which means that a saw with a protruding motor will be able to clear them, while it might not clear C clamps.

In general, start with two C clamps or two good spring clamps, whichever your saw will allow—get others if and when you need them.

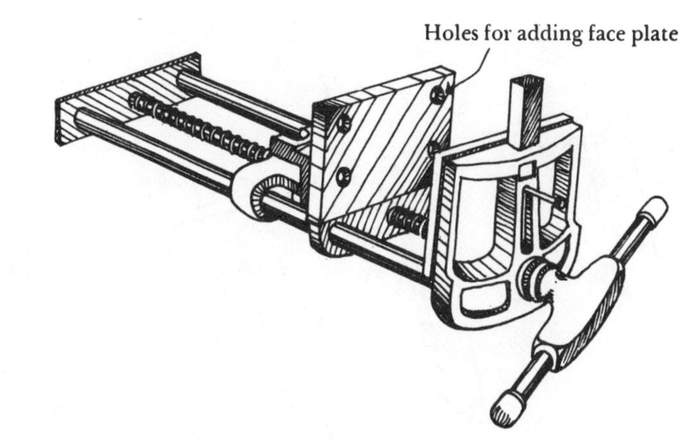

Fig. 50. Bench Vise

Tools for Cutting

These tools cut your work pieces—be they wood, plasterboard, metal, plastic laminates, or other building materials—to the sizes and shapes you lay out. They are not, however, always the best thing for precision cutting; sometimes you'll want to cut your pieces slightly oversized, assemble the project, then trim the pieces exactly with a finer shaping or smoothing tool (see next section).

The most important wood-cutting hand tool is the crosscut saw, which you can use to cut any lumber to length. For fine work involving mitered corners, you might also want to get a miter box and back saw. The utility knife, a pair of metal shears, and the hacksaw are also useful tools to have around for dozens of odd jobs. The utility knife is essential if you plan to put up any plasterboard.

Cutting Wood
Handsaws
Crosscut saw

The crosscut saw is used to cut wood across the grain, the most common carpentry cut (*51*). In other words, if you're going to cut a foot off a ten-foot 2x4, you use this saw or a power saw. If you're not going to buy a portable circular saw right away, you need a good crosscut saw. It's a good idea to have one anyway. It's sometimes handier than changing a blade in your power saw and finding an outlet; and it's the only choice in certain tight spots, or where there's no electricity.

Crosscut saws come in point sizes of 7, 8, 10, 11, and 12. The point size is one more than the number of saw teeth per inch. A smaller point number means bigger teeth, which cut faster and rougher. An 8- or 10-point saw is a good choice for most uses; get a 10 or 11 if you intend to do much fine work.

Fig. 51. Standard Hand Saw

Buy a good saw. It will save hours of time and trouble in cutting, and improve the accuracy of your cuts. A good saw can be resharpened and will last forever. Check that the handle is on tight. The blade should be straight when you look down its length, though it should also be flexible.

Saw use and care

Most problems in sawing occur because the wood is not supported correctly. The wood must be supported near the cut, or kerf, or else it will droop in the middle and "bind" the saw—the wood on either side of the kerf closes in on the saw. And if you don't support the outside ends, the weight will break the wood near the end of the cut, leaving you with a splinter on one piece. The ideal setup is shown in the illustration (*52*). Always try to use the principle of this setup—essentially putting full support under each of what will soon be two pieces.

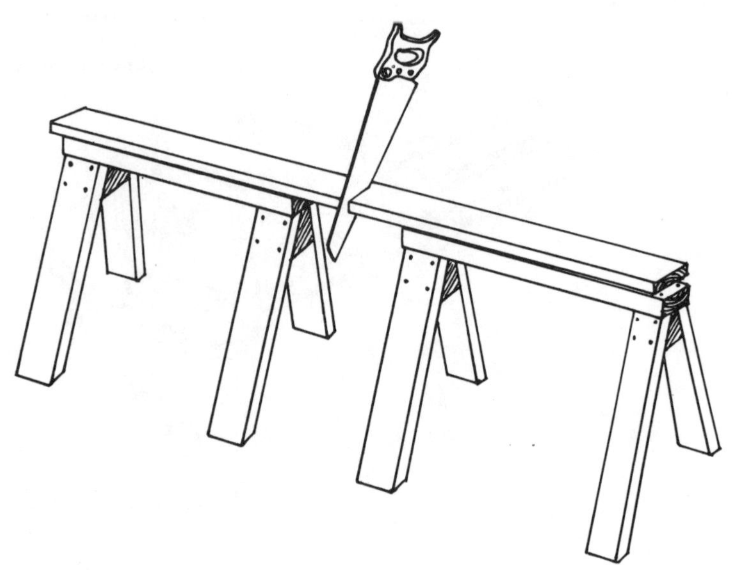

Fig. 52. Good Sawing Supports

If you are just cutting off a light piece—say two feet or less of a 2x4—you don't need the support under the shorter end. You can saw almost all the way through and then hold the little piece with one hand at the end.

Clamp or hold down the wood as you cut. Often you can hold a piece still with your hand or a knee. If you're using four supports, make sure they're all about the same height.

To saw a piece, first mark the line that you want to cut. Hold the handle with your index finger pointing straight out on the handle. Place the part of the blade near the handle right on the wood at the line, at a 45° angle to the wood surface. The blade thickness should be on the waste side of the line. Slowly draw the blade toward you, using the thumb of your other hand as a guide (*53*); as long as you pull slowly, your thumb is safe. Or you could hold a small piece of wood against the blade instead. Repeat once or twice, till the cut is started. Once the cut is started, take away your thumb and make longer, faster strokes, keeping the cut straight, of course. Remember that most of the cutting comes on the downstroke. Don't force the saw. Let it cut through the wood of its own weight, with just a little pressure from you.

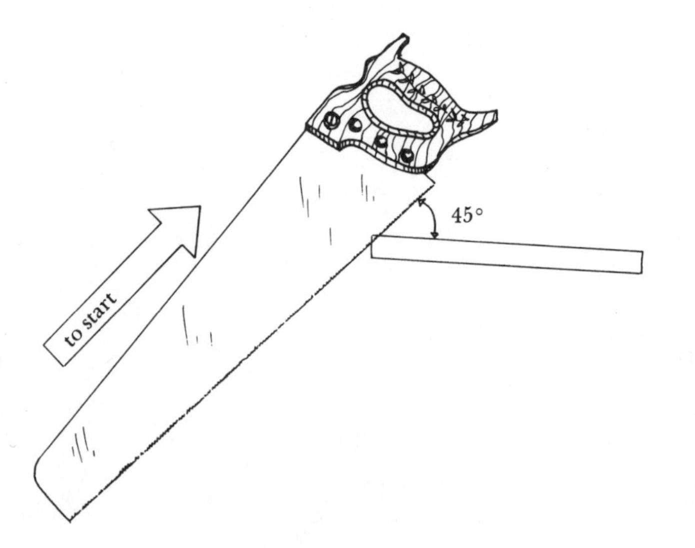

Fig. 53. Saw Use

To cut perfectly straight with a hand saw, improvise a guide for the blade. For instance, place a block of wood that you know to be square along the cutting line and against the saw blade. Or hold a try square, blade upright, near the saw blade, and eye the blade straight.

If the wood starts to bind, check the supports. Sometimes short quick strokes can loosen it. Another trick is to stick a small nail in the kerf behind the saw, spreading it enough to keep the blade free.

Practice cutting straight lines, supporting the wood. Watch someone who knows how; ask questions. Be careful not to cut yourself—the blade is sharp. Don't pick it up by the teeth.

When sawing, do not stroke down all the way and hit the handle on the wood.

Watch out for what's underneath the wood: your hands, feet, or something hard that the blade could hit and hurt itself against.

Go slow, never force a saw. You *can't* hurry a saw; it will just go off the line or the perpendicular.

If you start to saw off-course, back up a little. Carefully twist the handle a small amount to force the saw back in the right direction.

Particularly in old wood, watch for nails and other metal objects—they can ruin your saw blade.

Don't leave your saw on the floor, or any place where someone might step or drop some weight on it. If you don't have a case for it, hang it on the wall to store it.

Store your saws carefully, protecting the teeth and blade. Get covers for good saws, or make them out of cardboard or canvas. Keep your good saws oiled to protect them from rust. Use a light machine oil, wiping it on with a rag. Don't let water get on your saw.

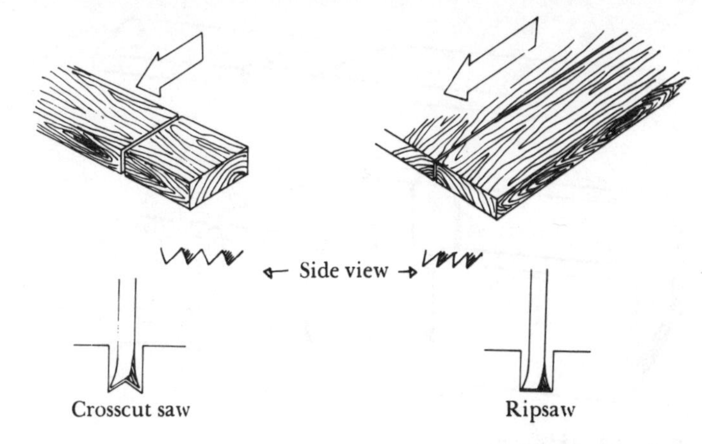

Fig. 54. Saw Types

Ripsaw

A ripsaw is similar in design and use to a crosscut saw (*54*). The only difference is that a ripsaw's teeth are designed to cut along the grain—for instance when you want to rip a 1x6 down to a 1x5. While crosscut teeth are beveled, ripsaw teeth are shaped more like little chisels. Ripsaws are rarely used these days, since portable circular saws do a much better job. You can also get pieces ripped at your lumberyard.

Compass saw and the smaller keyhole saw

The compass saw and the smaller keyhole saw have small tapered blades designed for making holes, notches, and other small cuts (*55*). A similar handle with several interchangeable blades, for different wood cuts and for metal, is called a *nest of saws* or a *nest saw.* Inside cuts can be started with a drill; then the blade is inserted in the hole. Inexpensive models are adequate, since the saw is often used for rough work—holes in paneling or plasterboard, for instance, that will later be covered over by switchplates, electrical boxes, and the like.

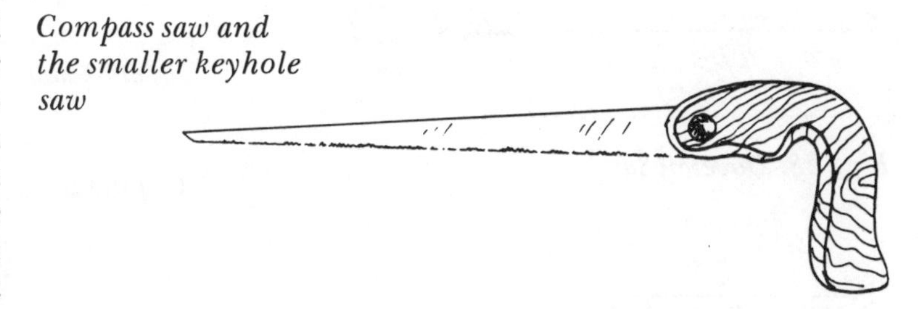

Keyhole or Compass Saw
Also with shorter blade, a plasterboard saw

The plasterboard saw is very similar, but designed to cut through plasterboard.

Backsaw

This is a saw designed for use in a miter box (*56*). Mitered, or angled, cuts have to be exact or they won't fit well. The backsaw has a reinforced back to hold it rigid, and a crosscut blade with very fine teeth to give a clean edge. It can make only shallow cuts; the reinforcement at the top of the blade is thicker than the teeth and doesn't let the whole blade pass through wood.

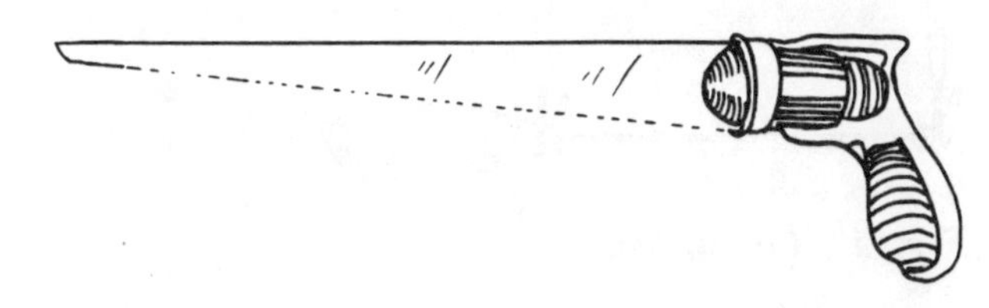

Fig. 55. Nest Saw

Miter Box

The miter box is a guide for making miters, or angle cuts, in wood joints, such as for picture frames, moldings, and finely worked furniture (*57*). Most angle cuts are 45°—half of the square corner. The cheaper boxes make only 45° cuts (from the left or right) and 90° cuts. Others can be adjusted to any angle.

The backsaw is used with the miter box because its extra rigidity gives a more accurate cut. Wood of any length is set in the box against a fence. A slot in the fence allows the blade to pass through at the desired angle. Boxes have varying limits as to wood thickness. Better ones have guides to hold the saw, and clamps to hold the wood.

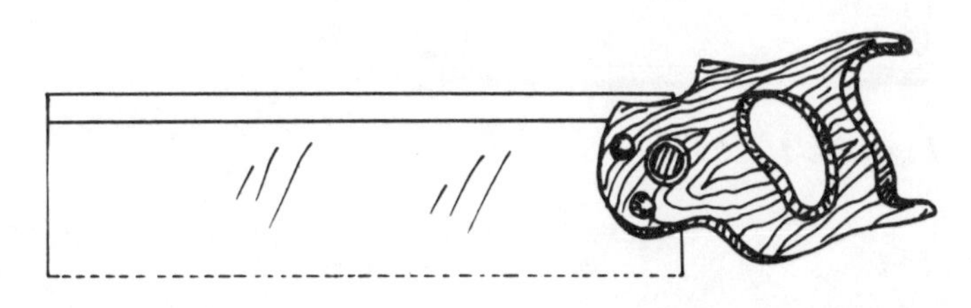

Fig. 56. Backsaw

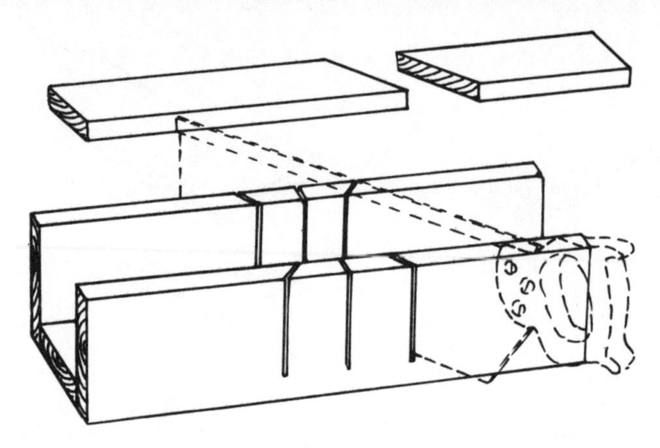

Fig. 57. Miter Box

Boxes range in price from a few dollars for those made of joined pieces of wood with saw kerfs cut in at 45° and 90° angles, to over a hundred dollars for a fully adjustable metal box with a fine backsaw included. A very good box can be bought, however, for around fifteen to twenty dollars. Buy one only when you need it, either for a lot of miter cuts (the box saves time besides being accurate), or when you want finely shaped miters. Buy a good saw—it can always be used for other jobs.

In the meantime, you can make roughly accurate miters by hand with a crosscut saw. Use the bevel square or adjustable square to lay out the angle. Or make your own miter box out of three pieces of hardwood, as shown here. Inaccurate miters can be sanded to fit better. Gaps can be filled in later with putty, particularly if they will be painted over anyway.

When cutting a miter, hold the piece well, clamp it if possible. Mark accurate lines with a fine pencil point. Then saw on the waste side of the line parallel to the surface of the board. With rounded or oddly shaped molding, practice on a few pieces till you're sure which side is to be cut which way. These pieces can be very confusing. Saw in short strokes so the blade doesn't slip out of the guide. Keep your miter box clean. With metal ones, oil the moving parts periodically.

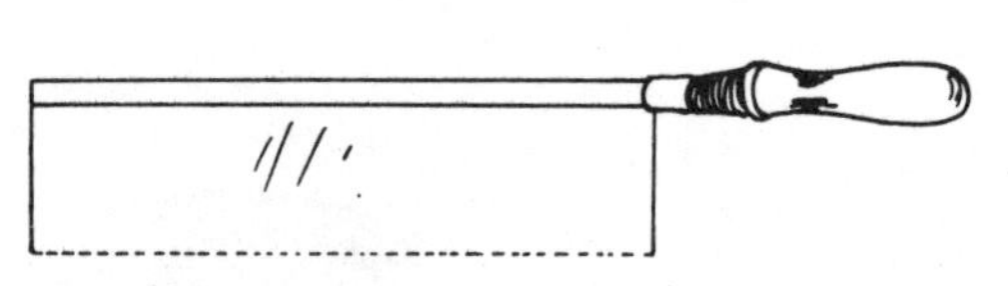

Fig. 58. Dovetail Saw

Dovetail saw

The dovetail saw is a smaller version of the backsaw, usually with even finer teeth (*58*). It's used for cutting dovetail joints, which must be done with even greater precision.

Coping saw

This has a bowed frame like a hacksaw, but is designed for thin wood (*59*). The blades are very fine, most of them flat, some almost like a wire, and used for precise curved cuts in thin wood. For interior cuts, first drill a hole, then insert the blade in the hole and *then* fasten it to the saw. Be careful not to force it, as it can break.

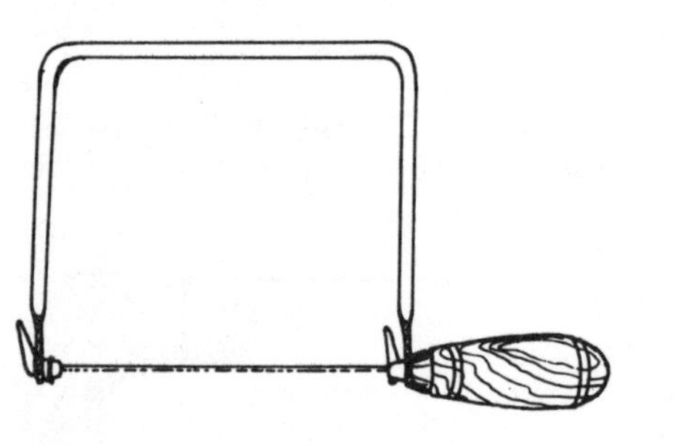

Fig. 59. Coping Saw

Cutting Metal

Metal-cutting tools are often needed in carpentry projects, particularly in general construction. You may have to cut off the ends of bolts, or cut away an old useless pipe that's sticking out of the wall; or you might have to cut a strip of sheet metal for a patch, or a couple of lengths of corner beading for plasterboard. Any wood saw would be useless, and also be ruined by trying to cut through metal.

Hacksaw

The hacksaw is a must for most carpenters (*60*). It has special blades to cut metal and plastics; it's also handy sometimes for quick rough cuts of wood scraps or thin stock like lattice. It's the best hand tool for cutting thicker metal objects, such as the ends of bolts, or nails stuck halfway into the wood. Use it for cutting J-molding, corner beading, and thin metal if you don't have metal cutters. The blade can be set to cut on the downstroke or upstroke, and with the teeth on the lower edge or up toward the inside of the saw. The frame prevents you from cutting into a wide piece of plastic, for instance, farther than the distance from blade to top of frame. Sometimes you will have to saw halfway through on one side of the piece, take the saw out, and finish the cut from

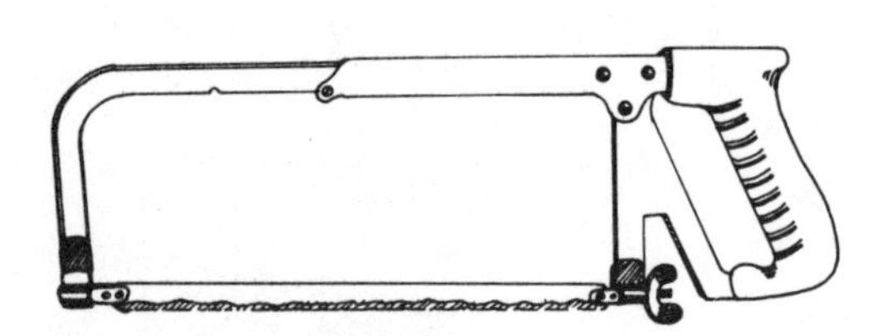

Fig. 60. Hacksaw

the other side. If that's not enough, a very inefficient, but sometimes necessary, solution is to detach and hold the blade in your hands by both ends and saw away.

Get a medium-priced or good hacksaw. The better ones hold the blades better, and are easier to adjust and to saw with (saving your hands and muscles). A good tool to have around—one you rarely plan on using, but often need. Keep extra blades around.

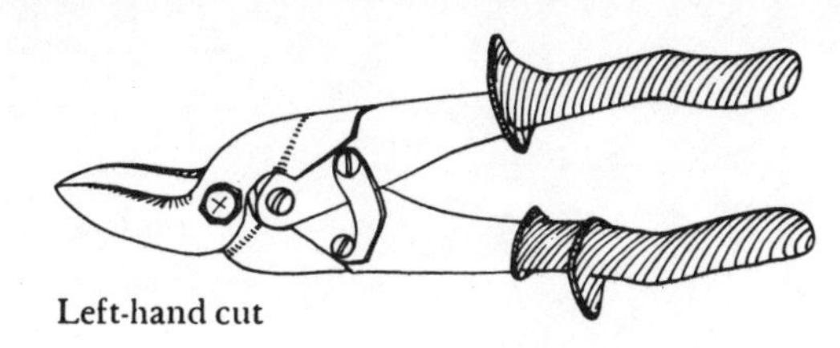

Left-hand cut

Fig. 61. Tin Snips

Metal cutters

Tin snips and shears are tools for cutting thin sheet metal. Most work like scissors. The cheaper simple-action cutters are adequate for only the thinnest of metals, such as corner beads. For anything thicker or longer, such as sheets of tin, or even J-molding, you will save your hands a lot of wear and tear by getting good shears, or the newer compound-action shears. Compound action gives as much cutting power with a short handle as normal shears give you with long handles. The compound-action shears also come right- or left-handed, which means the blade is twisted toward the right or left for cutting right- or left-handed curves (*61, 62*).

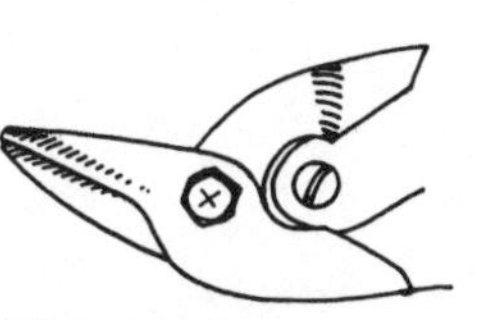

Right-hand cut

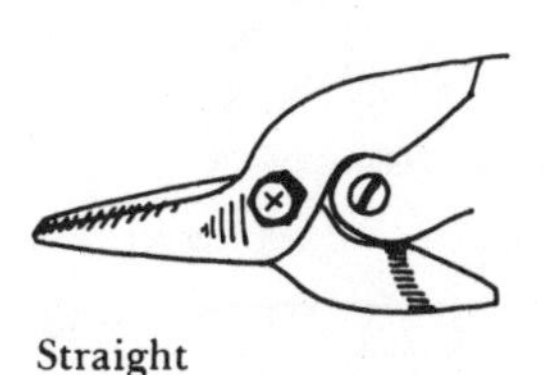

Straight

Fig. 62

Whenever you cut metal, wear heavy gloves. Be aware of where the edges and corners of the metal are at all times.

The most difficult part of cutting a sheet is fitting your hand and the tool in without slicing up your hand. Hold or bend one side of the metal up as you cut along, holding the other side down with your foot if you have to (*63*). Cut with short bites; don't try to close the blade entirely. The blades cut more easily toward the back where they are joined. Try to keep the metal that's ahead of the blades from crimping or bending. The blades don't shear properly unless the metal is flat.

Fig. 63. Using the Snips

Tubing cutter

The tubing cutter cuts even edges on electrical conduit, copper pipe, and other tubing. It is shaped something like a C clamp, only the adjustable jaw has a circular, metal-cutting blade in it (*64*). The pipe is clamped between the jaws and the tool is then rotated around the pipe. The blade spins about the pipe at the same time, cutting as it goes. After every few revolutions, you tighten the hand screw, which drives the blade deeper into the pipe. Some tubing cutters also have a file or triangular piece of metal that snaps out like a penknife blade and is used to smooth the cut edges.

Knives and Other Cutting Tools

For purposes other than cutting metal, the following tools are useful:

Mat Knife

The mat or utility knife is an indispensable carpentry tool. The blade is like a trapezoidal single-edged razor blade, only thicker and stronger. Several notches are cut out at the top, and the sharp bottom edge comes to a triangular point at each end. The blade fits between the two halves of the handle, the notches fitting into corresponding ridges inside. In most knives the notches fit in any of several positions, so the blade length can be adjusted. The two halves of the handle are then held together by a single set screw.

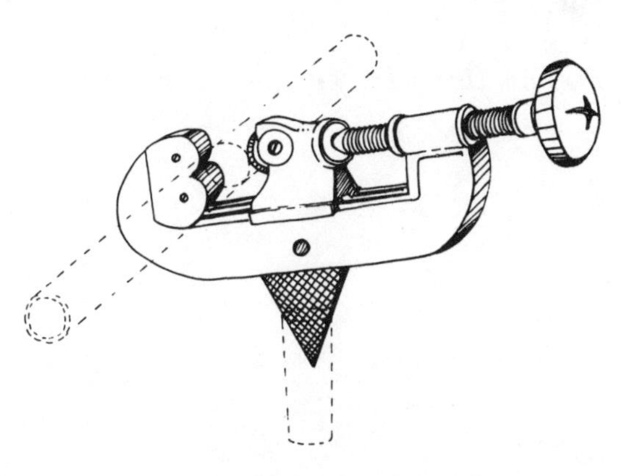

Fig. 64. Pipe-Tubing Cutter

When one end of a blade gets dull, open the knife, and

turn the blade around. Extra blades can be stored safely inside the handle to replace a dull one. Never work with a dull blade. It's slow and dangerous because it can slip more easily than a sharp blade.

Retractable Mat Knife

Retractable mat knives have a sliding switch on the outside of the handle to adjust the blade length without having to open the handle (*65*). Since you can totally retract the blade, it's much safer to carry around. The blade, however, tends to slip back in if you're cutting something very hard. The more common nonadjustable kind gives a firmer blade and is better for heavy-duty cutting.

Another kind has a blade designed to snap off in sections as each section dulls. The advantage comes when you're doing a lot of cutting and need a very sharp blade for each cut—you don't have to open the tool each time to change blades.

Mat knives are very cheap—a dollar or two—so get one of the better ones. Buy extra blades at the same time, and make sure they're the right kind. Blades come with different patterns of notches to fit certain brands of knives. Whenever you buy refills, it's a good idea to bring a blade with you, or to know exactly what the pattern is.

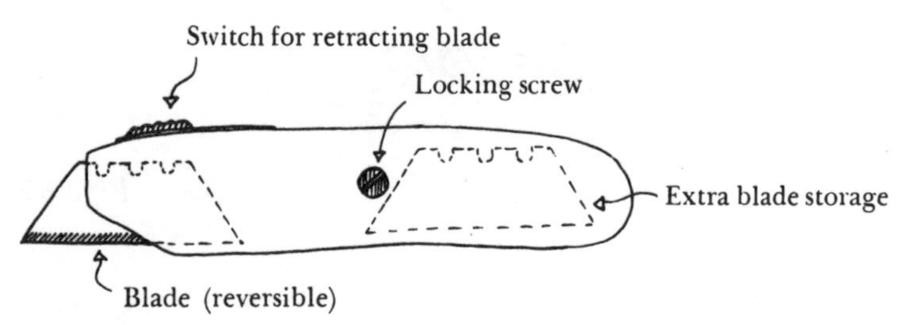

Fig. 65. Mat Knife

The main use of the mat knife in carpentry is in cutting plasterboard. You don't have to cut the plasterboard all the way through. You simply score a straight line on the white, finished side, snap the piece, and cut through the back. (See Plasterboard in "Room Renewal.")

The knife is also useful for cutting fiberglass insulation, trimming shims, sharpening your pencil, scraping drops of gunk off your tools, scribing lines, and many other things.

Heavy-duty plastic cutter

The heavy-duty plastic cutter makes it possible for you to cut plastic laminate (such as Formica) with hand tools (*66*). Similar to the utility knife, this tool has a stronger and sharper blade, and a unique design and cutting angle that help you apply much more cutting pressure.

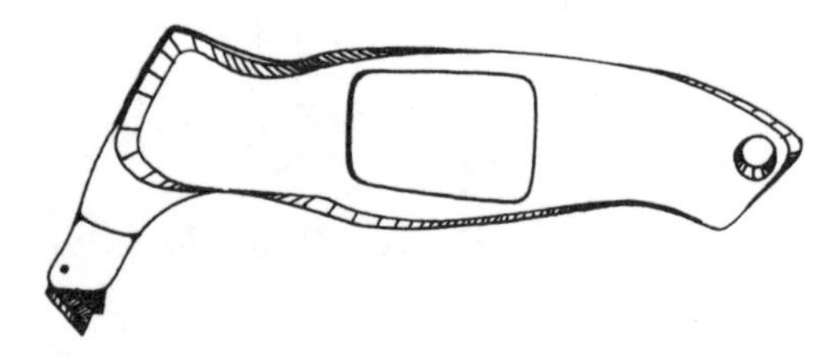

Fig. 66. Heavy-Duty Plastic Cutter

To cut a piece of laminate, first score through the top colored surface. Use a straightedge guide to get a clean line, and be sure the cut is continuous from edge to edge. Then hold the piece down on one side of the cut and, starting at one end, snap the other side up carefully. Then continue along the cut. It should break cleanly. Even a slight raggedness on the bottom side is all right, since pieces are usually cut slightly oversize anyway, and trimmed to fit *after* being applied to the project. Try some practice cuts first to get the feel. For long cuts, more than a couple of feet, try holding one side down with a straightedge along the cut as you snap up on the other side.

As you'll see later in the Built-in project, the final trimming is done with either a block plane and a file, or a router and a file, or just a file.

Pocket knife

The pocket knife is handy for rough cutting. The Swiss Army type is particularly useful around the house for small jobs when your big tools are put away, or when you are out of the house. The bigger models have several knife blades, screwdriver blades, an awl, scissors, a small ripsaw, can- and bottle-openers, a punch, a file, tweezers (handy for removing splinters), and even a toothpick.

Keep all knives sharp. They work better and are safer to use because they are less likely to slip.

Always cut away from or to the side of your body. Keep your fingers and other valuable possessions out of the path of the cut. Concentrate on the cut.

When you're done with the knife, close it if possible, or put it in a safe place. Never leave an open knife where it might fall on someone or be jarred loose—on top of a ladder, for instance.

Tools for Shaping and Smoothing

The tools that follow are necessary for the final shaping, smoothing, and trimming of cut pieces. The chisel is additionally useful in making small cuts and shapes *within* a piece of wood, such as notches or mortises for hinges.

Before you go very far in woodworking, you'll find a use for almost all of these tools. The chisel is essential for a number of shaping jobs. The plane and Surform tool are extremely helpful in removing extra stock and rough smoothing. Either a file or sandpaper is essential for any kind of finished look to your project. And if you do any kind of renovation or refinishing, you'll find steel wool and various scrapers ideal companions.

Chisels

Wood chisel

A wood chisel is a sharp-edged steel blade attached to a handle used with a hammer, or alone, to cut away pieces of wood (*67*). This is the tool you use to cut mortises for hinges and to cut notches. The blade is flat on one side and beveled, or slanted, on the other. The blade width ranges from 1/4″ to 2″ (in 1/4″ increments). A 3/4″ chisel is a good one to start with; or you can buy a standard set of four, 1/4″ to 1″.

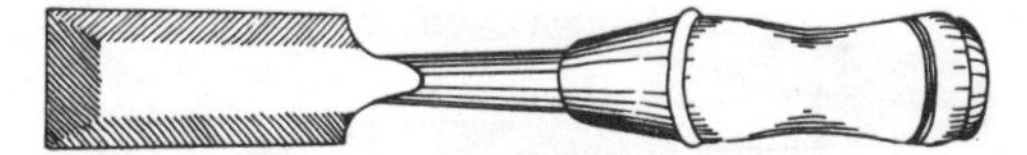

Fig. 67. Wood Chisel

The price of a chisel is a good indication of its quality. Cheaper ones are made of a softer steel that soon dulls or chips. Handles are wood or plastic; plastic usually won't split if you hit it with a standard steel hammer. Some plastic chisels have a steel cap on the end of the handle to take harder pounding.

Use and care

You are supposed to use a soft-faced mallet with a chisel, to protect both the handle and the wood you're chiseling. A soft face drives the chisel in slower, giving you more control of the cut. In most rough carpentry, however, carpenters just use a plastic-handled chisel and their nail hammers. It's handier, and the fineness and smoothness of the cut are not crucial. Today, power tools such as routers and saws with dado blades do much of the fine work once done by hand with chisels.

The chisel blade should be kept sharp. A dull blade will tend to slip off the cut and possibly into your arm, which it will still be sharp enough to cut. Wipe it off and oil it after

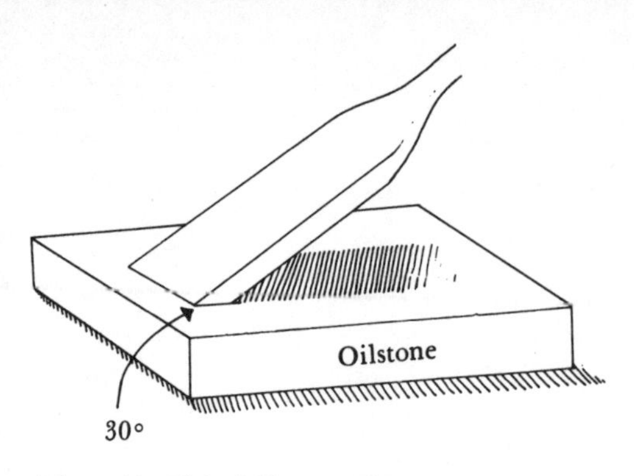

Fig. 68. Chisel Sharpening

every use. Use a light machine oil applied with a rag. Don't let the blade knock up against your other tools. Store the chisel in its original plastic sleeve, or find one for it—such as the kind that larger drill bits and saber saw blades come in.

Don't use a wood chisel on plaster, metal, stone, concrete—only on wood.

You should sharpen it periodically with an oilstone and some light machine oil. The bevel end is laid flat on the stone, then raised slightly from the back to about 30° (*68*). Push the blade back and forth, keeping the chisel at the same angle. When the edge seems sharp enough, turn the blade over, lay it flat on the stone, and rub it back and forth lightly a couple of times. This cleans the burrs off. As with any steel, you can tell if it's sharp by holding the edge near light. A dull edge reflects light, a sharp edge does not.

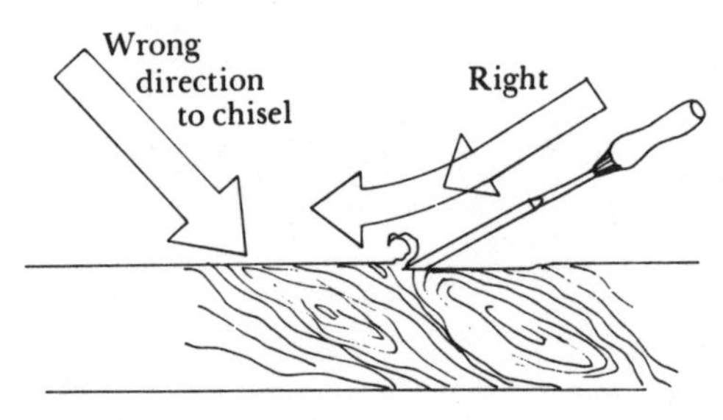

Fig. 69. Chisel usage with upward grain

Don't expect to master the chisel for a while. Using it correctly takes practice and familiarity with wood grains. You should be able to make rough notches and even mortises for hinges fairly soon, but any fine work will require practice. Here are a few hints:

Clamp or hold your work down firmly.

To start cutting, the bevel end should be down, the edge slightly pointed into the work to get it started. Chisel out a little at a time. The bigger the chunk, the less control you have, the more chance of the wood splitting.

Look at the grain on the side of the wood, figure out whether it is level with the top surface or at an angle up or down. Chisel *with* an upward grain (*69*). If you chisel against the grain, you run the risk of splitting the wood and going deeper than you want to.

Tap rather than hammer.

Don't chisel toward you, but to the side.

When chiseling straight down, as in outlining a hinge mortise, keep the beveled edge on the inside to insure a straight, flat side to the mortise. Work in from an edge, not out toward the edge.

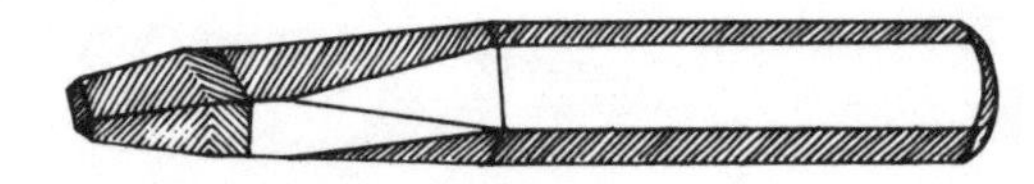
Fig. 70. Cold Chisel

Use the flat side of the blade down for fine cuts and trimming pieces.

If you push one side of the chisel ahead of the other, it will move more easily.

Always wear goggles.

Cold chisel The cold chisel is meant for cutting into plaster and chipping away at stone (*70*). You should use a heavy mallet, but a hammer is all right for plaster.

Brick chisel A brick chisel is a larger version of the cold chisel, used for cleaning and cutting brick (*71*).

Fig. 71. Brick Chisel

Planes A plane is basically a chisel blade supported in a structure that allows you to push it evenly along the surface of the wood. The depth and angle of cut are adjustable. All planes do the same basic job—remove stock and smooth rough areas. Power planes, sanders, and routers can be used instead; so can the Surform tool, described next.

Any decent plane you buy will come complete with instructions on how to adjust it, use it, sharpen it. The four standard planes are the smooth plane, jack plane, the fore plane, and the jointer plane. The big difference among them is length and width. The longer the plane, the easier it is to smooth down a long surface. The long plane spans the high points and cuts those first; the small plane requires *you* to keep it on the high spots.

Nevertheless, the smooth plane, around 8″ to 10″ long, is a good basic plane to start with (*72*). The only other you might find useful at first is the small *block plane,* which leaves a finer, smoother surface and which you can use with one hand (*73*). The design is somewhat different—the blade is set at a lower angle and the bevel is up, opposite to standard planes. It is also very useful for cutting end grains and for trimming edges of plastic laminates, lattices, etc.

Other planes available are the rabbet plane, for cutting rabbets or grooves on the edge of wood, and the router or dado plane, for cutting grooves along the wood surface. The electric router can do these jobs much more quickly.

Any standard plane can be adjusted for depth of cut by turning the adjusting nut. The blade should barely protrude beneath the sole of the plane. The lateral adjusting lever shifts the blade so it will cut level, or deeper on one side. The "frog" can usually be adjusted to open or close the mouth or space between the blade and the front of the opening. Just remove the plane cutter and lever cap and loosen the two screws that hold the frog; the frog can then move forward or backward. Retighten the screws and reassemble. This adjustment controls the smoothness of the shavings. In general, open it for rough cutting, and close it for finer planing.

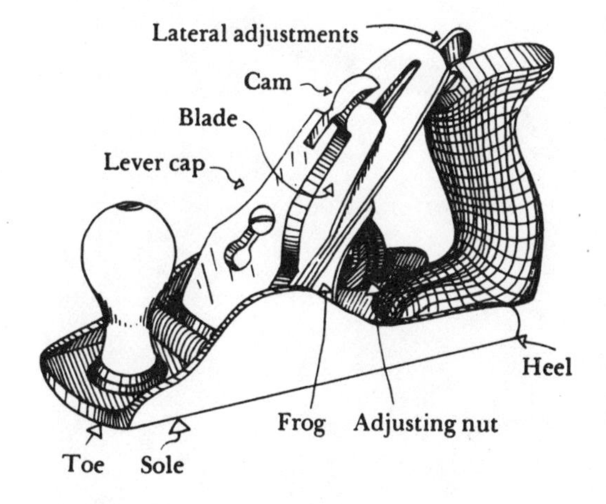

Fig. 72. Smooth Plane

Using the plane

To plane, clamp the wood down, or set it against a stop of some sort. (Plane *toward* the stop.) Don't make the cut too deep. This won't speed up things—it will only clog the plane and make it start "chopping" the wood. Set the plane on the end of the wood, straight and level. Push *down* slightly on the front knob and push it *forward* with the back handle. Move smoothly as you push along the wood. As you get to the end of the wood, put more downward pressure on the handle and less on the knob. Keep the plane level all the time, or you'll round off the wood edges.

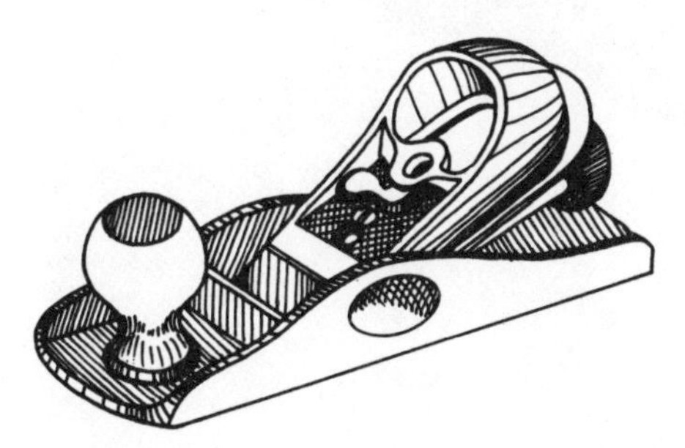

Fig. 73. Block Plane

When planing along the face of a piece, plane with the grain. Also look along the edge to see if the grain is running upward or downward; plane with the upward flow, as with a chisel. If you plane against it the wood won't split, but you'll get a very rough surface, because you'll be chopping into all the ends of the fibers.

To plane the end grain, you really need a block plane. But with either kind of plane, the trick is to stop the cut before the edge of the wood. If you cut to the edge, you will probably chip off a bit of it. Plane from one edge to just beyond the middle, then turn yourself or the wood around and plane from the other edge toward the middle.

Other ways to solve the problem of getting a smooth end grain are to make your original cut of the end of the board with a fine-toothed handsaw or a planer blade on a power saw, and/or sand it.

Never put your plane down on its base when not in use unless you've retracted the blade. Otherwise protect the blade by laying the plane on its side. Protect the smooth metal base also; don't let it get nicked by other tools. Keep the plane clean of shavings. Keep the blade sharp and clean. Oil it.

Surform tool

The Surform tool is a simple tool that is used more and more in place of a plane (*74*). It is cheaper, easier to work with, easy to keep clean, and has cheap replaceable blades. It's essentially a rasp blade placed flat at the bottom of a planelike casing. It has lots of very sharp open teeth, like a food grater, set in rows at an angle to the forward path. The shavings go right through the teeth, so the blade stays clean, unlike normal metal rasps without holes. There is a small model with a 5½″ blade, similar to the block plane, and a larger model with a 10″ blade in either regular or fine cut.

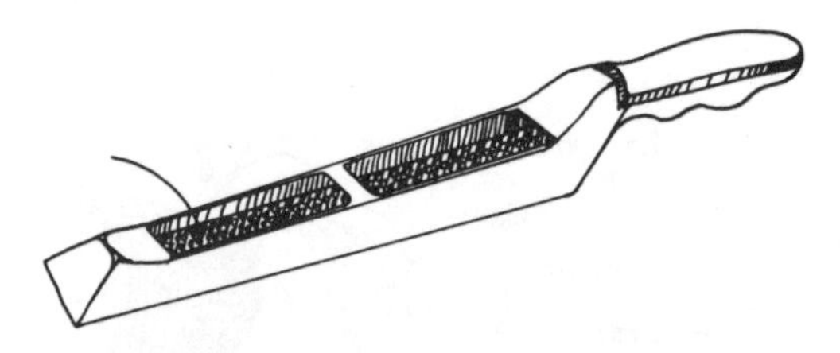

Fig. 74. Surform Tool

The Surform blade also comes on other tools designed to replace the file and rasp. Again, these come in various shapes and sizes, such as round and half-round.

The disadvantage of the Surform is that it does not leave as fine a surface as the plane or file, and it cannot be set to different depths like the plane.

Files and Rasps

Steel files, though designed for metalwork, are often used on wood to smooth out small areas and edges (*75*). Files come in varying grades of cutting ridges—from coarse, to bastard, to second cut, to fine or smooth. They also come single-cut, which has one set of teeth cut between parallel diagonal lines; and the faster double-cut, which has more teeth by having a second set of diagonal lines cut across the first set. You'll find a single-cut smooth file perfect for filing the edges of veneer tape, for instance, or the edges of Formica. While a double-cut file would not be too rough, you'd be safer using the slower single-cut till you have some experience with files.

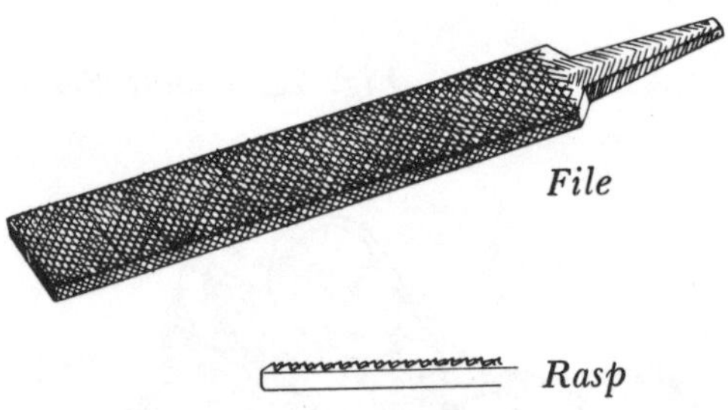

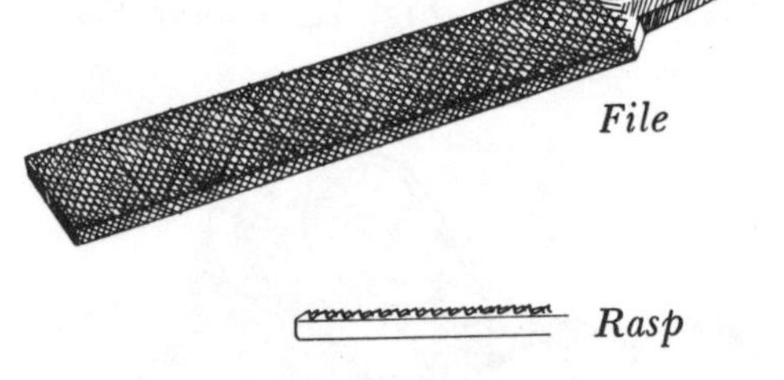

Fig. 75.

Rasps are designed for wood (*75*). They have bigger, triangular shaped teeth, and can remove more stock than a file, but they leave a rougher surface. A rasp should usually be followed with a light sanding of the stock to smooth the surface. Rasps are also used to shape curves and irregular surfaces. Rasps and files come in different cross-sectional shapes designed to fit into different spaces—round, half-round, tapered, triangular, square, and flat.

Fig. 76. File Card

Files usually come without handles. They are much safer and easier to use if you buy a cheap plastic handle to fit over the tang (the thin shank). If you're in a hurry, wrap tape around the tang, but be sure to dull it at the point. If the manufacturer put handles on files, they would cost a lot more. Be thankful and buy one or two cheap handles to use for all your files.

As mentioned, files are really meant for metalwork. When metal is cut with a hacksaw, shears, or tubing cutter, a burr forms on the edge that should be taken off by a file.

You will find odd uses for files and rasps. For instance, if you're trying to remove an old screw whose slot is worn away, a thin, fine file can cut a further slot in the screw so you *can* finally get it out. This also works for the new kind of one-way screw that comes with some door locks; the slot is shaped so that the screwdriver slips out in the counterclockwise direction. Supposedly, burglars can't take your lock off. Neither can you, unless you use a file or a screw extractor.

Use the *file card* to clean your files and rasps (*76*). This tool has rows of little curved wire bristles that can scrape the file teeth clean of dirt and filings.

Abrasives

Sandpaper

Sandpaper comes in a bewildering number of types. There are several things you should know to help you decide what kind to get. First, the four most common abrasives used to coat one side of the paper are *flint, garnet, aluminum oxide,* and *silicon carbide.* Flint (the cheapest and least fine) and garnet (more expensive, harder, and better for fine hand work) are the most common of these. Aluminum oxide is a little finer and most common on power sanders; silicon carbide gives the finest finish of all. Use flint for very rough hand sanding, like removing finishes or on very rough lumber. Use garnet or aluminum oxide for most other wood work, and the silicon carbide for exceedingly fine work, like smoothing paint or other finishes.

Papers also come in *closed* or *open grain.* The effect is similar to that of a file with many little teeth or a few big ones. The open grain has a few large grains and gives a rougher cut, but it will not clog as quickly. Thus it is good for taking off paint and such things that tend to clog the paper. Use a close grain for finer work and for speed, when clogging is not a serious problem.

Sandpaper comes in numbered grades designating its degree of roughness or fineness. It will be numbered in one of two systems. The most common system runs from about 20 for very coarse, to 80 for medium, to 150 for fine, to 220 or more for very fine, with a dozen or more grades between. The other system calls medium 0, and finer grades 1/0, 2/0, and so on to around 6/0. Coarser grades are 1, 2, 2½, etc. The latter system is often used for sanding belts.

It is good to have a supply from very fine through rough paper at hand. Sheets cost around a dime, so it's no great expense to buy extras.

Sandpaper comes in 9″x11″ sheets that are too big for orbital sanders or for hand sanding blocks. Cut them yourself to fit your tool. Most tools will take either exact thirds of a sheet—9″x3⅔″—or exact halves cut along the long dimension—4½″x11″.

Always use a block of wood or a sanding block when hand-sanding; see below.

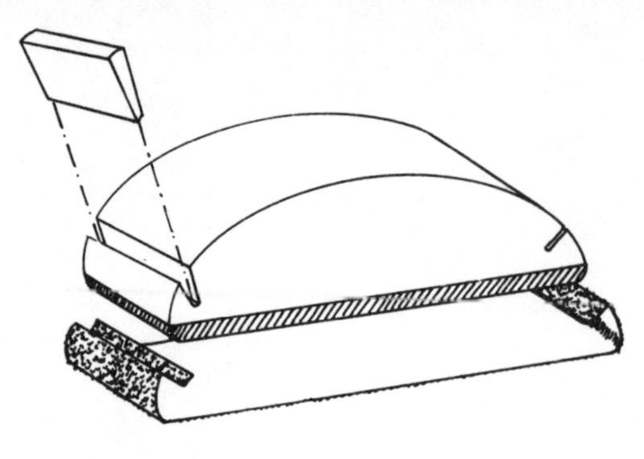

Fig. 77. Sanding Block

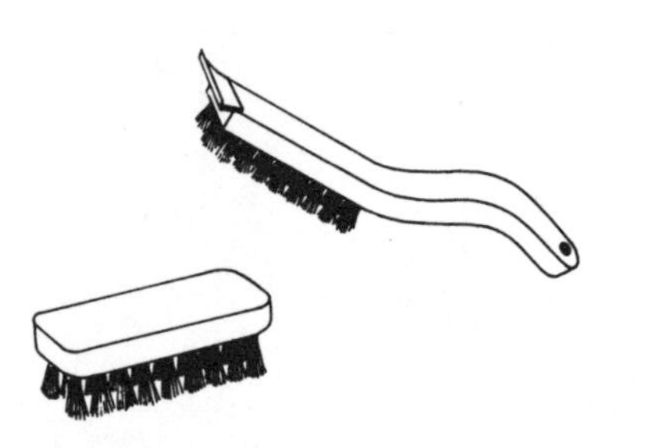

Fig. 78. Wire Brushes

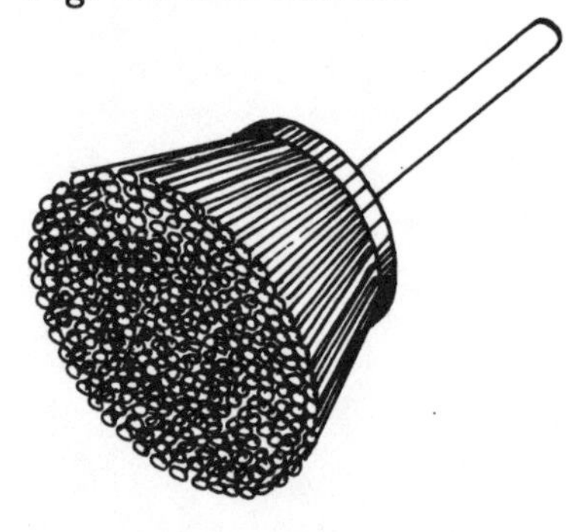

Fig. 79. Wire brush for electric drill

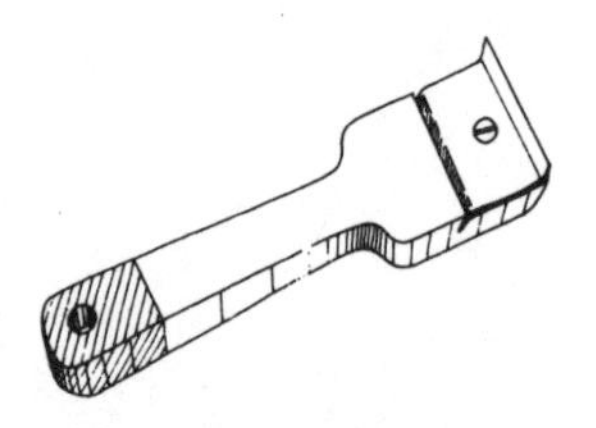

Fig. 80. Paint Scraper

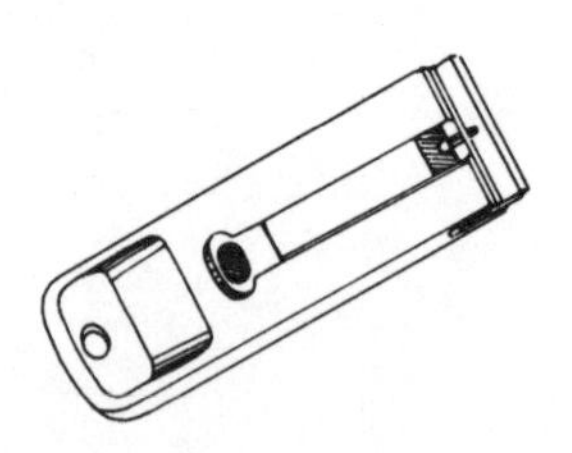

Fig. 81. Retractable Window Scraper

If paper gets clogged you can sometimes unclog it by scraping the clogged areas a little, or by "beating" the paper.

Always start with a coarser paper and gradually work down to as fine as you need. Never skip several grades at a time, such as going from rough directly to fine paper.

Sanding block

When you sand wood by hand, wrap the sandpaper around something other than your hand. A block of scrap wood is usually the cheapest and easiest thing. But a block of wood will most likely not be perfectly flat or smooth and will wear through the sandpaper unevenly. For a lot of hand sanding, you'll save sandpaper and energy by buying an inexpensive sanding block (*77*). You can cut large sheets of sandpaper into pieces that fit into clamps in the block.

Steel wool

Steel wool is the other major abrasive material. It is a tightly bound mass of fine steel wire—a scouring pad without the soap. It comes in grades 4/0 (super fine) to 3, very rough. It cleans rust off tools, removes old paint and finishes, smooths all sorts of surfaces, and finishes wood finer than fine sandpaper. It comes in individual pads or in bulk, which is cheaper. The bulk has to be cut or pulled apart to get usable pieces. You can use it in your hand, in which case it's a good idea to wear gloves. Or cut a hollow rubber ball in two and use one half as a holder.

Scrapers and Cleaners

Wire brush

A wire brush is a wood handle with stiff wire bristles (*78*). It is used to clean coarse rust from tools or plaster from brick walls. Cuplike wire brush attachments can be used with electric drills—a much saner way of cleaning a brick wall than with a hand brush (*79*).

Paint scrapers

Paint scrapers are used to scrape paint off such surfaces as old walls (*80*). The handle of the scraper holds a special blade so the entire cutting edge can scrape the work surface at once. On some models the blade protrudes straight out from the front of the scraper; on others, the blade is bent at a right angle so that you can pull rather than push the tool. Either model usually has replaceable blades.

Take special care when scraping not to dig into and mar the surface under the paint. To minimize gouging, some people dull the corners of the cutting edge with a file.

Window scraper

The window scraper (retractable-blade model shown) is a light-duty scraper for removing paint smears and drips off a windowpane, where you want to be careful *not* to also remove the paint from the sash (*81*). You could improvise one by holding a utility knife blade in the jaws of a pair of lineman's pliers; wrap the top of the blade first with some friction tape so the jaws can grip the blade better.

Putty knives (see next section), though they have duller blades than scrapers, can also be used to remove paint both from walls and windows. When using them as scrapers, it's a good idea to round the corners with a file to prevent gouging the undersurface.

Tools for Applying Gook

Much of carpentry involves the use of "gook" like glue, plaster, joint compound, and putty. The proper tool for applying each of these will make your job much easier and probably better.

Putty knife

A putty knife has a flat metal blade, 1″ to 6″ wide at the end and tapering down toward the wood or plastic handle. The blade is dull, not even sharp enough to cut you. The flexible blade bends to flatten putty, compound, or other fillers into holes and indentations in surfaces. (You can't lay it flat because your knuckles on the handle are in the way.)

Taping knives

The wider taping knives are used for applying joint compound and tape in plasterboard joints (*82*). They fill the tapered valleys between the pieces in one or two sweeps. The narrow knives are more suited for puttying and filling small holes. Always use a knife wider than the hole you are filling, so the blade can fill the hole even with the finished surface on either side of the hole.

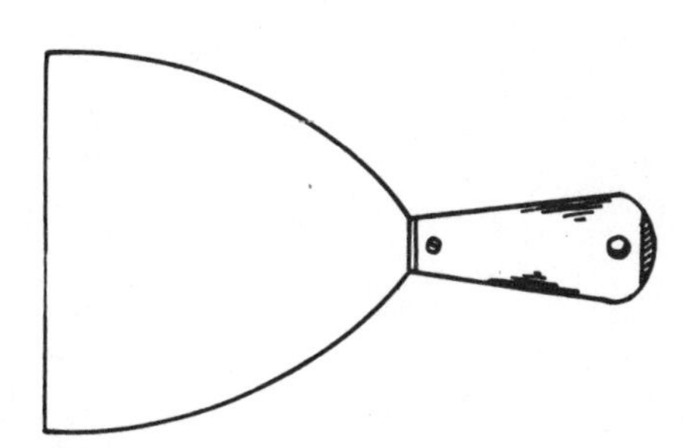

Fig. 82. Taping Knife

Window putty knife

The window putty knife has a special offset shape, with a flat rectangular blade at one end and a tapered curved blade at the other (*83*). The offset lets you press a flat surface of the blade against a windowpane without having to bend the blade flat by pressure and possibly break the window.

Fig. 83. Window Putty Knife

Always clean putty or taping knives after use. The best way is usually to scrape one off with another. If you are doing a lot of work with one, it's a good idea to clean it often while you work—otherwise the plaster or whatever you are using hardens into little pieces that can mar the surfaces you are trying to smooth. If you have to use water on stubborn stains, dry it immediately to prevent rust.

Trowels

Trowels are basically putty knives with large, thicker diamond-shaped blades (*84*). They are used mostly to lay on plaster, cement, and other heavy construction materials. The diamond shape lets them reach into corners.

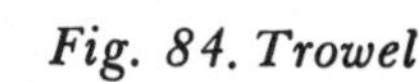

Fig. 84. Trowel

Floats

A float has a large flat base of metal, or wood, or wood and sponge, with a handle in the center of one side (*85*). They are usually square or rectangular, up to about a foot square. They smooth out large surfaces of whatever compound you are using. They can also just hold a pile of gook while you apply a rough layer with another tool in your other hand.

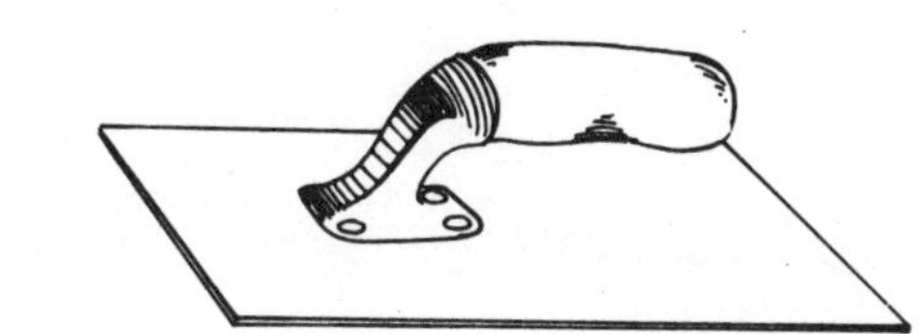

Fig. 85. Float

There are special floats and trowels for special jobs. For instance, there is a tooth-edged trowel for laying mastic and heavy flooring glues. The teeth give the glue a rough surface, which gives a better bond than a smooth flat surface, while also making it easier for you to spread the glue at a uniform thickness.

Caulking gun

A caulking gun (*86*) squeezes out the contents of special nozzle-tipped tubes of caulk or glue. (See Caulk in "Wood and Other Building Materials.") Attached to the trigger of the gun is a plunger that forces the bottom of the tube toward the nozzle, thus forcing the contents out the nozzle. To insert or remove the tube, rotate the plunger till its end piece

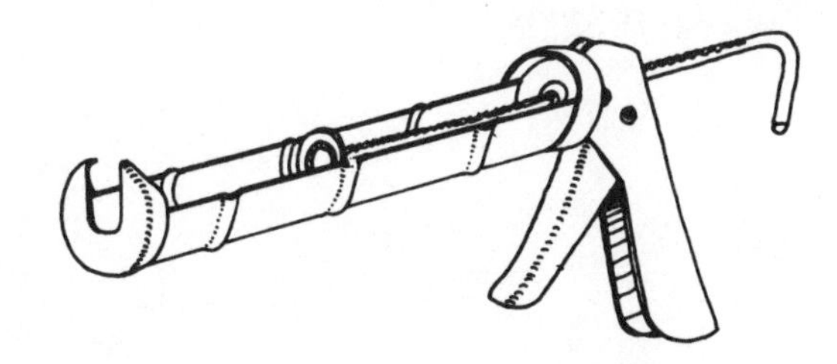

Fig. 86. Caulk Gun

is pointing down, and then pull the plunger all the way back.

A cheap caulking gun is fine—there's not much that can go wrong with it. Just scrape out the excess gunk each time.

Read the directions on any tube of caulk or glue before using, for any special directions. Before placing it in the gun, cut off the tip of the plastic nozzle with a knife so that the hole that appears at the tip is roughly the size of the bead you want. Cut the hole a little smaller than you think you need at first and test it. If you cut at a slight angle, you will probably find it easier to lay the bead.

When you first start squeezing the trigger, go very slowly. The caulk will feed slowly but very steadily into the nozzle. Squeeze a couple of times and wait until the caulk or glue stops surging forward. You want to get the tip just filled. Then place the tip on the work and again start slowly till you see how it flows. Most problems arise because the caulk is flowing too quickly, not too slowly.

With caulking tubes it's usually better to pull the gun along rather than push, so that the tip won't trail over the newly laid caulk and get clogged. Push only when you want to make the bead of caulk flat with the working surface. Let the tip flatten the caulk, and clean the tip often as you go to keep it from being clogged.

When you're done, if the tube is not empty, release the plunger. Otherwise, caulk will continue to flow out, very slowly. You can close the tube off by sticking a nail in the nozzle, point first.

Use tubes of glue in the gun the same way. They tend to flow the same way, only glue does not have to be laid so neatly.

Tools for Prying and Pulling

These tools will pry things apart—wood from wood, paneling off surfaces—and pull out nails and other fasteners. They will take structures apart, lift flooring up, jack heavy objects up slightly. They can be used wherever you need extra leverage and can fit them in.

Ripping bar

The ripping bar is made of steel and has a rough chisel at one end with a nail pull in it for nails with heads above the work surface *(87)*. The other end curves into a nail pull that can be hammered under a flush-mounted nail.

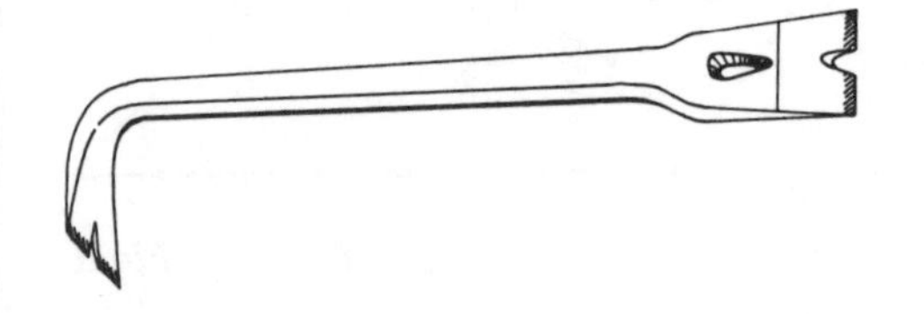

Fig. 87. Ripping Bar

It can also be used for wrecking or prying. Simply wedge or hammer one end between the two things you want to pull apart, then pull with both hands or push on the bar. To loosen some things you can push and pull quickly back and forth. The longer the bar, the more leverage and force you'll have. Be careful, keep your balance, and watch what you are doing. You are exerting a lot of force and things can come apart quickly. Also realize that the end of the bar will probably leave marks on the objects you're prying. Pry at the point where there is most resistance—right where the pieces are joined by a nail, for instance. If you place the bar between widely spaced nails, the wood might break before the nails pull loose. If you're trying to save the wood for something, that won't help.

Rip chisel

A rip chisel is a ripping bar without the curved end (*88*). About a 1½′ long.

Fig. 88. Rip Chisel

Nail pull

A nail pull is much smaller than either of the other pry bars (*89*). It can wedge under a nail that is hammered flush and pull it out without doing very much damage to the wood. The smaller sizes are best for finishing nails. For finely finished work, however, avoid pulling nails—try to sink them below the surface and fill in with wood-colored putty.

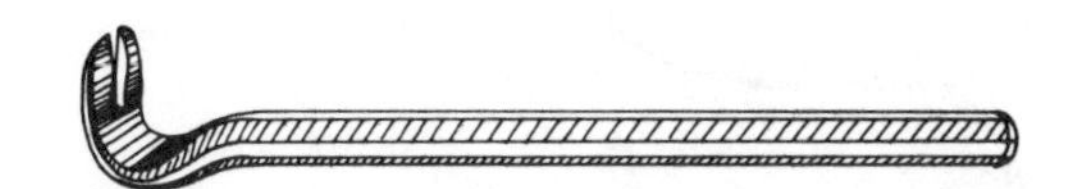

Fig. 89. Nail Pull

Remember that the handiest nail pull for nails not fully sunk is usually the claw on the head of your hammer. The rip-claw hammer is also good for prying smaller things, just like a small ripping chisel.

With all prying tools, blocks of wood can be inserted between the tool and the surface it's prying on for extra leverage, and also to protect the surface.

Tools for Safety and Convenience

A number of tools are designed for *you,* for your body, to make carpentry easier and safer. The most important and absolutely essential tool is a pair of goggles. Many carpentry operations shoot wood chips and particles into the air. If you want to keep your eyes open while you work, you'd best protect them.

Goggles

A good pair of goggles is an absolute necessity to protect your eyes when using any kind of striking or cutting tool (*90*). Nobody *wants* to wear goggles, but it's easier to keep them on when you have a comfortable pair that doesn't fog up easily or warp your vision. Some carpenters wear goggles for practically every tool they use. The smallest chip of wood or piece of metal can injure your eye if propelled with enough force. You *should* wear goggles *every* time you use a power tool. Electric drills are not usually as dangerous, but a drill bit *can* break and snap back at you.

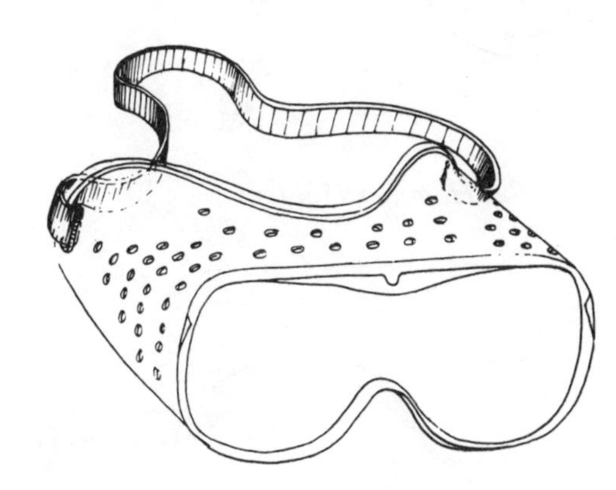

Fig. 90. Goggles

Hammering case-hardened nails such as brick nails and flooring nails also requires goggles. These nails are brittle and break off rather than bend. Some carpenters wear goggles for all nailing. Wear goggles whenever you use a chisel. Wear goggles whenever you are working over your head, particularly if it's an old ceiling.

The best cheap goggles to get are the ones with soft clear plastic sides and replaceable lenses. As the lenses get old and scratched up, you can replace them cheaply. Don't buy the cheaper one-piece goggles—they get distorted so quickly you'll never want to wear them. Goggles should have air holes on the side to keep them from fogging quickly, though all goggles will fog a little as you work. Make sure the goggles can be adjusted tightly, but comfortably, against your skin. Keep a pair of extra lenses handy. That way if the goggles get too scratched up while you're working, you won't be tempted to work without them—you can just change lenses. Keep your goggles clean and away from other tools.

If you want something better and longer lasting, there are more expensive goggles, particularly those like welder's goggles. You may even have an old swimming mask around that works fine (be sure it has a plastic lens and not glass).

Tool belts

One of the most persistently irritating problems in carpentry is mislaying a tool while you're working and then spending five minutes searching through the sawdust and other tools for it. Or realizing you need the hammer you left on the table twenty feet away. This is why somebody invented the tool belt.

You can spend up to twenty-five dollars for a fine leather tool belt with a dozen or so pockets for screwdrivers, chisels, your tape, hammer, nails, wrenches, and practically anything else you need at the time. This is really designed more for professional carpenters who are working throughout a whole building and need to have all their tools handy. If you do get one, it will be very stiff and uncomfortable to wear at first. Oil it as you would any leather to soften it and preserve the leather from cracking and moisture.

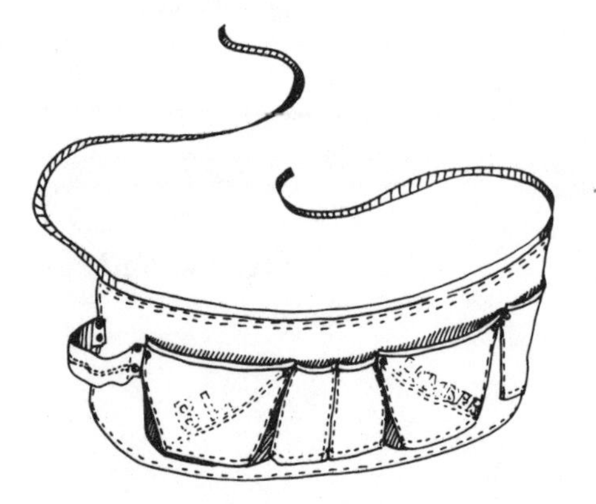

Fig. 91. Nail Apron

Nail apron

More useful for most people will be the canvas nail apron that ties around your waist *(91)*. It usually has two large pockets and one to three small ones, and maybe a loop on the side to hold a hammer. It's extremely useful and efficient for carrying the nails you need. You don't have to keep reaching over into a box to pick out nails. You can climb ladders with your hands free. You can also use these large pockets for other tools you might find handy for each section of the job—screwdrivers, pliers, nail sets, etc. And most important, probably a pencil or two. The loop is obviously handy for keeping your hammer with you all the time. Best of all, the price is right. Many lumberyards will give you one free for the asking, as advertising, since they imprint their logo on the front. At worst, you can pick up a good apron at a hardware store for a couple of dollars.

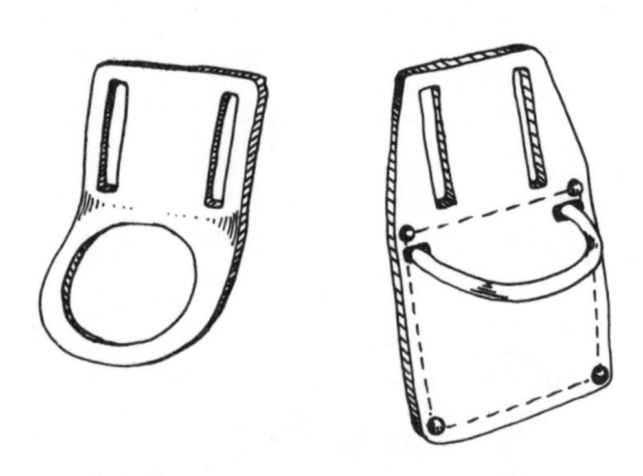

Fig. 92. Hammer Holders

If your apron has no hammer loop, you might try getting a separate *hammer holder* that attaches onto a belt *(92)*.

Gloves

Gloves are necessary for much of the work you will do. Cheap rubber dishwashing gloves are good for painting, staining, handling fiberglass or steel wool, and other light jobs harmful to skin. Workman's gloves are also cheap. Light cotton ones are good for carrying lumber and general light work; heavy ones with suede or leather palms are a little more expensive but good for more dangerous jobs like cutting and filing metal, working with threaded pipe, handling old rough lumber with slivers and loose nails.

Breathing mask

A breathing mask is cheap and a good thing to have around, particularly if you're susceptible to sore throats or have allergies *(93)*. The common kind consists of a replaceable piece of foam inside a plastic frame that fits over your mouth and nose. This mask works fine for filtering out sawdust, plaster dust, and fiberglass. It helps a little against the

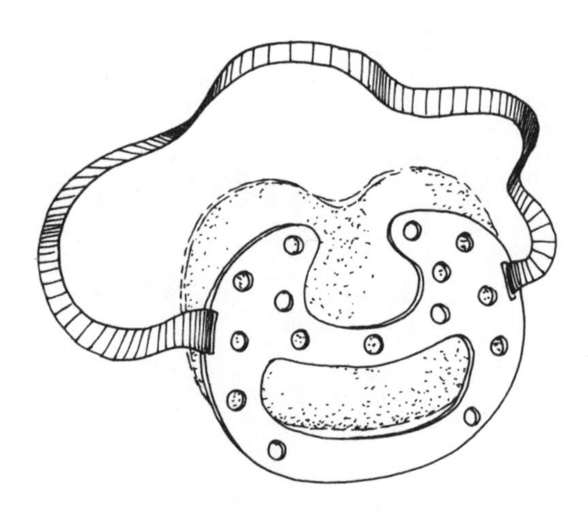

Fig. 93. Breathing Mask

fumes from paint and other finishing liquids; but the expensive masks with chemical filter pieces are the only ones that work totally against these fumes.

Always wear a mask when you work with fiberglass or mineral wool, or around a lot of plaster dust or dirt. The effects of such materials are not always noticeable until the next day, when you may have a very raw throat. If you have a cold or allergy, you might even wear the mask when you're creating a lot of sawdust, particularly with a sander.

Tools for Climbing On

Ladders

A folding ladder is all you'll need for work inside your home (unless you have cathedral-high ceilings). Get one about two feet lower than your ceiling. The aluminum kind is of course much lighter than the wood ladder, easier to move around and store. Most have a paint shelf near the top step to rest your tools or paint on. The kind of shelf that automatically folds open as the ladder opens is much stronger than the shelf you have to flip over yourself.

The folding ladder is in the multifinger-crunching category; make sure your fingers are not on the inside of the legs as you fold the ladder.

Never leave *anything* on top of a ladder when you're not up there with it. You'll forget about it, move the ladder and get hit on the head. Or someone else may carry the ladder away and get hit. Make sure the ladder is unfolded all the way, with the metal hinges locked straight, before you climb up. Make sure the legs are all even on the floor. Before going up, plan what tools or supplies you'll need so you don't have to keep climbing down for things. Don't reach too far while you're up the ladder; it's safer to move the whole ladder.

Scaffolding

Scaffolding is a large movable platform that can be raised to different heights. You can stand and walk around on it, and work directly on the ceiling. For large jobs, it's much more efficient than a ladder, which has to be constantly moved.

Scaffolding can be rented and delivered to your door. It comes in parts—pieces of steel tubing that fit together into a frame, heavy-duty wheels, and usually the wood (2x8s, for instance) that forms the platform. Be sure to lock the wheels when you use the scaffold. Unlock them only to move it.

Some people make a temporary scaffold by laying some boards across the steps of two ladders. This is okay for emergencies, but it's not really too safe because it's not very stable.

6/POWER TOOLS

How to Use Them, Which Ones You Need, and When to Spend the Money to Do the Job Faster

Much has been made of the old craftsmen who worked by candlelight with nothing but hand tools and their fingernails. True enough, some of these people were geniuses. Their work is fantastic. But the craftsmen of old didn't know about electricity. Most of them would have welcomed the chance at least to try power tools. There are people who prefer the esthetics of hand tools, who like to feel the total control of every stroke. If you are one such, more power to you. Technically, however, except for finely sculpted or hewn work, there are few jobs or techniques that a power tool can't do better and more quickly than a hand tool.

Electric tools are simply more powerful, modern variations on hand tools. Power tools as well as car engines are rated in terms of old-time horsepower; a saw motor is as strong as 2.3 horses, for instance. While a chisel is more efficient and faster than a fingernail, an electric router works faster and cleaner than a chisel. Power tools are potentially more dangerous, but are perfectly safe for anyone who knows how to use them—someone who follows the proper safety precautions and treats the tools with care and respect.

A power tool is also more expensive than a corresponding hand tool. It is worth the extra cost if you will be using it a lot. A power tool saves a great deal of time, and your work becomes easier and probably better. If you are just going to do one or two small jobs, a hand tool might be fine.

You may see the need for *some* power tools, such as a saw and drill, but prefer to plane and sand by hand. It's a very personal thing in many ways. The only way to know is to try. Borrow or rent tools, or try them out in some way at your own convenience. Or buy one, and use it for a while. You can always switch. An extra tool is no bad thing. You can use it when the other breaks down. Or sell it. Or give it to someone as a present. A good used tool is a wonderful present.

In "Hand Tools," the tools were grouped by function. The power tools here are presented in roughly the same order, even though there's often only one power tool that does the work of many hand tools. Certain hand-tool categories don't have powered equivalents—no powered tape measure or ladder.

You'll also find that we do not deal with stationary power tools at all, such as radial-arm saws or lathes. It's not that we don't like them; they are great tools. But you need a shop or some sort of permanent space for them, and they are very expensive. Most people are not ready for such a commitment till they've had years of carpentry experience. Portable power tools will be more than adequate for anything you want to do. They also have the obvious advantages of letting you work in any room, or in the garage, or outside (with an extension cord), or in another house. And you can store them away when you're done.

Once you decide you need a specific power tool, consider whether it would be worth while to rent (if it's outrageously expensive), or even to pay a lumberyard or a carpenter to do just the one or two difficult operations you need. Otherwise, try to select a model to buy.

Renting Tools

Large construction jobs often call for expensive power tools. You don't want to buy such a tool if you're never going to use it again. You can usually rent what you need cheaply from tool-rental shops, service centers, and sometimes from hardware stores. Among the more common tools you can rent are floor sanders, screw guns, reciprocating saws, electric hammers, flooring staplers (for nailing down tongue-and-groove flooring), Ramsets, sand blasters (for cleaning brick walls), disc sanders, and paint compressors. The Yellow Pages will also reveal places that will rent ladders and scaffolding. The rental people will be glad to show you how to use the tools.

Buying a Tool

If you follow the price ranges we suggest for each tool, and bring along a little common sense, you should be able to get a fairly good tool. As with hand tools, buy the best you can afford, but don't go overboard. The top-of-the-line power tools are not any more accurate than many of the less expensive ones; they're only more durable, for professionals who use them seven hours straight, five days a week.

Most people start cautiously with a low-to-medium-priced tool, then buy a good one if they find more use for it. But if you know beforehand that you'll be using the tool a lot, there's no reason to buy a cheap one. The good ones are just as safe when handled right—maybe safer, because you're not tempted to force them beyond their limits.

You never need to pay list price on power tools—you should be able to locate most any model at a minimum discount of five percent, possibly ten percent. Even higher discounts can be obtained through several of the reputable mail-order hardware companies, like Silvo or U.S. General. (See Mail-Order-Sources in Reference section.) Often you can find great buys in discontinued models; these tools were changed not because of flaws in the design, but usually because the company has a flashier design or new gimmick. The new tool might actually have a few extra features, but at the price difference it's not worth it. Buying old models is one way to beat the planned obsolescence built into our society, to turn it to your advantage. A "best" tool today may be off the market in three years; but it will still be as good then as now. And the same goes for tools that were good five years ago. Carpenters get a good tool, take care of it, and keep it till it dies. They don't rush out for the new model every year.

After deciding what tools you'll need, the next step is to go off to the hardware store to see what choice you have among the brands and different models. You may know the exact tool and model you want. Maybe it's been recommended by a friend. More likely you're open to suggestions. Tell the dealer the tool you want and either the range you want to pay or how good a tool you want. Now, of course, no matter how much the guy loves you, he will probably try to sell you one a bit more expensive than you're willing to pay. Don't let him knock you off course. (See Dealing with Your Hardware Store in "Hardware.") He may be right about what you need, or he may not. Also remember that a dealer may have a particularly profitable arrangement with one company, and may push that brand. The tool may well be the best one; it's up to you to judge from what he says and what you've read here or elsewhere about what that tool should have. A power tool may have a lot of great features, but you may not need them. Check the guarantee, find out where the service center is, and whether the store itself will take care of repairs.

A nearby manufacturer's service center is a strong plus for a power tool. It saves a lot of time and trouble with repairs and check-ups. You might take the trouble to call the center. Tell them, for instance, you have a saw that needs checking and a tune-up. Ask how long it'll take them to do it. Estimates may vary from same day to a week or more. This may also influence your choice.

Service centers are also good to have around because they may have a special attachment you suddenly decide you need. Hardware stores can't stock the more specialized accessories for every brand of tool. Centers are also good sources for used tools, or special buys.

A good store will let you examine tools right there, particularly power tools. You should at least ask. They may not let you actually work on a piece of wood, but you can at least get a feel for the tool. One thing you definitely have the right to do and should do—when you buy a tool, take it out of its box right in the store. Make sure all the parts, the instruction booklet, and guarantee are included. Have the dealer plug the tool in to make sure it works. If a store doesn't let you do this, don't buy there. There's no reason you should have to be frustrated at home and make another trip back to the store if it doesn't work right. Also the dealer can show you things about the machine right away, how it should be held and worked. It's always easier to learn in person than from an instruction booklet. But be sure *also* to read the booklet.

Living with Your New Tool

Before doing anything, read the instruction booklet carefully and also the appropriate entry in this book. Understand the tool's principle. Then practice on scraps, just as you would with a hand tool. No tool is going to do *all* the work for you—you've got to use your brain sometimes. If the tool doesn't work right at first, don't assume it is bad or poorly designed. Try to understand the way it's supposed to work, its own logic. Get the feel of it. Remember that all tools are logically made—think of them as simple extensions of your body.

General Information About Power Tools

Power ratings

All power tools carry some indication of their power ratings on their nameplates. Unfortunately, the ratings can be very misleading. Horsepower ratings are often determined by running the tool under no-load conditions, which is not much help. A high amperage rating, which is a measure of how much electricity the tool uses, is also not much help, since the tool might be an inefficient one. And revolutions per minute (rpm) can be equally misleading, also often being measured under no-load conditions.

What can you do to determine how powerful a tool is? First check through the manufacturer's catalogue in the store. Some of the companies are now giving torque ratings, which *are* a good indication of the tool's power. Torque is basically the amount of work a tool can do. You may also be able to find full-load rpm ratings.

Manufacturers' catalogues will also tell you if the tool has steel ball bearings, as opposed to nylon or bronze bearings. Ball bearings last much longer and are a definite sign of a better tool. If the type of bearing is not listed, assume it does *not* have ball bearings.

Similarly, a tool with nylon gears is not as good as a tool with steel gears. Good gearing, in fact, is the key to a good tool.

Beyond these possible checkpoints, your only source of information is the dealer, the price, and your common sense. A good dealer that you trust can cut through the technical jargon and say, "Here, this is good." Or with your common sense you should be able to arrive at a decent judgment through the relative prices of things. Remember that a tool with a lot of extra features and "consumer appeal" will cost more than a tool of equal value that performs only the basic functions. And the consumer-oriented machine will be the more likely to break down.

Double-insulated tools

Double-insulated tools have a simple two-prong plug; they do *not* have to be specially grounded with a three-prong plug or adapter. The inner components of the tool have the usual primary insulation, but there is also a secondary insulating layer that isolates the nonconductive outer casing of the tool. Thus, if the primary layer breaks down, the secondary layer protects you from shock.

On tools with a *three-prong plug,* the third prong is attached to a third wire which leads to a connection with the outer casing of the tool. You must always plug the tool into a three-prong outlet; if your home is equipped only with two-prong outlets, you must use an adapter with a grounding ear or wire, attaching the ear to a convenient grounding point. Usually the most convenient grounding point is the little screw in the middle of the outlet box plate, since, in a correctly wired house, the outlet box itself is the ground.

Care and maintenance

For maximum life and usage you must maintain your tools in good working order. If you want, you can simply take the tool to the service center every six months to a year (check your instruction booklet for the timing) and let them do everything. With most power tools, however, it's fairly easy to check two important things yourself—the brushes and the oiling.

Carbon brushes wear out periodically and have to be replaced. These are small, spring-loaded, carbon-graphite blocks that brush against and carry current to the spinning motor armature. As they wear down, the springs keep them in constant contact with the armature—firmly enough to maintain the current flow, but lightly enough to allow the armature to keep spinning. Your instruction booklet will tell you how to get at these brushes and how to check them. Usually when a brush gets less than 3/16″ long, it should be replaced. Your booklet may say how many hours of use you can expect before it's time to change the brushes; we'd suggest every six months or so if you use the tool a lot.

When you run your power tool, you can often see blue-white sparks in the motor housing. Do not be alarmed. This is normal. However, be careful not to start tools around flammable solvents like turpentine, or near gas appliances. If you start to notice a great increase in the sparking, however, and possibly a grinding sound, stop the tool. This usually means the brushes are worn down. Change the brushes before running the tool again.

You can buy replacement brushes at the hardware store or service center. Be sure to get the proper size.

Oiling is very simple. Your instruction manual should show where the points are to be oiled, how much to add, and how often to do it. Unless otherwise stated, use a light machine oil, like 3-in-1.

Some tools also need *greasing* periodically. This is a little more difficult. If the directions in the manual are not clear to you, let the service center do it.

Safety and General Tips

Always read the instruction manual that comes with your tool.

Ground all power tools, except for double-insulated ones.

Don't use power tools outside in the rain, or in damp or wet places. Wear rubber gloves if you work outside.

Do not wear loose clothing or hanging jewelry. Tie long hair behind your head.

Be sure the power tool's cord is out of the way; don't cut through it.

Keep fingers and other valuable parts out of the way.

Wear goggles with most power tools.

Watch out for obstructions while working in existing structures—old pipes, telephone cords hidden behind moldings, rollerskates, other people. Concentrate on using the tool as well and as accurately as possible.

Never force a tool. You won't get it to work any faster, and you'll just put it under extra strain. It's safer to let the tool work at its own speed.

Hold the work piece down, preferably with clamps. If you're working on the floor, you can kneel on it. Maintain your physical balance, however, at all times; and, of course, keep total control of the tool.

Disconnect the power from a tool whenever you are going to change a bit, blade, sanding belt, what-have-you; also whenever you are going to make an adjustment, such as depth, angle, and so on.

When you plug in a tool, be careful that the On/Off switch is in the Off position. Hold the tool by the handle anyway, just in case.

Before starting the power, be sure you've removed any adjusting key or wrench, and that all the depth or other setting nuts are tight. Of course, be sure that the blade, bit, belt, or whatever is in correctly.

Don't lend your tools. A person who borrows a tool probably won't be familiar with that tool. If something can go wrong, it will, and that person may not even realize it. Also, tools are often eccentric—particularly power tools. They get to know the feel of your hand, the speed with which you run them through the wood. Someone else will have a different feel, and the tool may get upset and start to work erratically. This may sound strange, but it's true. Think of the last time you lent your car to someone. Didn't it drive a little differently when you got it back?

Keep your tools in order in a place away from harm, dirt, animals, children, water, feet.

Clean the sawdust from the tool after every use. Make sure any protective guards are still working freely.

Sweep the work area periodically. Power tools create incredible piles of dust. Cleaning up clears your mind and also turns up lost pencils, bits, and other small tools.

Tools for Safety and Convenience

Extension cord

Buy a good three-wire extension cord and a *couple* of adapters (these are easily lost). This cord will work for either double-insulated or standard tools. If your outlets are the two-prong type, and you are using a double-insulated tool, use the adapter but don't bother grounding the ear. Of course, you must ground it for the standard three-wire tools.

Your extension cord must also be of the proper gauge wire for your tools; too thin a wire causes a drop in the voltage, a loss of power, and possible damage to the tool's motor. The smaller the gauge number, the stronger the wire. A twenty-five-foot #16, or a fifty-foot #14 will be fine for any tools up to around 12 amps. Note that the longer the cord, the stronger the gauge should be. If your tool is rated higher than 12 amps (unlikely), say up to 16 amps, use a cord one gauge heavier. The instructions with your power tool should have a table of recommended cord sizes for any situation.

Treat your cord well. Grab the plug to pull it out of a socket. Pulling by the wire puts pressure on the wire/plug connection and will eventually break it. Store the cord rolled up or coiled. Again this saves the end connection from strain, and also makes it easier to unravel the cord when you use it.

Trouble light

A caged lightbulb socket on a long cord. There's usually a hook on the back of the cage so you can hang it on a pipe or whatever is handy. Trouble lights are great for certain repairs—a leaking basement pipe behind the oil burner, for instance—and in construction areas when the new electric lines aren't put in yet.

Some of the older trouble lights were badly designed with an extension outlet right behind the light switch. This is dangerous, since you might accidentally stick a finger in the outlet when you turn the light off. Obviously, don't buy this kind. They are supposedly off the market now, but they're still sold in bargain shops.

Goggles

A good pair of goggles is essential for safety with many power tools—any saw, router, and sometimes with drills and sanders. It's vital that you not only protect your eyes, but be able to keep them open while you are running a power tool. See goggles entry in "Hand Tools."

Tools for Making Holes and Twisting

Electric Hand Drill

The power drill (*1*) is shaped and held roughly like a ray gun in the old spaceman movies. You squeeze a trigger and the motor turns the chuck at the front of the "muzzle"; the chuck holds the drill bit. It drills holes faster and more powerfully than the brace and other hand drills. Few carpenters are without at least one electric drill. With the proper bit it can put holes in wood, metal, plastic, brick, concrete, plaster, stone, and just about anything else. It can also handle many other operations that require a circular or rotary motion. With the right bit or attachment, the electric drill can drive screws, sand, polish, grind, wire-brush, mix paint, and rasp. There are even attachments to convert drills into circular saws, saber saws, routers, and even lathes. We're usually suspicious of any attachments that "transform" one tool into another. The attachments are never as accurate or efficient as

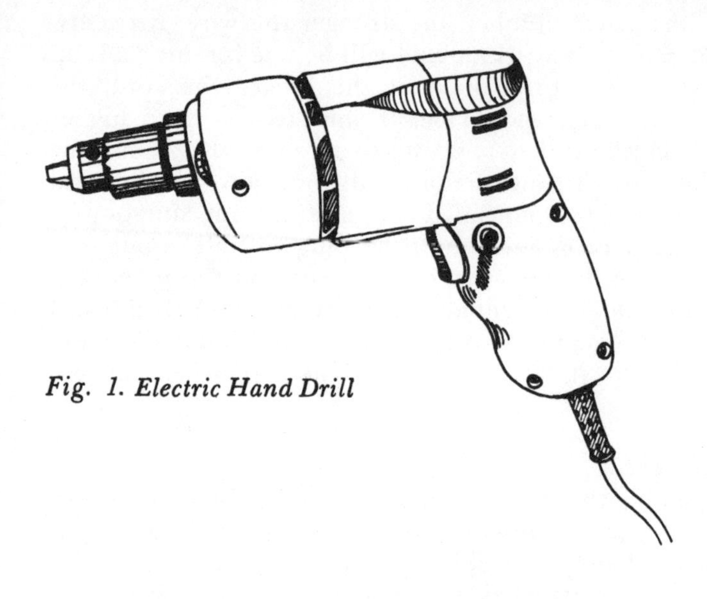

Fig. 1. Electric Hand Drill

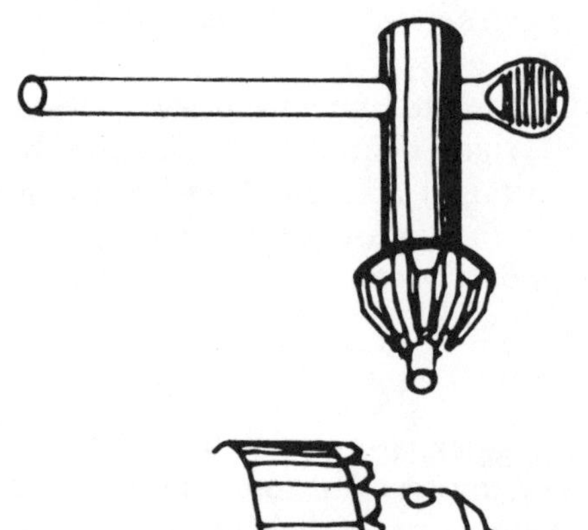

Fig. 2. Chuck and Key

Fig. 3. High-Speed Drill Bit

the tools they try to replace. And they may put extra strain on the tool's motor and inner mechanism, which are designed for specific operations.

A bewildering variety of drills are made. No one drill will be perfect for all operations. You will have to figure out the main uses you will have for your drill and base your buying decision on that.

The first consideration is the *chuck size* (2). Drills are labeled ¼″, ⅜″, or ½″. This refers to the maximum diameter of the chuck opening, or the largest size shank it will accept. The shank is the smooth, unthreaded part of the bit. The cutting part of a bit can be larger than the shank, so that both a ¼″ and a ½″ drill could drill, say, a 1¼″ hole, but the ½″ drill accepts a bit with a larger shank. More shank to grip gives more torque—turning power—to the drill. Therefore it is a more powerful and more efficient tool to use for very hard materials such as brick and concrete. The ¼″ drill turns much faster, however. While ½″ drills turn at no more than about 1000 rpm, and usually less, the ¼″ drill makes up to 2000 or 3000 rpm. Thus it's a much more efficient tool for drilling soft materials such as wood, where great torque is not necessary. It is also good for other operations where high speed is an advantage, such as sanding, mixing paint, and polishing. The ⅜″ drill is the middle-of-the-road choice.

The other main consideration in buying drills is *speed*—whether to get single speed, variable speed, or reversible variable speed. The single-speed drill turns only at its rated speed. This is fine for drilling most wood, thin light metal (when the hole is begun with a punch), and Plexiglass.

The variable-speed drill can rotate at any speed between zero and its maximum rpm, depending on the pressure you place on the trigger. Most of these drills can be locked in at any desired speed by pressing a small button. This type of drill is necessary if you want to drive screws. A single-speed drill would go too fast and probably strip the threads off the screw. Being able to vary the speed is also helpful for starting holes in metal, and for drilling through brick and concrete, where slowing down will sometimes help the bit to catch better.

Some variable-speed drills are also reversible, meaning that with a flick of the switch (while the drill is off), you can reverse the rotation of the chuck. This is great for removing screws and also bits when they get stuck in the work. It is of no help in making holes, since bits are designed to cut only one way.

Many carpenters have at least two different kinds of drills, to cover all the bases. If you are just beginning, though, the chances are that most of your work will be with wood, so you can safely choose a ¼″ drill. Unless you are building something with dozens and dozens of screws, you can get by without a variable-speed model. A cheap single-speed ¼″ model, even one as low as ten dollars, will be fine. You could even use it to drill a few holes in a brick wall.

If you plan to drill a lot into stone or brick as well as into wood—say you plan to hang a lot of shelves or cabinets on your brick wall—we suggest getting a 3/8″ drill as a compromise (unless you care to buy both the 1/4″ and 1/2″). At this point you will need a better drill. Hard materials like brick put a lot of strain on the tool. This means twenty to thirty dollars. You might also consider spending the extra few dollars to get a variable-speed model. Also consider getting a 1/2″ instead, but only if you don't mind taking twice as much time for every wood hole you drill. Then, later, you can buy a cheap 1/4″.

The only consolation to all this is that many good drills are sold at very large discounts, because their manufacturers like to bring out new models every few years. Look around for "obsolete" models. In fact, some of the newer models have cheaper chucks, made of a dull black metal, which tend to chip where the chuck key meshes with the sprockets.

If you buy a good drill with brick or concrete walls in mind, look for one with a D-shaped handle and/or one with an attachable handle on the side. These extras will give you extra leverage.

Basic Drill Bits

When you buy your drill, buy a set of high-speed wood bits in a case (*3*). The sizes should range from 1/32″ to 1/4″. These bits will make holes for any screw you might use, plus small to medium bolts. You'll need all these bits eventually, and it saves a lot of money and many trips to the hardware store to buy them in a set. Keep them in place in the case so you can tell at a glance where each size is. All bits are etched with their size, usually at the bottom of the shank—but sometimes the numbers wear off. They're hard to read anyway. When a bit breaks or wears out, replace it.

Bits up to 1/4″ have shanks the size of the bit. They will fit any drill. Bits slightly above 1/4″ may have a shank the same size as the actual cutting part of the bit or they may have a 1/4″ shank. If you have a 1/4″ drill be sure your large bits have shanks no larger than 1/4″. Get the biggest shank your chuck will take. A bit designed for a smaller drill will always fit a larger drill.

General Use of the Drill

Your drill should come with a chuck key, attached to the electric cord on a holder. Keep it attached so you don't misplace it. Choose the proper bit for the job.

Before changing bits, make sure the drill is unplugged.

Open the chuck wide enough to accept the bit. Do this by holding the drill in one hand and turning the thick collar of the chuck with the other hand in the direction that makes the jaws inside the chuck open up. Slip the shank of the bit between the jaws—no farther than the start of the threaded part of the bit, but as far in as it will go and still leave enough bit to go through the work. Tighten the chuck by turning the thick collar back the other direction. When it's as tight as you can turn it by hand, insert the chuck key into one of the holes on the side of the chuck and turn the key clockwise till the chuck is fairly tight. Repeat with each of the other two holes. This keeps the jaw pressure even. Tightening just one of the holes can damage the chuck.

Put on your goggles. If extreme accuracy is required, you can start the hole with a punch in metal or brick, or an awl in wood. You need just a slight indentation for the tip of the bit to rest in. This keeps the bit from "walking" away from the mark. Otherwise, start the drill with the bit right on the point; this works particularly well with a variable-speed drill that can turn slowly till the hole is begun. Unlike most power tools, a drill is best started with the bit or cutting element right on the work surface. Hold the drill straight in line with the hole you want to drill, with a slight pressure against the work. Squeeze the trigger, slowly at first if you have a variable-speed model. Once the hole is started, run the tool full speed.

Don't force the drill through wood. If the bit starts to smoke, burn, or bend, you are applying too much pressure. Slack up. Too much pressure will dull the bit, if not snap it, and also put added strain on the motor.

Once you've drilled as deep as you need, pull the drill out of the hole. Keep the drill running as you pull it out.

To remove the bit, reverse the process of putting it in. Put the chuck key in any one of the holes and turn it counterclockwise. Sometimes it will seem stuck. Give the key a hard rap with your hand to loosen it, or grasp it with a pair of pliers for more leverage. Don't hammer it—you could damage the chuck. As soon as you can turn the key, you can probably turn the chuck collar by hand and loosen it the rest of the way. Remove the bit and *immediately* put it back in its case. If you've been drilling several holes with it, the bit may be very hot. Be careful. If you don't have gloves or a cloth handy, you can slide the bit out with anything handy and let it drop into its case.

When you're finished using a bit, always remove it from the chuck. It's apt to be stepped on, bent, or otherwise damaged if it's left lying around in the drill. If you plan to use it again in a while, leave the drill in a safe place—not on the floor.

Tips on drilling

A bit that penetrates through the back of a piece of wood will splinter the wood as it comes through. If neatness counts, a scrap piece held tightly or clamped to the point of bit egress will stop the splintering.

To drill a hole to a certain depth, or many holes to the same depth, wrap a piece of tape around the bit at the proper distance from the point of the bit. When you see the tape reach the wood surface, the hole is deep enough. You can also buy a metal drill stop that fits over the bit.

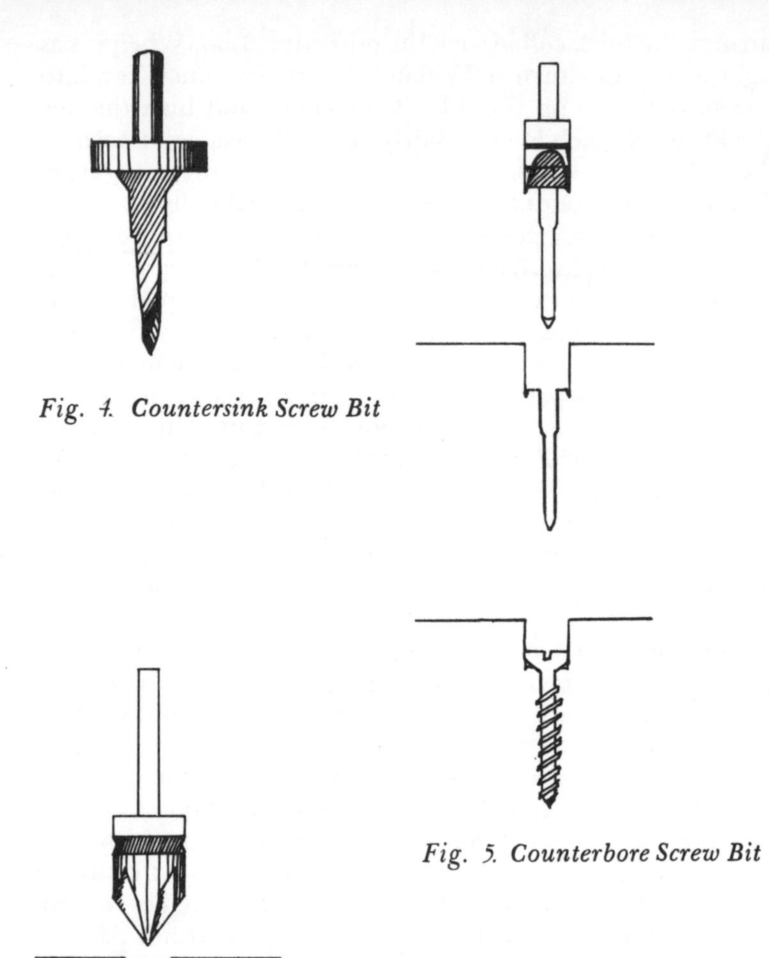

Fig. 4. *Countersink Screw Bit*

Fig. 5. *Counterbore Screw Bit*

Fig. 6. *Countersink*

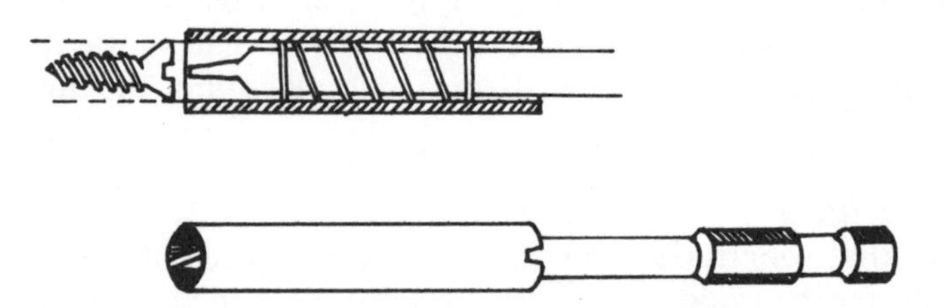

Fig. 7. *Screwdriver Bit*

If you're having trouble drilling through some wood, check to see if you're going through a knot. If so, move the hole. You should check that before drilling. If a hole *has* to go there, just keep at it, going slowly. The trouble could also be that the switch on your reversible drill is on reverse. A drill will make a hole in reverse, but much more slowly and with damage to the bit. This is a common oversight. Or your bit could be dull. Look at the point, compare it to a bit you know is good. If you have a variable-speed drill, the problem might be that the speed is locked in slow from a previous operation.

While drilling a deep hole or in hard wood, pull the bit out of the hole once or twice to clear out the sawdust. The bit will cut better when its spiral grooves aren't stuffed.

Sometimes a bit will strike something hard and get stuck. The motor may stall, or the drill handle itself may want to turn in your hands. For this reason you should always have a good hold of the drill. If the bit jams, turn the drill off immediately. Pull it out if you can. Use the reverse switch if you have one. In extreme cases you won't be able to get the bit out even with pliers and a stick of dynamite. Give up. Try to break off the protruding part of the bit, after detaching the drill. Or hacksaw it off.

If you are drilling a lot of holes and each hole requires two operations, do all the work that you can with one bit before switching to the other. Don't drill a hole, change bits and drive a screw, change back and drill a hole, change again and drive a screw. This is a simple but very important concept that can be used throughout all procedures in carpentry: when you are set up for one operation, do as much as you can before changing to another setup. It saves much time It also means greater accuracy and uniformity of like parts.

When you use an extension cord, tie a loose knot where the two cords meet—particularly when you are working on a ladder. This prevents the cords from coming apart and your having to shlep down the ladder to put them together again.

There are a cooling fan and open vents near the back of most drills. Don't hold the drill right next to your hair, a plastic drop cloth, or any flimsy material that can get sucked up and caught in the fan. And don't cover the vents, which help to keep the motor from overheating.

Always have the work piece held down well. After the bit goes all the way through, the piece will be likely to start spinning with the bit.

More Drill Bits

Most drilling is for screws or bolts. This can be done with the standard high-speed bits; choose the bit after you've chosen the fastener.

Screw bits

If you do much drilling for wood screws (see description under Screws in "Hardware"), buy one of the special screw bits that drill for the thread and the unthreaded shank, and countersink or counterbore for the screwhead all at once—a Screwmate or Screwsink, for instance (*4, 5*). Some of these bits will only countersink, because of a stop at the top; others will either countersink or counterbore. Some companies make adjustable screw bits, in which the bit sections for the thread and shank of the screw can be adjusted in length; therefore one of the screw bits can be used for screws of a given thickness but different lengths. These bits are very handy, but the nonadjustable types usually last longer.

Countersink bit

The alternative is to use two sizes of normal bits plus a separate countersink bit for the upper hole alone (*6*).

Screwdriver bit

If you have a variable-speed drill, you should also buy at least one screwdriver bit (*7*). Start with one with a standard blade, with a spring-loaded metal sleeve around it. The sleeve slips over the head of the screw and helps to keep the blade in the screw slot, so the blade is less likely to slip and mar the wood surface. Screwdriver bits come with different-size blades, and with Phillips heads and socket heads for bolts, among others.

Long wood bits

Buy other bits when you need them. Standard wood bits, such as the ones that come in sets, are proportioned—that is, the thicker the bit, the longer it is. If a certain thickness is not long enough, you can buy extra-long bits individually; $1/8''$ and $9/64''$ bits are useful to have in long lengths for pre-drilling nailholes slightly thinner than the nails when straight nailing might split the end of a piece, or for toenailing. Long bits bend easily—drill straight and slow.

Spade bit

To drill holes up to about two inches in diameter, get a spade bit (*8*).

Hole Saws

Hole saws also cut these, as well as larger, holes (*9*). A hole saw consists of a standard wood bit and an arbor to which is attached a piece that looks like a metal cup with sawteeth around the rim. Different-size cups, or saws, can be attached to the same arbor. The standard bit protrudes slightly farther than the hole saw and is placed at the center point of the hole to be cut. As it starts to catch, the hole saw cuts around the circumference. Because of its design, the saw will not be able to cut farther in than the depth of its cup. If the wood is thicker than the cup, drill as far as you can, remove the drill, and chisel out the chunk you've cut. Put the hole saw back in the hole, and continue. Place a backing piece underneath to protect the bottom surface.

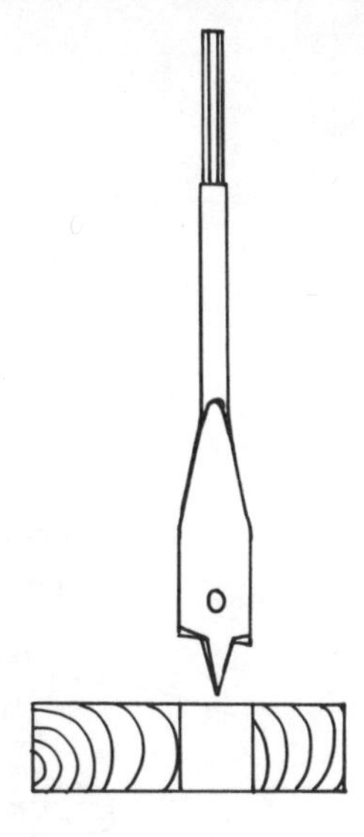

Fig. 8. Spade Bit

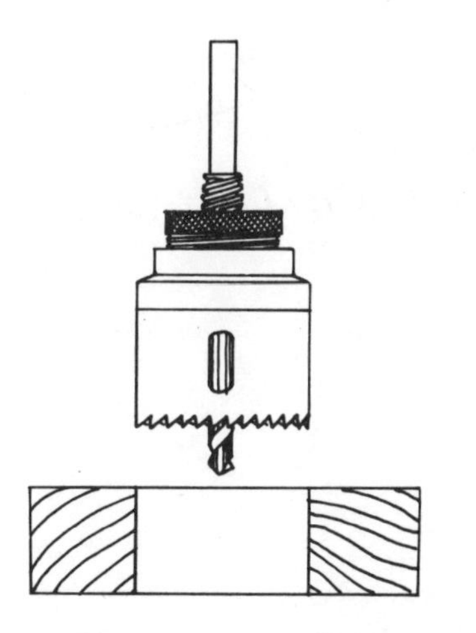

Fig. 9. Changeable Hole Saw

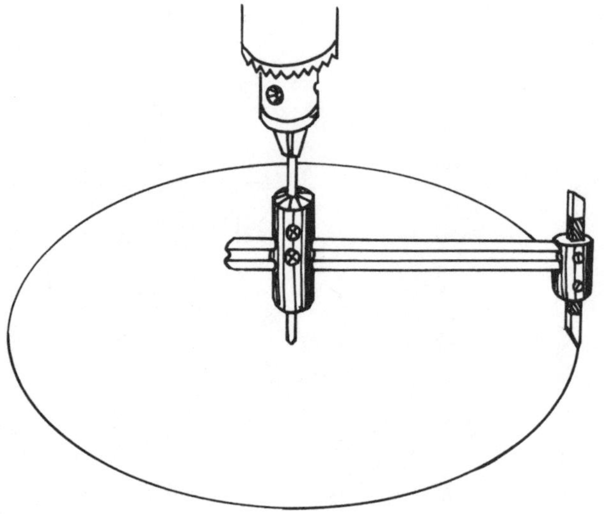

Circle cutter

The circle cutter can be used with a drill for even larger (or smaller) holes (*10*). It consists of a central drill bit that mounts in the drill, an arm extending from the shank of the bit, and a bit with a sharp chisel-like blade extending down from the arm. The distance between the drill bit and the chisel can be adjusted to any spot along the length of the arm. The bit is located in the center of the circle, and as it rotates into the wood, the chisel cuts around the circumference. This attachment is really designed for a stationary drill press, because it is difficult to hold the portable drill steady or vertical enough. Once you have some experience, however, it can be done. Hold the drill perfectly straight, go slowly—you definitely need a variable-speed model—and apply as little pressure as it takes for the chisel to keep cutting. Too much pressure will jam it into the wood. You can tell at the beginning if you are going straight by watching the cut the chisel makes— the depth should be even all the way around. This attachment is very useful for medium-size holes, as in speaker cabinets, where you might have trouble cutting a clean circle with a jigsaw.

Fig. 10. Circle Cutter

Drilling through metal

To cut through metal, use the high-speed bits in the set. Use a punch to start the hole. Vary the drill speed, if possible, to help the bit catch. Whenever you buy a set or individual wood bits, ask for high-speed. They cost a few cents more than regular wood bits, but they can also cut through metal. For thick metal, squirt some oil in the hole as you drill.

Carbide tipped bits

Buy carbide-tipped bits for brick, concrete, or stone. Buy one the size of the lead shield you will use, not the size of the screw or bolt. A cheap one will last for only a few holes in tough brick. If you expect to use it often, buy a good bit and take care of it. Don't force it, no matter how tough that brick is—you could bend the shank or ruin the tip by doing so. Store it in the plastic sleeve it comes in, and don't let any hard objects knock against it. To help drilling, take it out of the hole every twenty seconds or so and dip the tip into a lubricant to cool it off. If the brick is very hard and you have to drill a large hole, say $\frac{1}{2}''$ or bigger, try drilling a smaller pilot hole first with another carbide bit, say a $\frac{1}{4}''$ or $\frac{5}{16}''$ bit.

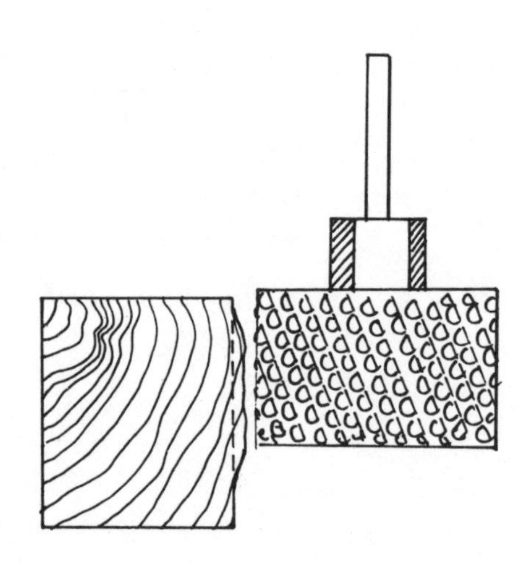

Fig. 11. Rasp Bit

Other bits and attachments

Other useful bits include *rasp bits* (*11*), for enlarging or evening the inside of holes; and *plug cutters,* for cutting out wood plugs to be used in filling in other holes, covering screwheads, and so on.

There are also wire brushes, pads for sanding discs and buffing pads, grinding wheels, paint mixers and so on. Check out your hardware store racks for interesting possibilities.

Screw Gun

The screw gun is basically a powerful drill with a chuck designed to hold and drive self-tapping screws without predrilling a hole (*12*). Its main use is in general construction—attaching plasterboard and other wall sheathing to wood or metal studs, or to metal sound clips (see Soundproofing in "Room Renewal"). It also screws wood to wood; the chuck can be set to drive the screw flush or to countersink it; an automatic brake stops the screw where you want it. Common bits are available for Phillips-head, standard-slot, or hex-head screws (*13*).

The screw gun is pressure-sensitive—you pull the trigger but it doesn't start till you press the screw hard against the work surface. Hold the gun straight as you start. Don't press the screw into anything else, such as your arm.

A good screw gun is the only kind to have—and it's expensive. There's no reason to own one unless you are doing a lot of general house construction, where it could save considerable time. Or if you're attaching sheathing to metal studs or sound clips, where you *have* to use one. Fortunately, you can rent a screw gun cheaply.

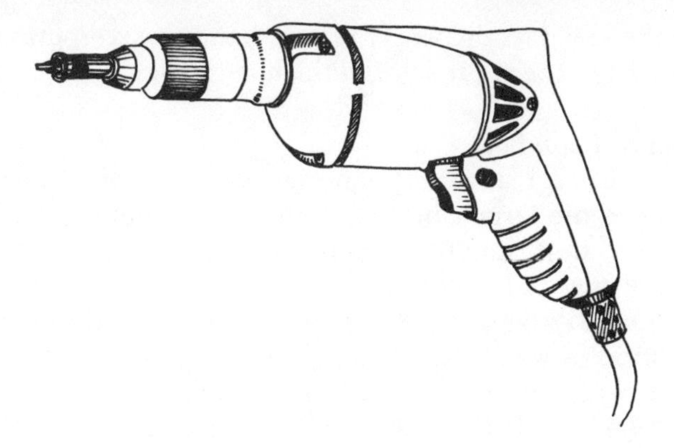

Fig. 12. Screw Gun

Hex Bit

Standard Screw Bit

Fig. 13. Phillips Screw Bit

Tools for Cutting (and Shaping)

Power saws are probably the most useful of all power tools. For every carpentry project, you have to cut wood pieces. Hand saws are slow, and accurate cutting takes experience and skill. Power saws are fast and, with a simple guide, very easy to use for cutting straight lines. There are different blades to make different types of cuts in different materials.

The portable circular saw can crosscut or rip lumber, and will easily cut plywood. With an aggregate blade, it even cuts masonry.

The saber saw also makes straight cuts, but it's not as powerful; it's really designed for cutting curves or holes within a piece, or for cutting into metal and plastics. The saber saw is a power version of the compass saw, the coping saw, and the hacksaw.

The main function of these tools is to cut, of course—to actually separate one piece from another. To a certain extent, they can also be used as shaping tools—for instance, to cut out notches or, with the circular saw, to cut dadoes and rabbets.

Portable Circular Saw

The portable saw has a circular blade rimmed with teeth (*14*). Pressing the trigger starts the motor that spins the blade. The baseplate of the saw rests on the surface of the work and keeps the whole saw steady, making the blade cut at the desired angle along the entire length of the cut. The cutting action and force are provided by the motor and teeth of the blade. Basically all you do is slowly push the saw along the line. There is, of course, much more to know before you try a saw cut, but the point is that you provide the knowledge

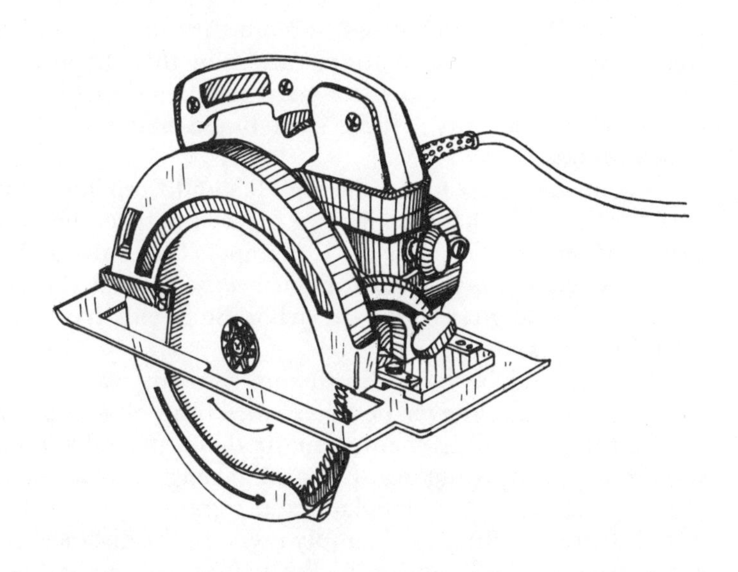

Fig. 14. Circular Saw

and the saw does the work. With a hand saw, *you* provide the knowledge, the accuracy, *and* the horsepower.

What to Look For

There is a great variety of power saws. First of all, a saw's size indicates the largest blade it will take; common sizes are 8″, 7¼″, 7″, 6½″, and 6″. Saws of smaller size, such as 3½″ and 4″, are meant for cutting thin sheets of laminate and plywood. A saw with an 8″ blade will not cut through an 8″ thickness of wood. Remember the blade spins on an axle, or arbor, which limits the cut to less than half the thickness, or 4″ for an 8″ blade. And there has to be room for the arbor, motor, and base plate—so an 8″ saw may cut to a maximum depth of less than 3″. Most saws list their maximum depth of cut at 90°, and also at 45°, which will of course be a bit less. The most common size for general-purpose work is the 7″ or 7¼″. It will cut through a 2-by at a 45° angle. Even an 8″ saw can't cut through a 4x4 in one cut; it has to be cut either in two cuts, one from the top and one from the bottom, or by a hand saw. Theoretically, you want as small a saw as possible that will do the job. It's cheaper and, more important, lighter. When you start to do a lot of cutting your arm will appreciate a lighter saw; or if you do any cutting above your head, or where you have to carry the saw from one house to another. A saw should, however, be heavy enough to give it some stability. Very cheap, light saws will have some trouble sawing straight through thicker pieces.

Hold the saw you're thinking of buying in your hand, put it in place on a piece of wood. Cut something with it, if possible, to test the balance and how easy it is to run and keep straight. The feel is a personal matter. A strong person will prefer a heavier saw.

The less power a saw has, the more time it will take to make a cut, and the harder a time you'll have with tougher cuts. Check the power ratings, though they may be misleading. (See General Information, earlier in this chapter.) In general, the price of a saw is a good indication of its power and quality—certainly within each brand and roughly between brands.

Look at the baseplate on the saw. It should provide a solid base for the saw. The stronger and bigger the base, the better it will sit on the work surface. It is impossible to use a circular saw to cut flush to a vertical surface because of the baseplate, the blade guard, and the arbor. Such cuts have to be made with a handsaw.

There should be two adjustments you can make on the saw. Neither should ever be made unless the plug is pulled. One is the depth adjustment. Usually this is made by loosening the depth-adjusting bolt by hand, raising or lowering the saw in relation to the baseplate, and tightening the bolt or nut. To measure the depth, simply measure the distance from the bottom of the baseplate to the bottom tip of the blade. Your saw should also have a miter adjustment, a similar bolt or nut that loosens to tilt the body and motor to any angle from 45° to the normal 90°. The degree of angle is registered by a measured scale along which the locking nut slides. This scale is not precise, particularly on inexpensive saws. Check the angle with an adjustable protractor or bevel square. Both these adjustments should move easily and tighten well by hand.

The saw should also be equipped with a retractable blade guard (*15*). The guard should extend around the back of the blade and below it when at rest; when you make a cut, the edge of the wood forces it up above the wood surface. The guard should have a flange or lever safely away from the blade that lets you pull it up by hand. This is useful for plunge cuts and clearing over obstructions. When the saw is taken from the work, the guard should spring back to place, protecting you from the spinning blade. Be sure the spring action works well, that the guard moves up freely, and springs back quickly. Sometimes a person will take the saw off the work, and place it down on the floor. Unfortunately, the blade keeps spinning for a while after you release the trigger. A good guard covers the blade so that you don't cut through your floor and so the saw doesn't jump forward.

Most saws come with a separate attachable rip guide. This is a metal bar that slides into two slots at the front of the baseplate. It is held in place by one or two set screws through the slots. The bar extends left or right with a lip at one end that slides along the edge of the wood. You can make long rip cuts by guiding the lip along the edge and cutting a line parallel to it at the set distance.

Among other features to look for are a special knob handle for your other hand to hold, to help guide the saw; a sawdust kick chute at the back of the saw, which shoots most of the sawdust out of the way of the cutting line; a safety switch that has to be pressed in before the trigger will work (it prevents you from accidentally starting the saw when just lifting it—a very good feature to have, but unfortunately found on very few saws); and double insulation. Some saws also have an automatic brake. This is a device that stops the blade within a few seconds after you release the trigger. This is safer than the normal gradual slowdown of the spinning blade, but no substitute for your own care and proper handling.

You will need a power saw when you begin to do a lot of cutting, particularly rip cuts and long cuts in sheets of plywood. You can get by with a handsaw for 2-by crosscuts, and you can even have plywood cuts done at the lumberyard. But after a while the price of the lumberyard cuts approaches the price of a good inexpensive power saw. And you can probably do a better job yourself. If you're going to get serious about a few projects, get a power saw. The only limitations compared to a handsaw will be you can't cut a 4x4 in one cut (you'll need two); you won't be able to cut flush to a vertical surface; and it will probably be safer to use a handsaw for cuts above your head.

Saws range in price from less than twenty to several hundred dollars. We can't honestly recommend the real cheapies. They work, but they work so slowly and inaccurately that you will be turned off carpentry forever. Get one if you must, but remember that even the best carpenters have trouble cutting straight lines with them. They go so slowly that you tend to push them beyond their endurance, which harms their bearings and motors and makes them veer off the straight line. We suggest buying as good a saw as you can possibly afford. Spend a minimum of thirty dollars for the saw, not counting extra blades or a case. One blade, the rip guide, and the wrench for tightening and loosening the blade should be included. If at all possible, just try a more expensive saw so you can feel the difference. Then you'll know what's your fault and what's the saw's fault when you use the inexpensive one. On the other hand the most you should spend on a saw, unless you will be using it every day for heavy construction, should be around eighty or ninety dollars. Anything higher than that will prove economical only for long-time endurance on heavy-duty work. You will gain a negligible amount in safety and accuracy and only a bit more in power.

If finances permit, a good starting saw would *list* in the forty-to-sixty-dollar range. Look around for discounts. Buy a brand name and be sure of the service and guarantee. If you really want to get serious, you can buy an even better saw later—and you can sell a medium-priced saw, or give it away, much more easily than a cheapie.

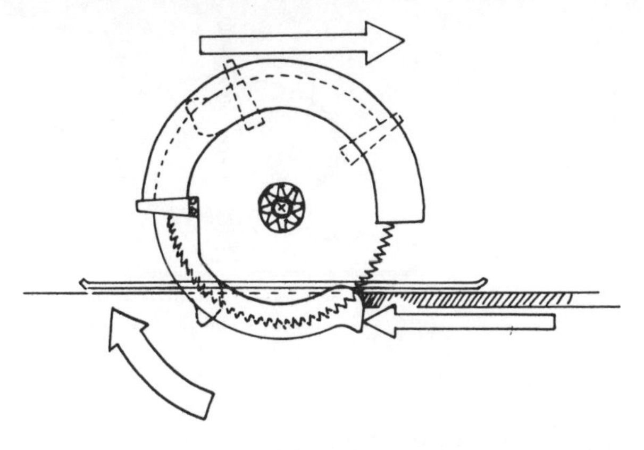

Fig. 15 Blade Guard Action

Getting the Saw Ready for the Cut

Once you become familiar with power saws, they are easy to use and fairly quick to set up. At first, however, if you've never used one, there is a lot to think about as you set the saw up for a cut. Read this section carefully, and also your instruction manual, before attempting any cuts. Understand as much as possible before cutting. The power saw is perfectly safe when you use it correctly and observe simple safety precautions. It can be very dangerous if you don't, just as an automobile can be. *Respect your saw.* After a while, most of the techniques and precautions involved with sawing become automatic, and things go much more quickly.

The first thing to do when preparing for a cut is to make sure you have the right blade *(16)*. For cutting solid lumber, the *combination blade* is handy because it will either rip or crosscut. There are also individual *rip* and *crosscut blades* for slightly cleaner work. The *planer blade* is thinner and leaves a smoother end grain. Because of its thinness, however, it will sometimes bend slightly at knots or on rips when the grain may carry it away from the cut. The *carbide-tipped blade* is much more expensive, but lasts much longer than any of these blades. It's worthwhile if you will be doing lots and lots of lumber cutting. For plywood you need a special *plywood blade,* which has extra fine teeth so as not to tear up the

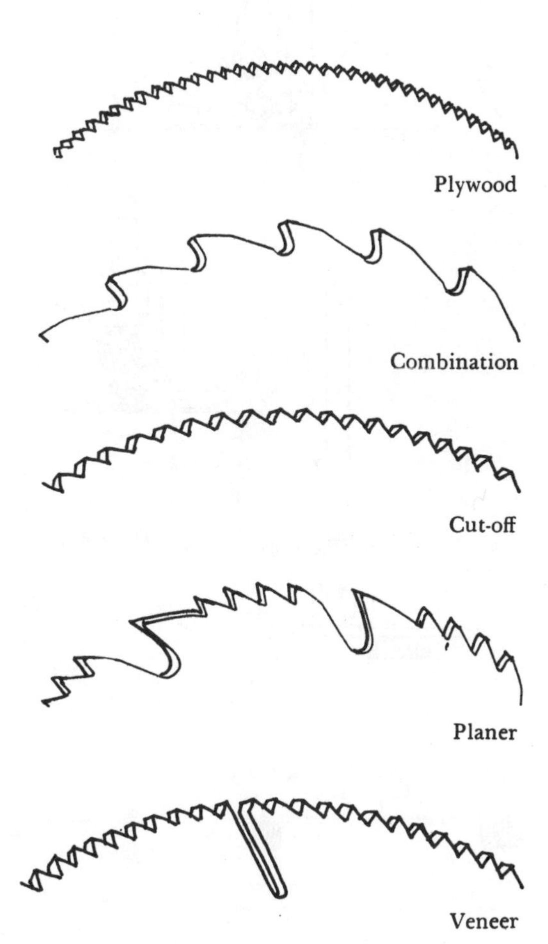

Fig. 16. Carbide-tipped

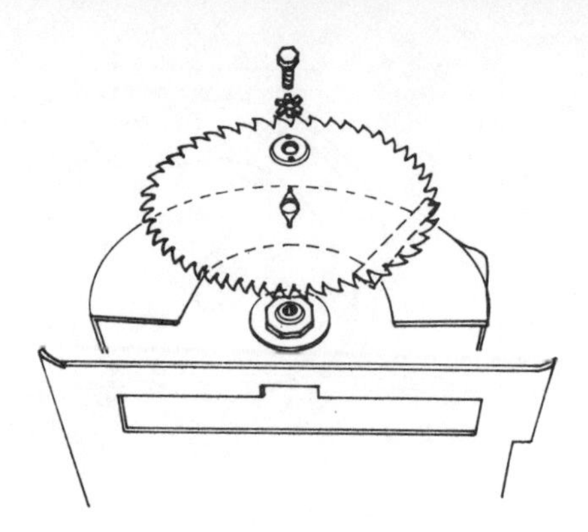

Fig. 17. Blade Mounting

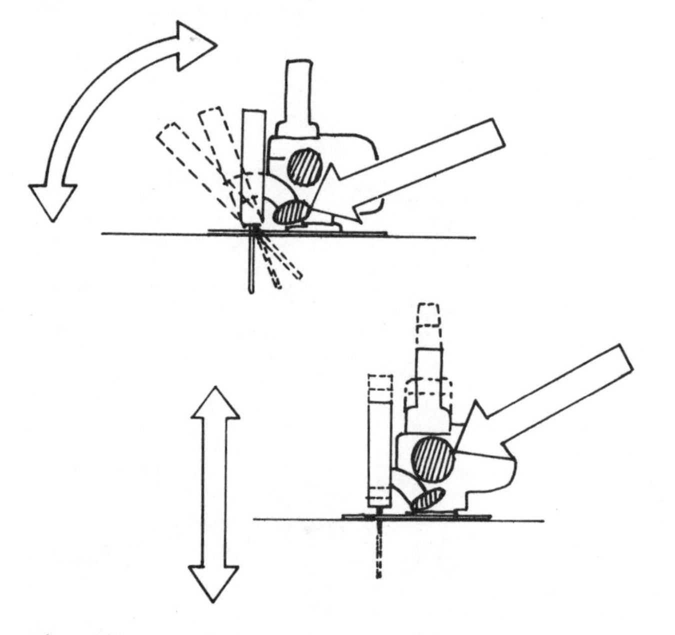

Fig. 18. Adjusting angle and depth of cut

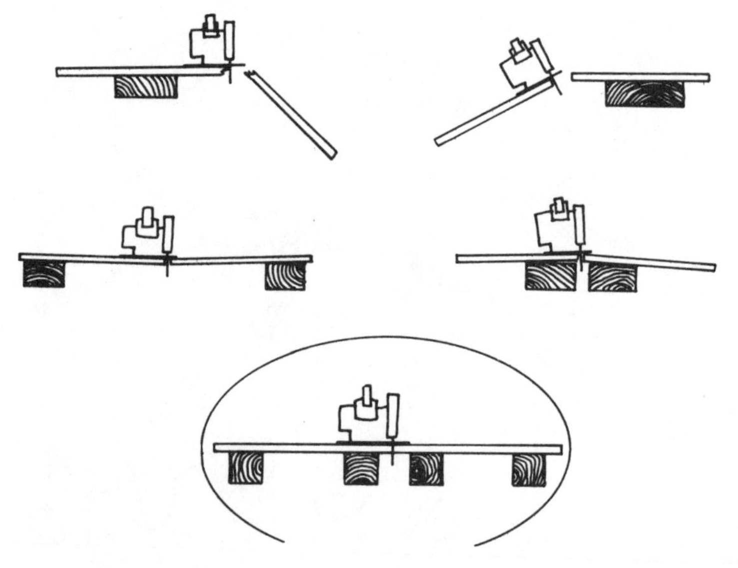

Fig. 19. Supports

veneer. Use a ply blade for composition board also—and don't expect it to last too long. There are other kinds for specialized uses—flooring blades for hardwood, and blades for cutting metal, stone, and plastic.

New saws usually include a standard combination blade. Buy a plywood blade when you need it. These two blades should be enough for a beginning.

When blades become dull, don't throw them away. Have them sharpened, and they will be good as new. It costs as much to sharpen a cheap blade as an expensive blade, so it pays to buy a good blade in the beginning.

Now you've got the proper blade. Put it into your saw (*17*). To do this, disconnect the power. *Always*. Follow the specific blade-mounting directions in your saw's manual. First, remove the bolt and washer from the shaft, or arbor. (Some saws have a nut instead of a bolt.) The side of the blade with writing on it is the side that faces out when the blade is mounted. Looking in that direction, the blade revolves counterclockwise. Holding the blade with the face out, slip it through the bottom of the baseplate into place on the shaft. You may have to raise the base plate to fit the blade in. Also, you may find that the hole in the blade is not big enough to fit over the shaft. Take the blade out. In the center of the blade you will see some "knock-out" rings. Knock out just the offending ring with an *old* screwdriver and hammer, or whatever is handy. Remount the blade.

Tighten the bolt over the washer by hand, and then tighten it further with the wrench that comes with the saw—about 1/8 turn, or whatever the specific instruction manual says. To tighten or loosen a blade with this wrench, it is usually necessary to hold the blade steady somehow. The best way is usually to wedge the front of the blade up against a block of wood. Some saws are designed to take two wrenches, one that fits between the blade and the inside of the saw, and holds the shaft steady, while the other wrench turns the bolt. When a blade slips on the shaft while cutting and doesn't turn while the motor is going easily, the blade is loose. Tighten the bolt.

The next step is to set the angle of cut (*18*). Most cutting is at 90° to the surface. Set the bevel adjustment by loosening the bevel nut or knob and moving the body of the saw in relation to the base until the knob lines up with the 90° mark. Tighten. This guide is not always accurate, particularly on cheaper saws. Check the angle with a square against the bottom of the plate and the inside of the blade below the base. Check miter angles with a bevel square. Later check the actual angle it cuts in a piece of scrap wood.

When the angle of cut is set, adjust the depth of cut. The saw should still be unplugged. The depth is determined by the thickness of the wood you are cutting. You want the blade to protrude below the wood no more than 1/8". This cuts down on splintering of the wood, and is also a safety measure. The less the blade protrudes beneath the surface, the less chance there is of hitting an obstruction you can't see. The easiest way to set the depth is usually just to loosen the depth nut, set the base of the saw on the work, with the inside face of the blade against the edge. You can then see how low the blade is in relation to the wood. Set it and tighten the nut. (You don't set the blade *flush* with the bottom of the wood because the blade wouldn't cut all the way through if the wood gets slightly thicker in the middle; or if it sags slightly, making the baseplate span slightly above the middle of the sag.)

Supporting the Wood

The saw is set. There are two more important parts to the preparation for the work. Supporting the work, and laying out the line to be cut. Supporting the work is probably what gives beginners the most trouble of all the techniques of sawing. Bad support can cause the saw to bind, which happens when the wood on either side of the blade closes in on the blade. Bad support can cause the wood to break unevenly at the end of the cut. It can cause the saw to "kick back." The principles of proper support are very simple. Follow them and you won't have any problems. Watch closely; we will show the ideal supporting system and then how to approximate and improvise along the same lines.

The ideal support for a piece of wood to be cut would be a flat surface at least as long and as wide as the wood itself. In that flat surface would be a slot right where the cutting blade will move along. Additionally there would be two clamps to hold the piece firmly to the surface, so that it wouldn't slide around or leap away from the saw blade. You could then run your saw along the wood over the slot and out of the wood. The piece would be right where it was before, only now it would be two pieces.

This system manages to support the whole piece of wood and the weight of the saw. It also supports the two halves that are the end product.

Look at the pictures (*19*) and you will see examples of bad support and the ensuing calamitous results. You have to think of the final pieces; place blocks to support them as if they are already in existence. Also consider the added weight of the saw in the middle. If the wood bends or sags, binding ensues.

When you cut a shorter piece off the end of a 2-by, you can eliminate some of the support. A piece of a few inches needs none—just saw steadily and let it drop. Up to a foot or two, depending on your reach and confidence with the saw, and assuming the piece is clamped or otherwise held down, hold the piece with your hand, well away from the saw, as you finish the cut, to prevent it from falling too sharply and splitting the end of the wood. It would be better to have someone else support the end—just to let it rest level in the hands.

Sawhorses are very handy devices. You can lay three or four 2-bys across them and lay the work on that. Or in some cases, like plywood sheets, you can put the work right on the sawhorses and saw from one to the other. Theoretically, it's bad form to cut through the top of a sawhorse. In practice, though, it's done all the time. If the top is at least a 2x4, and your blade extends no more than the 1/8" we suggest, all that will happen will be a shallow cut through the top of the sawhorse. After a number of jobs, sections might fall out between the kerfs so that this top piece would be too uneven to use. Take a few minutes and turn it over, or put another piece in. Most sawhorses come apart very easily. It's worth the time you save in setting up other supports for cutting. Just be sure of that 1/8" clearance.

The clamps in the ideal support method are mainly to hold the work steady against the forward force of the blade. If you use blocks on the floor, it's hard to use clamps. Usually you hold it down with a hand, a foot, a knee or else butt it up against something in front. The easiest method is to set the blocks on the floor, the work on the blocks, and one knee on the work (on the larger side). Go slowly, and you will feel if the wood starts to move. You may also have to hold it with a hand or weigh the end down, if it's very long.

Remember to keep your body away from the saw. Never stand or kneel behind or in front of the saw. Place yourself to the left, the side with the motor. If you hold the wood with your hand, keep it as far away from the saw as possible. Don't stick your nose down to get a close view of the cutting line.

Laying Out the Line to Be Cut

The second preparation is marking the line to be cut, or setting up a guide. If you are making a freehand cut on a piece, where accuracy is not necessary, you simply mark the length you want with your tape, mark a straight line perpendicular to the edge with your square, and cut by eye along the appropriate side of the line. Saws have a cutout in the front of the base plate that lines up with the blade. By making a few practice cuts you can see just how the blade relates to the cutout, and which part of it to follow. Just keep the line on that point in the cutout. This is not a totally accurate way to cut a straight line, but good enough for rough cutting. You may also carefully peer over the top of the saw and watch the blade as it hits the line. Keep your head as far away as possible, and as always *wear your goggles.*

Remember to consider the thickness of the blade. If the length you want to cut is on your left, cut just to the right of the line, so that the left of the blade is flush with it. If you cut through the middle of the line, or to the left of the line, the blade will take off some of the length you need.

You can get a perfectly straight line with no trouble or super skill by using a guide. Such a guide would be any straightedge fastened or held to the work piece against which the saw's baseplate travels; thus you don't have to watch the line or try to hold the saw straight. For instance, you could hold a try square on a 2-by as in the picture—with the blade

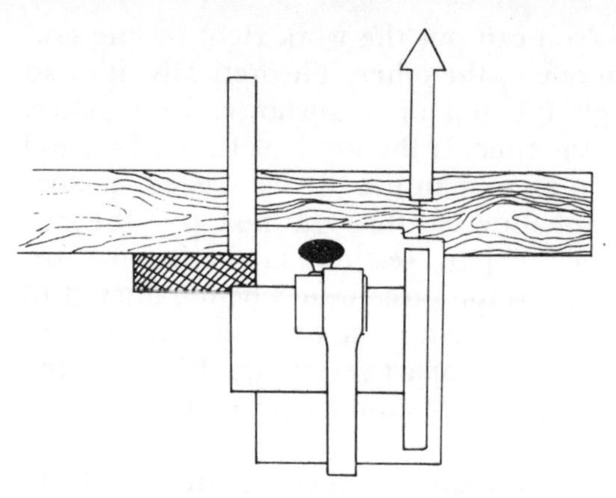

Fig. 20. Using a square for a straight cut

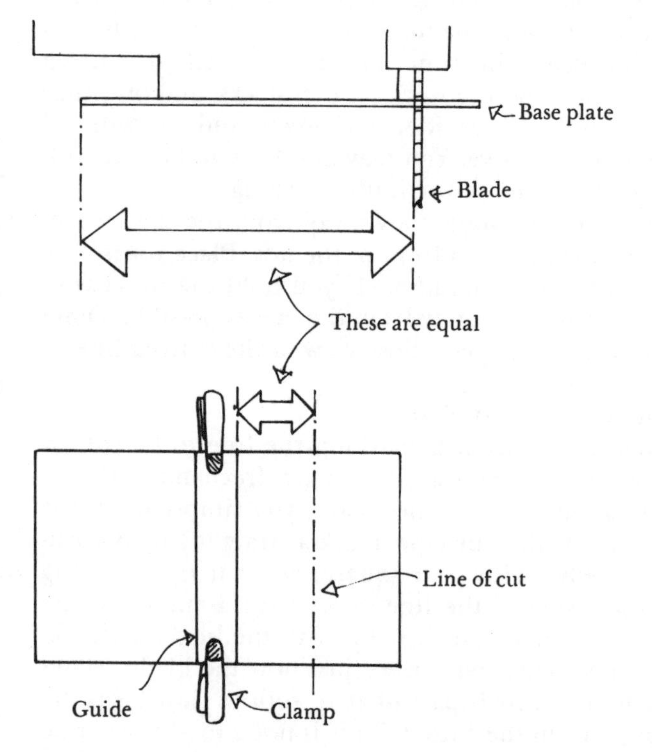

Fig. 21. Setting up the guide

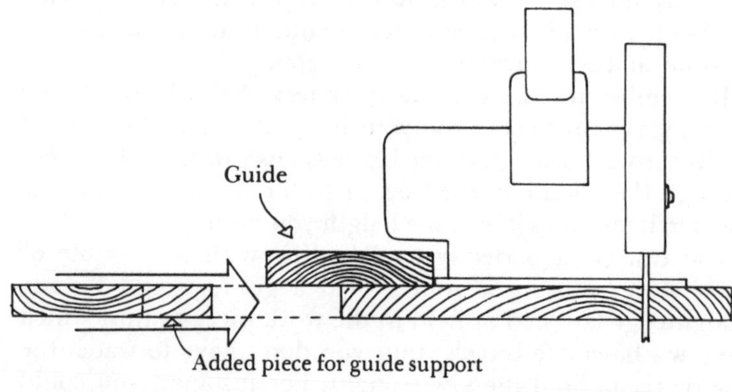

Fig. 22

perpendicular to the edge of the board *(20)*. If the saw baseplate runs up against it, the cut will be straight and perpendicular, just like the square. All you do is make sure the saw doesn't fade away from the square. This technique is handy for cutting 2x4s to length. To set the square in the right spot, you still have to mark the line to cut, set the saw blade at the line, then place the square against the plate and hold it in place well. This method of lining up is accurate enough for rough cutting, like wall studs. Remember to back the saw up from the wood a bit before starting it. Also move the saw to the side of the square till it reaches full speed. The bucking of the motor as it starts might push the square a little off. Start it, then move the saw against the square, then move it forward to the wood slowly.

A more accurate way of setting the square in the right spot is to measure the distance from the edge of the saw baseplate to the inside and outside of the blade. Do this by setting the square on a scrap piece. Draw a line along the blade of the square on the wood. Using the square as a guide, cut into the wood several inches, remove the saw, and measure from the drawn line to both sides of the kerf. Say the measurements are 4⅞″ and 5″. You then know that the square should be set 4⅞″ to the left of the line you want cut, if the desired piece is on the left. If the piece you want is to the right of the line you draw, then the guide should be 5″ to the left of the line, so that the blade thickness will not extend over the line.

A better, more stable, guide can be made from any straightedge—a long rule, or a straight piece of wood such as the factory edge of a sheet of ply—and some clamps or nails. This guide is good for long cuts, like cutting a sheet of plywood. The same principles of measuring apply. You might also measure the distance from blade to plate on the *right* of the saw. It's best to set the guide on the same side the finished piece will be. Holding a try square on the right of the saw is not usually a good idea because your hand is too close to the blade. If you use a metal rule, you will need either C clamps or spring clamps to hold it. Set it, if possible, so that some of the rule extends over both ends of the wood—so the saw will have a firm guide at either end *(21)*. Over a long cut, the middle of a metal rule sometimes bends to the side—the saw forces it. Tack a finishing nail into the work against the outside of the rule at the middle to stop this. If you don't have a rule, pick the straightest piece of wood you have—preferably a 1-by or ¾″ ply, to be sure the motor can clear the guide. Either clamp as with the rule or tack it down with finishing nails. The factory edge of plywood is usually perfectly straight. Check it to be sure by running the edge of your square or level along it.

Write down the measurements for guide distances on a piece of masking tape or such and attach it to the body of the saw in plain sight. The measurements are usually very easy to forget. Also note that the measurements may be slightly different for different blades, particularly between a plywood and a combination blade. Measure for each.

When you mark for cutting with a guide, you don't have to draw the line that will be *cut*. You might make a mark at the starting edge just so you can check the blade position as you begin. But all you have to do is get the measurement you want to be cut, adjust for the guide by adding or subtracting the proper distance, and then put down two or three points at that length. You don't even draw that line, just those points. Lay the guide on the points, clamp, and you're ready to cut. Lay the saw in, and check that the blade is on that single point you made earlier. If math is your mortal enemy, you can place points at the length to be cut, then mark back the guide distance with your tape, and continue as before. (No addition or subtraction needed.)

Spring clamps are the easiest to use for a guide—quick to put on and take off, and they don't jut up as high as C clamps. But they sometimes shift a bit if they're not put on just right. Watch carefully when the saw starts up—bring the saw against the guide only after it's running full speed. C clamps hold well, but take slightly more time to set up. They may get in the way of the motor if the guide is not wide, and they can leave imprints on the work if they're too tight. Use a thin shim between the clamp and the work if the surface is important. Try to put the clamps far enough away from the motor to give it clearance; use a wide enough guide.

If you are making a long cut with a guide and find that toward the middle of the cut the plate is slipping *under* the guide, you probably have not supported the piece in the middle and it is bending from the weight of the saw. Slip a support underneath (*22*). Check this beforehand if you can—it will be too late, once the saw strays off the cutting line. If you can't get a support under there for some reason, try to tack or clamp the guide down.

If you have trouble fitting the guide or saw on the work, you may be able to turn yourself or the piece around and work from the other side. Or clamp another piece right next to it to give the saw or guide something to rest on.

The Actual Sawing

Check that the wood is securely clamped or held; check that there are no obstructions under the wood where the cut will be; rest the front of the baseplate on the wood, with the blade just a little bit away from the edge. Make sure the blade depth is not too big. If you're using a guide, set the saw slightly away from the guide at first. Stand to the side of the saw. *Make sure your goggles are on*. Make sure the electric cord is clear of the cutting path, and also that there is enough slack and no obstructions so the cord will reach the full distance of the cut (important for long cuts). You have one hand on the handle and the other on the knob, or on the wood away from the saw, or holding the guide. Balance yourself. Start the saw in that position, slightly off the mark. When it reaches full speed, which you will know when the pitch of the motor evens out in a few seconds, slide the saw over till it's in place against the guide (there should be enough of the guide extending over the edge for the plate to have a good butt) and move the saw slowly forward the inch or so toward the wood. Let the weight of the saw rest on the wood. Hold it steady against the guide and feed it forward. Don't force it into the wood; it will move forward all on its own. With experience you will recognize when it's being forced. You can hear the pitch change drastically as the motor strains; you can smell burning, and see smoke in extreme cases. It definitely does not pay to gain a few seconds by forcing the saw. You will ruin the blade. And sawing with a dull blade puts a strain on the motor and can ruin *it*. Also, if forced, the saw will tend to stray off the straight line. If your saw is taking forever and driving you batty, try a new blade; or it may be the type of wood you're cutting; or perhaps the blade is slipping on the shaft—tighten the bolt.

As you continue cutting, maintain your balance and control over the saw. If it's a long cut and a wide piece that you have to maneuver about, you may want to stop in the middle of the cut and reposition yourself. To do this, while the saw is still going full blast, back it up an inch or so along the new kerf and then stop it. Hold the saw in place till the blade stops spinning. Picking it up while the saw is winding down or running could damage the wood, and it could be dangerous if the blade hits the wood and forces the saw to bounce up. When the saw has stopped, reposition yourself, set the blade carefully in place in the kerf or against the guide if you're using one, and start again an inch or so back Don't move it forward till the motor is on full again. Then move it ahead slowly.

As you approach the end of the cut, prepare yourself to support the weight of the saw. If the wood is well supported just cut through and, with the bulk of the saw still resting on the pieces, turn it off and let it stop, then pick the saw up. If the pieces aren't well supported, you may find yourself supporting the weight of the saw while it runs down. Only pick the saw up from the work if you have to, but always be prepared to. When the saw is lifted off the wood, the blade guard should spring back over the blade. Sometimes, though, fresh sawdust might clog the guard. Therefore never put the saw down till the blade stops spinning.

Experienced carpenters will sometimes saw 2x4s one after the other, lifting the saw off the wood at the end and setting the saw on the next piece before it has even stopped spinning. Their experience gives them control over the saw and a feeling for just how far they can stretch their safety precautions. Also they're in a hurry to finish the job on time. We strongly advise you not to get fancy like this. Take your time. The few seconds you have to stand around waiting for your saw to stop are well worth the safety insurance. When you're experienced and know the tolerance of yourself and your tools, you can start to stretch the rules at your own risk.

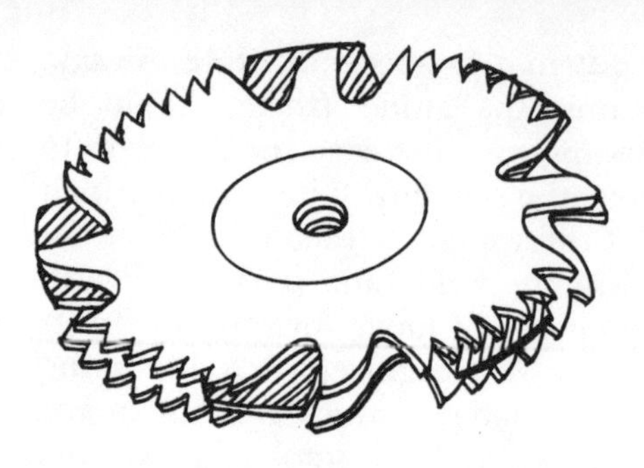

Fig. 23. Dado Blade

Other Sawing Techniques

To cut 4x4s, or other pieces too thick for a circular saw to cut through, square the cutting line on all four faces and make two cuts—one each on opposite sides. If the two cuts leave an uneven swirl on the cross section, rasp it flat.

To cut a 4x4 at an angle, draw the lines on all four sides as before. Make the cuts along the angled lines. If you make them on the level lines, you have to set the saw blade at an angle.

You can use the saw for other purposes than cutting a piece into two. By setting the saw blade very shallow, you can cut grooves or kerfs in wood. This is sometimes a desirable use, particularly for paneling. Or by cutting several kerfs next to each other, you can cut a groove or dado in wood. Dadoes, as explained in "Working Techniques," can be used to join one piece of wood strongly to another. A ¾″-wide dado will accept the end of a ¾″ piece of plywood or a 1-by, for instance. Dadoes can be cut more quickly and accurately with a router, but a saw will do if you don't have a router. (Use a chisel to remove the chips and smooth the bottom.)

Special attachments called *dado blades* can be bought (*23*). They can be adjusted to cut any size dado up to about 1″, or the inner thickness of the blade guard. They cost about twenty dollars, though, at which point you can start thinking of a router.

One great thing the power saw can do is to greatly simplify cutting notches and special joints. In the old days, notches were cut with chisels and maybe a handsaw. This called for great accuracy and skill, and great familiarity and experience with wood grains. Say you want to cut a notch 3½″ wide and 1″ deep on the edge of a 2x4. Simply set the depth of the blade to 1″. You can check it with a practice cut in some scrap. Mark off the 3½″ space on the 2x4 with your square. Make several cuts inside the space, say no more than ¼″ to ⅜″ apart. Be sure to make a cut flush with each boundary line, on the inside of the space. The cut sections are easily removed by knocking them sideways with your hammer. Any bits that remain can be cleaned out with a chisel. If the bottom of the notch is still rough or has some hills left, which often happens, chisel them out. Even easier is to make cuts on the end lines, turn the piece on its side, mark a line 1″ from the edge connecting the bottom of the cuts, and cut it out with a saber saw. The circular saw makes accurate depth cuts quicker than the chisel, and eliminates the risk of splitting the wood. (See Lap joints in "Working Techniques.")

Rip guide

A rip guide is included with your new saw (*24*). Remember you can use the rip guide for long cuts within 6″ or so of a straight edge of the wood. Set the guide in the saw, measure the distance from the inside of the guide's flange to the proper side of the blade (depending on which piece of wood will be the finished piece). Make a practice cut in scrap and measure the width to be sure. It is important that the edge it runs along be perfectly flat and as straight as possible. The blade will cut a parallel course and it will cut unevenly or bind if the edge is too uneven. Go slowly and keep the guide flange against the edge. It is usually more accurate and neater to set up a full-length guide, but not always practical. Use a ripping blade for a neater cut. In fact, if you want a really smooth edge, you rip to a 1/16" bigger size and then sand or plane the edge to size.

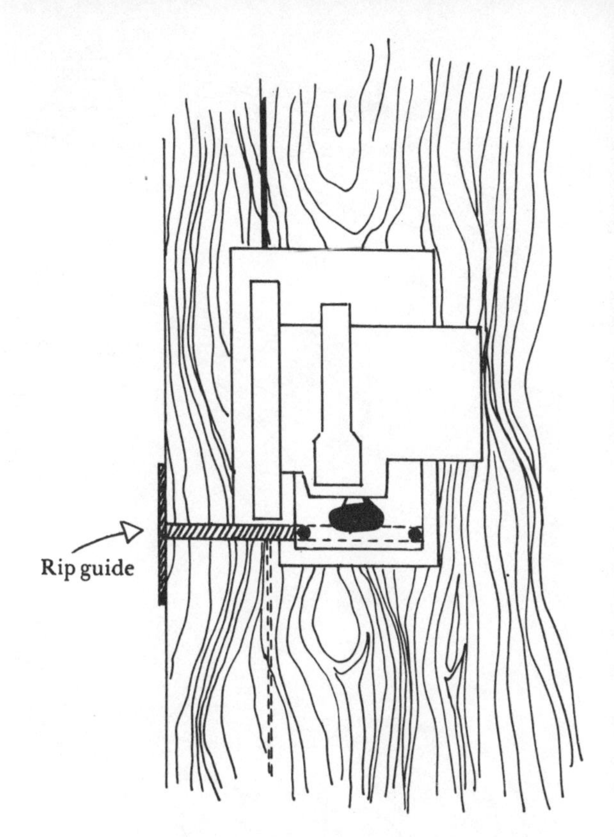

Fig. 24. Using the rip guide

Plunge cutting

The circular saw can be used to make a cut in the middle of a piece of wood, such as cutting a hole out of a sheet of plywood for a cabinet-door opening. This is called a plunge cut. The technique is to draw the line you want cut; lean the saw forward so it is resting on the front edge of the baseplate and the blade is an inch or more above the wood; open the blade guard all the way and hold it so the blade is exposed; start the saw on full and then lower the saw slowly onto the line, using the front edge as the fulcrum or hinge point. Go slowly and hold the saw firmly with both hands. The blade will slowly cut through the line. When it is all the way through, you can set it flat on the wood as usual, and continue the cut. Do not pull the saw backward.

When you lower the blade, it is best to have it a few inches ahead of the start of the line, to be sure you don't lower it too far back. Don't pull the saw backward when it's in all the way to cut those back few inches. Make the forward cut, stop, take the saw out, slip it back into the kerf facing the other way, and cut.

Free-hand plunge cutting is difficult to do with great accuracy. If accuracy counts, it is best to set up a guide. Then as the saw is lowered, the plate will butt against the edge of the guide and remain straight on the line.

Because the blade is circular, a plunge cut or any other cut that stops in the middle of a piece of wood will look like Figure 25A. In the last example, cutting out a door, you could make all the cuts with a circular saw, stopping the blade when it hits the end of the line on the *top* of the wood, then finishing the cut with a saber saw or a handsaw, which can cut perpendicular to the surface. (See the Captain's Bed project.) If you were to continue the cut beyond the line on top so the blade would cut all the way on the bottom, you would end up with cuts showing in the wood as in Figure 25B—very sloppy and ugly.

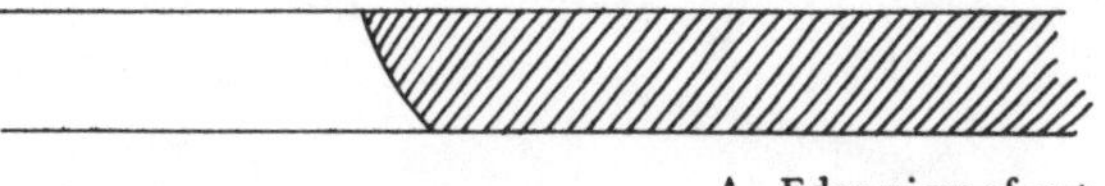

A. Edge view of cut

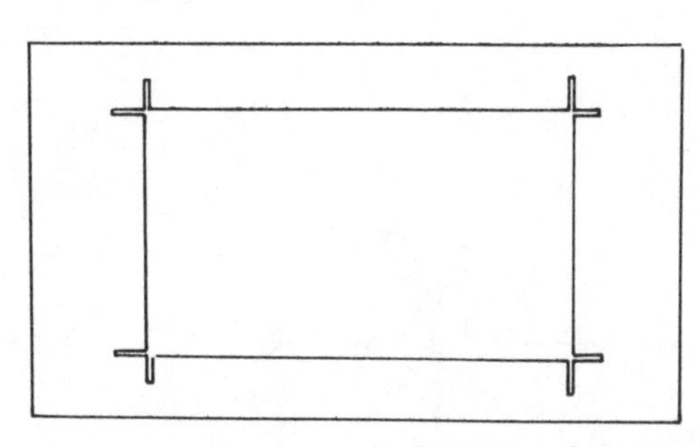

B. Top view of plunge

Fig. 25

Avoiding Splintering

Because the blade cuts in an upward direction, it tends to splinter the grain on the top of the wood. If you want one side to be the finished side, put that side down. Much less splintering occurs on the bottom. If you want both sides to be finished, you can eliminate a lot of the mess on the top by clamping a thin scrap over the line to be cut. This holds the top grain down when the saw blade passes through it. Mark your sawing line on the thin piece.

Miter cuts

Miters are easy with a circular saw. There are two kinds (*26*). The first (A) traverses the top face of the wood at an angle to the edge. The miter adjustment is not needed for this cut. Simply draw the line on the top face and cut. Use a guide for accuracy. A tool called a miter guide, or a combination square, can be used for 45° miters, just as a try square is used as a hand guide for straight cuts.

The other miter (B) is a little trickier. This is where the end of the wood is cut at an angle to the face. Set the miter adjustment on the saw to the proper angle. Check it with a bevel square. Make a practice cut and check that angle. If you want a 45° angle, make two practice cuts, put them together, and see if they form a 90° or square corner. Use a guide, preferably on the non-waste side. Hold the saw extremely steady, flat on the wood, and firmly against the guide. Any variations from the straight and narrow will make it very difficult for the miters to fit together.

Also note that the distances you've found for spotting your guide piece change when the saw blade is angled. You have to make a practice cut a few inches into some wood using a guide and measure the distance from the guide to the kerf again. You might note the distances for 45° cuts and tape them to the saw also, since that is by far the most common miter cut. For special plywood miters, see the Cabinet project.

Safety Measures and Things to Watch For

Always start the saw with the blade away from the wood. Know where the cord is. Look for obstructions underneath. Support the wood well. Extend the blade no more than 1/8″ below the wood. Don't force the saw. Don't use a dull blade. Check the wood for nails, knots, obstructions. Check the movement of the blade guard often. Hold the saw till the blade stops. Stand to the side of the saw, never in front or behind. Never saw backward—that is, never have the back of the blade cutting into wood (backing the saw up within a cut kerf is all right). Unplug the saw for all adjustments. *Always* wear goggles. Keep your hands out of the way.

Take care of the saw. Keep sharp blades in. Keep it clean. Empty the sawdust from inside the blade guard often. Lubricate according to the instruction manual. Check the armature brushes yourself periodically or take the saw to the shop to have them checked. Everyday use would require you to check them every three months or so. Store and carry the saw carefully. Dropping it or banging it against something could bend the baseplate, making it hard to cut straight.

Always concentrate on your work with a power saw, or any power tool. Don't rush. Watch out for other people, particularly children, and for curious animals.

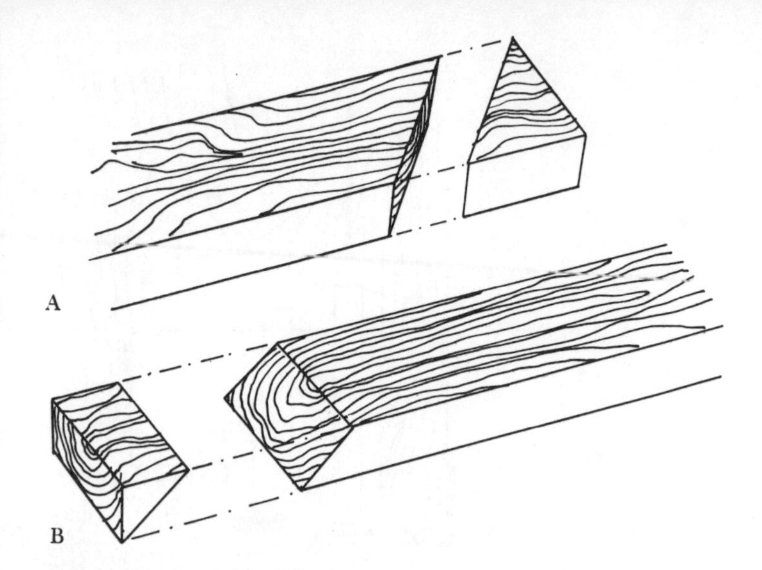

Fig. 26. Two types of miter cuts

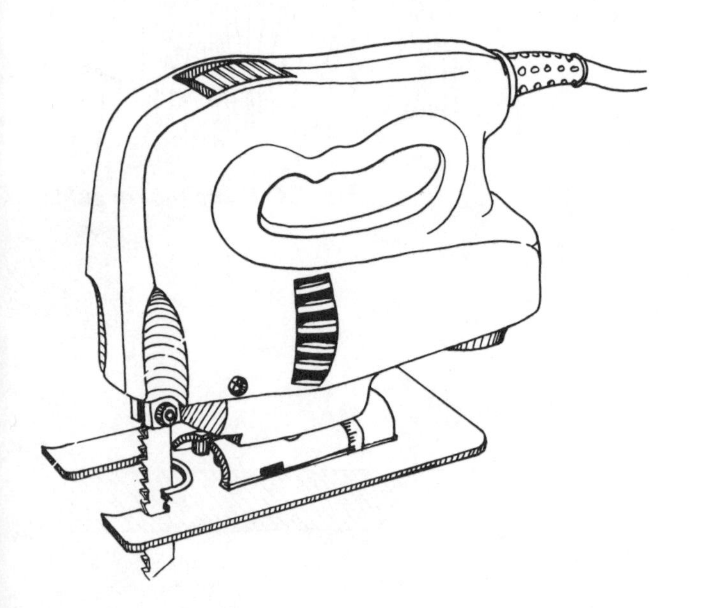

Fig. 27. Saber Saw

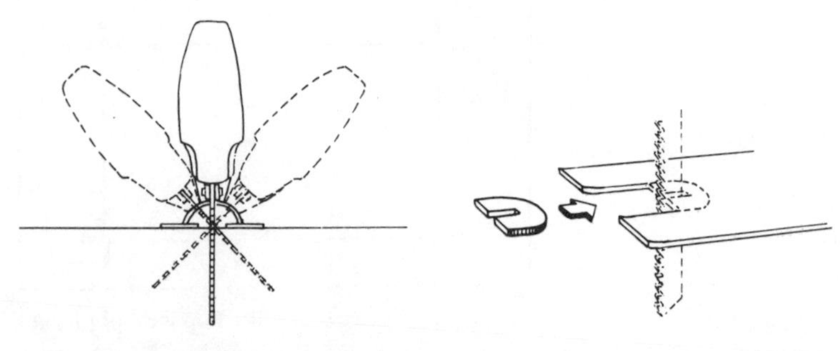

Fig. 28. Base Angle Adjustments *Fig. 29 Veneer Insert*

Saber Saws

The saber saw, portable jigsaw, or bayonet saw is designed to cut curves, intricate scrollwork, circles, and holes; it can cut out certain kinds of notches so that no chisel is needed; it can also, with the right blade, cut through plastic and metal (*27*). It has a straight, thin, narrow blade several inches long that moves in a reciprocating, or up-and-down, motion. The narrowness and flexibility of the blade are what allow this saw to cut curves.

One-speed, two-speed, and variable-speed saws are available. The speed is the number of blade strokes per minute, ranging from about 1000 to 3500 spm. If a saw has just one speed, it will probably be a high speed. The highest speeds are best for cutting wood. Theoretically the lower speeds are best for cutting metal and plastics. The blade catches better at a lower speed, and will last longer. In practice, however, high speeds will work well enough if you just push more lightly on the saw. The more speeds in a saw, the more gears and the more expense, and the more likelihood of something going wrong. Unless you plan on a lot of metal or plastic cutting, buy a single-speed model.

Most models come with a sliding switch on the handle top. This is easy to push on and off with your thumb, giving you good control as you work. Unfortunately it is also easy to flick on accidentally if you're just holding the saw. It can also get bumped into the "on" position when it goes in or out of storage, so be careful when plugging it in again. If the switch on the model you buy is not clearly marked "on" and "off," mark it yourself with tape or paint so you know at a glance. Toggle-switch models can also be located sometimes. Models with trigger switches and locking buttons are safer, but the button is much more likely to break down. It is best to get a sliding-switch model and to be careful with it.

Many saws have a swivel base that enables you to cut at an angle (*28*). The depth of the cut is nonadjustable; the manufacturer will say what the depth of cut is, with standard blades. It should be at least 2″.

Look for saws with ball bearings rather than nylon bearings. Nylon or plastic bearings just don't hold up. Check that the carbon brushes are easily accessible. Also look for one that has a sawdust blower—this is a great advantage in keeping the cutting lines visible.

An important feature to examine is the set-screw arrangement for holding the blade in the saw. The most common blade has two holes in the top of its tang. The tang fits into the blade holder of the saw and is held in by tightening the set screw through one of the holes. The cheaper models have just one set screw, which tends to loosen under heavy use. The blade will pivot backward as you saw, throwing off your accuracy and possibly grinding into the baseplate. The better saws have a second set screw at a right angle to the first, or one screw and a special blade with a right-angle jog at the end of the tang. Both designs hold the blade in position well. The set screws may have either screw heads or open hex heads for an Allen wrench. The latter last much longer.

We strongly suggest you buy a *good* saber saw when you decide you want one. Cheap ones start below twenty dollars and they don't last. They cut very slowly and the temptation is always to push them beyond their endurance. The plastic bearings wear out, and the single set screw begins to loosen. For another ten or fifteen dollars, on up to the fifty-to-sixty-dollar range, you can get a tool that will be a pleasure. Later if you start doing intricate cabinet work and detail you can look into the really expensive ones. An added feature on the finer saws is a knob at the top of the handle that turns the blade either direction. This helps in making sharp turns.

Various accessories are available for your saw. A combination rip and circle-cutting guide is very handy. Placed one way it works as a rip guide, just like a circular-saw rip guide. Turned upside down, it can be used as a circle guide. It has one or two holes at the end, through which one nail can be tacked at the center point of the desired circle. The saw pivots around that point.

The blade of the saber saw cuts on the upstroke, leaving a rough, splintered edge on the top surface. The work should be cut finished side down, as with the circular saw. Some saber saws have an insert (*29*), a small horseshoe-shaped piece of plastic or metal, that slides into the base around the blade and protects the top surface of the wood from splintering. It does the same thing that clamping a thin piece of wood along the cutting line does—it holds the grain on the work piece firmly as the blade slices through it—only this insert is much handier, and moves along with the blade. The plastic insert may come free with your saw, but buy a metal one anyway. It's better and just costs a quarter or so. Follow the instructions with the saw manual as to which blades to use with the insert. Some blades are too thick or wide.

Blades are available for cutting wood, metal, plastic, and many other materials; thin blades for fine scroll cutting (curves), thicker blades for heavy-duty use, extra-long blades for deep cutting. The more teeth per inch, the finer the cut. Extra-fine blades are used for metals. Your instruction manual or hardware store will have tables and lists of the best blade to use for your purpose.

You must, however, be sure to buy blades with the correct shape of *tang* to fit in the blade holder of your saw (*30*). There are several different designs. If you're not sure, bring an old blade or a tracing of it to the store. Or the model number of your saw. Or the whole saw.

Wood blades will last a long time if you don't force them. The most common death for a blade is to snap in two under too much pressure.

Using the Saber Saw

Skill with the saber saw is merely a matter of practice. Practice on scrap pieces before attempting new kinds of cuts.

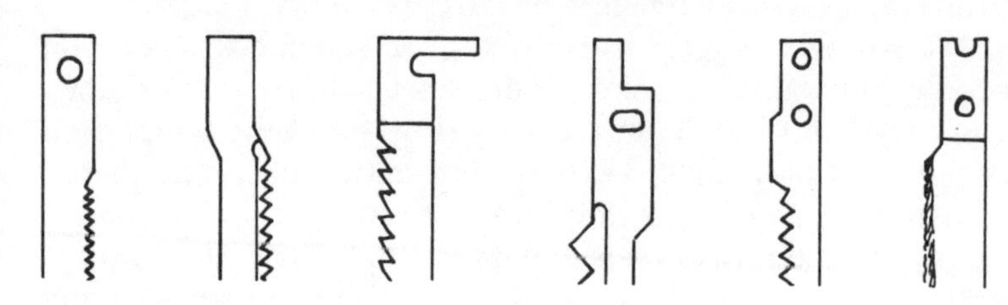

Fig. 30. Different Tangs

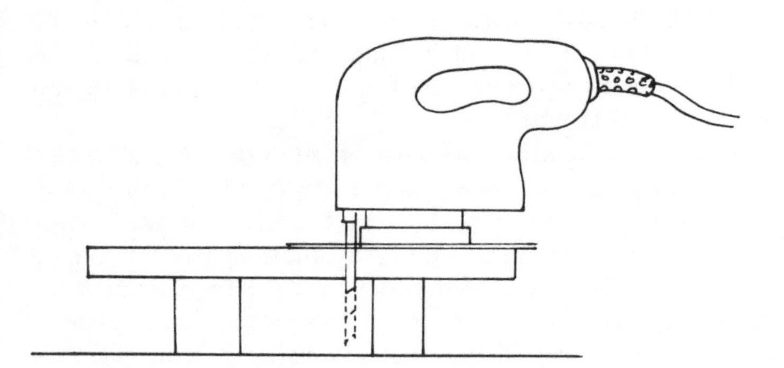

Fig. 31. Blade Clearance

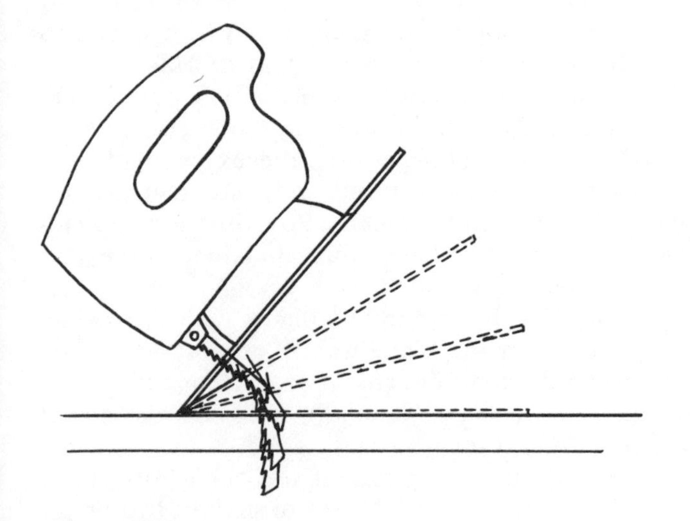

Fig. 32. Plunge Cutting

Freehand skill comes with care, patience, and concentration.

First of all, *always wear goggles*. Secondly, notice that the blade is exposed. There is no guard, so be careful. Always know where the blade is pointing. Hold the saw at top on the handle. Keep the other hand away from the blade. The saber saw is light enough for you to lift and move easily with one hand. For controlling the cut you might want to put the other hand on the top of the saw also.

Before installing the proper blade, be sure the saw is *not* plugged in. *Also, check that the switch is in the off position.* Loosen the set screw(s) and insert the blade, with the teeth facing the front of the machine. Tighten the screws.

Set up the work piece with good support, as you would for a circular-saw cut. Make sure the piece is clamped or well held down. If the cut is a very complex curve, you may have to stop midway and rearrange the supports so they don't get in the way of the blade. You must also leave adequate vertical clearance under the work for the blade (*31*). The depth of cut cannot be adjusted on a saber saw (except by using longer blades). Place the saw on the edge of the work and see how low the blade goes. You must also consider that the blade, while the machine is off, may be resting at the high point of its up-and-down motion. When the saw starts, the blade may go down another inch or so. So it would be a good idea to leave at least a 2″ clearance below the blade tip at rest. Were the blade to hit the floor or workbench surface, it could snap or force the saw to jump up.

If you are cutting into a wall, be very careful of obstructions behind it—nails, water pipes, electrical conduits. First find out what's back there, and/or saw *very* slowly, feeling with the saw for obstructions.

When the work is supported and the blade is set, check once more that the switch is off. Holding the saw in your hand, the blade pointed away from you, plug in the machine. It is a good habit to hold the saw rather than leave it lying on the table while you plug it in. You just might forget to check the switch and the saw could start up on its own.

Be sure your goggles are on. Turn the saw on before placing it on the work. It takes a moment for the saw to reach full speed, and the resistance of the wood before that moment could short the motor, ruin the tool, and make an awful noise. You could rest the edge of the saw on the work if you want, the blade away from the edge; but the tool vibrates so much you can't hold it too steady that way—and you run the risk of the tool sliding into the wood off the cutting line.

For freehand cutting, watch the blade itself along the line. Bring the tool slowly to the work line, with your eyes in a good spot to see what's going on, maybe a foot or two above and just to the side of the blade. Start the cut carefully, till you're sure you're right on the line. Hold the saw down tightly on the work. This will dampen the vibration, let the saw cut more efficiently, and give you more control over the cutting. If your saw does not have a sawdust blower, you'll find the cutting line ahead of you getting covered with sawdust. You'll have to blow it away yourself, which means backing up the saw or stopping (another vote for the better models). To stop in the middle of the cut, back up an inch or so, turn the saw off, and hold it in place till the blade stops moving. *Never* pick the saw up from inside the cut while the blade is moving. The blade could catch on the inside edge and start the saw shaking up and down in your hand. To start it again, turn on the switch, holding the saw firmly an inch or so back. When it reaches full speed, bring it forward again and continue cutting.

You can get a very accurate cut if you go slowly. Let the saw go at its own pace. The sharper the curve, the slower you should go.

If the end of the cut is the end of the wood, continue the blade out of the wood and lift it clear while you turn it off. Do not put the saw down till the blade has stopped moving. Many people even pull the plug before laying it down for any length of time. If the end of the cut is in the middle of the wood, do as described above—do not pick up the saw till it is completely off.

Sometimes the saw will stop moving forward, as if it's caught on something. It may *be* caught, particularly if you are cutting *with* the grain. Back up the saw and stop. Check the wood. It could be a splinter along the path that has hooked over the baseplate. Or it may be a nail or a particularly hard knot. Remove the obstruction and continue. If it's a knot, you can usually get through if you go very slowly. Best to avoid knots earlier when you lay out the lines. Certain grain patterns might also give you trouble, causing the blade to bend to the side at the bottom. Again, go very slowly, giving the blade a chance to cut through the patterns before getting caught in them. Sometimes it also helps to tilt the saw forward a bit, using the front of the baseplate as a lever and attacking the grain at a different angle.

On circles, particularly, the bottom of the blade will tend to bend out away from the line. Again, go slowly.

Plunge cuts

You can start a cut in the middle of a piece of wood in one of two ways. You can drill a pilot hole wide enough to slip the saw blade in, and start from there. This is fine if you can put that hole in the part of the piece that will be waste. If neither piece will be waste, however, you can make a plunge cut, much as with the circular saw. Place the saw along the cutting line, tilted forward on the front edge of its base (*32*). Be sure the blade edge is well above the wood surface, almost parallel to the work surface. Remember, when the saw starts, the blade may move an inch or so lower, so leave a lot of room. Hold the saw firmly and turn it on. Slowly tilt the saw back to the normal position. As the blade hits the wood, the saw will chatter a little. It may want to push or pull. Hold it firmly and keep tilting it very slowly. The blade will cut through the line and eventually through the thickness of the wood. Then you can right the saw and cut as usual.

Line up the initial plunge cut well into the line—that is, ahead of the beginning line. That way, if the saw starts to pull backward while plunging, you will not cut farther back than you want. After you finish the plunge and forward cut, stop the saw, pull it out, place it back in the kerf going in the other direction, and cut that part of the line you missed.

Plunge cutting seems difficult at first. Practice on some scrap pieces till you have it down. The only trick is to hold the saw firmly, control it; and to guide the blade carefully into the cutting line. To help your aim, you can set up a guide as with the circular saw. Lower the saw using the front end as the fulcrum or hinge, and the guide to keep it on the line. Find the distance between the guide and cutting line the same way as with a circular saw—clamp a guide on a scrap, cut it a few inches with the saber saw, and measure the distance between both edges of the kerf and the guide.

A guide is almost a must for perfect straight-line cutting with the saber saw. The blade is so narrow that it's extremely hard to keep it on the straight line by hand. It *wants* to cut curves.

Finishing inside saw cuts with saber saw

Use this saw for finishing cuts inside pieces that a circular saw can't make—for instance, cutting an opening in a piece of plywood. Note that the saber saw blade is thinner than the circular-saw blade, so it will leave a smaller kerf. If you finish a circular-saw cut with a saber saw, run the blade flush to the edge of the kerf on that side that is to be the finished piece. If both sides are to be finished edges, run it flush to the side that will allow easiest removal of the extra stock later. For instance, in cutting an opening, cut flush to the outside, and then trim the extra off the inner piece, which could be the door, when it's removed. If the extra were on the inside corner of the opening, you would have a much harder time.

Be careful with your cuts if your saber saw is one with the bottom of the blade set farther out front than the top. It is set this way supposedly to give a more powerful cutting action; but when you try to cut out a neat corner, you'll find that the bottom of the wood has saw cuts extending beyond the corner. The only solution is to make the last part of the cut with the saw tilted forward a bit so that the blade is approximately vertical.

Rip and circle guides

You can use the manufacturer's rip guide with your saw, but cut slowly and carefully. The guide will hold the saw and top of the blade on the line, but a slanted grain might pull the bottom of the thin blade away from the line, giving you a slanted edge. Rip with the circular saw whenever possible. Most saber saw rip guides can be turned upside down and used as a circle guide. The guide will have one or more holes in its flange end. Place as thick a nail as possible through the hole into the center of the circle.

The best way to cut a circle with the guide is to find the center of the circle and draw the circle on the wood with a compass. Or improvise with string and a pencil. If one side of the circle is to be waste, cut or plunge through that side and bring the blade right up into the line of the circle. Stop the saw. Slip the guide into the saw as it sits there. Don't tighten it yet. Place as thick a nail as possible through the hole in the guide flange and into the center point of the circle. Straighten out the saw so it is on the line. Tighten the rip guide. Now cut, slowly. Watch that the saw stays on the circle line. Slowly, so the bottom of the blade doesn't bend outward. If the saw goes off the line much, check that the nail in the center hasn't come loose.

Some guides are designed with the flange almost an inch above the wood rather than on it. With such guides, the nail may bend between the flange and the wood. Two remedies: drill a larger hole in the flange and use a thicker, stronger nail, or stick a shim between the flange and wood, and hammer the nail through it. Perhaps do both. Some manufacturers provide a little peglike device for this situation that works fine (while it lasts).

If neither side of the circle is to be waste, you have a problem. It's almost impossible to plunge-cut neatly on a curve. You can try, and later sand the pieces at that point. It won't be perfect, but it may not be noticeable. On large circles, bigger than the rip guide, slow freehand cutting can be fairly accurate. Some sanding will probably be necessary. For very small circles, the saw may not be able to make the turning radius. A hole saw or large spade bit on a drill is best for making holes, and a circle cutter on a drill is best for cutting out a circular piece.

General Tips

Always remember to keep the finished side down and/or use the antisplinter insert.

On plastic laminates and fine veneers, you may want to protect the surface from the bottom of the saw plate, particularly from an old plate that might be scratched up. Lay some masking tape on the surface right up to the cutting line.

Remove the blade before storing the saw. Store it carefully, since the plate or blade guard can be damaged. If it doesn't have its own case, store it away from other objects that might knock against it. Check the carbon brushes. Keep the saw lubricated as described in its manual. Keep the plastic sleeves that most blades are sold in. Store the blades in them; label the package if necessary. (When blades break, save the sleeves for other tools—very handy for chisels, for instance.) Keep the set screw hex key attached to the electric cord. It should come with a little rubber or plastic holder that slips onto the cord. It's always handy that way, and hard to lose. If the holder breaks, buy another one for a few cents.

Tools for Shaping

The Router

The router is one of the most versatile and enjoyable tools ever made (*33*). It lends itself to all sorts of creative endeavors. Find the right bit and you can use the router to cut grooves and notches (straight or curved), round off edges, shape all kinds of molding patterns, trim laminates flush to edges, plunge cut, etch, emboss, dovetail, rabbet, dado, and mortise.

The motor has a direct-drive shaft (meaning the shaft comes directly out from the motor—no gears) with a collet at the end, into which you place the router bit (*34*). The motor rests on a round, flat base. The bit protrudes through and below a hole in the center of the baseplate. The top or motor part is encased in a cylindrical housing that threads into the base and can be adjusted infinitesimally to get the desired bit depth. Thus you can easily rout a slot for such things as a 1/12″-thick veneer inlay. Most routers can be adjusted accurately to within 1/64″.

The most important feature of a router is its power. All that the tool really does, since it works on a direct-drive shaft, is plow a cutting bit through the work. The more powerful a motor, the easier time you're going to have—and the longer your bits will last, because you won't be tempted to push them so hard.

Other differences between routers are a matter of taste and the type of work you do. The better a router, the more accurate its depth adjustment will be. The differences will show up, however, only if you plan to do very intricate inlays. A fifty-dollar router should be powerful enough and fine enough for most work, and one for eighty or ninety dollars will be more accurate than you could tell.

Routers come with different-size collets, much like drill chucks—1/4″, 3/8″ and 1/2″. Some models are made so that the collet size can be switched, allowing you to use any bit on the same machine. As with drills, the larger collets can take larger bit shanks, providing more torque and power.

Usually the inexpensive routers have a toggle switch near the top of the motor, and near the base, two knobs with which you can hold and guide the machine. The expensive models will have either this arrangement or one knob and a large D-handle with a trigger switch inside. This latter arrangement gives better leverage and easier access to the on/off mechanism. But it's really a matter of taste and feel.

Some models have motors that can be detached to drive other tool attachments, such as planes, jointers, sanders, and grinders. You may have no plans for such accessories now, but if you're going to spend the money, you should get something that will offer you as much versatility as possible. If you do plan to use such attachments, choose a more powerful router than you might pick just for routing purposes.

Also inquire whether the router has, or can be easily

adapted to, routing attachments such as rip guides, circle guides, and dovetail and butt-hinge templates.

Installing Bit and Setting Depth

First read your instruction manual carefully. Routers differ in many details. If the adjustments on yours differ from the general description we give, follow your instruction manual, not us.

Never make any kind of adjustment on the router unless the plug is pulled. The bit, when it is in the router, is not guarded. It protrudes through the bottom plate and could be very dangerous.

To install the bit, check that the router is unplugged and the switch turned off. Loosen the key or locking device that holds the motor in the base. Unscrew the motor from the base and lay it down on its side. Check the tool's manual as to the proper method to install the bit. The most common arrangement requires the collet to be loosened or tightened by two wrenches (included with the router). One wrench holds the collet shaft from turning. Rest the wrench end on the bench for leverage. The other loosens the collet lock nut on the shaft. Once the wrench loosens it, the collet lock nut can be turned by hand till the opening is wide enough to accept the shank of the router bit. You should leave about 1/8" of the bit shank showing, or enough space to slide the thickness of one of the wrenches between the collet and the top of the cutting edge of the bit. You may have to carefully pry the bit out later with the wrench in this position. Bits are often hard to pull out by hand. Slip the nut back up by hand and then tighten well with the two wrenches.

Now place the router base on a flat surface. Mount the motor back in it and screw the motor down slowly till the bit just touches the flat surface. Tighten the motor in this position. The bit is now at ground zero, flush with the bottom of the base. Now locate the depth-adjusting ring that encircles the motor, with a numbered scale inscribed on it. You will also notice several vertical lines inscribed in the motor body above this ring. Rotate the ring till the 0 point is lined up with any one of these vertical lines. If there is no vertical, scratch one or draw one yourself at any point on the motor, up from the ring. This ring does not move the bit or the motor—or anything for that matter. It is simply a reference point. Now unlock the motor again and rotate it down some more, tilting the router so the bit can come through. The vertical line on the motor will move along the numbered scale, showing the exact depth of the bit. Stop the line at the depth you want the bit, and lock the motor in again.

Once the bit is set, lay the router down on its side for adjustments or when you are just putting it aside. The bit is unguarded, so be sure the tool is unplugged to protect yourself from accidents. You should usually check the depth with a practice cut in a scrap. This will also give you a chance to check the accuracy of the more intricately shaped bits.

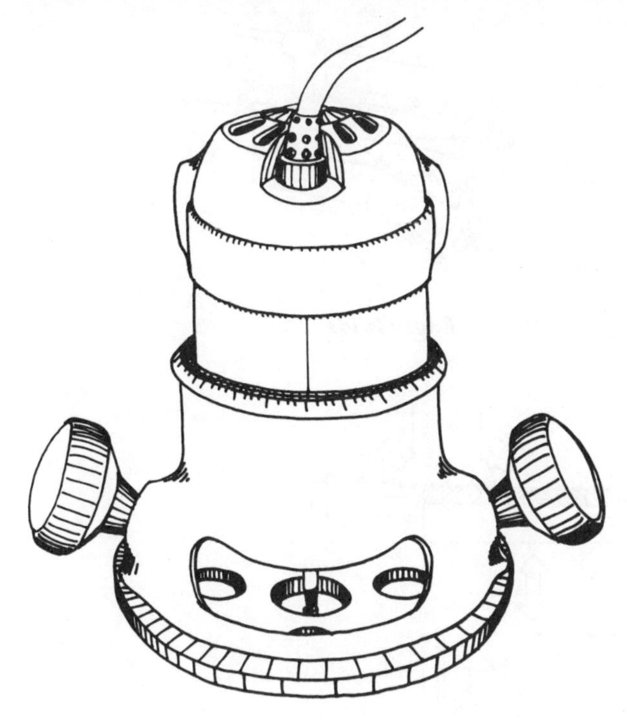

Fig. 33. Router

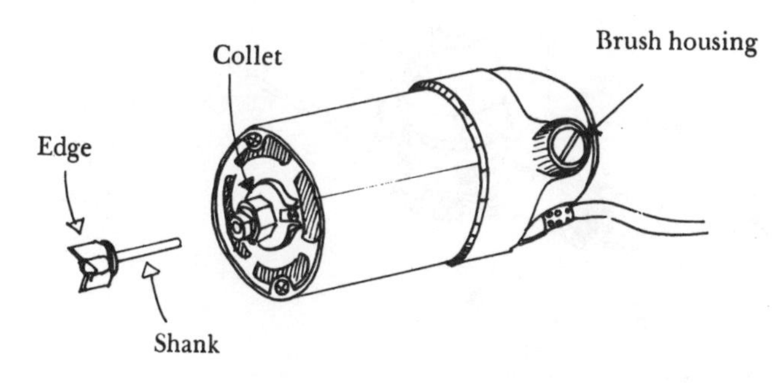

Fig. 34

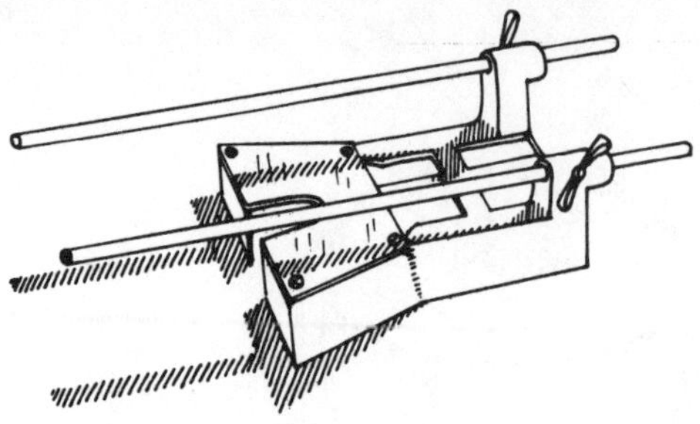

Fig. 35. *Router Edge Guide*

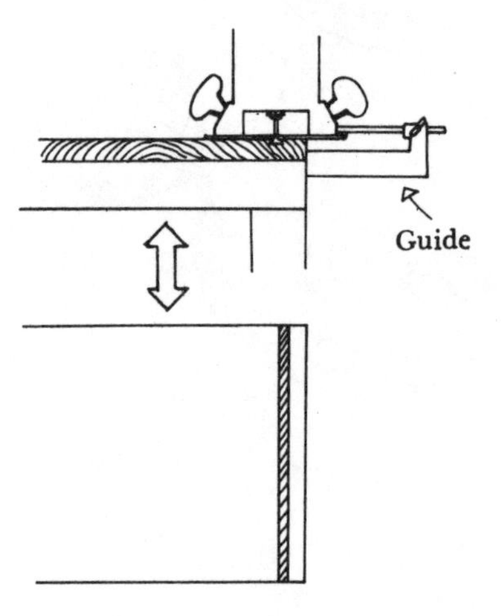

Fig. 36. *Using the edge guide*

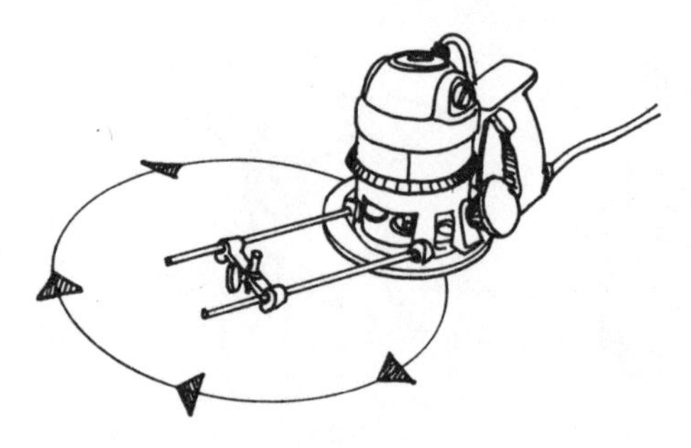

Fig. 37. *Circle Guide*

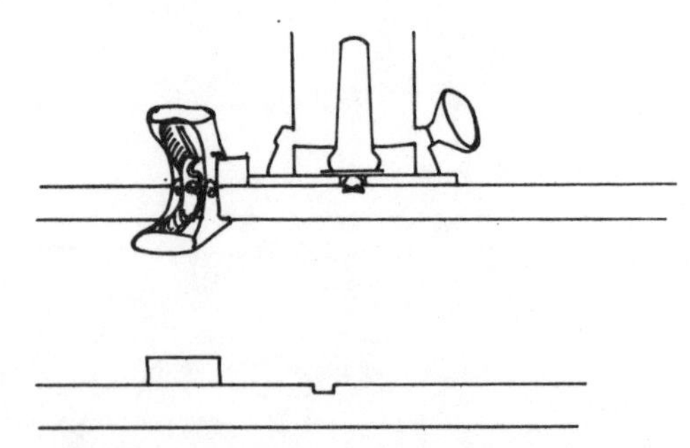

Fig. 38. *Straight rout guide set-up*

Routing with the Router

Make sure the work is well held down. Put your goggles on. Place yourself or the work so that the cut will be either away from you or from side to side. Never cut toward yourself. Hold the router firmly with both hands on the handle, rest the base on the work with the bit just a little away from the edge. Start the motor.

When it reaches full speed, bring the bit into the work and start your cut. Maintain firm control of the router, since the rotating motion of the bit tends to make it dance away on its own paths. If the cut is to stop in the middle of a piece, back the bit up from the end of the cut and turn off the motor while the router is still in the wood. When the bit stops completely, lift the router off the wood.

In cuts that go from one end of a piece to the other, it is often a good idea to stop the router an inch or so from the other end, lift the router out (after it stops) and finish the cut in the other direction. Router bits coming *out of*, as opposed to *into* a piece of wood, tend to splinter the edge.

There are four ways to make a cut or pass with your router —freehand, with a guide, with a template, or using a self-piloting bit.

Freehand Method

The *freehand* method is used for such things as cutting out your name or following some intricate design you've drawn out. It takes skill and experience to make accurate cuts freehand, but it's a lot of fun.

Guide

A guide should be used for all straight-line cuts within a piece. There are combination edge and circle guides, used the same way as those for the saber saw (*35*). Simply attach guide to the router base at the desired distance from the bit and run the guide along the edge of the wood (*36, 37*). For crosscuts, use a homemade guide of a straightedge and clamps, used just as with the power saw (*38*). To find the proper distance between guide and bit: Clamp guide to a scrap and make a practice cut, running the router base against the straightedge of the guide. Note that since the base is round, and the bit is round, the whole router can rotate in your hands as you move it and still stay on line, as long as it rotates tight against the guide. Measure the distances from the guide to both the inside and the outside of the rout. Note them down with the size of the bit. Remember to be careful not to use the inside measurement for an outside cut, or vice versa. Remember also that a different-size bit will change the distance from guide to bit.

Template

A template is a guide for a definite pattern or enclosed shape. It is a negative cutout for the positive shape you want to make, like a cookie cutter. The router travels around the template, cutting out the exact same shape on the wood below. Templates can be used for dovetail joints, or butt hinge mortises, for instance. Manufactured templates are expensive.

To use such templates, you must first insert a special *template guide* (they are inexpensive) into the center hole of the base plate (*39*.) A flange protrudes slightly below the guide and surrounds the bit shank. This flange is what moves along against the template and guides the router. The router base itself sits on top of the template. You can make a homemade template for your own patterns with some time and care. Use the special guide and find the distance from the outside of the guide flange and the outside edge of the bit you will use. The template you make will have to be that much bigger than the actual pattern. Take a piece of plywood or hardboard at least as thick as the depth of the flange, and cut the patterns out with a jigsaw. This is a lot of trouble, but you may find it easier to cut out the exact pattern with the jigsaw than trying to cut the pattern freehand with the router. At least your mistakes will be on scrap, rather than on the finished piece.

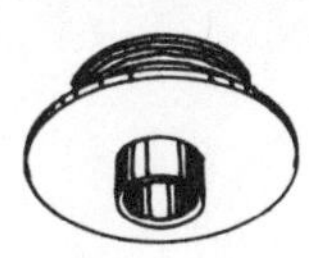

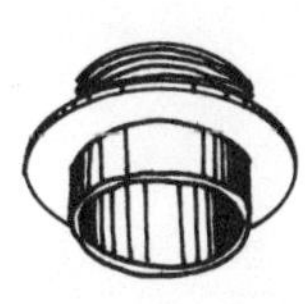

Fig. 39. Template guides for routers

Self-piloting bit

The fourth method of cutting is to use a self-piloting bit, such as the quarter-round bit in Figure 41. This can be done only on cuts along an edge. The bit has a part that extends *below* the cutting edge, and runs along the edge of the work piece. It thus guides the cutting part in a path parallel to the edge. Laminate-trimmer bits are self-piloted, for instance. They work equally well on straight or curved edges. A *threaded arbor* will take a variety of cutting heads and pilots.

When using a guide or template, there is always a preferred and most efficient direction in which to move the router. The reason has to do with the rotary motion of the bit. Looking down from the top, the motor and bit rotate clockwise. As the bit moves forward in the work, the front half of the bit receives a resistance from the wood in the opposite or counterclockwise direction. The back half of the bit receives *no* resistance because that part of the work piece has already been cut away. Thus the router will tend to move away from that resistance. The idea is to set up the guide or template so it will counteract that tendency (*40*). If you set up the work to cut away from your body, the router will want to go to the left. Set the guide on the left. If you set up the work so you'll be routing from right to left, the router will want to move toward you. Set the guide up between you and the router. Routing from left to right will require the router to be between you and the guide.

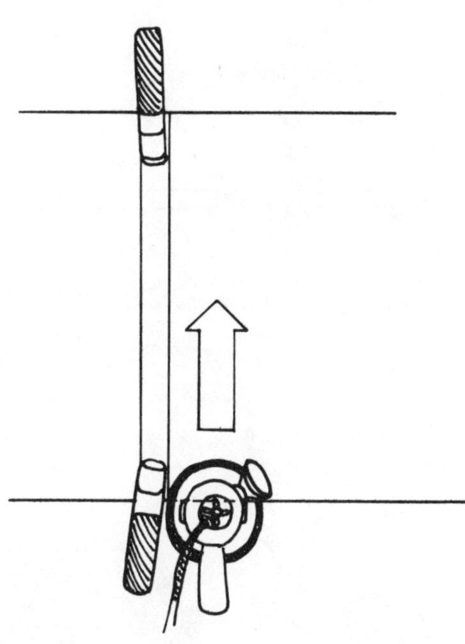

Fig. 40. Routing Direction

Take these considerations into account when you decide where to put the guide. Sometimes it may be impossible to set the guide the proper way because of space limitations. You *can* make your passes on the wrong side of the guide, but rout very carefully and slowly and keep an even firmer grip on the tool. In fact, when you make a cut from one end to

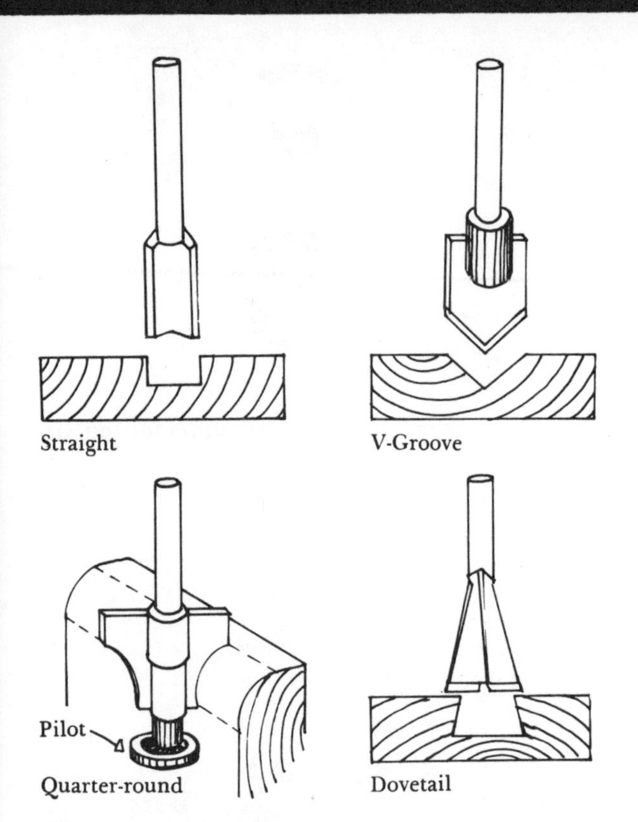

Fig. 41. Router Bits

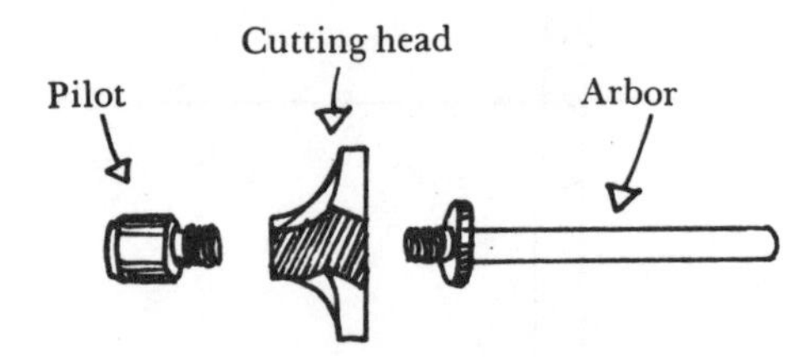

Fig. 42. Threaded Arbor Router Bit

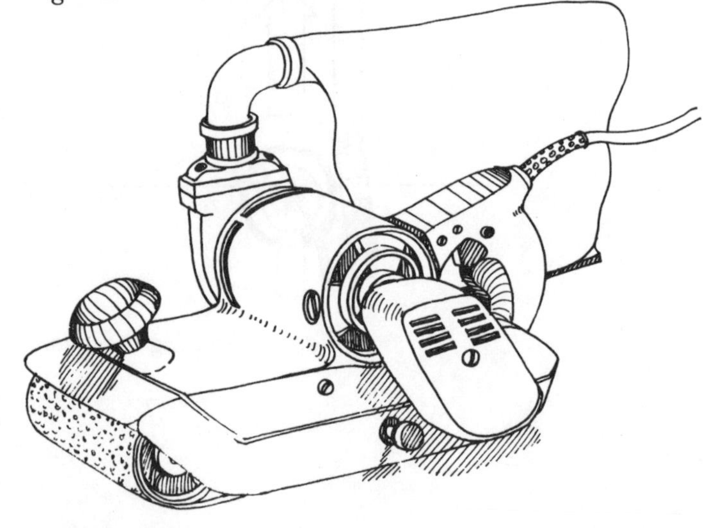

Fig. 43. Belt sander

another with a guide, you should stop before the other end, remove the router, and come back the other way—"the wrong way"—to prevent splintering of the edge, as we mentioned before. There's no way around this. Just leave as little routing as possible at the other end and work carefully. All we can do is maximize our opportunities for perfection.

The same principles of resistance apply to templates. When working on the outside of a template, move the router in a counterclockwise direction to counteract the wood's resistance. When working on the *inside* of a template, run the router in a clockwise direction (unless the instructions with the template specifically give other directions). Similarly, work counterclockwise around the edges of a piece with an edge guide or a self-piloting bit.

The router is shaped so you can't make a pass flush to a vertical surface. The base will meet the surface before the bit does. The best way to solve this problem is to avoid it. Plan ahead. Rout that piece before it's attached to the vertical piece. Always watch out for blind alleys when you plan your projects. If it can't be avoided, or if you're working on something already built, rout as far as you can and finish the rout by hand with a chisel.

Router Bits

Here are drawings of some common router bits (*41*). Manufacturers and hardware stores will have foldouts with pictures of many more and the shapes they create. Your instruction manual may also have a complete listing. If you are still unsure of what kind of bit to use to get a certain effect or configuration, ask your hardware store or even service center. Ask someone who knows, who's not just guessing.

Here are a few things you should know. Bits will be made of either steel or carbide. Steel bits are fine for cuts into soft woods like pine and fir (pine being a common wood for 1-bys and thus bookcases). They are also easy to sharpen yourself on a small sharpening stone. Carbide bits are about three times as expensive and will last much longer. They cut better and cleaner, and are better for harder woods and particularly plywood. Invest in a carbide bit if you plan to use it for more than one big job—it will be well worth it. For instance, the ¾″ straight bit is used to rout dadoes for ¾″ plywood and 1-by lumber to fit in bookshelves and cabinets, and much furniture is made with this lumber. A carbide bit of this type would be a good investment.

Most bits come either as one-piece units, shank and cutting bit molded together, or else as just the cutting bit with a threaded hole through the center that accepts a separate shank or arbor (*42*). This latter type usually can have three or even four cutting edges rather than two because of its design, and is thus stronger and faster. The separate arbor can be used for different bits. Buy the bit or shank with a diameter that fits your collet. Some bits come in odd sizes and are meant to be used with an adapter. Adapters can also be used to adapt a bit to a different-size collet.

You can tell that a bit is dulling if it burns the wood, creates a lot of smoke, leaves a lot of stringy fibers on the edges of the rout, or simply cuts too slowly. Take into consideration, however, that you may be cutting deeper routs than before, which would slow the tool down; or that the piece of wood might be a harder wood than you're used to.

Don't force a bit to cut out too much at once. With a 3/4″ straight bit, for instance, don't cut out more than 1/4″ depth in one pass. If you need a deeper rout, make one pass at 1/4″, then lower the depth of the bit and make a second pass.

New bits usually come in plastic cases, with a special form-fitting piece of plastic over the cutting edge. Save both plastics and store the bit in them. Never store a router with the bit still mounted in the collet.

Keep your router clean of sawdust and dirt.

Again—pull the plug when making any kind of adjustment.

Be careful not to touch a just-used bit with your bare fingers—it will probably be very hot.

Get a listing or diagram of the manufacturer's different bits—it will give you ideas.

Planer

The planer is an electric plane. It's not a tool that you will need too often, particularly if you own a belt sander. Its main uses would be to plane the face of very rough lumber, such as wood salvaged from the streets, or to plane down doors. If you don't have much of this type of work, the hand plane or belt sander will suffice.

If you do want a planer, however, and you also own or want to get a router, consider getting a planer attachment for the router motor. This happens to be one of those "transforming" attachments that actually works well. The result is a good planer and a good router, but with the expense of only one motor between the two.

Tools for Smoothing

Sanding wood by hand is a long, tedious way to spend the day. Power sanders can sand, plane, or smooth your work in very little time.

The *belt sander* is the work horse, able to remove a lot of stock quickly, but also able to do relatively fine sanding. The finishing sander, either the orbital or reciprocal type, is a bit impractical for heavy stock removal, but it can finish wood to a glasslike surface. If you plan to lay a new floor, the large and powerful floor sander is indispensable and easy to rent.

There's also the disc sander (including the disc attachment for your drill), a tool we don't particularly care for. It's too easy for someone who isn't being extremely careful to leave gouges and swirl marks in the wood.

In general, we suggest you begin sanding your projects by hand, with a sanding block. Buy a cheap finishing sander when you get tired of this, or when you start to have a lot to sand. Buy a belt sander only when you have to, when the only other way to do the job requires just too much time and energy. Get a good model. Use your finishing sander till it dies and then decide whether it's worth it to you to get another or switch to an expensive model.

Belt Sander

The belt sander (*43*) is used mainly for heavy and rough sanding. It can take off a lot of stock, but it can also be used for finishing solid wood with fine sandpaper. It requires great caution with fine veneers—you might sand right through them. It also takes off paint and other finishes. It's *not* meant for sanding floors.

The tool looks sort of like a locomotive or tank. The sandpaper is a continuous belt looped and stretched around two roller wheels. The motor turns the wheels and thus the paper at a very high speed. The bottom of the loop is held flat against the work by a thin metal plate backed with rubber or foam. Most sanders are started by a trigger switch, which usually has a locking button. The tool runs only at top speed, there being no reason ever to sand slower. Finer, more careful work is accomplished by switching to a finer-grade belt.

Many sanders have a chute that ejects the sawdust and a bag that collects it. This feature adds about ten dollars to the cost of a machine, and will not collect *all* the sawdust; but we think it's well worth it in terms of breathing ease, cleaning up, and keeping the work clean and clear as you sand. If you buy this model, check it in the store. Be sure the fan works when the motor goes on, and that the bag blows up like a balloon.

The size of a belt sander refers to the size belt it takes. You must always use that size belt. The sizes range from 2″x21″ to 4½″x26″, the first number being the belt width and the second number being the length of the loop. The standard size for most home sanders is 3″x21″.

An important design feature to look for is whether the sander can sand flush on one or more sides. This is necessary for sanding inside a bookshelf, for instance, close to the back piece, or any kind of inside corner.

A sander should also have a built-in key that adjusts the sideward motion of the belt on the roller. A new belt may slowly slide off the rollers as it runs or, worse, into the inside of the machine, cutting into the housing. The key adjusts the tension of the belt between the wheels and keeps the belt centered. You might also check this feature in the store when you buy your new machine. If the belt cannot be adjusted to stay in the center, the rollers may be imperfect or incorrectly mounted.

The belt sander is designed for heavy-duty work. You will also tend to work it steadily over long periods of time. Therefore it is one of the tools that you should spend a little more

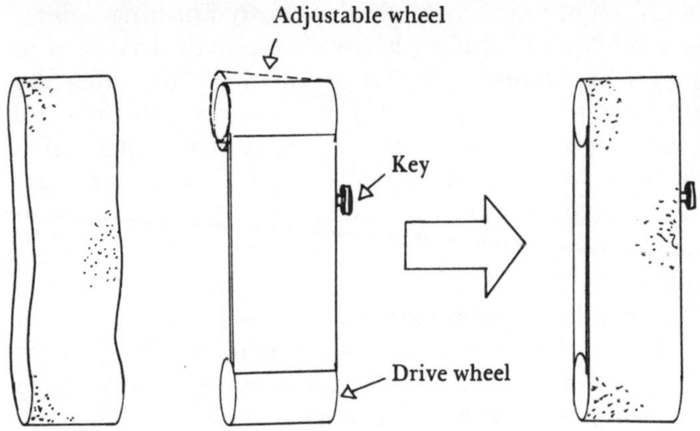

Adjustable wheel

Key

Drive wheel

Fig. 44. Mounting the belt

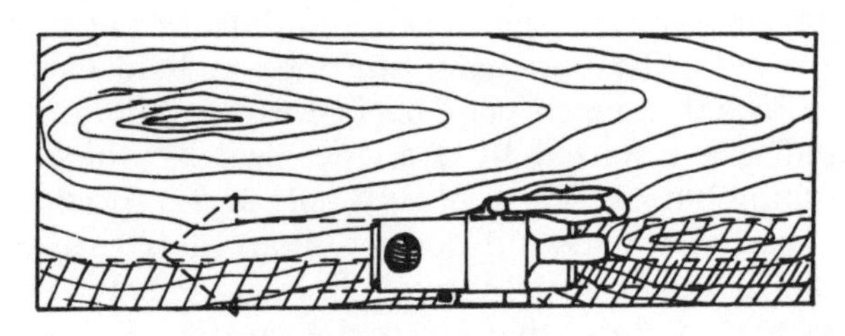

Fig. 45. Sanding Direction

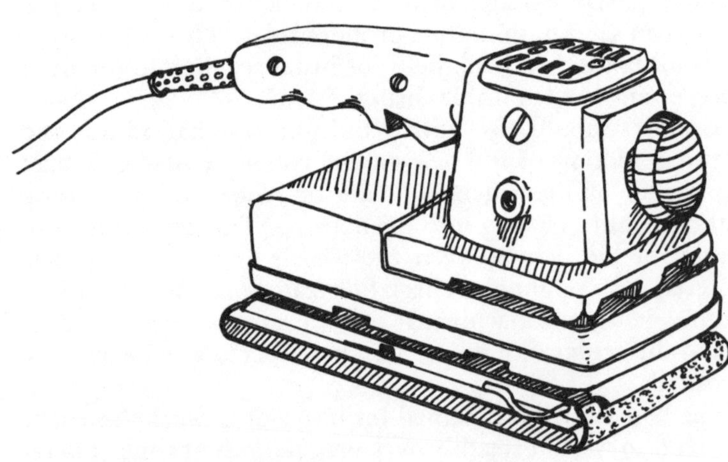

Fig. 46. Finishing Sander

money on. Cheap ones won't last long enough to make you happy. Better ones will also shorten the time it takes to complete a basically tedious step in your project. You will probably have to get a model that lists for at least fifty dollars.

Using the Belt Sander

Always wear goggles. (You may also want to wear a dust mask.) Never change belts or fiddle with the machine unless the plug is pulled.

First thing you have to do is load the belt on the machine. Different sanders use different systems. Follow the specific instructions with your machine (*44*).

You must put the paper on in a certain direction. There should be an arrow on the sander to indicate the direction of rotation of the rollers. Looking from the left side of the sander, the rollers rotate counterclockwise. The top half of the belt is going forward and the bottom half backward. You will see an arrow on the inside, or nongrit surface, of the belt. The belt should be put on so that this arrow corresponds to the arrow on the machine. If the arrow is on the top of the roll, it should be pointed forward and to the back if on the bottom. This is very important for the belt's life. The arrow is oriented by the direction of the seam's overlap; running the belt in the wrong direction will split the seam.

Before sanding the work, you have to center the belt. Put on your goggles. Plug in the machine. Turn it upside down, leaning the tool on its body, so the belt is not touching anything. With one hand on the adjusting key, press the trigger and watch the belt on the rollers. Adjust the key so the belt goes to the center and stays there. Note which way the key turns to make the belt move in each direction. Now you're ready to sand.

Make sure the work is firmly held down. If it's a large piece like a cabinet, its own weight should be enough. Always start the machine off the work. Let it reach full speed. Then lay it on the work and sand back and forth in the direction of the grain (*45*). It will tend to move forward. Don't press down if you're on a horizontal surface; let the weight of the machine itself determine the pressure. All you do is control its movement and direction. On vertical surfaces, you have to support the machine and also supply the pressure. With experience you'll gain the feel to give as much pressure to the sander as its own weight would give it on a horizontal surface.

Hold the sander flat on the work. Tilting it gouges the wood. Sand in a straight line all the way back and forth, not in a circular motion. To cover a wide area, move the machine to the side about half the width of the belt at the end of each back-and-forth motion. Be careful at the edges to keep the sander flat, or it will round them off. The idea is to sand the whole surface gradually, rather than one part a lot and then the next. Hills and valleys are difficult to smooth out once you make them.

To stop, lift the machine off the work while it is still running. Release the trigger. Hold the machine in the air until it stops completely, then put it down off the piece. If you put it down while it is still running, it will take off like a tank.

As you work, check now and then that the belt remains centered on the rollers—particularly when you start and the belt is adjusting to the work.

If you have a lot of stock to remove, run the machine diagonally to the grain. This removes stock faster, but leaves a very rough surface. In such cases, another tool might be a better choice. A plane, or chisel, even a saw could remove the bulk of the stock, leaving the final smoothing to the sander.

Keep the sander away from your body and clothing. Stand to the side and behind the tool. If you use the locking button, keep your finger on the trigger so you can release it quickly. Always hold it firmly, with both hands. Skill and a feel for the belt sander come with practice.

Sanding Belts

Know what size belts your sander takes and buy *only* those. Jot down the size, e.g. 3x21, on a piece of paper and take it to the hardware store if you want. If you don't know the right size, jot down the make and model number of your machine.

The sandpaper is graded the same as sandpaper for hand sanding. The heaviest grades are used to take off a lot of stock, such as sanding off overhanging edges, trimming doors, cleaning old dirty wood, and removing old finishes. Open-coat belts are for soft woods and removing finishes; closed-coat for hard or soft woods. As with hand sanding, work from rough to fine in gradual steps. The exact order will vary with the kind of wood and type of finish you want. In general never skip more than one or, at most, two grades within a numbering system. The best method is to take a scrap of the wood you are using and test it with different grades.

The belt sander can give just about as fine a finish as you could want. To get that glasslike smoothness that hardwoods can have, you will need to switch to an orbital sander. Be very careful using the belt sander on thin veneers, however. Rough and medium grades can sand right through the veneer. Most fine veneer plywoods need only light sanding with very fine papers. Fine veneers should really be finished with orbital sanders.

Belts do wear out. If a belt *seems* to be sanding much less efficiently than it was an hour ago, it probably *is*. Tracks or smears start to appear on worn belts. Belts will clog especially fast on green woods.

Another sign of a worn belt is that it won't stay centered on the rollers. If this happens quickly to *all* your belts, however, have the rollers checked by your service center. They might be unevenly worn.

Orbital and Reciprocal Finishing Sander

Finishing sanders are designed for light sanding jobs, for giving superfine finishes to wood and veneers (*46*). They will not take off much stock, even with the roughest grade papers. A sheet of sandpaper fits over a base of thick felt padding on a metal plate. It is held on by a clamp or roller attachment above the plate. On the orbital sander the base vibrates in an orbital or circular motion. On the reciprocal or vibrating sander, the base vibrates back and forth. Some finishing sanders can do both. In any case you have much more control than over a belt sander. The motion is much gentler and the tool itself is much lighter.

Because this tool requires a less powerful motor and is used for much finer work, a good finishing sander will be much cheaper than a good belt sander. You can pay less than twenty dollars and get a tool that will work well. The extra money you pay will show in the speed at which it sands, its longer lifespan, and a more even finish.

Use aluminum-oxide sandpaper for finishing sanders, or garnet or silicon-carbide, but not flint. It is much cheaper to buy larger sheets and cut them to size; precut pieces are exorbitantly priced. You can cut them with a straightedge and a mat knife along the nongrit side. Many sanders are sized to get exactly three sheets out of a standard 9″x11″ sheet—3⅔″x9″. This is the same sandpaper used for hand sanding, so the grades will be the same. As mentioned above, if you have a belt sander, it is better to use for the rough work through to the fine, then use the finishing sander for the extra fine work. If you have no belt sander, rough paper on the finishing sander can be used for work that is not too messed up. It will take a while.

Always pull the plug before switching papers. Always check that the switch is off before plugging it. The switch is usually a slide switch, similar to that on the saber saw, and equally prone to turning itself on when you're not looking. Oil your sander often, as directed in your manual. Wear goggles and, if you like, a dust mask.

Using a Finishing Sander

Start the tool off the work. At full speed, place it on the work. The vibrating type should be moved along the grain. The orbital kind can be moved in any direction. As with the belt sander, don't sand in one place too long. Hold it steady and flat; let its weight determine the pressure. Most finishing sanders can sand flush to a vertical surface on one or two sides. Remove the sander from the work before turning it off, and let it stop before putting it down.

KNOW-HOW

7/WALLS AND CEILINGS

What's in There and How to Fasten Things to It

A continuing source of mystery to many people is just how to attach objects to a wall or ceiling. "What kind of fastener do I use?" "Are you sure there are studs back there?" "How do I attach to brick?" "Will one bolt be enough?" The inner structure of a wall or ceiling is the solid part, the part to which you should fasten. There's no reason why these wall innards should remain a mystery to you. There are, after all, only a few basic possibilities, and it's very easy to find out which kind you have.

In this first section on walls, we will initially describe the four basic types of inner wall structures; then the different surfaces covering these structures, how they are attached, and how to discover which inner structure is underneath. We'll then describe the fasteners to use for each structure, under different load-bearing and wall-covering conditions.

Walls

Inner Structures

There are four basic types of structures underneath the wall surface that meets your eye.

Wood stud frame

This consists of vertical wood studs, usually 2x4s, spaced either 16″ or 24″ on center—that is, the *centers* of adjacent studs are either 16″ or 24″ apart; the spaces between the studs will be slightly less. The spacing is unfortunately not always consistent, but you can use those numbers to start with. See Walls project for illustrations and more detail.

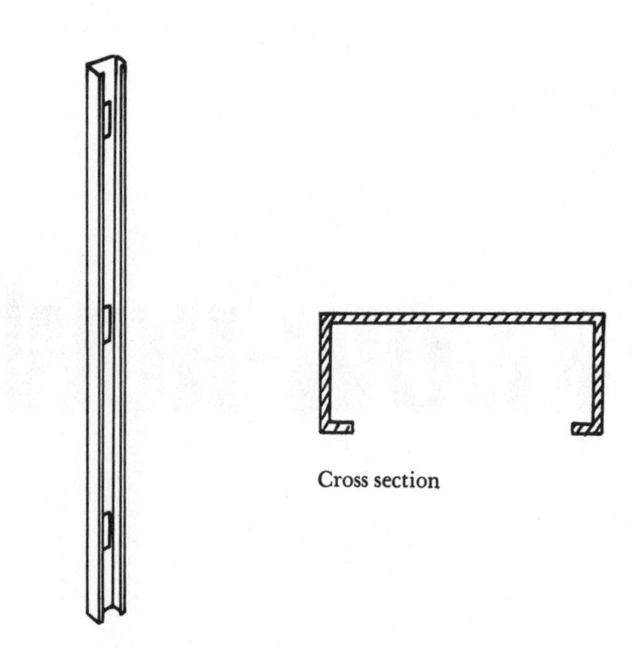

Fig. 1. Metal Stud

Metal studs

Modern buildings often use these for interior partitions instead of wood, as a fire-preventive measure. The studs are hollow, and are not meant to have heavy loads hanging from them. They are also spaced either 16″ or 24″ on center.

Masonry

A masonry wall may be made of common brick, cinder block, plaster block, or stone. They are somewhat difficult to fasten into, but hold well once the fastener is in.

Concrete slab

Often found in modern high-rise buildings, this is a poured aggregate concrete, reinforced with steel rods. It is extremely hard, practically impervious to normal drilling. If this is your basic wall structure, you're going to have a lot of problems fastening into it.

Once you identify the structure of one of your walls, don't sit back and expect all your walls to be the same. You may, for instance, have exterior brick walls and interior partitions of wood studs.

Wall Types

Following are the five common wall surfaces you'll find attached to these inner structures. This list should cover at least 99 percent of the houses and apartments in this country. If your wall is covered with wallpaper, by the way, there will still be one of these surfaces under the paper.

The *furring strips* mentioned below are lengths of wood, usually 1x2s, attached to a rough surface like masonry to give a level nailing surface for the covering pancls. They may be attached horizontally or vertically, on 16″ or 24″ centers. They are sometimes used just to create an air pocket between the structure and covering, to insulate the covering from moisture.

Wood lath are strips of 1/4x1 rough-cut wood used as a base for plaster. They're usually applied horizontally, with gaps of about 1/4″ between strips.

Metal lath is a heavy screenlike material used as a base for plaster.

Plaster

Usually about 3/4″ to 2″ thick, applied on top of metal or wood lath. The lath is attached in turn to either wood studs, directly to masonry, or to furring strips attached to the masonry. Or you may find a thin coat of finish plaster directly on concrete-slab or masonry construction.

Plasterboard

Anywhere from 3/8″ to 1 1/4″ thick, on wood or metal studs, or on furring strips attached to bricks, blocks, or studs.

Wood

Often on studs, or covering an old wall. The wood could be hardwood panels, solid wood sheets, or planks, probably tongue-and-groove.

Paneling

May be applied either directly over an old wall, or on furring strips attached horizontally to studs or vertically on masonry. Sometimes attached directly to studs, particularly in attic and basement renovations. Usually a veneer or plastic fake wood surface on a hardboard base, usually 1/4″ thick.

Masonry

If the surface is masonry, the structure is the same.

Finding Out What's Behind Your Wall

You want to find out what type of wall structure you have and, if there are studs, to locate them.

If you suspect your walls have wood studs, there are two simple approaches that may work. First try thumping the wall with your hand or gently with a hammer. Where it does not sound hollow, you're at least close to a stud. On plasterboard walls, there are usually slight bulges along the studs.

Hold a light against the wall and look for bulges, seams, or nail heads.

Otherwise, one approach is to pry off a piece of the base molding. Plaster walls are sometimes unfinished at the bottom, making it easy to actually *see* what's back there once the molding is gone (*2*). If you can't see, try drilling through the wall surface with a thin, extra-long bit to see if you can hit something solid. Use an old wood bit, since it will be dulled by the plaster; or use a masonry bit through the plaster and switch to a wood bit when you're through the surface. Drill in an inconspicuous spot—for instance, in the space behind the molding piece or along the area where you want to fasten the object. Start anywhere, drill holes in a horizontal row about an inch apart until you hit something solid.

Fig. 2. Removing base molding

You should be able to tell by the feel of the drill, by how easily it goes through a layer, what that layer is. A wood bit will go easily through plaster, slowly through wood, and hardly at all through masonry. Also, the dust on the tip of the bit tells a story. White means plaster; light yellowish sawdust is, of course, wood; fine red dust means brick. No dust and total resistance means concrete slab. If you hit wood, drill through to see how thick it is. It may be just lath, about ¼″ thick and of little structural use. You can tell the thickness of layers, and where they begin and end, by measuring how much of the bit is in the wall at each point. Then you'll know how long your fasteners have to be.

The one thing to be very careful about is plumbing or electrical lines hidden in the wall. Try to trace them first, if you can; or check your house plans, if there are any. Drill in an area that seems unlikely to have such lines. Drill slowly; stop if you hit something that feels like hard metal. If, however, you find thin metal about an inch and a half wide, running vertically, you've found a metal stud.

Another way to see what's behind the wall surface is to remove the plate covering a switch or outlet. This also helps in determining which direction the power lines run.

If all else fails, chop away a small piece of wall from behind the molding for a good look, and fill it later.

Fastening Loads into Wall Structures

Ideally, you always want to fasten into the solid supporting structure of the wall, not just the wall covering. The fasteners we suggest should work for the corresponding structure no matter what the outer wall surface is.

First, some arbitrary definitions:

Light load—less than five pounds. A small painting.
Medium load—five to fifteen pounds. A mirror.
Heavy load—over fifteen pounds. A bookcase.
Heavy dynamic load—A loft bed.

Proper fasteners for each structure

When the wall has wood studs, use nails or screws for the light to medium loads, and screws or lag bolts for the heavier loads. Try to fasten into the *center* of the stud thickness. Nails, the weakest of the fasteners, should be driven at a downward angle, so that any compression stress will tend to tighten them into the wall.

Metal stud walls are not meant to support heavy loads. Do not try to hang a loft bed from such a wall, for instance. You can support heavy static loads like cabinets if you can get several thick molly bolts (at least 3⁄16″ thick) through a couple of the studs. Use toggles, mollies, or sheet-metal screws in the studs for light and medium loads. Try to design any heavy load to be supported mainly by the floor.

For light loads on masonry walls, you can sometimes get away with cut nails or special masonry nails. For heavier loads, use lag bolts or screws with the corresponding shields (*3*). Always try to get at least 3⁄4″ of the shield into the solid masonry. For instance, if you have 3⁄4″ of plaster on brick, and a 1 3⁄4″ shield, you can sink the shield flush with the plaster surface. With 1 1⁄4″ of plaster, you would have to sink the *front* of the shield into the plaster at least 1⁄2″.

The concrete-slab wall is a big problem. Most likely, you will not be able to drill into it to set a lag shield. If you can, it may take hours. And then you might run into an impenetrable reinforcing rod. The best method is to hire someone or rent a power drive tool like the Ramset. (See "Hand Tools".)

Be sure to get full instructions on how to use the Ramset; tell the people the type of wall and fastening you want to do. Wear ear plugs and goggles when you use it. First trace your plumbing and electrical lines so you won't hit them. Your superintendent may know their locations.

The only other alternative is to design another type of construction that eliminates the need for hanging your project. For instance, you can wedge 2x4s in place between floor and ceiling with "timber toppers" (see under Joints and Joining in "Working"); then attach the object to them. Timber toppers are not good for heavy loads, however.

On any hollow wall—a wall with a cavity behind the outer surface—you can fasten light and medium loads right to the wall surface with molly or toggle bolts. This is true no matter what the inner structure is. Examples would be plasterboard walls on wood or metal studs; also a plaster on lath on furring strip on brick wall.

In general, nails are not good for hanging heavy loads, particularly dynamic loads, because they bend, and tend to work loose. Toggles and mollies are also inadequate for heavy loads; they are designed to get the maximum holding strength out of the plaster, but the plaster has limited strength. Mollies are the stronger of the two. We've used several thick mollies for a medium-heavy load, but only when there was no choice. Screws and bolts are strongest.

Number of fasteners

A light load usually needs only one or two fasteners. With medium and particularly with heavy loads, you have to take

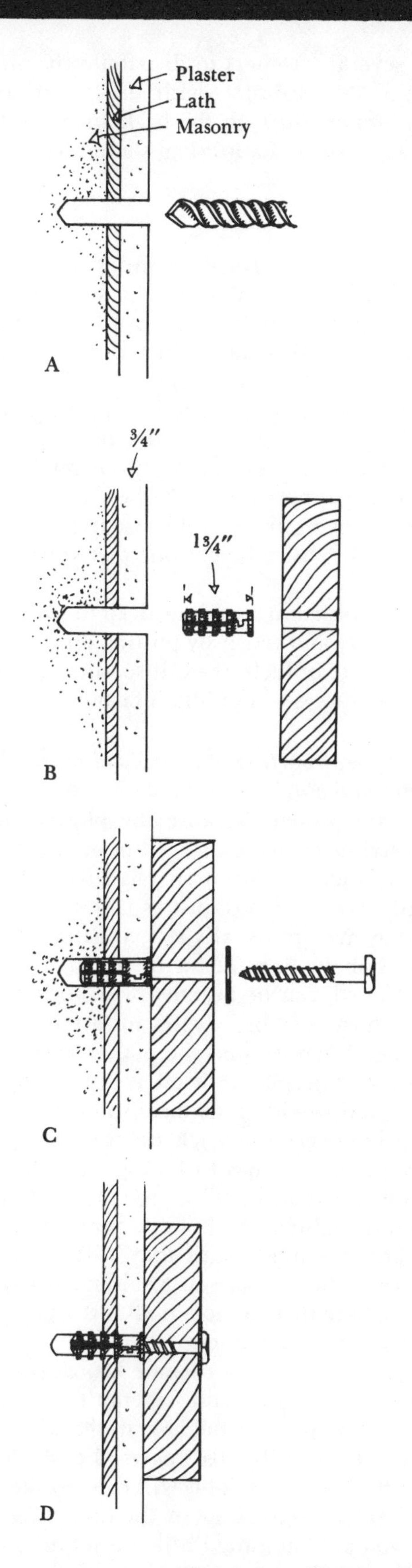

Fig. 3. Installing lag shield and bolt

care to space several fasteners to distribute the stress evenly. For instance, a 3′x3′ cabinet, with back securely fastened, should have fasteners through the back a few inches in from *each* corner. The beam of a loft bed should have a bolt every 16″ or so.

Size of fasteners

The heavier the load, of course, the stronger the fastener should be. Dynamic loads require the strongest joints and fasteners of all. (See "Structure.") For the loft bed, lag bolts 5⁄16″ in diameter would be fine. Used with lead shields, they should reach at least 3⁄4″ into masonry; without shields they need to run at least 1″ directly into wood studs. Unless you're building a whole house, there are few occasions that would require a larger bolt or screw. Most objects smaller than a loft can be adequately supported by 1⁄4″ or even 3⁄16″ lags.

If you're ever uncertain as to the strength of a certain size bolt or screw, simply use a bigger one that you *are* sure of. It can't hurt.

If you're ever uncertain as to how deep to sink the fastener, put one in part way and test it by pulling on it, even hanging on it. See if it's loose or feels weak. If so, sink it farther, or get a longer one that allows you to sink it farther.

Special problem—supporting a heavy load when there's a gap between plaster and a brick structure

Gaps often are present because the plaster lath may be attached to furring strips, either 1-bys or 2-bys, which are attached to the brick. If the furring is 2-bys, you can fasten into them with screws or lags, just as if they were studs. Unfortunately, they are not usually spaced with any consistency and you'll really have to hunt for them. Light and medium loads, as mentioned, can be fastened with toggles or mollies.

If you can't find any thick wood to fasten into, you will have to get a lag shield beyond the plaster and gap, and sink it into the brick. The plaster itself is not strong enough to support a bolt and shield. This means that the lag bolt will be unsupported between the brick and the outer wall surface. (The hole in the plaster has to be big enough to allow the shield to pass through, so it will be too big to support the lag bolt.) Therefore tighten the bolts as much as possible—the tighter they are, the more rigid they will be. Use stronger bolts, and more of them, than you would normally use for the situation; do *not* use this method at all if the gap between the brick and the outer surface is more than 2″.

Drill your hole first. Use a carbide bit, of course. Put the shield on the end of the bolt several turns and stick it through the plaster end into the hole in the brick. Pound the bolt a couple of times lightly to set the shield and then tighten the bolt a few turns. This will expand the shield some so it will stay in the wall. Loosen the bolt, take it out, and put up what you are hanging. Put a washer, of course, under the bolt head, and the bolt through the hole in the object. When the bolt goes back into the hole in the wall, it will grab the shield and you will be able to tighten the object to the wall.

You may have problems drilling through the lath, whether it's wire or wood, if there's no support directly behind where you're drilling. The lath is flexible, even though the plaster is on it, and it may bend back from the bit and start to vibrate and rattle. This weakens the plaster all around it. Sometimes additional pressure on the drill will help; this flexes the lath to its limit, so it doesn't vibrate as much. A variable-speed drill is also very helpful. Slowing the speed allows the bit to catch easier on metal lath. Sometimes switching to either a smaller or larger bit will help. Use your high-speed bits, of course, to get through wood or metal.

Ceilings

Once you get above the surface, all ceilings are much the same. The basic structure consists of joists running across the narrower dimension of your *building* (not your room or apartment; Figure 4). Joists are usually at least 2x8s and often thicker. It is the joists that you must find and fasten into for any heavy objects. Only a concrete-slab ceiling has no joists.

Finding the joists can be a bother. If you know where you want to hang your object, drill some test holes at that point through the surface. You may be lucky. Otherwise, start drilling at 1″ intervals to either side of the first hole till you find the nearest joist. Run your row of holes along the long dimension of the house, so you'll intersect the joists running across the short dimension. Also be sure to use a long bit, since the joists may be as much as 4″ to 5″ above the ceiling surface. Use a *thin* bit so you'll have less patching to do later.

If there is already a heavy object hanging from your ceiling, chances are that it is below a joist. Measure off from this joist in standard 16″ steps till you get to where you want to hang your object. Test-drill. Unfortunately, joist intervals are not as dependable as stud intervals. While 16″ is popular, the actual spacing may be anywhere from 12″ to 24″. Also the intervals may vary across the house. So you're usually forced to hunt and peck.

Mollies and toggles are not as effective on ceilings as on walls. Gravity works directly against the thickness of the surface covering, and will tend to break the fragile plaster. Use these bolts for lighter loads, and make sure to use a large washer to help disperse the stress.

For heavy loads, use screws or lag bolts in the joists. Avoid nails when hanging something from a ceiling; gravity and the load work directly to pull them out.

To hang things like flowerpots or lamps, you can use large screw eyes or ceiling hooks. These come with lag-screw threads, or regular bolt threads for molly or toggle attachments. Braided picture wire is often used with these fasteners.

Ceiling Types and How to Fasten to Them

Plaster

As with wood stud walls, the plaster is usually on wood or wire lath. The lath is either attached directly to the joists, or to 2x4s in turn attached to the joists (*5*). Toggles and mollies are fine for light objects, barely adequate for medium loads. Put lags into the joists for heavy loads, and preferably for medium loads also.

If the plaster is on concrete-slab construction, you've got problems again. Drilling into a ceiling is even harder than drilling into the wall, since you have less leverage. Also slab ceilings have a greater concentration of water pipes, electrical lines, etc. Either use a Ramset, or design a construction that doesn't need to be fastened to the ceiling.

Plasterboard

Mounted on furring, 2x4s, or directly to the joists. Use toggles and mollies for light loads, lags or screws for medium or heavy.

Tile

These may be the small 9″ or 12″ acoustic tiles, or anything up to the large 2′x4′ panels. The larger ones are probably part of a suspended ceiling, in which case remove a panel and see what's above.

The smaller tile is usually mounted right on an old ceiling surface, or on 1x2s crossing the joists.

Tiles are not strong enough even for mollies or toggles. All fastening must be done with screws or lags into the joists.

Tin

If your ceiling has 2′x4′ panels with a raised pattern on them, the panels are probably tin. They are usually used to cover up old, crumbling plaster ceilings. Use a high-speed wood bit to drill through the tin. Treat the ceiling as a plaster ceiling.

Wood

The other common ceiling is wood—panels, planks, even plywood. It may be attached directly to the joists, or to furring, or 2x4s on the joists. The layers will run at right angles to each other. For instance, the joists run north-south, the furring east-west, and the planks north-south again. If the wood surface is 3/4″ thick or more, screw light, medium, and the lesser heavy loads right to it. Otherwise, use screws or lags to the joist.

Remember that gravity is more of a factor in hanging things from ceilings than from walls. Therefore, don't economize on the strength or number of your fasteners. If you have any specific problem with a ceiling or wall that you can't figure out with the help of our examples and guidelines, ask your friends at the hardware store for advice.

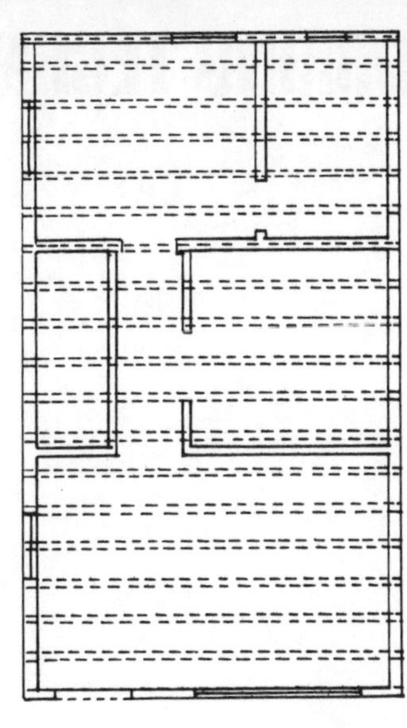

Fig. 4. Normal Joist Run

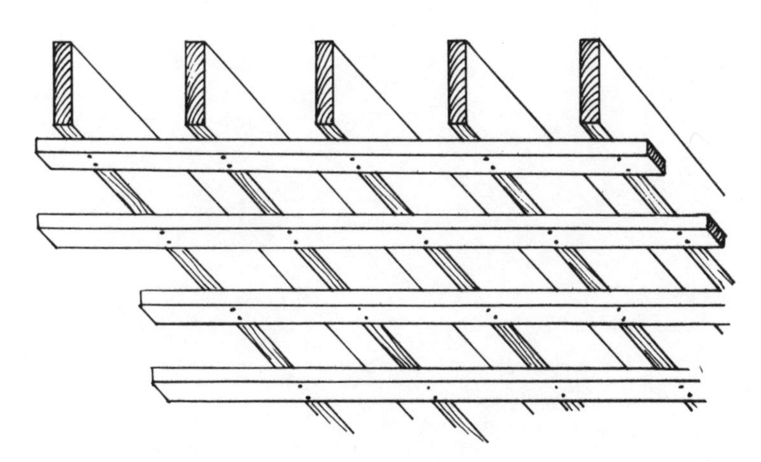

Fig. 5. The way additional surface supports are applied over ceiling joists

8/WORKING TECHNIQUES

Using Your Head, Hands, and Feet to Put Everything Together

Any project, no matter how complex, is put together with a lot of simple techniques. If you walk into a large auditorium with fifty-foot ceilings, marble columns, and red velvet curtains, you know that someone had to hang the curtain rods, that the molding along the floor was nailed in with a hammer and a couple of finishing nails.

You can learn techniques. Some you learn instantly, just by watching or being told how, such as turning a screwdriver to drive a screw. Some take a little more time, such as hammering without bending the nail or hitting your thumb. Other techniques, like planing, take years to perform skillfully, although you can do them adequately with a little practice.

The point is they all can be learned. There is nothing otherworldly about carpentry. Don't be discouraged with what you're doing; you'll improve. We'll try to show *how* to do things, and *why* and *when* to do them. These are ways and techniques we've found to work well. But don't follow our words as gospel. Follow your own ideas. And whenever you can, learn firsthand from someone who knows. Watch. Ask questions, the dumber the better.

You now know how to plan the design and structure of your project, how to use the tools you need, and how to fasten to walls and ceilings. This chapter describes the basic techniques that you can use to tie all this knowledge together to actually build a project from start to finish—from how to select the right lumber, how to carry and store it, how to lay out the pieces to be cut or the area you'll be building in, to the actual joining of the pieces, the final assembly, and the preparation of the project for finishing. Some of the more specialized construction techniques—for example, cutting plastic laminate—will be described later in the Project chapters (Part IV), since they are clearer when seen in the overall context of a structure.

Since strong joints are the heart of any structure, we have lengthy sections on fasteners and joining. In Fasteners, you'll learn how to pick the right fasteners for the job, what size you need, and how to use them. (Also check Fasteners in "Hardware.") In the Joining section, we describe the common and not-so-common wood joints possible to use, when to use each, how to cut and form them, and how to put them together. Remember to check back to Chapters 5 and 6 for all the basic tool techniques that you will need to apply in making joints, attaching fasteners, measuring, leveling, and the like.

Try to assimilate as many of the techniques and guidelines that will be important to your specific tasks. Keeping all of these general work tips in mind, and knowing how things hang together, you will soon be ready to begin work on your own project.

General Work Tips

Planning ahead

If you're building a piece in a work room and planning to carry it into another room later, be sure the door and hallways are wide enough to let you pass. If not, design the piece in parts that can be put together in the final room.

When you plan the construction of a piece, go over the steps carefully. Be sure that doing step B now doesn't make it impossible to fit step E in later. For instance, once you attach the shelves in a cabinet, you may not be able to fit a power tool such as a drill or sander in between the shelves to do a necessary job. If you plan ahead, you can sand or drill before the shelves go up.

Never trust a floor to be level or room corners to be square. Always find out the reality before making your plans.

Be careful not to nail or drill into hidden pipes or wires within the walls and ceilings. Study Chapter 7 if you're attaching anything to walls.

Keep tools sharp as you work—watch for dulling saw blades, chisels.

Be aware of the grain direction in the wood you're working on. This is obviously necessary with sanding and planing, but just as true with nailing, screwing, and other operations. Watch the grain as you nail near the end of a 2x4, for instance, for warnings of any stock split. Avoid fastening into knots and the surrounding grain, which is very hard and dense.

Grain is strong, tends to carry stress, nails, planes, and chisels along with it. As grain runs lengthwise, it may tend to travel downward slightly also; nails and chisels should be worked counter to this downward slant.

Keep your workspace clear; clean up periodically as you work. It will make a great difference psychologically, not to mention turning up lost pencils and drill bits.

Don't waste time and labor being more exacting than you need to be. Wall paneling needn't be cut perfectly straight to 1/32" when the seams are going to be covered by molding. You can cut a little extra off and it won't matter. Wall studs don't have to be cut as accurately as the parts for a cabinet. Spend your time on the things that deserve it.

Consolidate any operations that are repetitive or similar or require the same setting on a tool. Do all your cutting at the same time; all the routing; all the countersinking. Work efficiently.

Concentrate on the work. Plan ahead as much as possible, leaving only the tool operation itself to center your mind on. If you make careless mistakes, or if things get too confusing, stop and take a rest. Figure out why you messed up. Come back in a little while, try to plan or organize things on paper one step at a time. Simplify.

If you make a mistake and blow up, try to smash a waste piece, not the project.

Plan as much as possible, but don't be rigid about it. Something always pops up that can't be foreseen. Be flexible enough to change plans when the situation calls for it. The important thing is the final project, not the original plan.

Keep yourself well fueled with whatever gives you energy. Some people like yogurt and wheat germ; the most manic carpenter we know eats lunches of Dr. Pepper and Drake's Cakes.

Selecting and Handling Lumber

What Size Wood to Use

In general, use 2-by lumber for structural framing. Vertical members can be 2x4s—unless they are used as posts, in which case use 4x4s. For proper beam or joist size, consult the span charts in the Reference section. One-by lumber and 3/4" plywood are used for surface coverings and for combined structural/finished surface members in boxlike structures and furniture. Thinner lumber and ply are used mostly as surface coverings or skins over framing.

These are all general rules, of course, just to give you a rough idea. The Project chapters will show you more exactly when to use what. Using the projects as parameters, along with the structure principles in Chapter 1, you should be able to figure what you need for any project.

Wood Selection

Don't pay for clear grade when inexpensive common or construction grades will do. Construction grade is fine for walls and most general construction. Clear lumber is mainly for looks, or where perfect straightness over the entire length is required. Try to match the needs of the project with the wood. A stud wall needs relatively straight wood, not clean wood. Some warps are acceptable (see Stud Walls). For finished furniture, wood with knots may appeal to you.

However, if you need particularly good wood, say for a finished tabletop, don't try to save by getting a cheap grade, thinking you can plane and sand it down. The extra work is not worth it.

As you gain experience you'll know what you can get away with. You might get a few pieces of different grades to experiment with. See how much trouble it takes to sand each smooth, how easy each is to work with, what kinds of marks sand off easily and what kinds don't. Some wood may look pretty bad when you get it, but come clean in a flash.

You've undoubtedly noticed a door or drawer in your home that sticks in the humid summer but works like a dream during the winter, when your heating system dries the air. Lumber shrinks across the grain and very slightly in thickness (but hardly at all in length) as the natural moisture in its fibers evaporates. You should always buy kiln-dried lumber, which has the lowest moisture content. Even this will dry out somewhat, and tend to shrink. Therefore, whenever possible, let your new lumber sit inside the home for at least forty-eight hours to become accustomed to the temperature and humidity conditions; expect a bit of shrinkage anyway over the years. Woods for fine cabinet work are available that have been dried even further than standard kiln-dried lumber.

A big advantage with plywood is that it is less likely to warp and it will not shrink. Thus it is particularly good for furniture and large flat areas where laminated solid wood might shrink.

Store all wood flat. Lay lumber flat on floor, in neat piles. Lay plasterboard and plywood sheets flat on the floor, being sure no nails or dent-causing objects are under them. Proper storage prevents wood from warping or getting damaged.

Carrying lumber

This is a technique vital to your happiness. Don't exhaust yourself and don't smash up the materials. The lumberyard should deliver the materials to you in good shape. It's your job to keep them that way.

First of all, wear work gloves for rough lumber and plywood. Splinters are annoying, and they can also make you drop what you're carrying. Have someone help you carry—it's much easier and quicker.

Boards such as 2x4s are easy enough to carry. Pile several on top of each other and (if you're doing it alone) lift the pile at the middle. If you tilt the pile a little toward you while it's on the floor, you should be able to get your fingers under it. Take only as many boards as you can comfortably carry, without letting any swing out from the pile. Cradle them in your arms, as if you were going to smash in the door with a battering ram.

Wide lumber like 2x12s is easier to carry if all the pieces are the same dimension.

If you're carrying things upstairs, take a little less than usual. You have to angle the pile upward to maneuver it, and you don't want any piece sliding down behind you.

Also angle the pieces upward to turn corners. Angling "shortens" the horizontal length; you fool the corner into thinking you have a shorter piece. Go slowly and the corner won't notice you. Go fast and the corner gets nervous and sticks a wall in your way.

If you're on the left side of the pile, as you walk forward, your left arm should be under and your right arm over the pile. Otherwise you're twisting your body too much to see where you're going. Reverse arms if you're on the right side.

If you live in an apartment house, there may be a trapdoor in the elevator ceiling that lets you stick the ends of long pieces through. Be careful if you're going to the top floor—make sure there's enough clearance under the roof and pulleys. If there's no elevator and you can't maneuver the lumber up the stairs, an alternative is to cut the pieces to size out on the street. If you need them full length, you can sometimes pass them, one by one, straight up the middle of the stairwell. One person passes it up to a second person on the second floor. The first person runs up to the third floor while the second person holds on, then passes it up to the first person. Don't laugh. We've had to do this more often than we'd like to admit.

When all alternatives are impossible, cut the pieces, carry them up, and laminate pieces together (glue, nails, staggered joints) to make the long ones you need.

Full 4′x8′ sheets of ply or plasterboard are difficult to carry because they are so unwieldy. Most people should be able to lift a sheet of ½″ ply, and many a ¾″ sheet without too much trouble.

The trick is to lift and hold in the right spot, and to get the proper hand under the sheet. If you are going to carry the sheet on your *right* side as you walk *forward,* you want to lift and hold it with your *right* hand under the *center* of the bottom edge. Alternatively, if you carry it on your *left* side, you want the *left* hand under. Otherwise you're twisting your body too much, and the sheet seems twice as heavy.

Say the sheet is on the floor, and you want to lift it to your right side (*1*). Stand along an 8′ edge, pivot the piece up till it rests on the other 8′ edge. Now, grab the top edge with both hands (grab the left 4′ edge with one hand if you can reach it), tilt the front corner up and slip your *left* foot under the sheet as a wedge. Then slip your *right* hand under the sheet, hold the top with your other hand and lift. Tilt the top of the board a little onto your shoulder and head as you lift to take up some of the weight. For best balance your left hand should be just a little ahead of the center and your right hand a little behind the center of the board.

The best method, of course, is with another person. You each take an end. Arrange your hands so you face forward. Your being on opposite sides of the sheet makes it harder for the piece to fall over, but also harder to squeeze around corners.

Always be careful not to damage the edges and corners of ply and plasterboard, or any large sheets. Carry the finished side facing toward you, so it doesn't get the brunt of protruding pipes, fixtures, parakeets, and the like.

Layout

Layout is the vital first step in any carpentry. It's the measuring and squaring of things, the outlining of shapes—whether it's the location of an object to be built or the cutting lines on your wood.

General Construction

Before any job, check that your measuring tools agree with each other. No amount of skill will help you if you measure a space with your carpenter's rule, and build a piece to fit it with a tape that reads a quarter of an inch larger.

Take care of your measuring tools. Don't let the hook at the end of your tape get bent. Don't leave your tape or carpenter's rule lying open, waiting to be stepped on.

Most layout problems are fairly straightforward. Use your plumb bob to find a point on the ceiling directly above a given point on the floor (or vice versa). Lay out long straight lines between two given points on a surface with the chalk line. Find perpendiculars with your rafter square.

Walls are often rough and uneven. To accurately lay out a perpendicular to a wall on the floor, lay a long straightedge against the bottom of the wall and butt the rafter square up against it. To extend the perpendicular, first make your mark at the wall; make several marks up to 4′ or 5′ from the wall by laying another straightedge against the square. Start your chalk line at the wall, run it directly over the other marks, and continue out as far as you want. Check (or have someone else check) the positioning, and then snap the line.

When obstructions such as pipes or loose molding prevent you from laying a straightedge against the original wall, measure out beyond the obstructions at several points, lay a straightedge on those points, and measure from there with your square (*2*).

Always check your floor level before constructing anything on it. Lay a straight 2-by along the floor with the level on top of it. You can run your carpenter's level along the floor to find specific hills and valleys. Depending on what you are building, you can plan to shim the bottom (see discussion of shims under Joining, later) or to cut the bottom edges of the piece to fit the floor contours.

To find a point on one wall that is level with a point somewhere else, use your line level. Measuring up from the floor the same height at two different spots is no good if the floor isn't level. If the distance between the two points is small enough, it's more accurate to place a straight 2-by between them with the carpenter's level on top.

Helpful Tricks

In any true rectangle, the two diagonals are equal in length. Therefore if you lay out lines *AB* and *BC* perpendicular to the walls, and you find *AC* and *BD* are not equal, you'll know that at least two of the corners are *not* square (*3*). Either you laid the lines out wrong, or more likely, the room corner is not square.

This trick is also helpful in squaring the assembled frame of a cabinet or other rectangular object, when the frame is still subject to sway. Just keep adjusting the cabinet till both diagonals are equal; then you know the frame is square.

Another way to check if two lines are perpendicular is to set a 3–4–5 triangle on them. Any triangle with sides in that proportion (6–8–10; 1½–2–2½; etc.) is a right triangle, with the 3 and 4 sides perpendicular. Mark off three feet on one line, four feet on the other. If the distance between these two points is 5′, the lines are perpendicular (*4*).

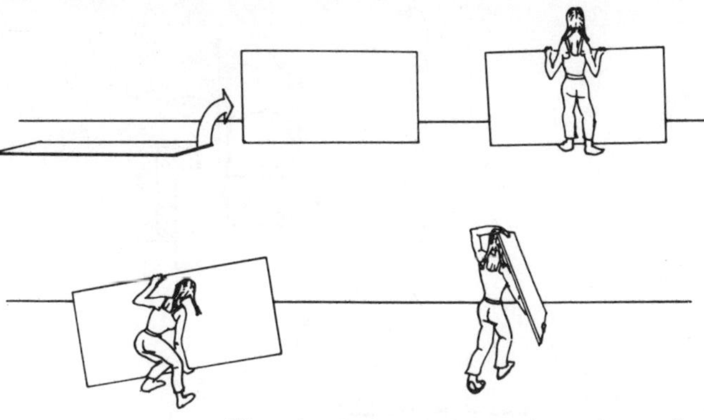

Fig. 1. Lifting and carrying a large sheet

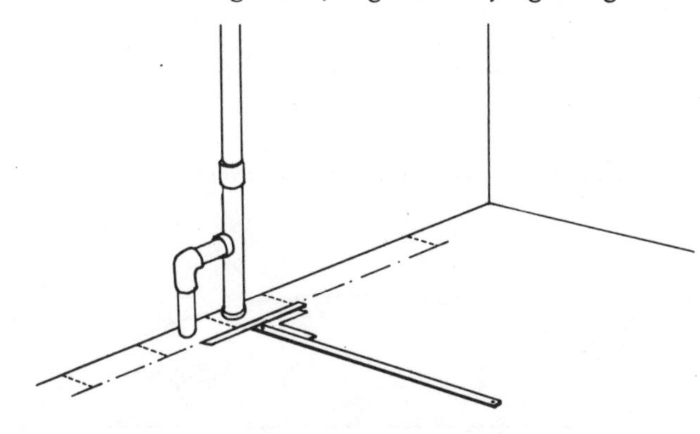

Fig. 2. Layout of a perpendicular from an obstructed wall

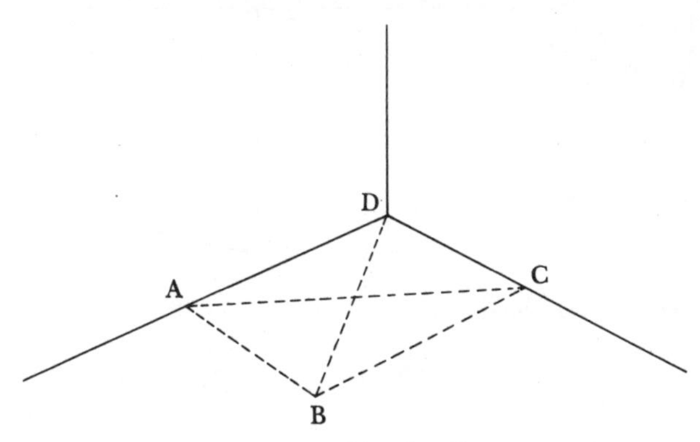

Fig. 3. Finding if a corner is square

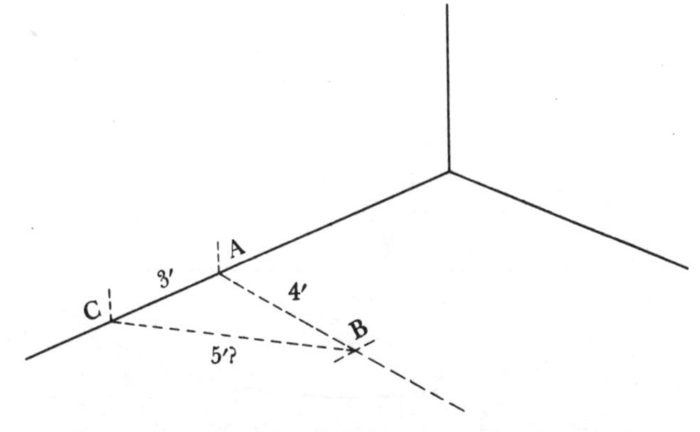

Fig. 4. Finding if a line is perpendicular to the wall

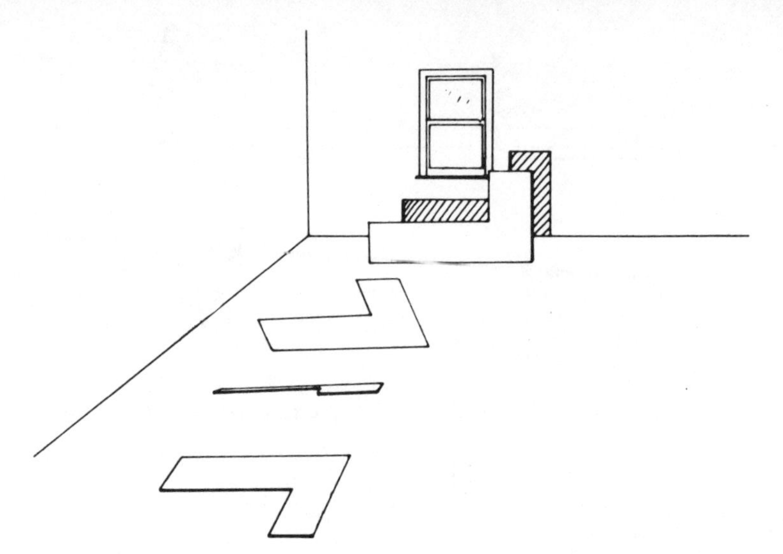

Fig. 5. *Correctly fitting an irregular piece*

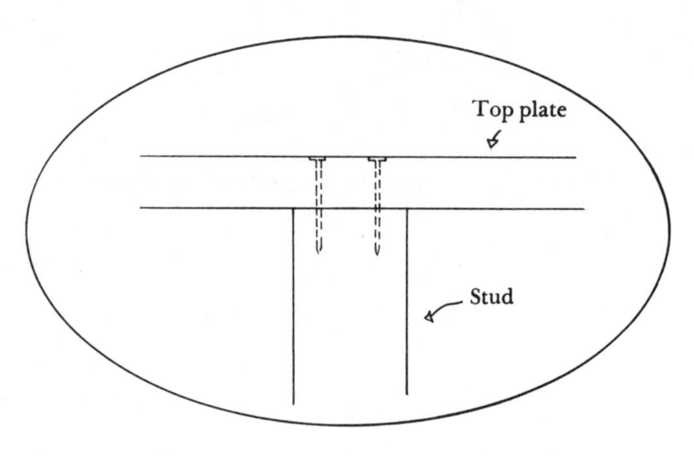

Fig. 6. *Standard End Nailing*

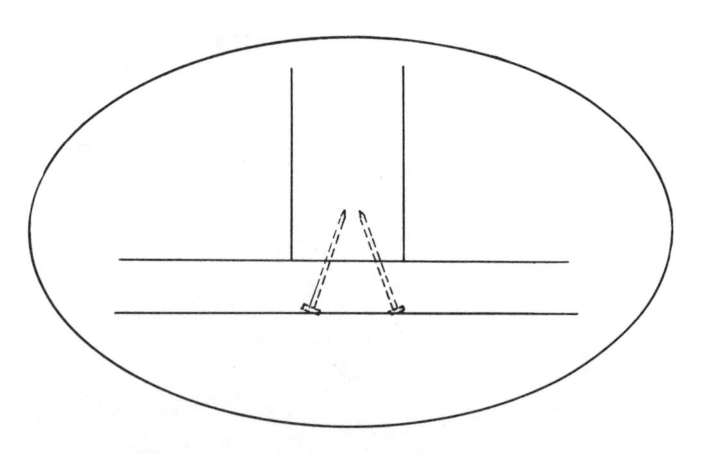

Fig. 7. *Adding strength by opposing nails*

Do not use felt-tip pens for marking. The ink is difficult to remove and bleeds through paint. Use pencil or chalk.

Selection and Layout of Pieces for Cutting and Assembling

In any selection of lumber, some pieces will be straighter than others. Use the best pieces where they count—on the ends, or around openings in walls, for instance. Use the worst lumber to cut out the smallest pieces. Select and arrange the wood before cutting.

Check the actual dimensions of the wood—a 2x6 could be easily anywhere from 5⅜″ to 5⅝″ wide. Check the square of the face and edges when it's important to your project.

Make all layout marks on the wood with a sharp pencil, or with an awl. Check your measurements. Always be aware which side of the cutting line is the waste side and thus will have the thickness of the saw blade.

Be careful laying out the cutting lines for odd-shaped pieces of plywood. Since plywood is marked and cut on the bad or back side, you must draw the mirror image of the outline, if you want the bottom face to end up as the front of the piece (*5*).

If you're cutting a number of pieces at once, label each as you go for quick reference later.

When possible, always check the fit of pieces before actual assembly.

Plywood Cutting Plan—Creating Your Own Jigsaw Puzzle

Whenever your construction calls for a lot of different plywood pieces, a cutting plan will tell you how many sheets to order. It also organizes the sawing of the pieces.

Sketch several 4x8 shapes to represent full sheets of ply, and one by one fit the dimensions of each piece within these shapes in the most efficient manner—you want the least waste. The easiest method is to draw on graph paper, using the squares as size references. You don't have to draw to scale, however; just keep track of how much of the 4x8 sheet is left after each piece is drawn in. Fit the biggest pieces in first and then try to fit the smaller ones around them.

Remember that each saw cut wastes about ⅛″ of wood, so you won't be able to get four 2′ pieces out of 8′. And don't forget to account for the grain direction as you plan. The grain is easy to keep track of if you list all measurements across the grain first, and with the grain second.

It usually takes a couple of sketches to find the best layout pattern. Just stay organized and it'll come easily. Generally you'll know you've done well if the total waste from all sheets seems less than one full sheet (32 square feet). Sometimes a project requires a number of large pieces, which can cause a lot more waste. For instance, if you need just five pieces, and each one at 2½′x5′, you will need one full 4′x8′ sheet to get each piece. The waste will be almost two-thirds of each of the five sheets. In that case, you should consider changing your plans, unless you're sure you'll be able to use the waste later.

Cutting Plywood

Most cutting information is covered in the Cutting sections of "Hand Tools" and "Power Tools." A few plywood cutting hints, however, can save you a lot of time:

Lay the good side of the plywood down for layout and cutting.

Always align the blade thickness on the waste side of the cutting line. Practice on scraps with your guide on either side of the saw.

The closer you cut parallel to an edge, the more chance of splintering the veneer. Try not to cut within ½″ to an edge.

Remember in your planning to leave space for the saw kerf. Don't leave yourself with a 2′-wide piece out of which you have to cut two 1′ pieces.

Include the factory-cut edge as one side of a piece whenever possible. It's probably straight. Similarly, use the factory corner when you can, although you should check it with a square to be sure.

Immediately after you cut each piece, label it on its edge (not on the face). This saves time later. You might even draw arrows showing which way it will be positioned, if you have any preference.

Order of Cuts

Make the most efficient cuts, those which eliminate other cuts; they will leave you with less awkward pieces to handle and move about. A big problem in working with plywood is that it's such a major undertaking just to move it. Generally you want to make a cut that will go from edge to edge, preferably along the longest dimension.

Look at your cutting plan and pick which sheet to cut first. Pick an efficient order of cuts.

Don't leave yourself with a narrow piece that has to be cut narrower—a piece so narrow you can't rest both the saw and the guide on it. For instance, don't leave a 5″x6′ strip that has to be ripped down to 3″x6′. (It's all right if it had to be crosscut down to 5″x4′.) Plan ahead. Cut that thin strip to exact size while it's still attached to a bigger piece.

When you have several sheets of plywood, you can use one sheet as an 8′ straightedge guide to make a cut on another sheet. Make the long cuts first, before you cut up the piece you might have to use as a guide. Even better, cut a piece about 8″ wide from the waste area of one sheet and use its factory edge as a guide. This is much easier to maneuver than a full sheet. Check the edge again *after* the cut—sometimes it warps slightly after being cut off the full sheet. A second cut will stay straight, however.

Fasteners and Joining Technique

"Should I use nails or screws? How many do I need? Where do they go? Should I notch the joint?"

It all depends—on the type and amount of stress the joint will undergo, on how sturdy you want the object to be, and on how much trouble you're willing to take. This section is designed to give you an understanding of the many different joining techniques, how strong each is, and what its uses are.

To choose an appropriate joining method, first try to understand the stress the joint will be under, then choose a method that is effective under such stress. Use the information and examples in this section as guidelines for your specific problem. Or you may be able to find a situation similar to yours among the Projects. If you're still unsure if our method will work for your specific problem, use a stronger method. It never hurts to overstructure. We do it all the time.

Remember that you want to build the object so that the whole structure works as one unit. One weak joint out of ten weakens the whole structure. Good joints make strong, long-lasting work. A good joint, as we said in "Structure," is one where the two pieces are held together so tightly they act as one piece. They do not move apart under stress.

Nailing

Nailing is the quickest, easiest way to join two pieces of wood. (See also under Hammer in "Hand Tools.") Nails, however, do not grip the wood very well (except for screwlike flooring nails). Thus they do not make very strong joints when used without glue or specially notched joints. Nails are strongest when the stress is in the direction in which the nail is driven. For instance, the top plate of a stud wall is nailed down to the tops of the studs (*6*). The load on the wall is mostly downward, so the nailed joint is actually tightened by the load. If you were to nail a plate to the ceiling, however, an equal load would pull the nails down and out. A piece nailed to a wall will support more of a load than the one nailed to the ceiling, but it's still weak. Most of the downward stress is taken up by the thickness of the nail itself; there is really not much of a bond between the plate and the studs. The stress will not be passed efficiently from the piece to the wall; the nail will want to bend and pull out of the wall. You could strengthen this joint and many similar ones by driving the nails in at a downward angle, more along the path of the stress.

Because nails work well only in one direction, they are not good under dynamic loads, which set up stresses in several directions.

Nails are good for framing in general construction, when a skin such as plasterboard or plywood goes over the framing. The sheathing binds the pieces together and disperses the stress to all the pieces.

One way to make a simple nailed joint stronger is to drive the nails at opposing angles (7). This makes it very difficult for the joint to loosen since any pull on the nails that would loosen one nail tends to tighten another. It's also a great help if the pieces are cut accurately so they meet flush.

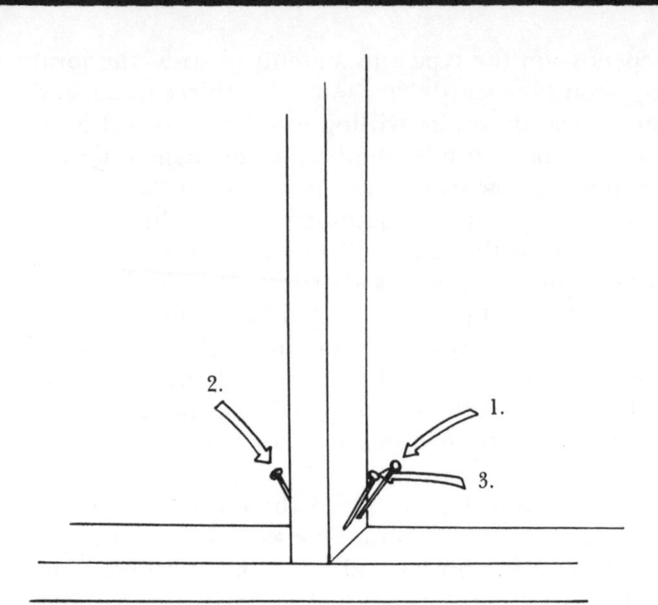

Fig. 8. Toenailing

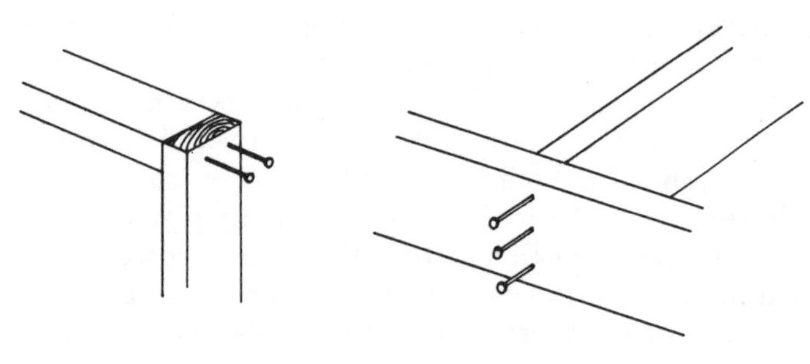

Fig. 9

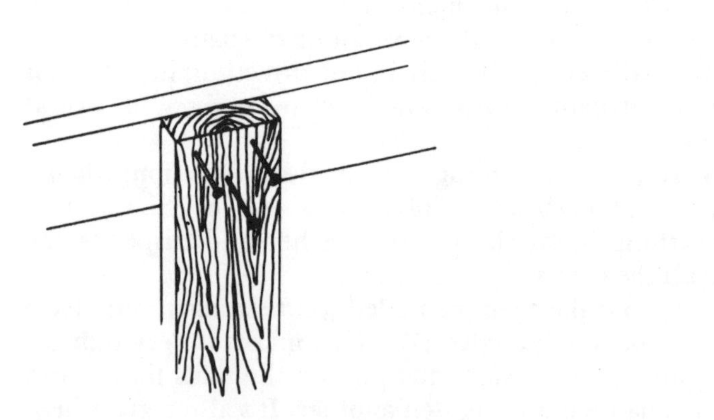

Fig. 10

When nails are used with glue and notched joints, they act partly as joiners but more as clamps to hold the pieces together till the glue sets. Modern glues, applied and set well, form a joint as strong as the wood itself. Notches offer a wider surface area for the glue, besides "holding" onto the wood.

Tacking or temporary nailing

Many times you will want to join two pieces temporarily and then take them apart before permanently joining them. You might have to make some measurements with the pieces in place, or hold them while you drill for bolts or screws. Nail the pieces together, but leave ½" or more of the nails above the surface. Then you can remove the nails easily with your hammer claw. This "tacking" in place is particularly useful in finishing furniture. Put in all the finishing nails part way, check the position of the pieces; if you have to readjust, you can pull the nails without damaging the finished surface.

Toenailing

This is a very weak joint, but sometimes it's the only way to get a nail in. It's fine when used to hold pieces in place till they're covered with a skin, such as with plasterboard over studs. The nails are driven in at an angle, as in Figure 8—one on one side, one on the other, and a third nail in the first side. The actual hammering is tricky because you are hammering partly against the side of the first piece, thereby knocking it off its mark. Set your toes up against the opposite side to resist the movement and you can toenail more accurately. Or wedge a block of wood against it with your foot.

It also helps to drill a hole for at least the first nail (a hole slightly smaller than the nail thickness, of course). This prevents splitting the grain of the first piece, and makes it easier to drive the nail accurately. Start the hole just a little with the drill bit perpendicular to the wood, then angle it through both pieces the length of the nail. (You may need an extra-long bit.) Drive the nail through the first piece till you feel it grab the second. Adjust the pieces till you feel the nail point drop into the hole drilled in the second piece, and hammer it all the way. Drill the other holes and hammer those nails in.

Some other nailing techniques

If you have to hold a heavy piece of wood in place and also start a nail in it, try starting a few nails while it's down on the ground. Let the nail points just poke out the back. Press the piece and the protruding points into place, hold it up with one hand, and hammer the nails with the other. Once the first nails are in, both hands are free to start others. This technique is also helpful if for some reason you want to do as little pounding on the second piece as possible.

When you're nailing in a spot too small for the hammer head to fit, use a nail set.

If a nail bends while hammering it, you can sometimes straighten it with your fingers, and continue nailing carefully. If it bends again, though, take it out. In general, don't waste your time—nails are cheap.

Hammer flooring nails straight and steady. The spiral type are brittle and will break rather than bend. Wear goggles.

Be careful nailing near the ends of pieces. Blunt the point of the nail so it's less likely to split the grain. Put the nail upside down on the spot to be nailed, and give it a tap on the point with the hammer, then turn it over and nail. The blunt point tends to cut the grain rather than spread it. Also angle the nail in slightly, away from the end; sometimes the grain can carry the nail out the end. Drilling a thinner hole for the nail almost always solves the problem.

Don't try to nail through knots or the surrounding area. When you cut the wood, try not to leave knots at the ends where you will have to nail.

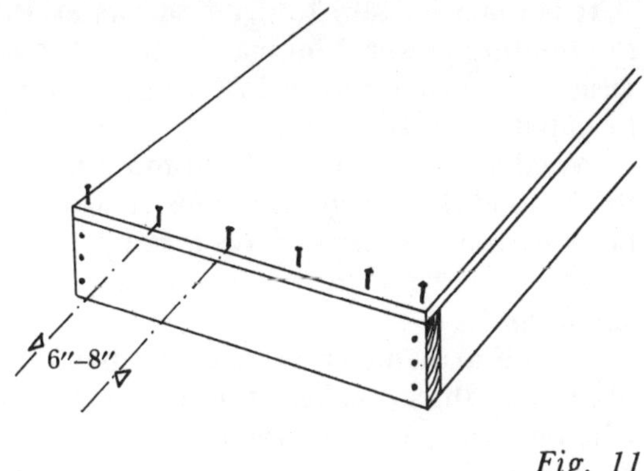

Fig. 11

Nailing patterns

The number of nails you use is largely a matter of common sense. Too many nails placed close together actually weaken the grain. For instance, you only need two nails to butt 2x4s together *(9)*. Generally, you can tell if there are too few nails just by testing the joint.

Figures 9–12 show some examples of nailing patterns that you can use as guidelines. You'll find others in the Projects section.

A few hints: Nails placed close to each other along the same grain line can split the grain; that's why the triangular pattern in Figure 10 is better than four nails in a square pattern.

The end grain does not hold nails very well; try to drive the nails through the face or edge when possible.

Always try to nail the thinner piece to the thicker piece.

Drill first in hardwoods, and wherever a nail might split the wood—such as near the ends. Try to place nails in the "meat" of the wood, rather than along the edges. The more wood around a nail, the better the grip.

Nail size

A nail should be long enough to grip the second piece without going all the way through it *(12)*. Be careful not to make it so long that the increased thickness splits the wood. Test on scraps.

Screws

Screws are thicker and stronger than nails and they grip much better. Their threads bite into the wood and hold on tight. Screws are particularly effective with glue because they hold the pieces tight while the glue sets. Screws without glue are still strong, and have the advantage over nails of being easily removable. Screws are a much better choice when the joint is under any kind of pulling-apart force.

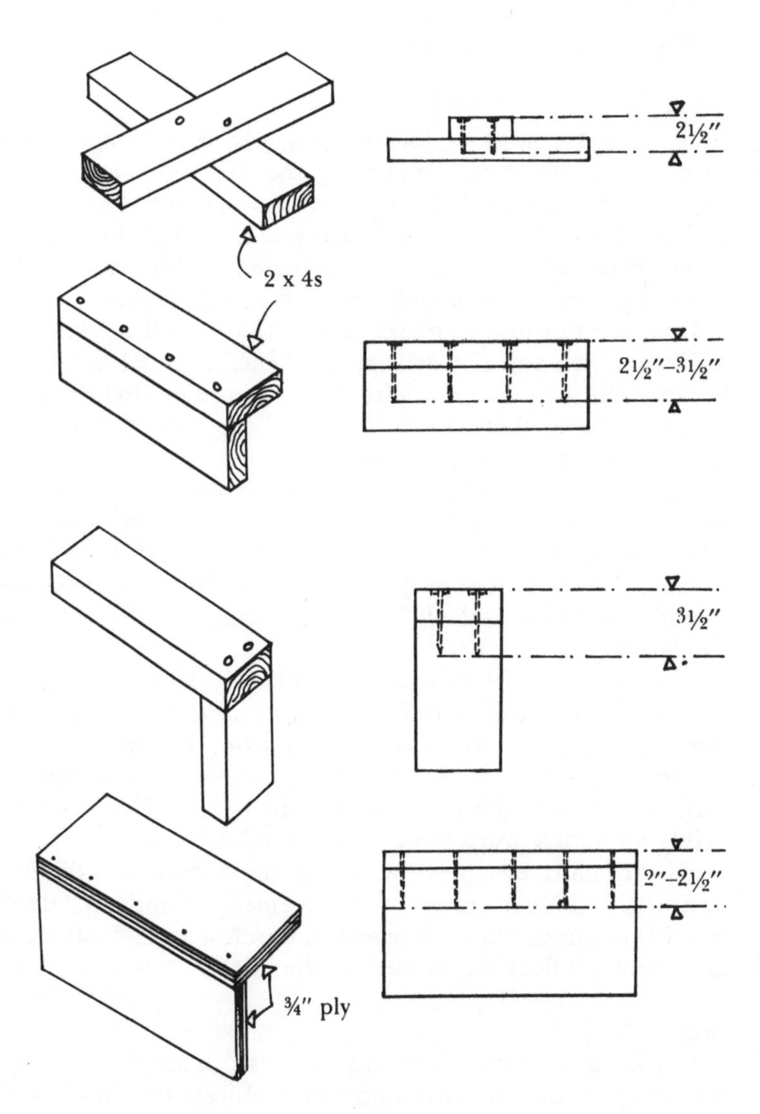

Fig. 12. Suggested Nail Lengths

It is not necessary to drive screws at an angle to increase the holding power. You may be forced to at times, but otherwise it only causes grief in trying to countersink the angled head below the surface.

Wood screws are used much more in furniture and on 1″-thick wood than in general construction, where lags or other bolts are much easier to drive through a couple of 2-bys.

Screw choice

Choose a diameter or size number only as thick as you need. Too thick weakens the wood. No. 10 or No. 12 screws will be strong enough for most uses. The screw should be long enough so that half to two-thirds of its length enters the second piece; ideally you want all the thread (about two-thirds of the screw length) into the second piece. If two equally thick pieces are being joined face-to-face, use a screw at least 1/8″ less than the combined thickness.

The type of screw head is unimportant structurally, so long as it is screwed tight against the wood.

Screwing tips

If the screw must be located precisely, as in hinge placement, mark the center point for the drill with an awl or punch. Then the drill bit will not tend to "run away." Drill straight. Drive the screw slowly till you're sure it's in straight. Never hammer a screw in, even a little bit. This defeats the whole purpose of the threads, and damages the screw.

Drill for and insert one screw at a time, till the pieces are well set. Then you can drill all the remaining holes. This is because the first screw might shift the pieces slightly as it enters its hole; then the two parts of any other previously drilled holes might not line up.

Use the proper screwdriver or screwdriver bit—one that is about as wide as the screw head, but not wider, and one whose thickness fits the slot well. The longer and thicker the handle, the easier it is to drive the screw. Soap on the screw threads also helps. If the fit is still too tight, drill the hole one size larger.

Use the screw bits that combine all the drilling operations in one. They save a great deal of time and trouble. For driving a lot of screws you can use a ratchet screwdriver or a power drill with screwdriver bit. Be careful with the power drill, to avoid marring the finished surfaces. Use a screwdriver bit with a protective sleeve around it.

As with nails, be careful when you must screw into the end grain. Use longer screws than normally. Similarly, don't screw into knots; don't place screws too close to each other; as always, fasten the thinner piece to the thicker.

Bolts

Lag bolts are very strong, and are used mainly in general construction and for fastening heavy objects into walls and ceilings, directly into wood, or with shields or anchors in masonry. They may also be used anywhere it is impractical to use a bolt with a nut (*13*). For the construction projects in this book, the kind most people will be doing, you will rarely need a bolt thicker than 3/8″. In the Loft Bed project, for instance, 3/8″ lags are more than adequate for attaching and supporting the beams to the walls.

Always use a washer with the lag bolt to keep the head from sinking into the wood and to help spread the compression stress of the bolt head. You can counterbore for the washer and bolt head for a more finished appearance.

Carriage bolts are also strong joiners. Use them when you have access to both the front and the back sides of the pieces and particularly when you wish the pieces to be easily taken apart (*14*). Be sure to use a washer with the nut. Run the nut very tight, but don't break off the thread; the tighter it is, the stronger the joint will be. It gains its strength from the compression between the bolt head and nut/washer end. Drill the hole the same size as the bolt thickness; you may have to lightly hammer the bolt through the hole. Hammering is okay for the carriage bolt since its threads are not meant to catch on the wood.

Carriage bolts are easier to work with than screws or lags—they go right through their holes, and the nuts tighten quickly.

Glues

Wood glues are so strong that they can make super-strong joints without nails or screws, so long as the surfaces meet flush and are well clamped till the glue sets. In general, though, we always use nails or screws with glue, if only to help clamp the pieces till the glue sets.

The end grain of wood does not glue well. Thus, simple butt joints with the end grain should be avoided when you need strength. Notched joints, cleats, or even the miter joint will be stronger. The miter cut gives an angle that approaches the surface grain, so the glue adheres a little better.

Follow the directions on the glue you buy as to setting and drying times. Spread it well, covering the entire area to be joined, but try to limit excess that will squeeze out. Excess is messy and a waste, and with finished surfaces causes extra work. Dried glue spots and film do not take a stain and have to be sanded or scraped off.

Joints and Joining

Now that you understand the uses of the different fasteners, it's time to learn about the different types of joints you can make. You'll find that the stronger and fancier a wood joint you make, the weaker a fastener you can use. Joints like the dovetail or the mortise and tenon need nothing but glue to keep them tight for instance. The second part of this section covers special joining hardware that eliminates the need for specially shaped joints in certain situations. The final part offers some solutions for several joining problems commonly encountered in carpentry.

All-Wood Joints

To help understand the use and importance of joints and fasteners, let's take a look at several different ways of attaching shelves to the sides of a bookcase. This will give an overview of some of the basic joints and their structural properties.

Butt joint

The simplest and quickest way to build a bookcase is to cut out the pieces, butt them together, and nail through the side supports into the shelves (*15*). The load on each shelf is transmitted to the sides of the structure only through the nails. The strength of the bookcase then is only as strong as the nails and their grip on the wood. With a heavy load, the structure could fail in any one of several ways. The shelf could start to sag and pull out from the sides, or the weight could cause the wood of the shelf immediately above the nail to split and eventually fall down; or the bookcase could, with time, start swaying or leaning to one side and slowly come apart.

You can improve on this structure in several ways. You could use screws instead of nails (*16*), which grip better and hold the pieces together more tightly; but there is still a slight danger of the wood splitting above the screw. You could glue the joint, which would help somewhat. But it's unlikely that the pieces would fit tight enough for the glue to really set strongly.

Cleats

A simple and effective improvement would be to use 1x1 or 1x2 cleats as shelf supports (*17*). The cleats are glued and screwed to the sides from the inside, and the shelves are similarly attached to the side supports, with each shelf resting tightly on its cleats. The screws from the outside into the shelves set up opposing forces to the screws in the cleats; together they can disperse lateral stress much more efficiently than if there were just screws going into the shelves. The tight fit of the shelf on the cleat also helps to disperse the load to the whole structure.

Dadoes

A stronger support method than the cleats, and also a cleaner-*looking* method, is to cut out grooves or dadoes on the side supports (*18*). The joint is glued and can then be nailed through the side with finishing nails (which would not be strong enough for a simple butt joint). The bottom of the dado acts as a support, like the cleat, only it can't come loose from the side like a cleat might, since it *is* the side. The joint gives three different surfaces to glue; and when done correctly, the tight fit alone makes a good joint.

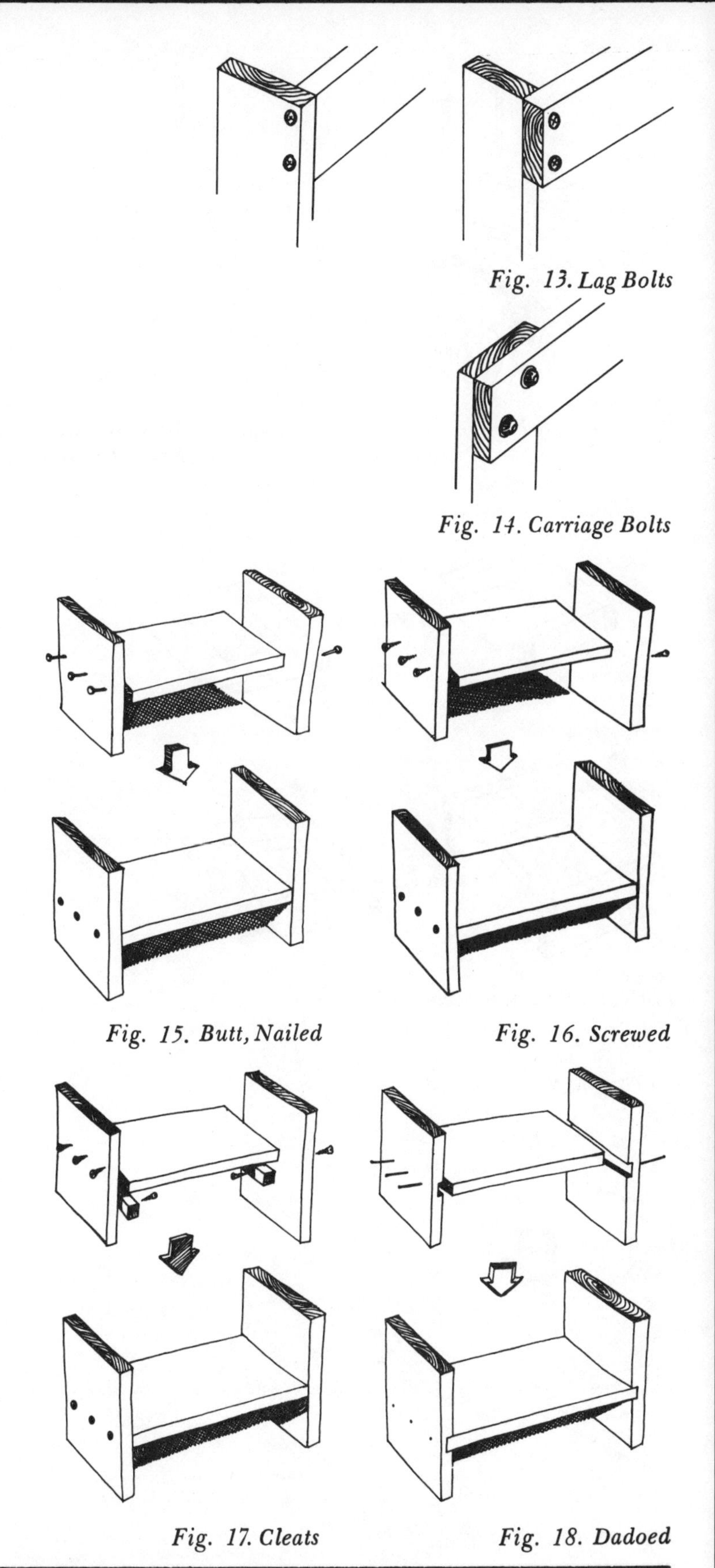

Fig. 13. Lag Bolts

Fig. 14. Carriage Bolts

Fig. 15. Butt, Nailed

Fig. 16. Screwed

Fig. 17. Cleats

Fig. 18. Dadoed

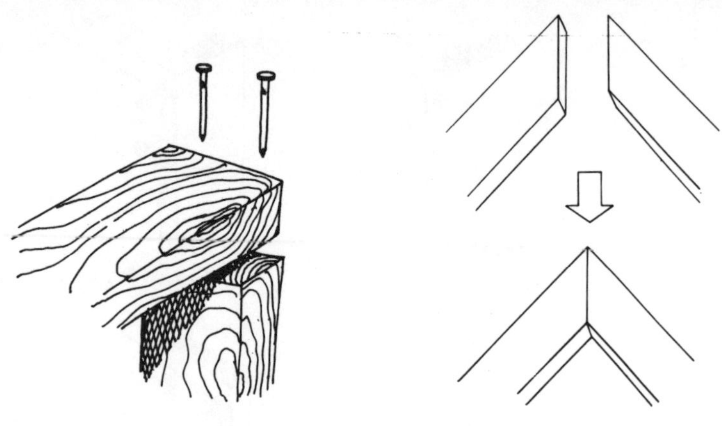

Fig. 19. Butt Joint

Fig. 20. Miter Joint in a Frame Corner

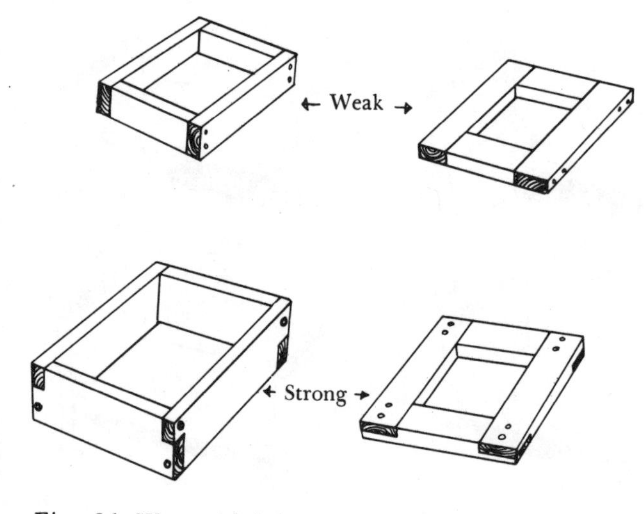

Fig. 21. Ways of joining

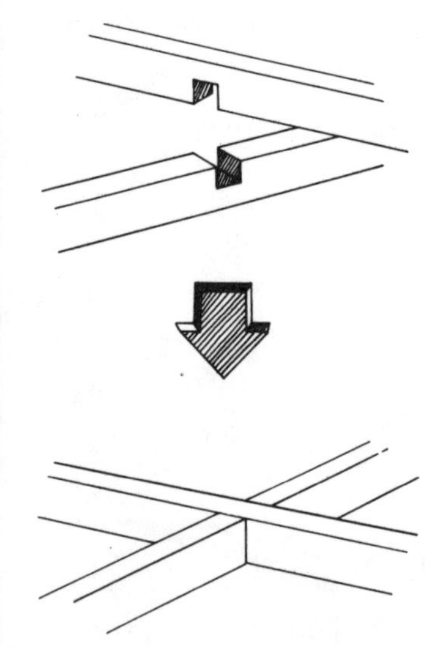

Fig. 22. Cross Lap Joint

Besides the dado, there are other specially shaped joints—the lap, dovetail, and mortise-and-tenon joints—in which the parts interlock. The fit is what gives these joints their strength; the nails and/or glue simply hold the pieces tightly in place so the joint can function.

Making the Joints

Butt joint

The butt joint, where the end of one piece butts up against the side of another at a 90° angle, is the simplest; its most common form is the nailed butt joint (*19*). It has no inherent strength or locking effect, so it's not particularly strong, especially if the stress tends to pull the pieces apart. However, when one piece is butted on top of another (instead of into the side), it is adequate. It's used mainly in general construction where sheathing will later tie the pieces together better. Otherwise, screws or bolts with glue can be used to strengthen the butt joint.

Cleated joint

A cleated joint is essentially a supported butt joint. Cleats are a simple, handy way to join all sorts of furniture when you don't own a router to make dado cuts; they don't look as neat as dadoes, but structurally they are fine.

The usual cleat construction is to screw and glue the cleats—frequently 1x2s—into the supporting side piece. The horizontal member is then laid on top and held tightly against the cleat while it is fastened to the side piece (not into the cleat). In a bookcase, you would screw into the shelf through the side; where you can't get to the outside of the side piece, toenail the horizontal member in place. (Cleats are used in the Bookcase and Captain's Bed projects.)

Miter joint

The miter joint is used at a corner where each piece is cut to an angle equal to half that of the corner (*20*). For a regular 90° corner, each piece is cut at 45°. A miter box makes accurate cuts a snap; but you can lay out your own accurately with a little care. Just use a bevel square to lay out the 45° cutting line. The joint is more decorative than strong, since it hides the end grain at the corner. Usually glued and nailed, it can also be fastened with dowels. It is used mainly for moldings and frames.

Another type of miter cut is made by setting your bevel on your power saw to the desired angle. In this way long edges can be mitered for the finished edges of furniture, for instance, as in the Cabinet project.

Lap joint

The lap joint is very easy to make, helpful in many situations, and always stronger than butt joints (*21*). For a cross lap joint, notches are cut out of each piece so they can interlock and the surfaces will remain flush (*22*). Usually a piece of wood is weakened if you cut a notch out of it; cut a 2″-deep notch out of the middle of a 2x4 joist, and it will be no stronger than a 2x2, or the thickness of the wood left above the notch. But when a second piece is notched and lapped tightly into the first, the joist is strong again. Each piece fills the gap in the other piece. But if the joint is loose—not cut accurately, or not nailed or glued—it will be weak.

The middle (*23*) and end (*24*) laps with the pieces on edge (nailed and glued) are very useful in all sorts of joist and header (support beam) situations. (See Loft project, for instance.) Simple nailed butt joints are weak; placing the joists on top of the headers is weak, awkward, and space-wasting. Lap joints with the pieces flat are useful, strong, and attractive for all sorts of furniture framing and bracing. Screws and glue are often used for such joints.

When the pieces are of equal thickness, the lap joint is made by removing half the thickness from each piece. If the pieces are unequal such as 2x4 joists supported by 2x6 beams, notch the smaller member halfway, and the thicker member that same distance from the top.

To make the lap joint, mark the width of each notch on each piece in the desired location. Make accurate measurements and lay out the lines with your square and a sharp pencil. For precision measurement, clamp the pieces together exactly as they will be and trace the edges with a pencil or knife. Mark the depth of cut on the faces. The cutting is easy with a power saw. Set your depth to the notch depth (make some practice cuts to check it). Then carefully make your cuts just on the inside of the notch outline. Make several more cuts within these kerfs, about half an inch apart (*25A*). The little pieces between the kerfs knock out easily with hammer and chisel (*25B*). The ridges and chips left on the bottom of the cut you chisel out to the marked depth (*25C–D*). Basically, however, the depth of the cut is taken care of by the depth setting of the saw blade.

With a handsaw the process is the same, only it's up to you to saw to the proper depth.

Work carefully, as the more precisely the pieces fit, the stronger the joint will be. It's better to cut the notches too narrow or short rather than too large. You can always enlarge the cut to fit, but all you can do with a loose notch is stick tiny shims in it.

Be careful that your two pieces are not slightly different sizes. If so, notching half of each will not give flush surfaces on top. Decide which surface is most important to be flush, and measure your depths from that surface. Let's say you have two 2x4s to notch across the faces. One is 1½″, one 1⅝″ thick. Remove ¾″ from the bottom one, and *leave* ¾″ in the second one above the notch for a flush top surface.

On end laps, eliminate the chiseling by making one cut just within the edge of the outline, and then with a hand saw or saber saw make a cut into the end of the wood along the depth line.

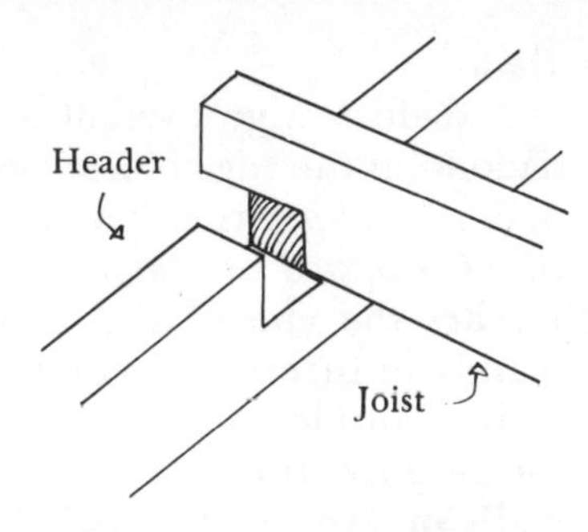

Fig. 23. Middle Lap Joint

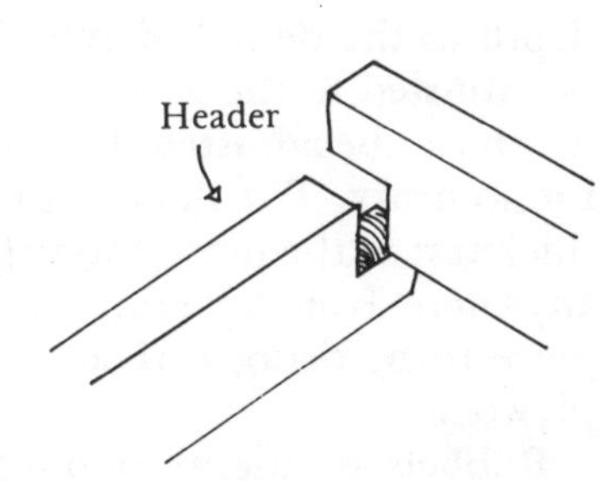

Fig. 24. End Lap Joint

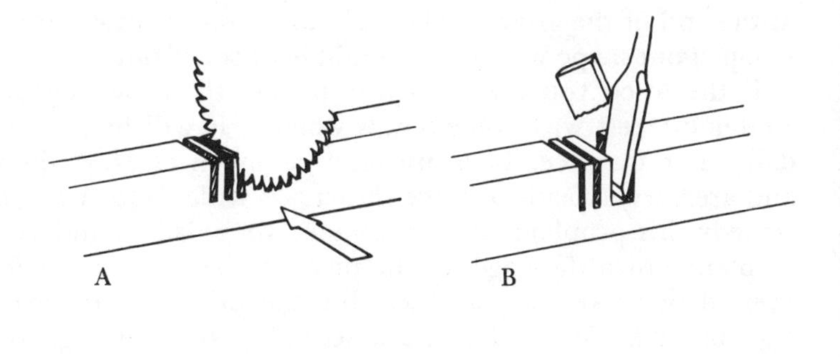

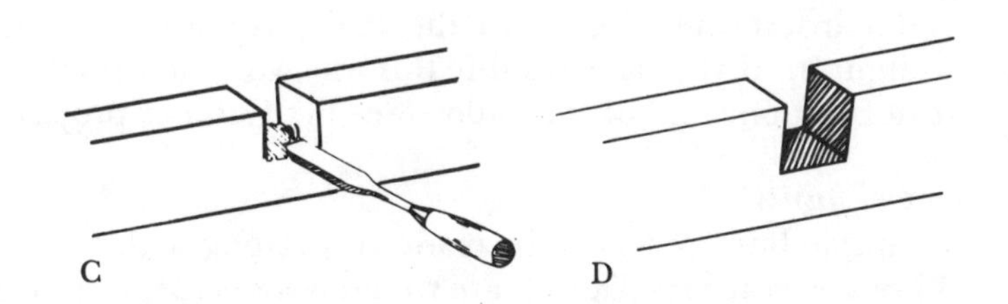

Fig. 25. Making a notch

Dado

A dado is a groove cut across the face of the wood; in a dado joint the edge of a second piece is inserted into the dado (*26*). A *rabbet* is a dado along the edge of a piece (*27*). A *blind, stopped* or *closed dado* is a dado that stops before it reaches the end of the wood (*28*). These joints are used mainly in furniture, as opposed to general construction. The dado is an ideal way to join shelves to the sides of a cabinet, for instance. It is very strong when glued and fitted properly. Nails or even screws are often driven through the dadoed piece into the edge of the second piece.

A router or dado blade on a power saw makes this joint with ease. Use a straight router bit with a diameter equal to the desired thickness of the dado, and set the router cutting depth to the desired depth. The dado blade for the saw can be adjusted to the exact thickness, and of course the cutting depth can be adjusted. In either case use a straightedge guide for accuracy. The thickness of the dado is determined by the thickness of the piece entering the dado; the depth should be anywhere from one-third to one-half the thickness of the piece to be dadoed, if solid wood; and about one-third for plywood.

Rabbets are useful to join the ends of frames and cabinets; they are also commonly used to install the back of a cabinet (*29*). Since the cut is along the edge, note that you can rout a narrow groove with a larger-diameter bit, if necessary; for instance, you can use your standard ¾″ bit to cut a ½″ rabbet. Just set your guide to let ¼″ of the bit overhang the edge. This trick is very handy, since good router bits are expensive, and you don't want to buy any you don't need.

The blind dado made with a router has to be squared off at the end with a chisel. The bit is round and leaves a semicircle at the end of the groove. The dado blade on the saw leaves a scooped-out shape which also should be chiseled out.

If the wood you use is slightly thicker than the standard router bit, test with some scraps whether it will fit into the dado. For instance, 1-bys are often as much as 13/16″ thick, and are very difficult to force into a ¾″ wide dado. You can remedy this problem two ways. Make your dadoes and sand or plane down the edges of the pieces to be inserted; or for every dado make one pass, then shift the guide over the extra 1/16″ or whatever, and make a second pass to enlarge the groove.

To insert the piece into the dado, try first to push it straight in. If you have trouble this way, you can usually slide it in from one end of the dado. (See the Cabinet project.)

Dowel joints

Simple butt joints can be made very strong with the use of dowels as fasteners. Dowels are round wooden pegs, anywhere from ⅛″ to more than 1″ in diameter. They can be inserted and glued into a predrilled hole like a screw (*30*); or, in the blind dowel joint, inserted into holes on the inside of each piece, each hole equal to half the length of the dowel (*31*). Dowel rods are sold in 3′ or 4′ lengths. You cut your own pins. Bevel the end that goes into the wood (or both ends for the blind dowel joint) with a file. This makes it easier to knock the pin into the hole. It's also a good idea to make a thin, shallow cut with a knife or saw along the length of the pin to allow excess air and glue out of the hole as the pin goes in (*32*). You'll find it easier to make this cut along the three-foot rod and then cut the pieces up. Easier, but more expensive, is to buy special dowel pins already beveled, with a spiral fluting around them for the air and glue.

Dowel joints are useful in joining pieces end to end, in joining the corners of frames (*33*), and in joining the rails of tables and other furniture to the legs.

The thicker the dowel, the stronger the joint—as long as there's enough wood left around the dowel. A dowel should be no more than half as thick as either piece it is joining. The length can vary, depending on the wood, which edge the dowel is entering, and so on; in general, ⅛″ doweling should enter each piece *at least* ½″, while thick ¾″ doweling should enter each at least three inches when possible. For other size dowels, interpolate from these figures. The hole you drill should be about 1/16″ longer than the dowel, for sure clearance.

Dowels that go through the top of the first piece are simple to install. Just clamp or hold the pieces tight and drill your hole the size of the dowel rod. Blind dowels are much trickier, because you have to drill once in each piece for the dowel and make sure the holes line up perfectly.

One way to line up the holes when joining pieces is shown below in Figure 34. Clamp the pieces together, with their eventual top surfaces out. Square lines across both edges at once wherever you want dowels. Then mark the center of these lines carefully. Measure equally from the outside of each piece, if they are slightly different thicknesses and you want the eventual top surface flat. Before drilling, start the holes at these center points with an awl.

Another method for lining up holes for blind dowels is to use dowel centers (see under Layout, in "Hand Tools"). Drill your holes in one piece; insert dowel centers in the holes. Locate the second piece accurately in place and press it against the first. The points on the dowel centers will mark the second piece. Drill your holes at those points. Be sure to drill straight in both pieces.

If you plan to do a lot of blind doweling, invest in a doweling jig (see Layout in "Hand Tools"). It speeds things up and improves accuracy.

Mortise and tenon and dovetail joints

The mortise and tenon (*35–36*) and the dovetail joints are two very sophisticated joints used by experienced woodworkers. The first is very difficult to make till you have some experience with the chisel and saw. Fortunately, the dowel

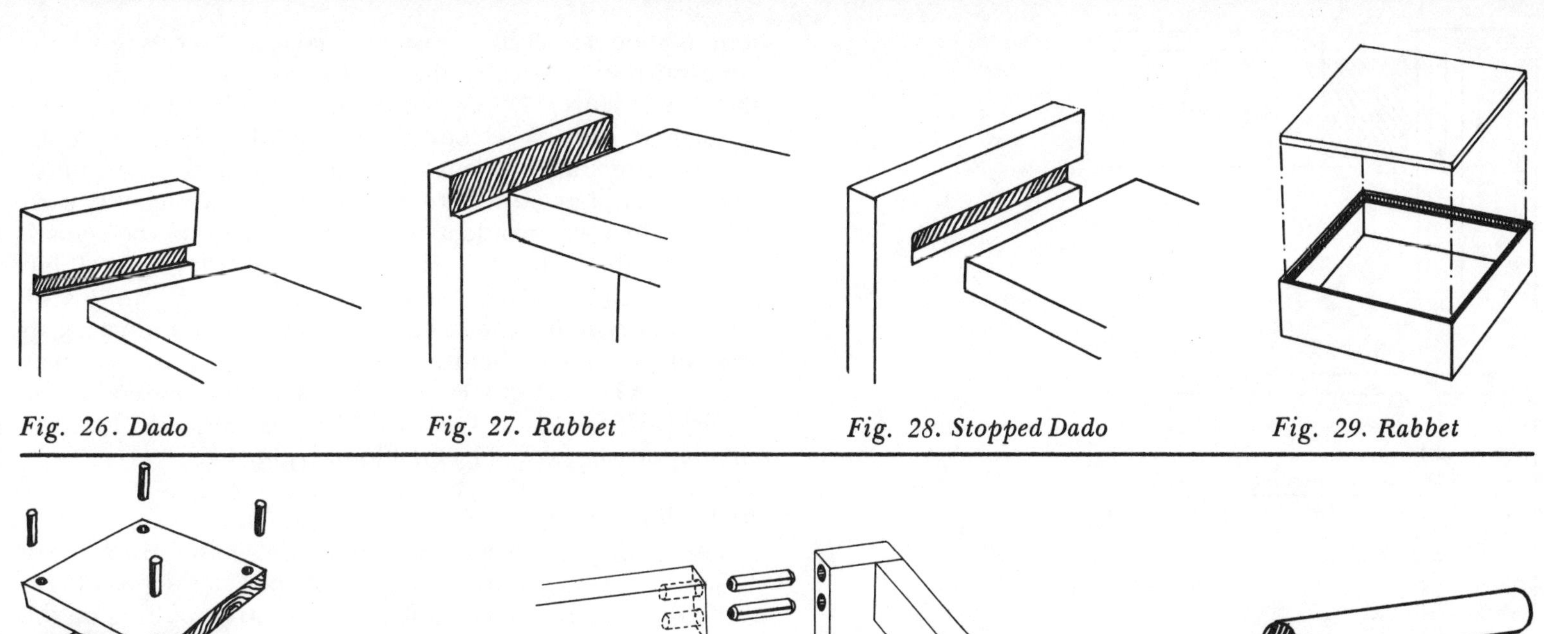

Fig. 26. Dado *Fig. 27. Rabbet* *Fig. 28. Stopped Dado* *Fig. 29. Rabbet*

Fig. 30. Dowel Joint

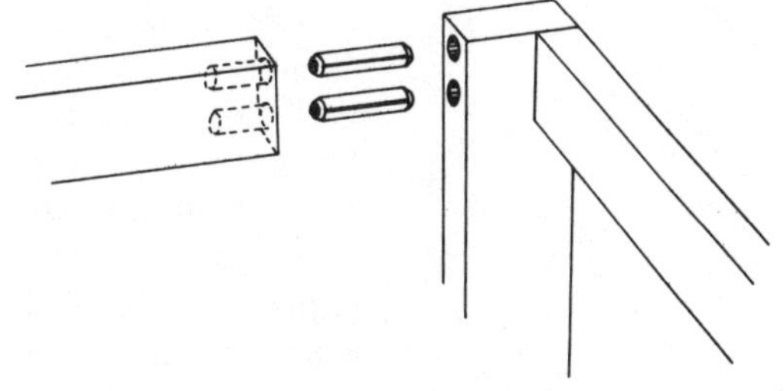

Fig. 31. Blind Dowel Joint

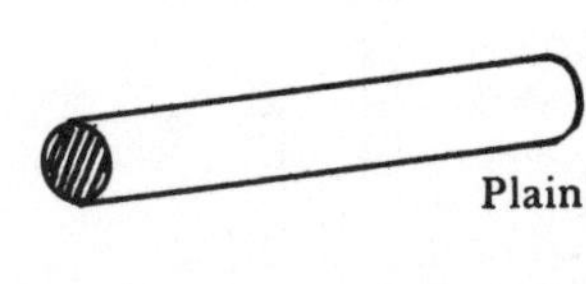

Fig. 32. Dowels

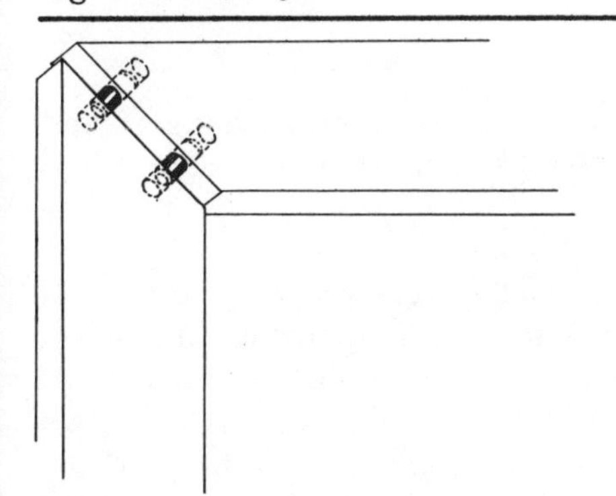

Fig. 33. Dowel Miter Joint

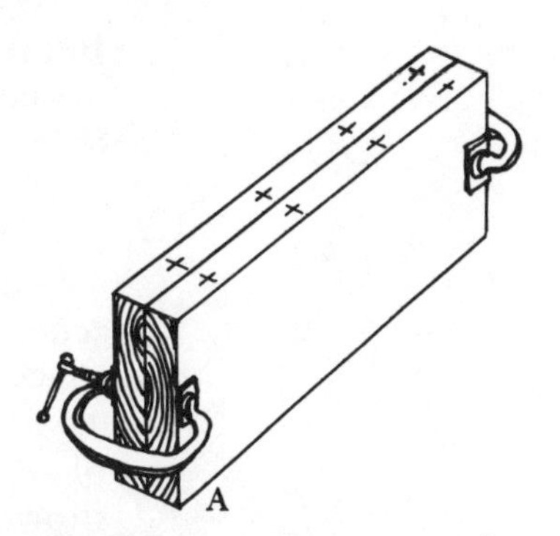

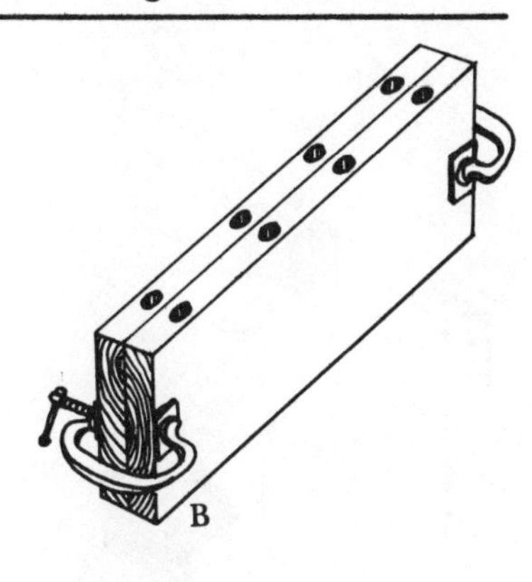

Fig. 34. Doweling Set-up (A, B, C, D)

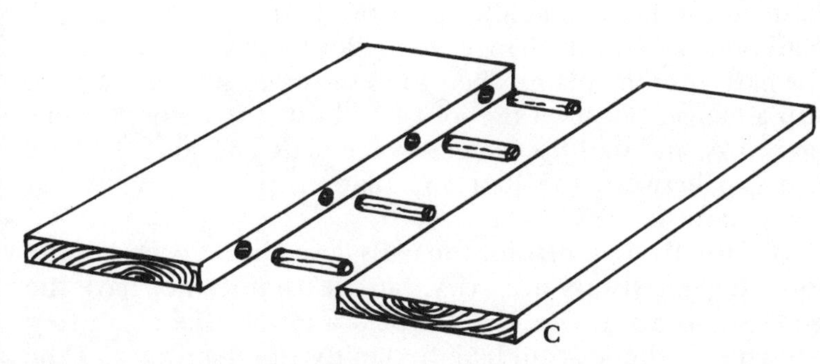

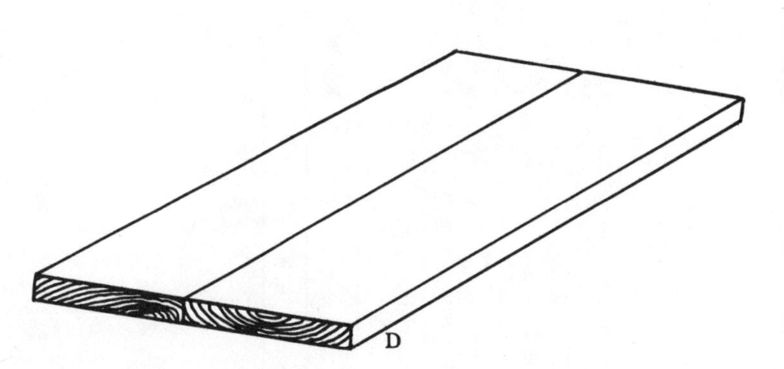

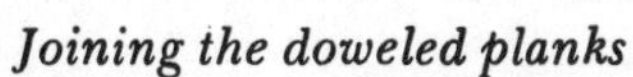

Joining the doweled planks

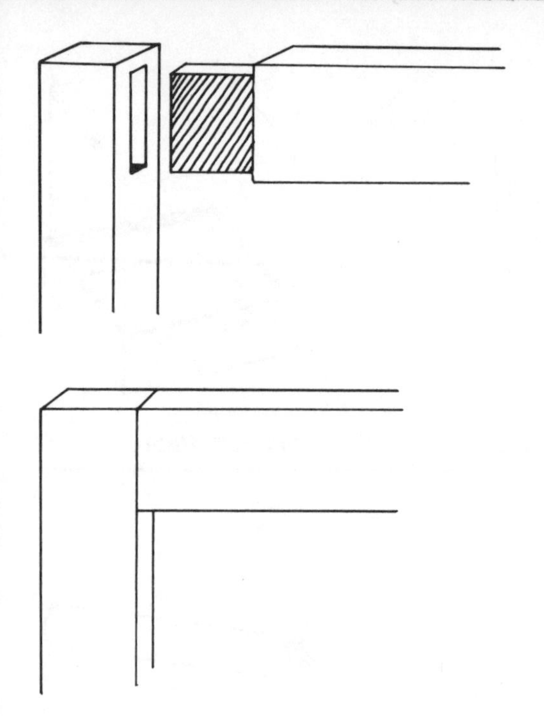

Fig. 35. Blind Mortise and Tenon

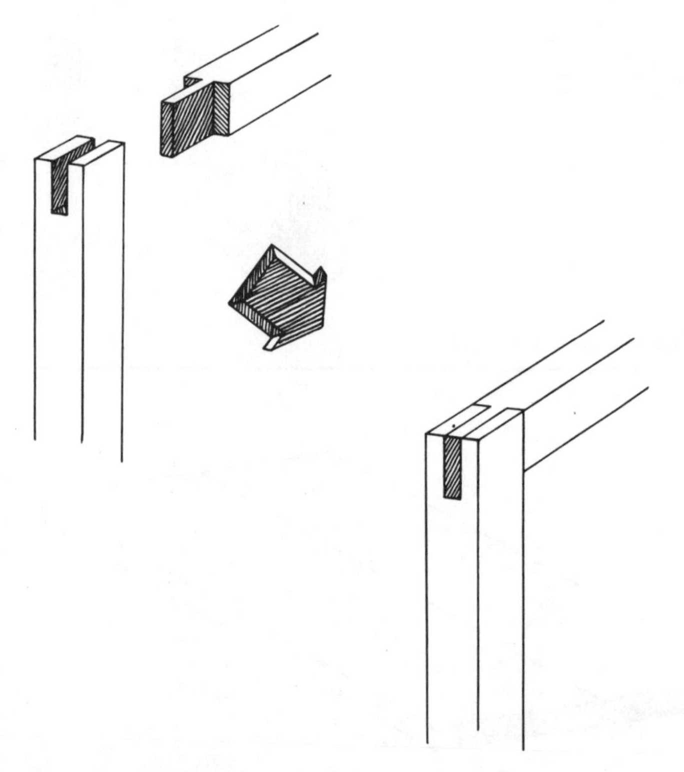

Fig. 36. Open Mortise and Tenon Joint

joint is about as effective, easier to make, and serves much the same purpose. Basically the mortise and tenon is made as shown in Figure 37. Lay out the mortise, drill several holes through it, and chisel out the rest of the wood. An open mortise can simply be cut out with a saw and chisel, much like a notch. Lay out the tenon and cut the waste away carefully, again as you would a notch. Cut with a saw, then chisel. It sounds simple, but the trick of course is in laying it all out and cutting precisely. If the final fit is loose, the joint loses a lot of its strength. Always cut the mortise first and check its final measurements before laying out the tenon. You may find it easier to cut the tenon with a hand saw, preferably the firmer back saw (also known as a tenon saw), than with a circular or saber saw. The final joint is glued.

Multiple dovetail

The multiple dovetail is a beautiful joint, used today mostly in drawer construction. It is shaped so the pieces grip each other like interlocking fingers. Any stress pulling on the pieces actually *tightens* the joint. Glue is all that is needed—no nails or screws. You can make this joint by hand only with much care and trouble and a lot of skill. There are, however, expensive dovetail templates (thirty dollars and up) for routers to make these joints fast and simple. The *single dovetail* joint is much simpler to make and works extremely well (*38*).

Joints with Hardware

Remember that there are all sorts of metal hangers and fasteners for wood joining (see "Hardware").

Joist hanger

You'll probably find the *joist hanger* the most useful in general construction. It gives an extremely strong joint when attached tightly. Always use the special nails that come with it. The logistics of getting the joist in place between two plates are these: with an assistant holding the other end up, slip your joist end into the hanger and hold it in place at the correct height. Fasten a side of the hanger to the plate with one nail. Squeeze the hanger tight against the joist; nail the other side of the hanger to the plate. Make sure the joist is flush to the plate, and sitting in the bottom of the hanger. Nail both sides of the hanger to the joist and drive the rest of the nails into the plate. Then go to the other side of the joist, slip a hanger on, level the joist or adjust it to your previous markings, and nail the hanger to the plate and joist. If there is a gap between the joist and plate, slip in a shim when you're done.

We usually drill first for the nails going into the end of the joist. It's not always necessary, but we'd rather not have the end split as a surprise. You've spent a lot of time measuring and cutting the joists to length; don't waste the time and the wood just to save a few seconds of drilling.

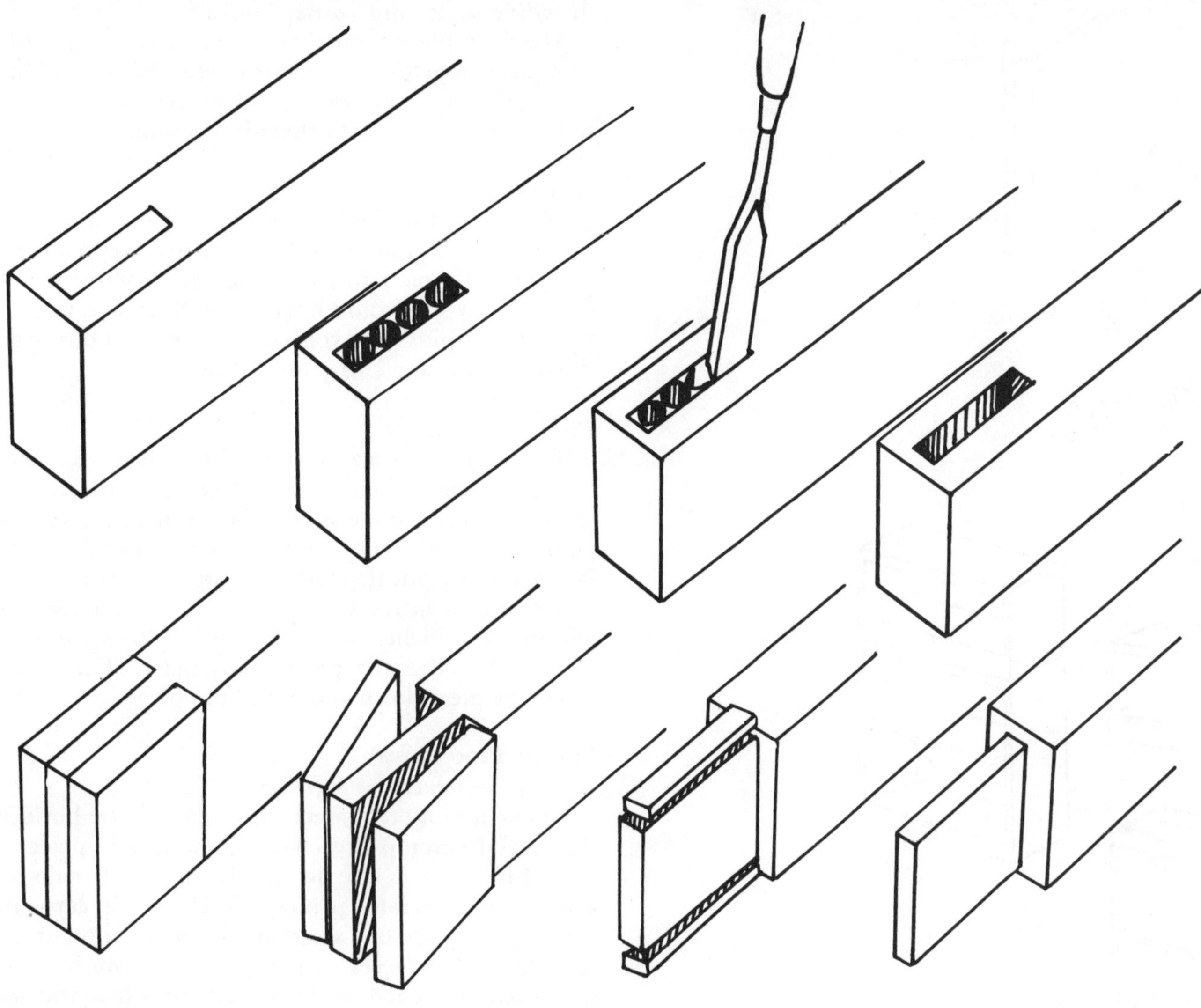

Fig. 37. Making a mortise and tenon

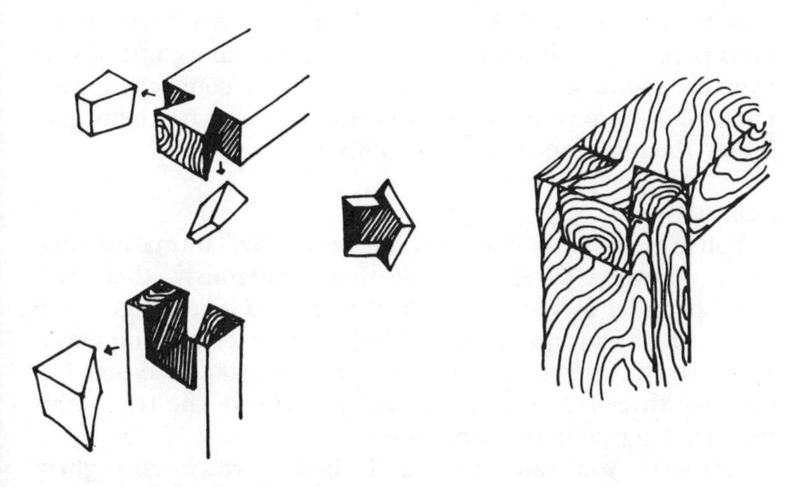

Fig. 38. Single Dovetail Joint

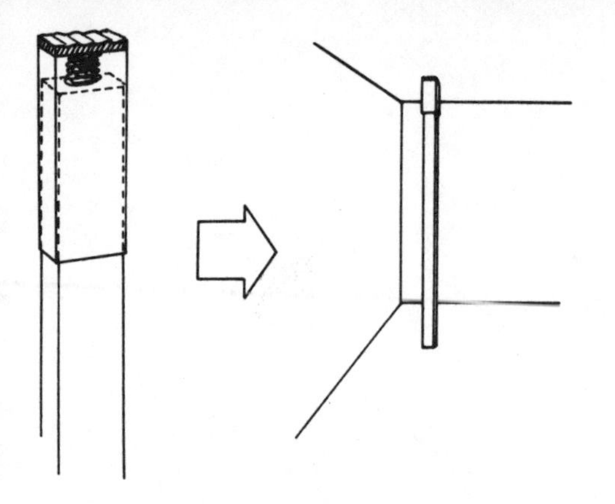

Fig. 39. Timber Topper

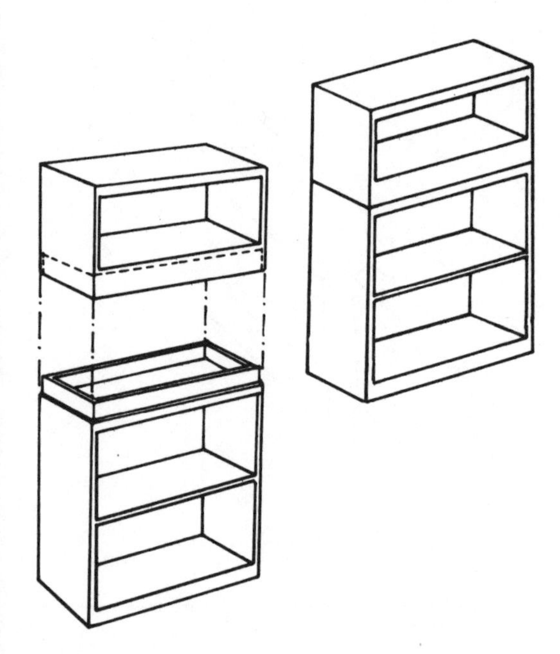

Fig. 40. Male-Female Joining

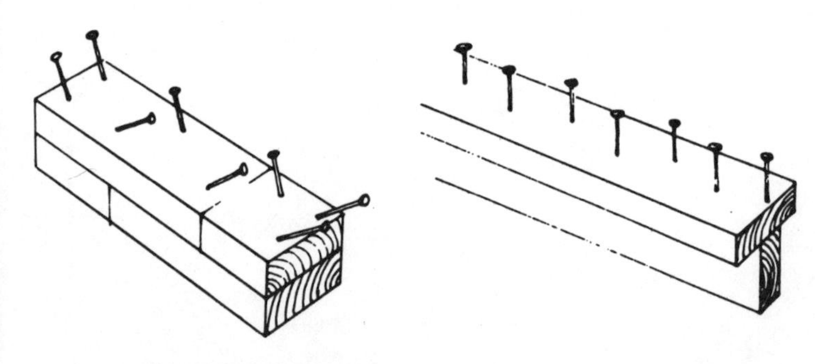

Fig. 41. Lamination Nailing

Mending plates and corner braces

Mending plates and corner braces are commonly used to join pieces together or reinforce weak joints—wobbly ladders, tables, chairs. They come in many sizes and shapes. Mortise to hide them, putty over them if you want.

Hinges

Hinges are wood joiners also. We think of them as ways to make a door swing open, but actually they join the door to the jamb. If you wanted to take the trouble and spend the extra money, you could build most of a house with hinges. By pulling the pins out, you could take the house apart. Keep them in mind for collapsible structures.

Timber toppers

Timber toppers are cheaters. Very clever. Say you want to run some studs from floor to ceiling in the middle of the room, and hang some shelves between them. Your ceiling is poured concrete (very difficult even to scratch). How do you fasten those verticals to the ceiling? The timber topper is a sleeve with a heavy spring in it *(39)*. Cut the verticals to within a few inches of the ceiling, slip these sleeves over the tops, and wedge the pieces into place. The spring action forces the piece down and holds it in place.

Joining Procedures

Joining constructions together

You've undoubtedly seen children's plastic building blocks that fit right on top of each other and lock in place. The male part of the bottom one fits snugly into the female part of the upper one. This same principle will work in carpentry, when you want to place one structure *on top* of another *(40)*. This joint keeps the pieces in place, lined up flush. This type of construction is good for large structures that you want to be able to take apart, or move around, or possibly have the option of using as two separate pieces.

For horizontal connections between pieces, carriage bolts may be all you need. For instance, if you want three bookcases perfectly flush to each other and not to move out of line over the months, a couple of carriage bolts connecting each pair holds them tight. When you want to separate them, just loosen the nuts and remove the bolts.

Laminating

You can make a double-thick beam, by laminating two pieces of wood together. Glue them generously, then nail them together as shown. Angle the nails to give a tight grip *(41)*. This technique is particularly useful for making extra-long pieces. Say you have only 8′ 2x6s and you need one 11′. Cut two three-foot lengths, then join them to the two eight-footers, staggering the seams as shown.

Similarly you can make a T beam, which is slightly stronger than a double laminated beam (see "Structure").

You're only as strong as what you are joined to.

Always try to fasten into the strongest element of a structure. If you attach something to a wall, try to reach the studs; don't just attach to the plaster (see "Walls and Ceilings"). A loft bed is fastened to the wall through its beam, not through its plywood floor. The posts support the beams, not the plywood floor.

If you want to hang a cabinet by its back, be sure the back is a strong-structured member of the unit (it usually isn't). If it is just 1/4″ ply tacked to the back edges, the first load of books will tear the cabinet from the back (*42*). (See Box Construction, Chapter 9.)

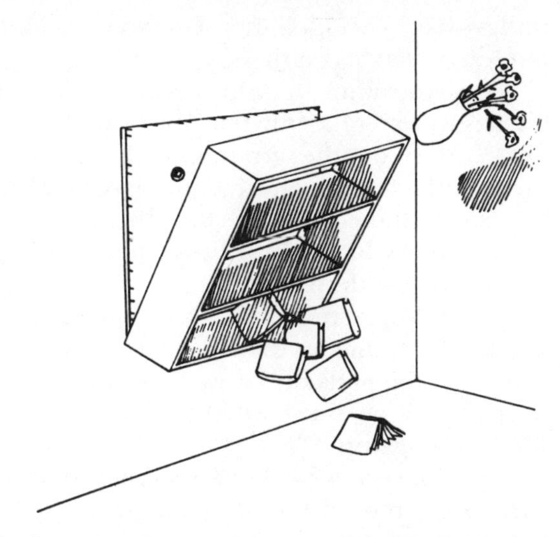

Fig. 42. What could happen without proper joining of back and frame

General Construction Aids

Shimming

You shim something by wedging a small piece of wood behind or under it to make it level or bring it out to a certain point—for instance, behind furring strips to bring an old wall out level; or under the bottom plate of a stud wall if your floor is not level; or under the base frame to level a built-in. Cheap cedar shingles make the best shims because of their tapered wedge shape. The wedge shape eliminates the need for measuring the gap and cutting blocks to fit. Just take two shims and slide them under the piece from both sides at once, the thin ends first (*43*). Together they maintain a level surface on top—just one shim would be slanted. The farther in you push them, the thicker they are in the middle. Push them in till they fill the gap tightly, or the piece is level. If the shim part that still sticks out is in the way, chisel it off after nailing down through the leveled piece and shims.

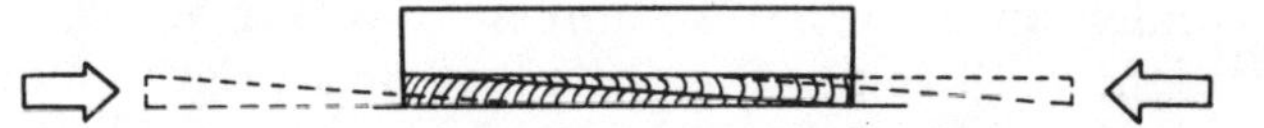

Fig. 43. Leveling by sliding shims under object

Shimming is difficult without shingles, but it can be done. Use layers of something thin, such as 1/4″ hardboard or ply. For thinner sections, chisel or plane slivers and shavings off a piece of wood. For large gaps, say several inches, start with a section of 2-by and then put the thinner pieces in.

When shims seem like they might fall or slip out, glue them in place. You can nail them also, but sometimes this will just split them if you don't drill first.

For any large construction job, such as a number of walls, or laying a new floor, it's worth it to buy a bundle of shingles. They'll save you much time in leveling.

Assembly

Always check the fit of pieces before fastening them permanently. You're going to make a mistake sometime, and it will be hard to correct once it's glued tight.

The big problem in assembling pieces is holding them in place till you get the first few nails or screws in. One of the advantages of an articulated joint is that the shape itself pretty much holds the pieces in place for you. In other cases you have several alternatives. An assistant is usually the best answer. If you're working alone, use your whole body to hold things—feet, knees, elbows. Use your head. Clamp pieces or put heavy weights on them.

When you want to hold pieces up till you can drill them for screws or bolts, put temporary nails in. Prenail the first piece down on the ground, with the points just coming through the back, then lift it in place and hammer the rest of the way. Pull the nails once the other fasteners are in.

Temporary blocks in strategic positions are often helpful—particularly with miter joints, which tend to slip as you nail them (see suggestions in Cabinet project). When you're assembling a frame on the floor, and want to keep it square as you drive the nails or screws, a couple of blocks nailed to the floor on one outside corner provide a square mold to press the frame against.

If you have any kind of assembly problem, stop and think. What's the root of the problem? If you're hammering on an area that is not well supported, stick something under it. Support it.

When you work on one area of a piece, be aware of the other areas that are already assembled. You don't want to damage them—scrape them on the floor, or knock them off square or out of their dadoes.

If something doesn't fit right, do something about it. Trim it or cut another piece. File it, rasp it. Be wary of pieces that will fit only when you have to twist them or bend them in some way. Such unnatural posture puts the whole structure under an inner stress, forces a weaker structure to bend slightly. Of course the piece may be all right—your structure may be a bit off. In that case it's probably too late to fix it. So don't worry about it.

Many times you have to knock pieces in. If the surface will show in the finished project, never hit it directly with your hammer. Place a block of wood on it, near the point of most resistance, and hammer the block of wood. Move the block around as you hammer, so it won't leave its imprint. Or use a soft-headed mallet.

Similarly, never drive nails all the way with the hammer on a surface that will show. Use a nail set for the last little bit.

Be careful not to let the screwdriver blade slip on the wood. Clean off excess glue as you go; have a roll of paper towels and some warm water handy. Make your pencil marks lightly or in places that won't show. Use thin blocks between clamps and the work surface.

Preparing Wood for Finishing

Finishing Plywood Edges

Most people don't like the look of plywood edges in a finished piece. You can cover them with veneer tape to give the look of solid wood. Or you can glue and tack 1/4″x3/4″ lattice stripping around the edges, which gives more protection to the ply edges.

Veneer tape

Veneer tape comes in 8′ rolls in many different wood types, 1″ or 2″ wide. To apply it to a plywood edge, cut a length of tape a couple of inches longer than the edge. Spread contact cement on the plywood edge, and on the back of the veneer tape. The back is the inner side of the roll, usually with a paper backing. Now wait ten or fifteen minutes till the cement is dry to the touch. If the tape is a particularly long piece that keeps wanting to roll up on its own, tack it face down to a board. Put the tacks at the ends, so the marks will be cut off later. When the cement is dry, lay the tape in place, starting at one end with a half-inch or so overlap. Apply it carefully, since it's hard to adjust once the cement surfaces catch each other.

Press the tape down firmly everywhere. Run something over the surface—the side of a pencil or, what we prefer, the plastic top to a cat-food can. Press down all the air bubbles. Any stubborn bubbles can be punctured with a pin to let the air out.

Trim off the excess with a fine file. Use the rougher side first, finish with the fine side. File downward at an angle, so that you actually bend the excess veneer down. Do the ends first.

When different lengths of tape intersect, as in the front edges of a bookcase, position the seams to give the appearance of butted joints; or use the fancier variations suggested in Figure 25 of the Cabinet project.

Lattice

Lattice comes in 3/4″ widths (among others) so it will fit 3/4″ ply edges exactly. Simply cut it to length, apply with regular wood glue and tack down with 2d or 3d finishing nails. Use your nail set to sink the nails. Lattice does not give plywood the appearance of solid wood, because you see the lattice thickness. Since it *is* solid wood, itself, however, it offers more protection to the plywood edges. This is useful where the edges get a lot of wear, such as the top edge of a captain's bed. Lattice is also useful if you are attaching hinges into the plywood end grain; it gives the hinge screws something better to grip than the plywood core.

Sometimes the 3/4″ lattice may be slightly smaller than the plywood edge. To get a perfectly flush fit, we often use 7/8″ lattice, and then belt sand the overlapping edges.

Filling Screwholes with Wood Plugs

If you counterbore for a screw, you can fill the hole above it with putty (see Chapter 3) or a wood plug. Drill the hole with a bit that drills and counterbores. Buy a plug cutter of the same size and manufacturer as the bit. Cut the plug out of whatever kind of wood you want. You can match the grains or contrast them. Glue and insert the plug; sand off the excess. Try not to leave too much excess on hardwood plugs—they are difficult to sand.

Sanding

If your project will have a lot of interior corners and areas hard to sand, sand the pieces as much as possible before assembly. At the end, you'll still have to sand a little, but a lot of the work will have been done.

To sand an edge without rounding it off, clamp it between two other pieces, the top edges flush as in Figure 44.

You usually use a sanding block to hold the paper, but be imaginative for difficult areas. Wrap the sandpaper around a thin strip of wood or a metal rule to get at inside corners and other hard-to-reach places; or on the tip of a pencil eraser for small spots.

To sand small items by hand, bring the mountain to Mohammed. That is, tack the sanding paper down, and rub the item on the paper. Also see Sandpaper in "Hand Tools" and Sanders in "Power Tools."

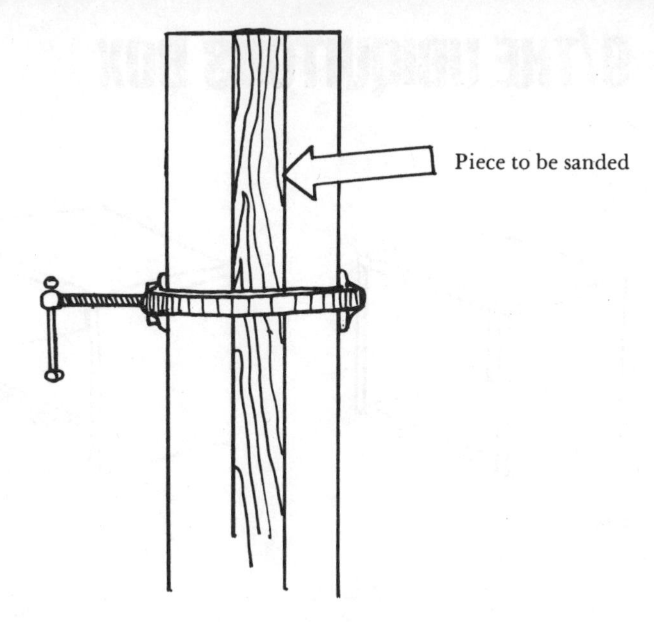

Fig. 44. Set-up for accurate end sanding

9/THE UBIQUITOUS BOX

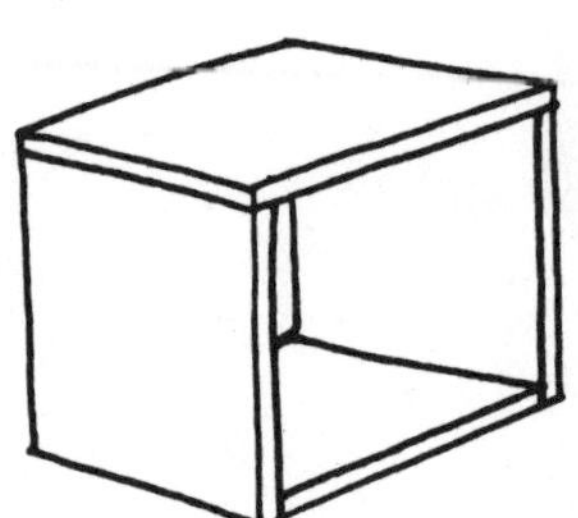

Fig. 1

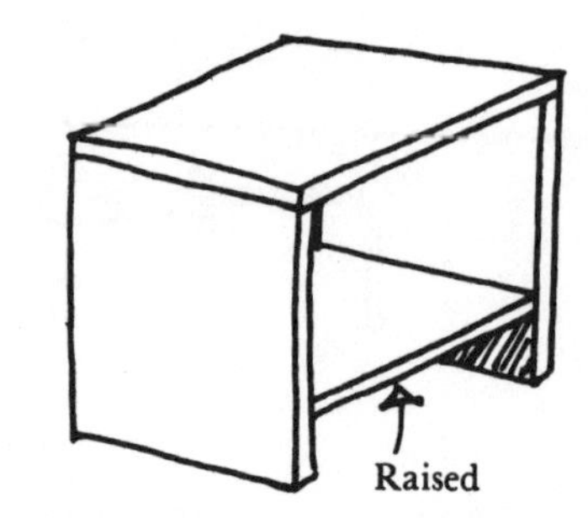

Fig. 2

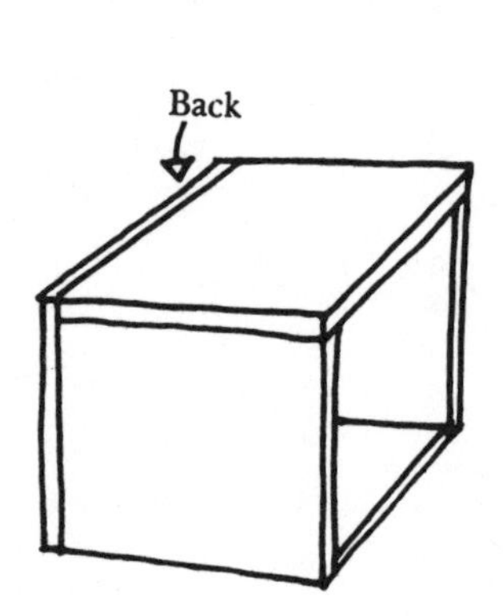

Fig. 3

Fig. 4

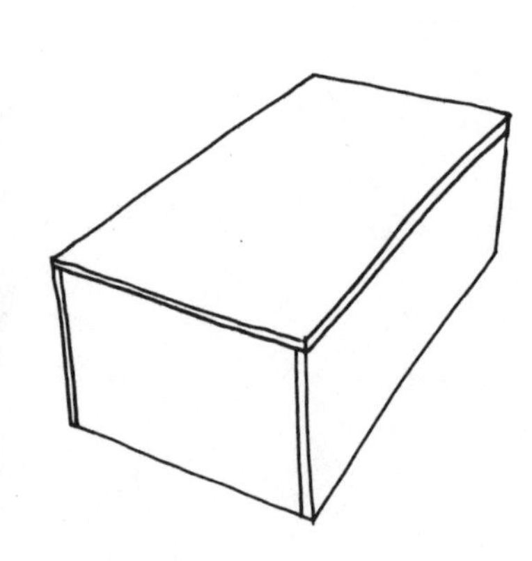

Fig. 5

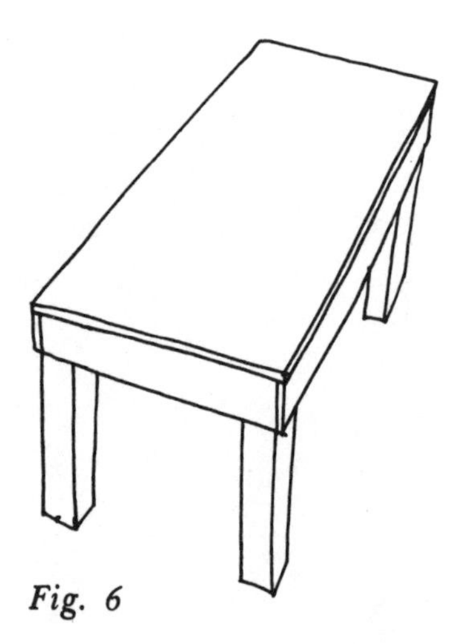

Fig. 6

Box Transformations

Transforming a Simple Box into Cabinets, Tables, Chests, Beds, Drawers, and Other Forms; and Other Useful Constructions

Basic Box and Variations

The box is one of the fundamental structural patterns in carpentry. The square or rectangular shape makes it easy to work with, since all the cutting and joining can be done at right angles. The box is also relatively strong and stable when it's joined well and has a back. The drawer is another element common to many projects; basically it, too, is just a box. If you familiarize yourself with these elements, plus a third one—a simple cabinet door—you'll have the basic elements for 90 percent of the projects you may want to build. Chests, dressers, desks, tables, cabinets, closets, and beds are nothing but forms or combinations of these elements.

When you try to design a piece, and you feel lost, try drawing a box and playing around with it on paper. See if you can make it resemble your vague notion. Can you stretch it out, squeeze it, add to it, cut it up, hinge it, or adapt it in any way to what you want? Most likely you can.

Let's try playing with a box and following it through its transformations. First we have a simple four-sided box made of 1-bys, no front, no back, as shown in Figure 1. This is not particularly useful as is, being a little weak and prone to sway under lateral stress. (See explanation in "Structure.") It's easy enough to put together, however, and could serve as a utility stand or table, or even as a storage piece for records, magazines, books.

If you check Figure 2, you'll see we could make it look more like a table by raising the bottom horizontal a few inches.

To help the box resist lateral stress, we can add a back, as shown in Figure 3. This binds the sides together so they can't sway or slide into a parallelogram shape. This type of box is very useful as a storage piece; it could even serve as one unit of a modular storage system. Stretch the box vertically, put another horizontal piece or two in the middle, and—as seen in Figure 4—you have a bookcase.

Lay the box with its back up, like Figure 5, for a platform. Or a bed. Mount it on some legs, as shown in Figure 6, and you have a table. Or a loft bed.

Or, as in Figure 7, lay the box with its back on the floor and you have an open chest. Hinge a door to the top of the chest (see Figure 8), and you have a hope chest. Or a trunk. Or a bench with storage room inside it.

Squash the open chest and you have a drawer. Stick the drawer in the bookcase, add some doors, and you have a cabinet, as drawn in Figure 9. Or fill the bookcase with drawers and you have a dresser.

Put a door in the middle of a cabinet, hinged at the bottom with folding supports, and you have a secretary. Or a minibar. Put a similar but larger bottom-hinged door on a large box, fasten it well to the wall, and stick a mattress inside for a fold-out bed.

Playing with and sketching boxes in these ways can give you your basic design. Then it's just the question of choosing an appropriate inner structure and adequate joints and fasteners. A loft bed, for instance, has to be stronger and sturdier than a storage box. Use your knowledge of structure principles (see Chapter 1) and joining techniques (see Chapter 8); also consult the following examples plus the projects as guidelines.

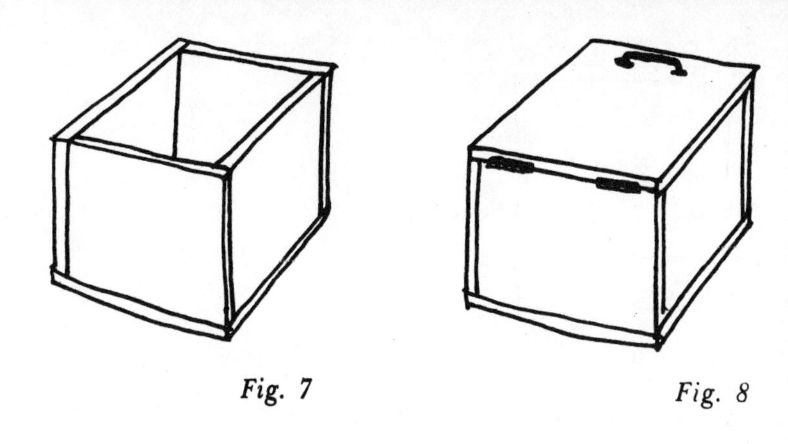

Fig. 7 *Fig. 8*

Box Construction

The original four-sided box frame *could* be just nailed together. Note that the top horizontal piece is joined *on top* of the sides, not between them, for maximum strength against loads on top. As explained in "Structure" and "Working Techniques," however, screws and glue make the box much stronger than nails. And better yet would be to use special joints like dadoes or rabbets, as discussed in Joints and Joining in "Working Techniques."

All you need to make a box structure, such as a cabinet, rigid is a 1/4″ thick back (plywood). You can just nail and glue the back right to the back edges of the project. If you want to hang the cabinet or box on the wall, the back should be thicker—3/8″ or even 1/2″—and you should definitely insert the back piece *into* the back of the frame—either into rabbets or against cleats (see Figure 8–29 and the Cabinet project). The usual hanging method is to bolt or screw through the back into the wall. If the back were just joined *to* the back edges of the cabinet, the load would be mainly on the fasteners and tend to pull the frame away from the back (see Figure 42 in Chapter 8). Inserting the back means the load is *on* the back piece and thus more directly dispersed from the frame to the back to the wall. If you use the rabbet method, glue the back on and drive 1″ finishing nails through the back into the rabbet. With cleats, glue and screw the cleats to the sides, then glue and nail the back to the cleats.

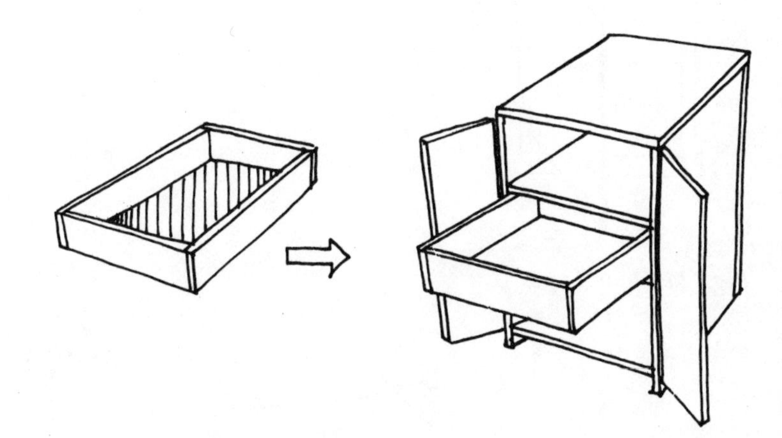

Fig. 9. Simple Cabinet

Another reason for inserting the back inside the frame is design—it looks better because you can't see the edges of the back piece from the side. Just remember in your design that an inserted back shortens the depth of the inner shelves. For instance, if the outer frame is 12″ deep, and the back is 1/4″ ply, the inner shelves can be no more than 11 3/4″ deep.

The frame of the open chest does not need very strong joints because none of the sides is under very great stress. The bottom, however, which carries most of the load, should be well joined to the frame—at least nailed and glued. Also the bottom should be thicker ply, at least 1/2″ and preferably 3/4″.

In the platform structure, the top piece need only be nailed to the box frame *(10)*. The downward stress of any load *helps* keep the nails in.

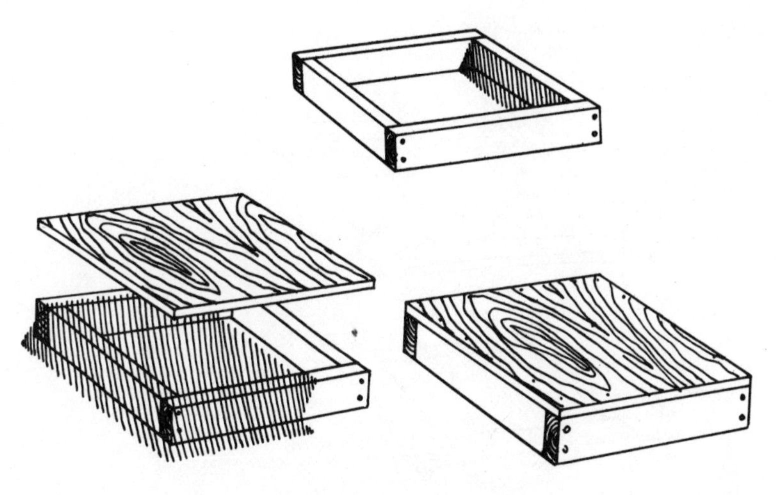

Fig. 10. Simple Top Application for Platform

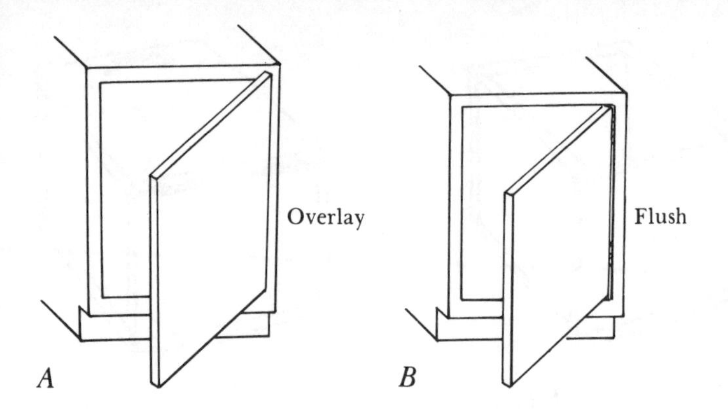

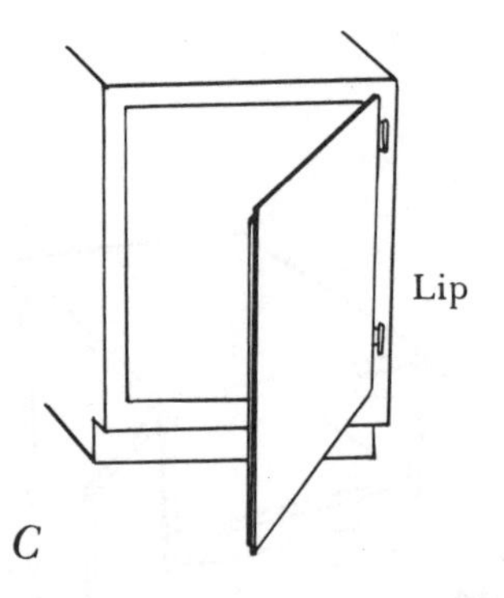

Fig. 11. Door Types

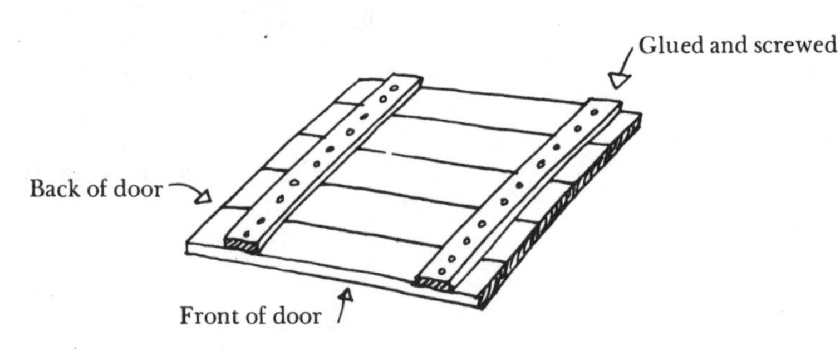

Fig. 12. Constructed Door

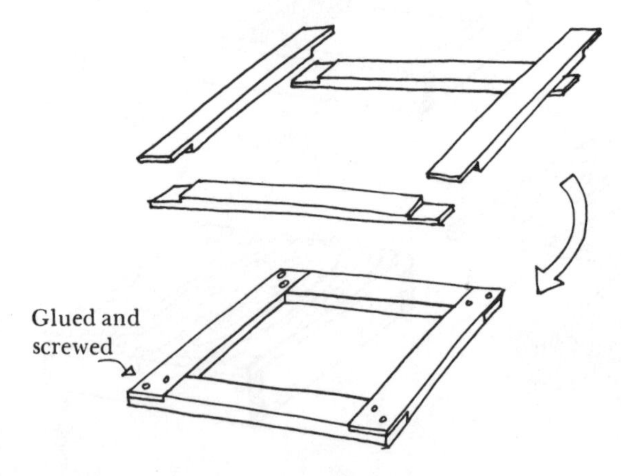

Fig. 13. Lap jointing a frame

The type of joint between the sides of the frame depends on the use of the platform. If it's a small piece—say 1′ square, meant to support a static load—nails are enough. Glue would be helpful, but not necessary. Any piece much larger, particularly if it supports dynamic loads, should be at least nailed and glued and preferably screwed and glued.

A large platform, such as a bed or loft structure, also requires a thicker top piece, since the top is not only holding the frame rigid but actually supporting the load. Use at least ½″ ply, preferably ¾″, with adequate supports underneath. (See Captain's Bed and Loft Bed projects.) The platform is essentially a floor, and a floor has joists supporting it.

Cabinet-Door Construction

The top to our hope chest in Figure 8 is really a very common type of cabinet door—an overlay door, one that closes against the front edges of the frame (*11A*). Cabinet doors are usually made of ¾″ stock; anything less tends to warp over the years. Narrow butt hinges or a length of piano hinge joins the door to the frame. The hinge is really a type of joint that allows the door to swing in relation to the box.

The flush and the lipped styles of cabinet doors are more difficult to make and hang, but they usually look better—more a part of the whole piece.

The flush door (*11B*) is mounted *inside* the opening, like the door to a house. The door should fit precisely, with about ¹⁄₁₆″ clearance all around. Narrow butt hinges, piano hinge, or special cabinet hinges (see "Hardware") can be used.

The lipped door (*11C*) requires more work to make, but it's easier to fit since it covers the opening and any irregularities. Neither the lipped nor overlay door has to fit exactly, for that reason. To make a lipped door, rout a rabbet around the door's edge ⅜″ deep and ⅜″ wide. The inside, thicker portion of the door should fit within the opening, with about ¹⁄₁₆″ to ⅛″ clearance all around, plus room for the hinge, of course. Thus the outer part of the lip extends about ¼″ to ⁵⁄₁₆″ over the front edges of the door opening. This clearance needn't be as exact or consistent as with the flush door. (See Built-in Kitchen Cabinet project.)

In many manufactured cabinets, lipped doors close against an *apron,* a facing frame of 1-bys. It's not necessary and we don't particularly care for the look; but if you like it, it's easy to make. Just fasten the 1-bys (usually 1x2s or 1x3s) to the front edges of the opening (flush to the outer edges), and then hinge the door to the 1-bys. Special hinges are used for lipped doors (see "Hardware").

With flush and lipped doors, remember that any shelves behind them must be indented so the door can close. The idea is similar to inserting a back inside a cabinet and decreasing the depth of the inner shelves accordingly.

It's usually easiest to cut a cabinet door out of ¾″ plywood. Veneer-tape the edges or, if you like, nail and glue lattice to them. Or make a door out of solid wood, as shown in Figure 12.

You could make another type of door by stretching a fabric over or inside a 1-by frame. Make strong corner joints, such as lap joints, so the frame will be rigid (*13*). Some of the materials possible are burlap, caning, printed fabrics, bamboo mats, and so on. The fabric can be stapled to the inside of the frame, or sandwiched between two frames.

A cabinet door should of course have some kind of catch to keep it closed. Check "Hardware" for the various kinds.

Sliding doors

Some people like sliding doors in cabinets. We don't, but there's nothing wrong with them. You can buy sliding door tracks cheaply and install them easily (they usually come with complete instructions). They can be mounted right on the surface of the door opening, or you can mortise for the tracks and recess them out of sight. Or you can make your own tracks by routing grooves along the bottom and top of the door space. The grooves only have to be 1/4″ wide for a 3/4″ door. You can rabbet the bottom and top of the doors to fit the grooves (*14*).

Make the top grooves and rabbets twice as deep as the bottom ones, so you can lift the door into the top to install and remove it. Make the grooves slightly wider than the part of the door that fits into them. Clean the grooves well and wax them before installing the doors.

Use special inset door handles for sliding doors, so they can slip past each other.

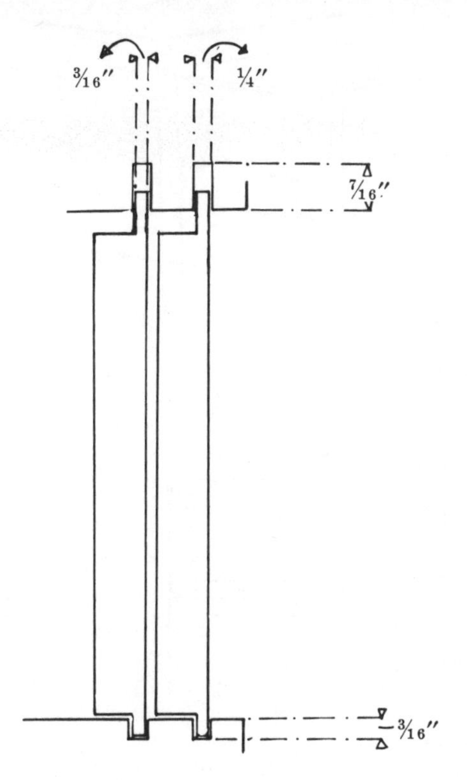

Fig. 14. Sliding Door Set-up

Drawer Construction

Just as we can hinge a door to any box frame, we can also add a drawer. The front of the drawer can fit the opening in any of the three ways that a cabinet door fits—overlay, flush, or lipped. The basic structure of all these drawers is nothing more than another box, sized to fit within a larger box.

The simplest drawer has an overlay front and only butt joints (*15*), glued and nailed (finishing nails are fine). This structure is adequate, particularly for light loads like clothing.

Good drawers have better joints, however. Every time you pull open a drawer, you put the joints, particularly at the front, under stress. In finely made drawers, the bottom is inserted into dadoes cut into the frame, and the four sides are joined with dovetail joints. The dovetail actually tightens as you pull on it. You need either a lot of time and skill to make this joint, or you need a router and expensive dovetail template.

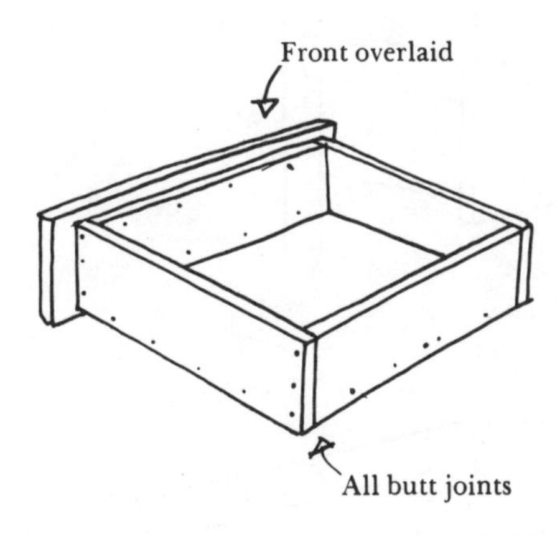

Fig. 15. Simple Drawer

Fortunately there are other good ways to construct a strong drawer. Obviously, what you want to do is to make all the joints stronger, either by supporting them with cleats or by using interlocking joints such as dadoes and rabbets. For construction details, see the Cabinet project and its variations.

The thickness of wood for the drawer depends on the use

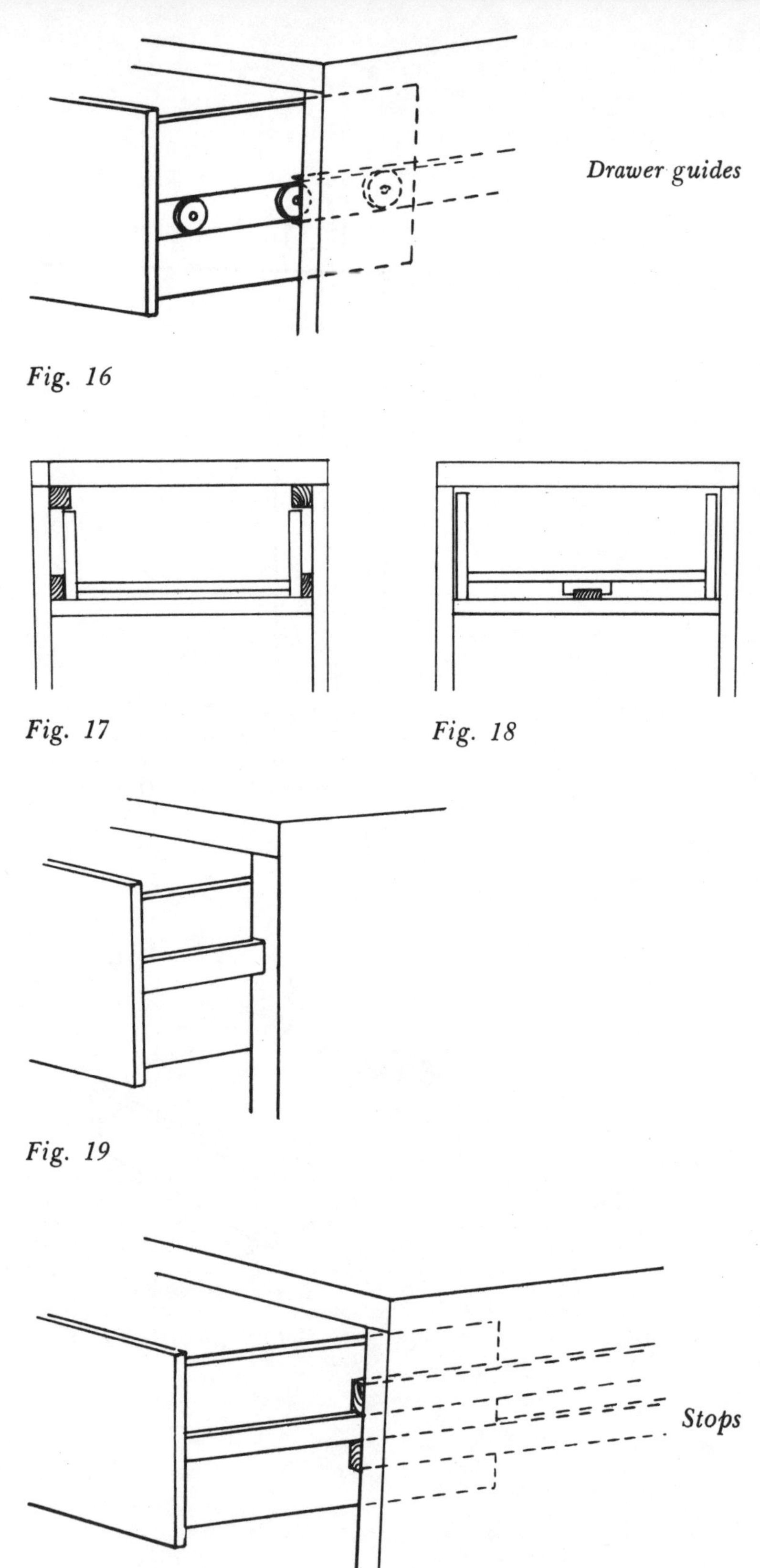

Fig. 16

Fig. 17

Fig. 18

Fig. 19

Fig. 20

of the drawer and on your own taste. The bottom and sides can be as thin as 1/4″ for light loads; for heavy loads like tools, use 1/2″ or 3/4″. It's best to use 3/4″ for the drawer *front,* however, because of the added stress at the front.

A drawer has to slide smoothly in and out of the cabinet, of course. Allow about 1/16″ on each side for sufficient play between the drawer and the cabinet.

You could simply have the bottom of the drawer sitting on a shelf, with the sides of the drawer guided just by the sides of the cabinet. This would work, but the large areas of facing surface would provide too much friction for a smooth action. Some system of guides is usually used to help the drawer slide smoothly.

You can buy ready-made *roller hardware* cheaply, complete with easy installation instructions; one type is illustrated in Figure 16. Just be sure to read the instructions before constructing the drawer; they tell you how much clearance you need for the hardware.

Or you can make your own drawer runners or guides with 1x1 or 1x2 cleats. There are several possible types. The *sleeve guides* (*17*) are attached to the cabinet interior; the *bottom guide* has a dadoed guide on the bottom of the drawer riding on a cleat attached to the cabinet (*18*). Another method is to attach *side cleats* to the drawer to fit into dadoes in the side of the cabinet (*19*). Or you can put single cleats on the sides of the drawers, to fit between double cleats on the cabinet walls (*20*). A little wax on the runners helps the action.

All of these systems are simple; you just have to be very careful in laying out the measurements for the placement of the runners. You want them to be straight and placed so the drawer fits neatly in its space.

Let's examine the placement of runners attached to the sides. The first runner can be placed anywhere in the lower half of the space; the other two are placed in relation to the first. Say the drawer front is to fit flush in the opening, with 1/16″ clearance at the top between it and the shelf above it; and that the drawer sides and front are all the same height. We attach the bottom runner to the cabinet side so that the runner's top edge is, say, 4″ below the above shelf. The drawer runner rides on this runner, and is attached to the drawer side so its bottom is 4″ minus 1/16″ below the top of the drawer side, or 3 15/16″. The upper runner on the cabinet is placed above the first one by a shade more than the thickness of the drawer runner between them. Again, see Cabinet project for an example of side-runner placement.

You also need to make sure that the drawer won't go in too far. Overlay and lipped drawers are stopped by the edges of the drawer front contacting the edges of the drawer opening. A flush drawer needs a stop on the inside of the opening to keep it from sliding too far in. If you are using cleat runners, just indent the ones attached to the cabinet by the depth of the drawer front (*21*). Another method is to fasten little blocks or a strip to the cabinet behind the drawer (*22*).

Other Building Components

Any furniture that you stand at—such as cabinets, bookcases, even captain's beds—needs a space at the bottom for your toes. Otherwise, every time you stand at the piece to reach for something you would kick the bottom. Standard toe space is about 3″ deep and 3½″ high. Toe space can also be obtained at counters by extending the countertop 3″ out. (See Cabinet and Captain's Bed projects for examples.)

Diagonal members or bracing can be very effective in dispersing stress. A diagonal brace is more economical than a whole sheet of plywood used as a back. Other uses include a temporary support for a stud wall, permanent supports for railings, or counter shelf supports.

Used as a support, the diagonal bracing should be no less than 45° to the horizontal (*23*). Always join the brace to the framing, not to the skin. Diagonals can be incorporated into the structure; a railing, for instance, will resist lateral stress better with a truss web, rather than vertical studs.

A cantilever is an unsupported overhang. The joists or framing of a horizontal structure can overhang their supports by as much as half of their supported length and still be strong. The advantage of the cantilever, obviously, is the extra space you get underneath the front (*24*).

Note that the structure is somewhat like a seesaw. If you stand on the front of the cantilevered loft, the stress is down at that point and *up* at the back. The beam and posts are like the middle or lever of the seesaw. Therefore you need to support the back of the loft against a small amount of upward stress—if the joists are notched in, glue and opposed toenails would do it.

Toe space

Triangular structure

Cantilevers

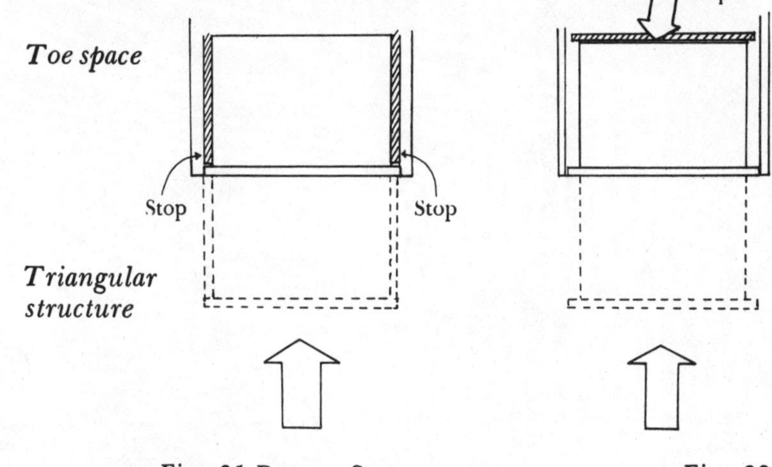

Fig. 21. Drawer Stops

Fig. 22

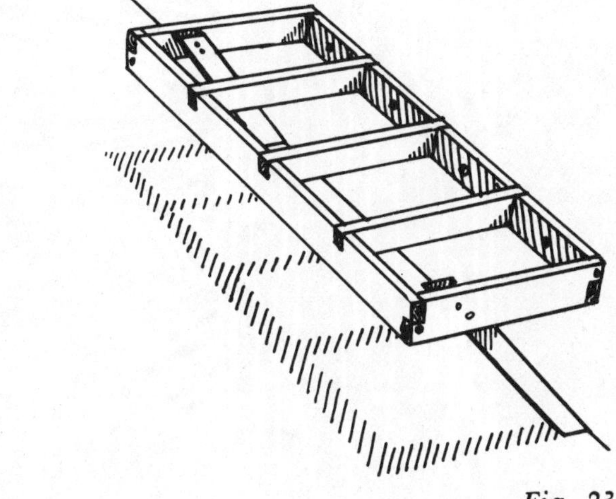

Fig. 23

Modular Construction

When possible, plan to build in a modular system—that is, where certain elements are a uniform size or even multiples of a given dimension. Your work will be much easier, and there will be less chance for error. For instance, if you want to build a number of boxes for storage cubes, make them all the same size or multiples. The layout of the cutting is easier; most of the pieces will be interchangeable, so your assembly will be simple; the finished product will be uniform, and able to be neatly stacked.

If you have a choice as to the cube measurements, make the most efficient use of your wood. If you're using plywood, plan to use almost the full sheet. Rather than sides that would be 2′2″x2′2″, make them 2′x2′ (or a shade less, allowing ⅛″ for the saw cut).

Most house construction is designed around the module of the 4′x8′ size of plasterboard and plywood. That's the reason studs and joists are put at 16″ intervals. If you plan for the maximum use of full sheets, you minimize your labor in measuring and cutting down the sheets to fit.

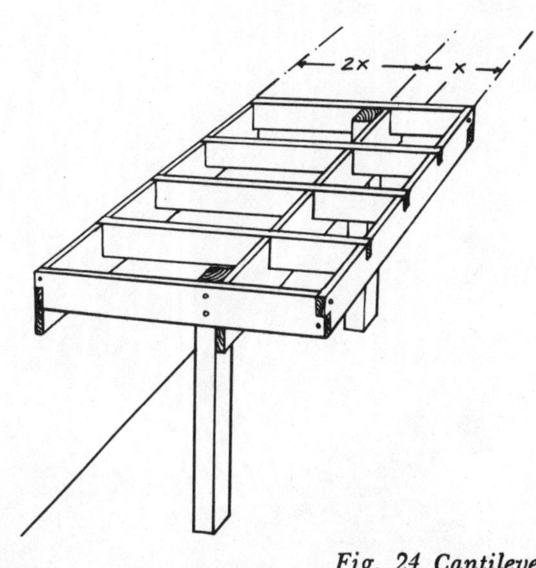

Fig. 24. Cantilever

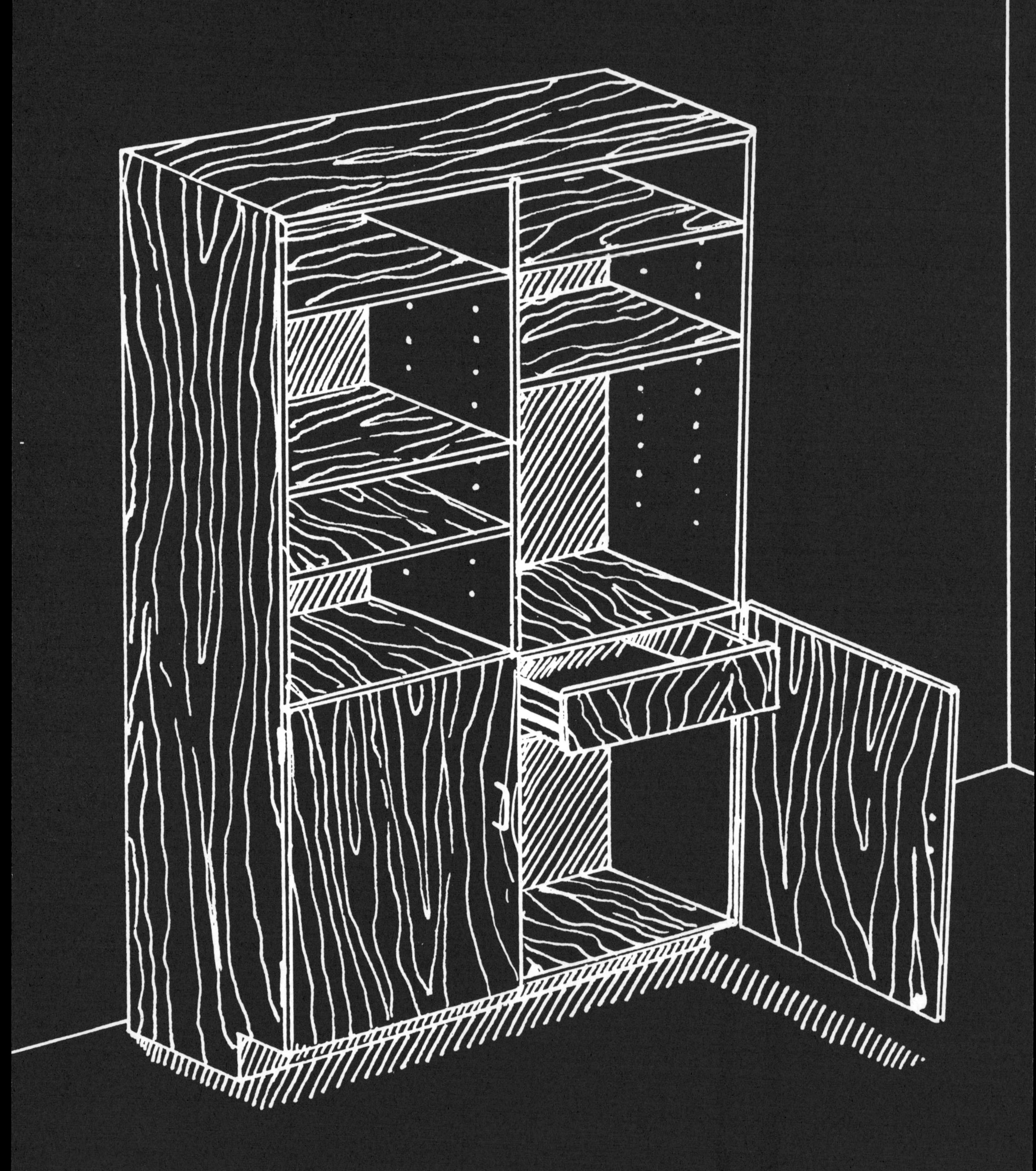

MANIFESTATIONS

10/FURNITURE PROJECTS

Step-by-Step Instruction from Planning to Completion and How to Modify Our Designs to Suit Your Own Needs

We've chosen the projects that follow for several reasons. First, they seem to be the ones people are most interested in having us build. Every project is complete with exact construction plans, materials, and tools needed. Variations in the design or construction, ways to work without the more expensive tools, are often included.

None of these projects requires a workshop setup or powerful table tools. We've built all of them in homes with nothing more than a floor to work on and with the listed tools. With the plans, you should be able to build any one of them, or any suitable variation.

More important, we chose these projects because they illustrate so many principles and techniques of carpentry. We are trying to present an organized *approach* to carpentry—a way of thinking and problem-solving—that will enable you to design and build projects of any kind. We don't want you to be dependent on our designs, much as we love them.

We include measurements so that, if you want, you can build a project exactly to our plan. But we explain everything along the way so that you'll see how to design a similar project to your own shape or size, with additions or deletions. That is really our intention—not to present you with a bunch of projects to build by rote, but to give you an understanding of how to design and build from your own ideas.

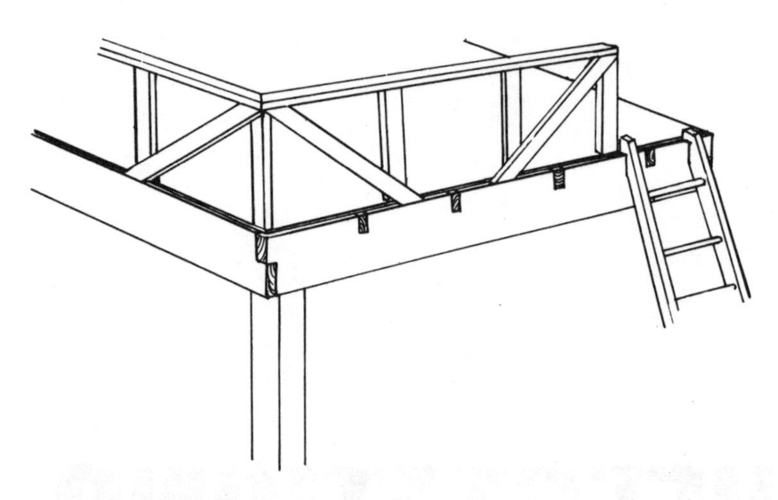

A. Diagonal Bracing

Any planning of a project necessarily involves a certain previous knowledge of construction and of the proper order of steps. This may be the hardest part for you—trying to plan ahead when you've never used those techniques before. So we also try to explain things very carefully in these long projects, showing just where and how to look for possible problems, and what they might be.

You may get a little confused in your own planning. Particularly at first. But stick with it. Any plan involves lots of wasted paper and erasers.

In some of the projects we delve less deeply into the designing and construction, though we still explain the basic structure. These are simple projects, whose construction should be fairly clear to you after reading "Working Techniques" and studying the longer projects.

Three Step Approach to Creating a Project from Scratch

First, without worrying yet about structure or how to make it, come up with an ideal visual design for what you want.

Second, try to give the piece a well-integrated structure, one suited to the type of stress and loads it will undergo. Make the whole structure work for you.

Third, make full plans for its construction, figuring just how you will make each part and the proper order of steps to follow. This will all take a lot of time. Be patient. Use lots of sketches throughout to help you understand things.

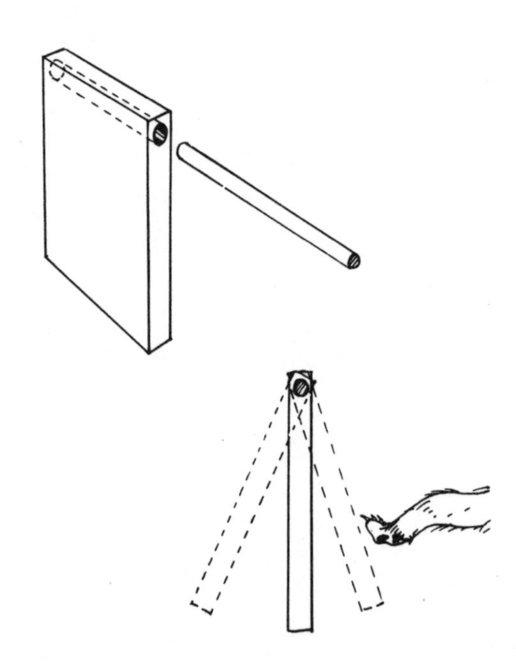

B. Cat Door

The steps will inevitably overlap at times—for instance, you may choose a certain design because you know that a certain structure will work well with it. Also, you may find your ideal design gets a bit mangled by the realities of structure or economics.

"Fine," you moan, "that's all very nice and general, but exactly *how* do I come up with this ideal design and all the structure?" Here are some suggestions.

First, you should familiarize yourself with the structural principles and working techniques in this book. Use those and any other related sections as reference. Don't forget the full explanations of tool usage in chapters 5 and 6. See if any of our projects are similar in some way to what you have in mind. Look around at similar furniture or structures for ideas—in homes, magazines, other books. Make some sketches. Use your common sense. These are all bases for creating your own solutions.

It will also help to ask yourself a lot of questions, particularly in the design area. The more answers you get, the more you'll define your design. If you start with nothing more than a vague notion of what you want, try to define it. What is it? What is it for? Storage? Display? Work space? Eating, sleeping? Noise abation, liquor libation? If it's for storage, how much space do you need, what kind of access? What size or shape area will the piece be made for? Standard dimensions for some furniture can be found in the Useful Dimensions table or the Dimensions of the Human Figure table in the Reference section. Or figure what's comfortable for your body; check out similar furniture around the house. Do you want the piece free-standing or built-in? What particular requirements does your home have? Do you have to worry about pets or little children? Noise? Electrical outlets? Water? What about ventilation? A loft bed, for example, should be near a window, since the air near the ceiling gets very stuffy in the hot months.

The answers will begin to define your project. Then, make sketches. Fool with them till you like the way something looks. Remember that the piece won't be in a vacuum, that it should relate somehow to the area around it. Do a scale view to see the proper proportions, and any other kinds of views that help you. Write down the measurements you want.

Make your design or structure in the most obvious way first. Then see if you can streamline it, modify it. Pare it down, combine, eliminate.

Another useful approach, particularly in figuring the structure and construction, is to break down the complex whole into its simple parts. Try to understand what's happening in those parts. Is it a joint? Is it something that has to be cut? Is there a dynamic stress at work? Once you see the underlying principle, you'll know what related areas of carpentry can help you.

You want to build a railing for a loft, for instance. What is a railing? It's like a wall, only shorter. So you can structure it like a wall. As you'll learn, however, a wall can sway if it's not joined to the ceiling, or at both ends. How to strengthen it? What strengthening or bracing techniques do you know? Diagonals are good. Join the railing to the loft floor with diagonals (*A*).

A more subtle example of finding the underlying principle: you want a cat door in your back door. It has to swing in *and* out, so the cat can go either way. Regular hinges won't work. (There *is* a special double-swing hinge, but it doesn't swing freely.) How can you construct a cat door that swings freely? Well, you need a hinge-type device, right? What is a hinge? What's the principle? A hinge is a flat plate that swings on a pin. The pin is inside the plate. Rather than trying to adapt a hinge to your cat door, you could make the door itself the actual hinge (*B*). That is, drill a hole through one edge, stick a dowel in, and insert the dowel into your door.

Learn to adapt the principles and techniques you're familiar with to unfamiliar situations. Every project is new, different from the last one. But the basic principles remain the same. We often feel a moment of panic when we first try to figure out a new design. We rarely know how we're going to build something when we start. We just assume that there will be a way. And there always is.

Planning the construction—step by step—is extremely important, particularly if you're not very experienced. For one thing, it minimizes mistakes. You may, for instance, assemble some pieces with nails and glue and then discover you have no room left to fit your drill in for the next operation. Advanced planning tells you to drill first, and then assemble. Any mistakes appear on paper, rather than in the project. Full plans mean economical lumber and hardware orders. Planning also lets you concentrate more on the actual work. Your carpentry will be easier, more fun, and you'll produce better work for it. The final product will be a more integrated design and structure if all the bugs can be worked out on paper.

Make final building plans and sketches of all details. Include all the dimensions. Calculate the amount of wood required, and make an organized list. Decide on the tools and hardware.

The Projects

The following projects are concrete examples of our approach. The longer sections in the beginning show how and why a design is formed, how (and sometimes how not) to structure it, and at least one way to put it together. Meticulous attention is paid to many tiny problems that can arise, the kind of problems most books don't mention, but that we know from experience can happen. Things like uneven floors, wood that's a little thicker or thinner than it should be, room corners that aren't square. The shorter projects are complete as to final design, structure, and general construction steps; but much less attention is paid to the process of design and to the tiny details of construction, which should all be much clearer to you by that point—if you've read the earlier projects carefully.

Final Tips

A few final tips about planning and carpentry in general.

Don't be afraid to experiment with new techniques, new tools, new designs. That's where the fun is. When possible, practice on scraps first.

When you can't decide which of two or three similar choices is the right one, probably they all are.

Insurmountable problem? Sleep on it. Carpentry problems are *always* smaller tomorrow.

Stay flexible of mind as you work. You should have complete work plans, but something may come up that requires a change in midstream. You can't foresee everything.

Good carpentry is using the techniques and principles as basic forms from which you shape your ideas. Different pieces or different environments call for different combinations.

As you work, be aware of the flow of the project, its changing shape, how strong or weak it is. Is it staying square? Watch your tools, their condition, how best to use them under different circumstances.

Understand the principles of each tool and you can stretch its functions, use it more fully and accurately. Always use tools properly and safely; stay friends with your tools.

The good carpenter is the person who manages to get into that flow. He is part of the construction, flexible and open to what is happening with the wood and the relationships of the parts.

BOOKCASE

A bookcase is one of the simplest pieces of furniture to build. It's also a useful piece that people often seem to need one more of. It's a good beginning project to build, or even to read about, since it illustrates a number of basic construction principles, without being complicated.

Our approach to this project is pretty much the same approach we use for all the projects. We start with discussing how to design for your own needs, how to structure the piece to fit the design, how to plan the materials, and continue right through the actual construction showing the (or one) correct order of steps to follow. This approach takes all the complicated problems that tend to intimidate people and breaks them down into clear, easy-to-deal-with steps.

Planning

Designing your own

Your first step is to decide *where* you're going to place your finished bookcase. You may find, for instance, that you have very little floor space for more furniture, in which case you should plan to build a hanging bookcase. Or you may find you have a specific space of, say, 23½″ between two other cabinets, which would define the maximum width you could make this project. Let's assume, however, that you have ample space for any reasonable-sized bookcase. Therefore your design will depend pretty much on the number of books you have, plus the size of the books. And we'll also assume that you want to make this project as simple as possible.

Start by drawing the rough outline of what you want (*1*). This looks like the bookcases in "Drawing" and also in "The Ubiquitous Box." There's nothing particularly complex about designing this piece. The top shelf should be within your reach, so make the bookcase 6′ high.

What about the width of the bookcase? Since we want to make the project simple, we don't want to have to put in center supports for extra-long shelves. Bookcases and most cabinetwork are made of 1-by stock (or ¾″ ply), and the suggested maximum span for a 1-by shelf supporting heavy books is 32″. (See Useful Dimensions in Reference section.) Adding 1½″ for the thickness of the two sides gives us a maximum width of 33½″. Let's make it an even 33″.

The depth of the shelves and the vertical height between them depends on the books you have. Most paperbacks fit on shelves 6″ deep, with as little as 8″ vertical space. Large hardcovers, particularly art books, might need 12″ depth and 14″ or more height. The point is that you should look at *your* books and see what *you* need. Since you may have only a few very tall books, you might give one shelf 14″ vertical space, the rest about 10″. The latter accommodates most hardcovers and also allows an inch or so of finger space—so you can reach in to pull the book out. We'll say, for our purposes, that a 10″ depth suits your needs.

Write these measurements down on a sketch, trying to fit them into the 6′ height. Allow ¾″ thickness for each shelf. You'll find you can fit in six shelves, plus a top piece, and have about 2¾″ left over at the bottom for toe space, a sufficient amount (*2*). So this will be the design you work from. The measurements worked out fairly easily here; on another design, say with a 6′6″ height, we would have had to adjust the heights of some shelves to make everything fit neatly.

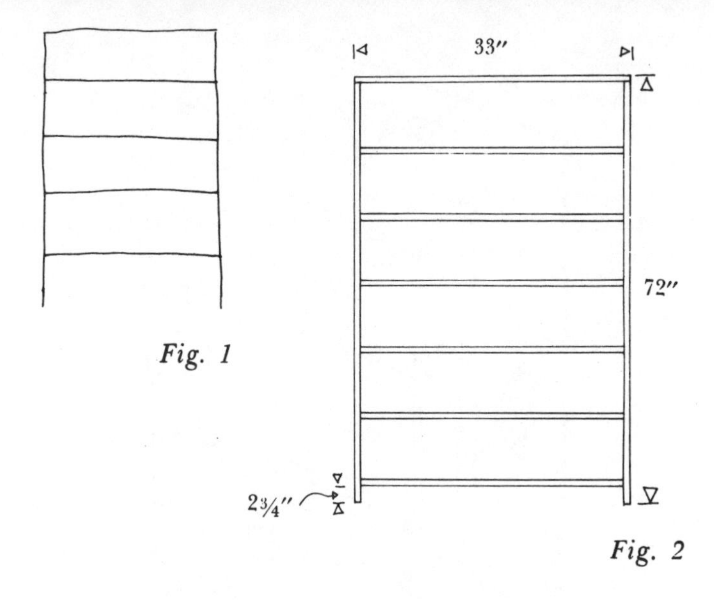

Fig. 1

Fig. 2

Structure

How do you put this piece together to obtain your ideal design? Let's keep in mind that for this project we are mainly interested in economy and ease of assembly; we don't care about building a masterpiece. After all, it's just a bookcase. We do want it to be strong, however. We will also assume that for this beginning project you do not have a vast arsenal of tools.

Two standard and structurally sound ways of putting bookcases (and cabinets) together are with dadoed joints and with cleats. Dado construction requires a router, or an expensive dado blade for a power saw—neither of which most people buy till they're well into carpentry. In the Cabinet project, we show how to rout and dado. Here, cleats provide the strongest simple method of joining the pieces together (*3*). The cleats are screwed and glued to the side pieces, and the shelves are glued on top of the cleats and then screwed into through the side pieces. (See Joints and Joining in "Working Techniques.")

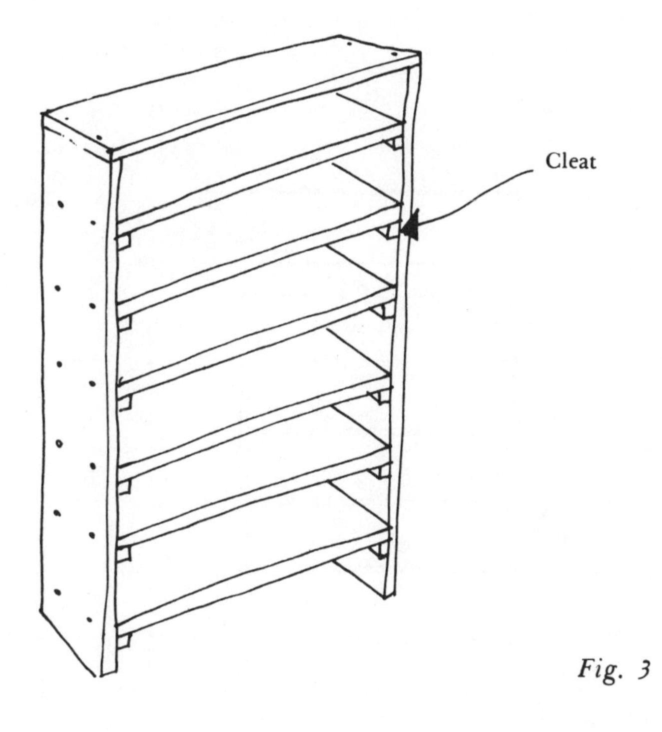

Fig. 3

Your ideal design has to be changed somewhat in looks. That's the way it is sometimes between the ideal and the reality. You could maintain the pure lines of your original by simply gluing and screwing the shelves to the sides without cleats, but as we've stressed before, this type of joint is a bit weak. There's too much stress right on the screws; you are not making the whole structure work as one piece.

The wood is also a structural element. As mentioned earlier, standard bookcase stock is either 1-by or ¾″ ply. Solid 1-bys are easier to work with in this project, particularly if you're not going to worry about perfection and exact measurements. Plan the shelf depth to be the actual width of the 1-bys—1x10s are actually 9½″ wide; let that be your shelf depth. Then you only have to cut the pieces to length. With plywood you would have to cut to length *and* to width. Unfortunately 1-bys may vary slightly in width, often up to about ⅛″. But the slight difference shouldn't bother you much for a simple piece like this. All it means is that the shelves will not all be perfectly flush at the front. (If you do want perfection, see the Variations at the end of this project.) Another reason to use solid-stock 1-bys is that you don't have to cover the edges with veneer tape or lattice strips, as you do with plywood.

Since we're still not interested in super-fine work, common-grade 1-bys, with some knots and imperfections, should be

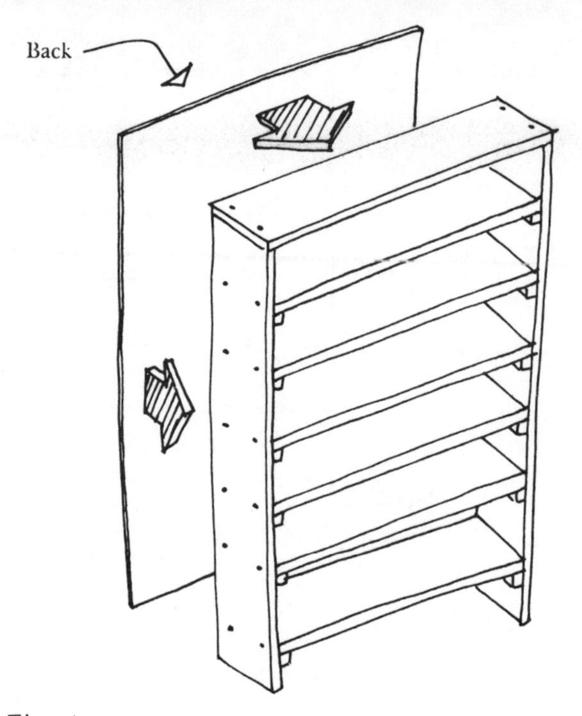

Fig. 4

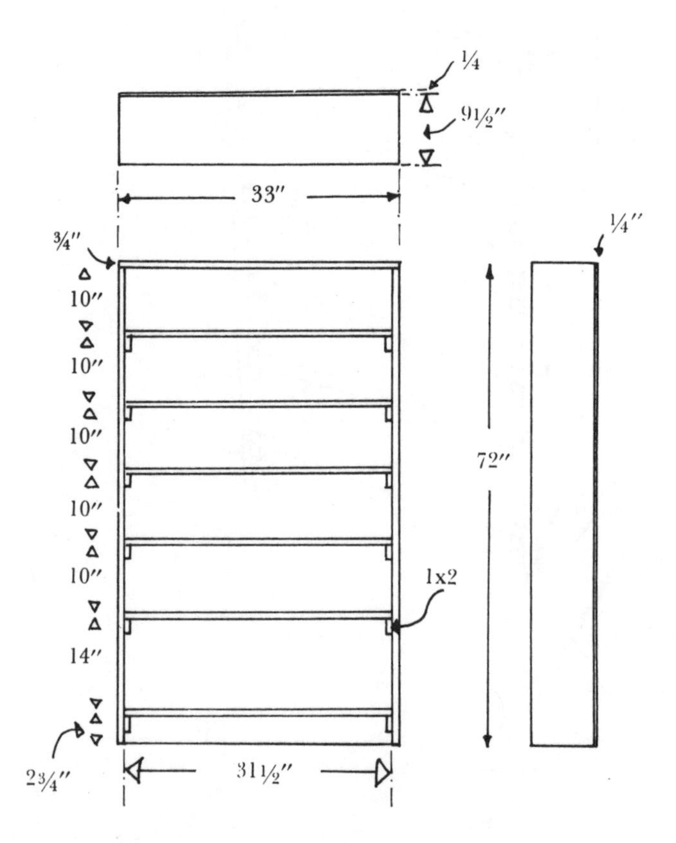

Fig. 5

fine. It's half the price of clear lumber, and some people even prefer the rustic look. Just try to pick the lumber yourself at the yard, to be sure the pieces aren't grossly warped or split. To be sure you'll have enough good lengths, get about 15 to 20 percent more than you need.

Standard stock for cleats in cabinetwork is 1x2. You could use 1x1s, but 1x2s are of course stronger. You should order *clear* 1x2s—not for looks, but to be sure you get straight pieces. Common-grade 1x2s are often in pretty bad shape; they are used mainly in furring out walls.

This bookcase is, of course, a basic box structure. With all the shelves acting as bracing, it will be more rigid than just a box frame. But we'd suggest adding a back to really firm it up. In fine work, the back is inserted into the frame so it doesn't show from the sides. Since that would mean reducing the depth of the inner shelves and adding extra cleats for the back piece, we can just fasten the back against the rear of the frame (*4*). It's certainly strong enough, just not as clean-looking. Again, ease and speed of construction are the deciding factors. Make the back out of ¼″ ply. Glue and fasten it to the frame and shelf edges with ¾″ finishing nails.

Finish

A bookcase does not need much protection in the way of a finish. One or two coats of stain will be sufficient. The wood may need some sanding prior to the stain, depending on your taste. If you want to really protect the surfaces, give it two or three coats of varnish or polyurethane.

Working plans

It's important to have a working plan of some sort, a blueprint that will be your construction guide. Draw the project in whatever style you find easiest to work with—oblique, front and side views, orthographic, etc. It can be a rough sketch, but include all the measurements—and don't forget to account for the thickness of the wood (*5*). Do enough sketches to show *all* the pieces and *all* the joints. The more you draw, the less chance there is of making an error in the actual construction.

List of pieces

Consult your plans and make a list of every single piece of wood (*6*) with its measurements. It's helpful to group the pieces by type of lumber as you go.

1x10s

Top shelf—one at 33″
Other shelves—six at 31½″
Sides—two at 71¼″

1x2s

Cleats—twelve at 9½″

¼″ plywood

Back—33″x72″ (you may end up cutting this a shade shorter in either direction)

Lumber order

From your list of pieces, calculate the total lumber you need in the specific lengths that will give you an efficient use of the wood. For instance, you could get three 31½″ shelves out of an 8′ length; but that leaves almost no room for eliminating bad knots, splits, or warps that occur in common grade. If you pick the wood out yourself, you might be able to get a perfectly good piece. Otherwise, it's best to plan for waste. One possible arrangement is three 10-footers and one 8-footer. One 10′ piece gives three shelves with about 2′ waste; the other 10-footers each give a shelf and a side with about 1½′ waste. The 8′ piece gives the remaining sixth shelf plus the top, with about 2½′ waste. This arrangement leaves a very generous amount of waste, probably more than you need. But it saves you an extra trip to the lumberyard in case you need just one more piece. Also remember that any good leftovers can be saved for the next job.

The cleats come from clear stock, so the only waste you have to consider is what you lose through saw cuts. (Usually figure about ⅛″ per cut.) Therefore a 10′ length of 1x2 would suffice, with practically no waste. You might get a 12′ length instead, however, just to insure against a mistake on your part—you might accidentally cut a piece too short, for instance.

The back, of course, can be cut from one 4x8 sheet of plywood. You want it good on one side. Quarter-inch Philippine mahogany (or lauan) is clean and cheap, and a good choice if you don't mind the contrast between it and the pine 1-bys.

Thus, your order looks like this:

1x10s (common grade)
Three 10′
One 8′
1x2s (clear)
One 12′
¼″ *plywood,* good one side (mahogany, if you like)
One 4′x8′ sheet

When the lumber is delivered, check it carefully before you accept it. Make sure all the pieces are there; that they are the proper lengths; that you've got 1x10s, not 1x8s; that the 1x2 is indeed clear; that the ply is not damaged; and that the common 1x10s are not grossly warped or split (expect some flaws, however—that's why it's cheaper).

Hardware order

You need screws to join the cleats to the sides, and the sides to the shelves. Screws 1¼″ long are the biggest you could use without the cleat screws breaking through the outside surface of the bookcase. No. 8 or No. 10 screws are both strong enough. They should have flat heads so they can be countersunk below the surface. Otherwise you might scrape your hand reaching for a book.

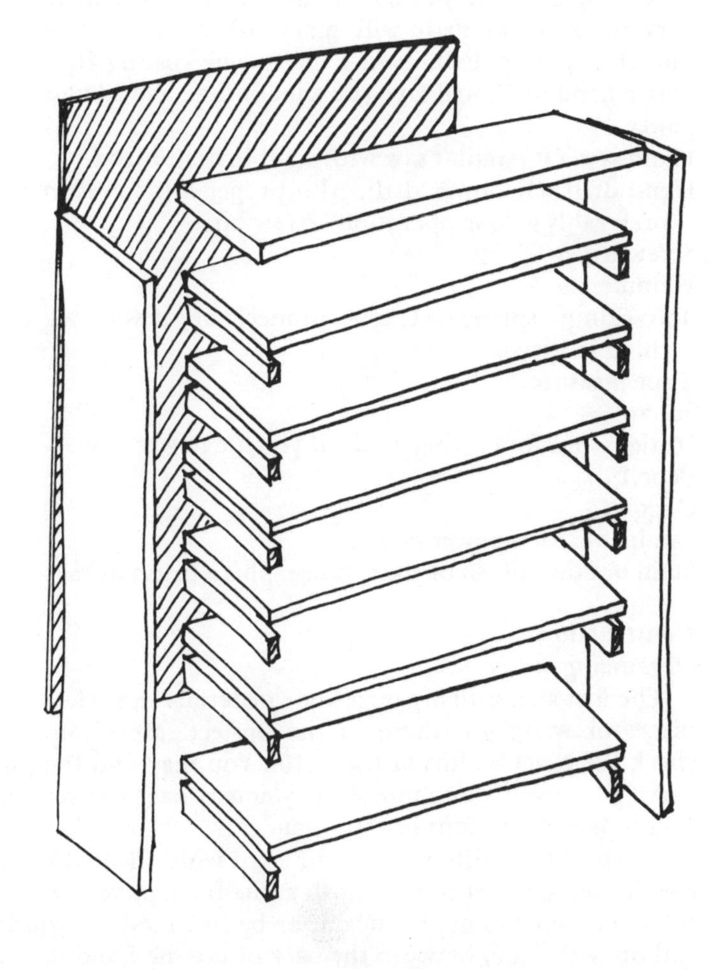

Fig. 6. View of Whole Structure

You need the 3/4″ finishing nails for the back piece; you could also use some 1 1/4″ finishing nails just for tacking the pieces together while you drill and screw them.

60 1 1/4″ No. 8 wood screws, flathead
1 lb. 3/4″ finishing nails
1 lb. (or less) 1 1/4″ finishing nails
Wood glue—any white carpenter's glue like Elmer's, or a liquid-hide type
Stain and/or varnish or polyurethane

Tools

Planning the tools you need should be part of your original working plans. You may have to adapt certain projects because you can't afford a certain tool. We're using cleats instead of dadoes, for instance, because most people don't buy routers till they have more experience.

This project could easily be built with nothing but a few hand tools. Except for the plywood, all the pieces can be cut to length with simple crosscuts. All you need is a handsaw and a minimum of skill. You can even cut the plywood with a handsaw if you like; it just takes a little extra care to make those long cuts. Or you could have it cut to size at the lumberyard. A hand drill will make all the holes you need, though a power drill, even a cheap one costing little more than a hand drill, will save you a lot of time and a lot of muscle power.

Hand saw. Or circular saw with combination blade
Hand drill. Or power drill. Also proper bits for screwholes, preferably a "one-operation" screw bit
Screwdriver
Hammer
Two clamps, spring or C type (unnecessary if you tack everything in place)
Tape measure
Square
Straightedge for cutting guide, if you use power saw
Pencil
Goggles
Sandpaper, or a power sander
Stain or other finish of your choice, plus a brush or rags

Construction

Organizing materials

The first step is to organize your materials and select which pieces of wood go where in the project. Most important, check the exact widths of the 1x10s. You may find the pieces vary 1/8″ or even a bit more. If so, you may have to make some decisions and revisions in your plan.

The problem with pieces of different widths is that some of the shelves may not line up flush at the front or rear with the sides. The shelves might jut out or be indented; or you may end up with a gap between the back of the shelf and the back piece of plywood. Since the difference will probably be no more than that 1/8″, however, we don't feel it's too crucial to even out the pieces for this simple project. It's not worth the trouble. Just try to choose the pieces so that the elements of the outer frame—the sides and the top piece, plus the bottom shelf, if possible—are all the same width. This looks better in the front and, more important, insures that the back will set flush against the frame for a good joint. You can choose the frame members to be the narrowest pieces or the widest pieces, or even the in-between dimension—just so long as they are all about the same. It's up to you whether you want the shelves to overhang, or be slightly indented, or a little of each. If you want to spend the time actually evening all the pieces to the same width, see the suggestions in Variations.

As you select which pieces go where, also look to be sure there are no bad splits or warps within the needed lengths, particularly on the sides and top piece. Mark off the areas that are too flawed to use, then mark off and measure where each of the bookcase elements will be cut from. Always start with the longest pieces first—the sides in this case—then try to fit in the smaller ones. Remember that a 10′ piece may look pretty warped, but when it gets cut into three smaller pieces the warp may be hardly noticeable on each.

Cut the pieces

Now cut all the pieces out, including the cleats. It's always best to do as much of one operation at one time as possible. Setting up the tools and space takes time; you don't want to repeat the setting up more often than you have to. Nevertheless, we prefer to wait to cut the back till after the assembly. Then, if the final real dimensions are slightly different from the original ideal design, the back can be cut accordingly.

The cleats should be cut to the *real* width of the frame; or, if the shelves are less deep than the frame, to the depth of the shelves (so that the cleats don't show in front of the shelves). Use your original plans as guidelines, not gospel. Change them when necessary to keep the project unified and sound.

Check the dimensions of the pieces you've cut. Be sure you have five equally long shelves plus one that's 1 1/2″ longer. At this point you might select how you'll arrange the wood in the bookcase. If any pieces have particularly nice grain patterns or particularly ugly flaws, you'll probably want to mark them to be placed accordingly. For instance, a piece with a bad gouge on one side could become a low shelf with the bad side facing down so it will never be noticed.

Pre-assembly work—sanding

We know you're eager to start putting all the pieces together, but you should always stop and think first, "Are there any jobs easier to do now than later, when the piece is assembled?" In this case there is one thing you can do now. If the wood is very rough and you plan to sand it before staining, sand it now. You'll still have to do a little finish sanding later, but you can get the bulk of the work done while the pieces are separate. It's much harder to sand the inside of a bookcase—particularly the joints and corners—than it is to sand individual pieces of wood.

Always try to think ahead, save yourself trouble. In this case you *could* do all the sanding later. In other projects, the step you forget may not be *possible* to be done after assembly.

Layout cleat positions

For the shelves to be joined to the sides, the cleats must first be attached to the sides. Then the shelves sit on the cleats.

If the side pieces are bowed, the bow should go outward in the middle when the frame is assembled; therefore the cleats should be attached to the inner or concave side of the piece (7). The bows are arranged this way because it's easier to pull a bowed piece *in* to meet the shelves than it is to push the bow out. Also, in the finished piece, it's better structurally for the middle of the side piece to want to bend out and separate from the shelves, than for both ends of the side to want to pull away at the frame corners.

The top of each cleat should be flush with the bottom of the shelf it supports, obviously. Therefore on the inside of the side pieces, lay out these lines for every cleat. For instance, the bottom of the bottom shelf is to be 2¾" off the floor (8). Use your square to draw a straight line across each side piece at 2¾" from the bottom. Set the two sides together and mark them at the same time.

Attach cleats

Your aim is to glue each cleat, place its top edge on its layout line while keeping its ends flush to the edges of the side piece, drill two holes through the cleat into the side piece, and fasten the cleat down with two screws.

The main problem is how to *hold* the cleat in place while you drill the holes, particularly since the glue makes things slippery. One way to hold the cleat is to use clamps. C clamps or spring clamps work well here. Use a thin scrap piece between the clamp jaws and the wood to protect the bookcase from marks.

You could also tack the cleat in place with some 1¼" finishing nails. Start the nails into the cleat before gluing it, letting the points protrude just a bit through the back of the cleat. Apply the glue and press the cleat in place. The points should catch on the side piece just enough to hold the cleat steady while you hammer the nails in a little more (not all the way). If the cleat still slips off the line, remove the nails and try again.

When the cleat is in position (either clamped or tacked) drill two holes through it for the 1¼" screws. Each hole should be about 1½" in from the ends. A screw bit is best here. If you don't have one, and have to use several bits for each hole, it's best to tack all the cleats in place first; then drill all the holes with one bit before changing to the next.

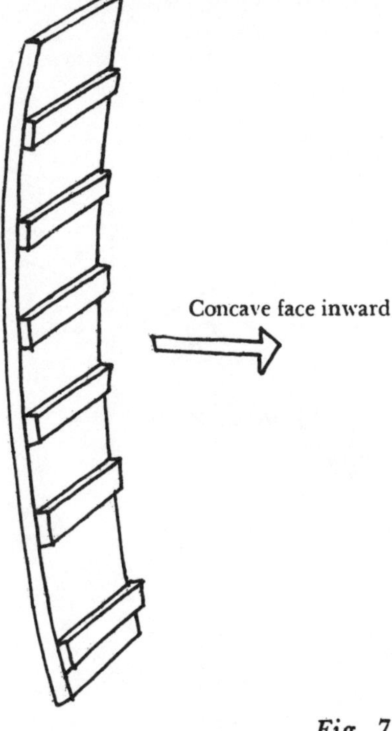

Fig. 7

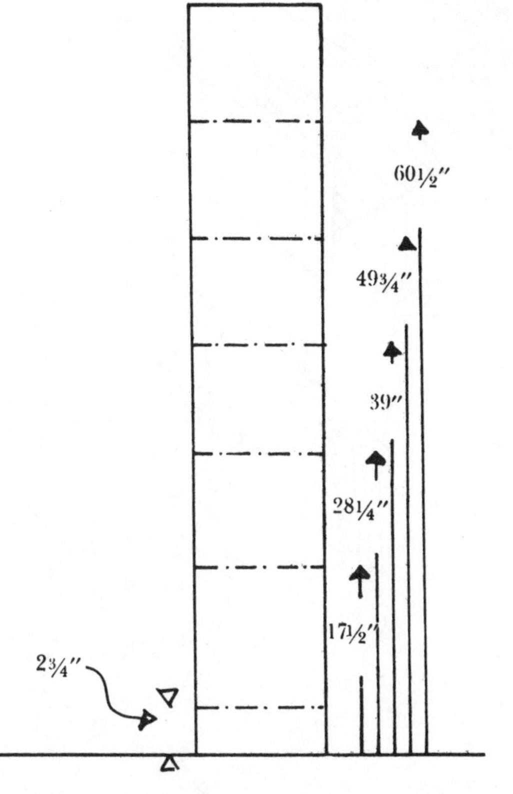

Fig. 8. Correctly finding cleat heights

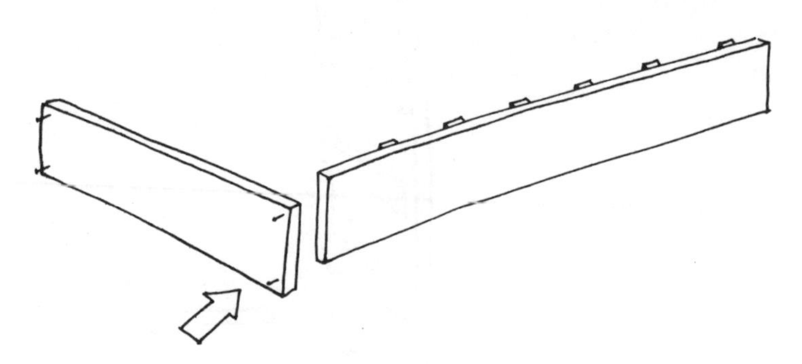

Fig. 9. Joining top

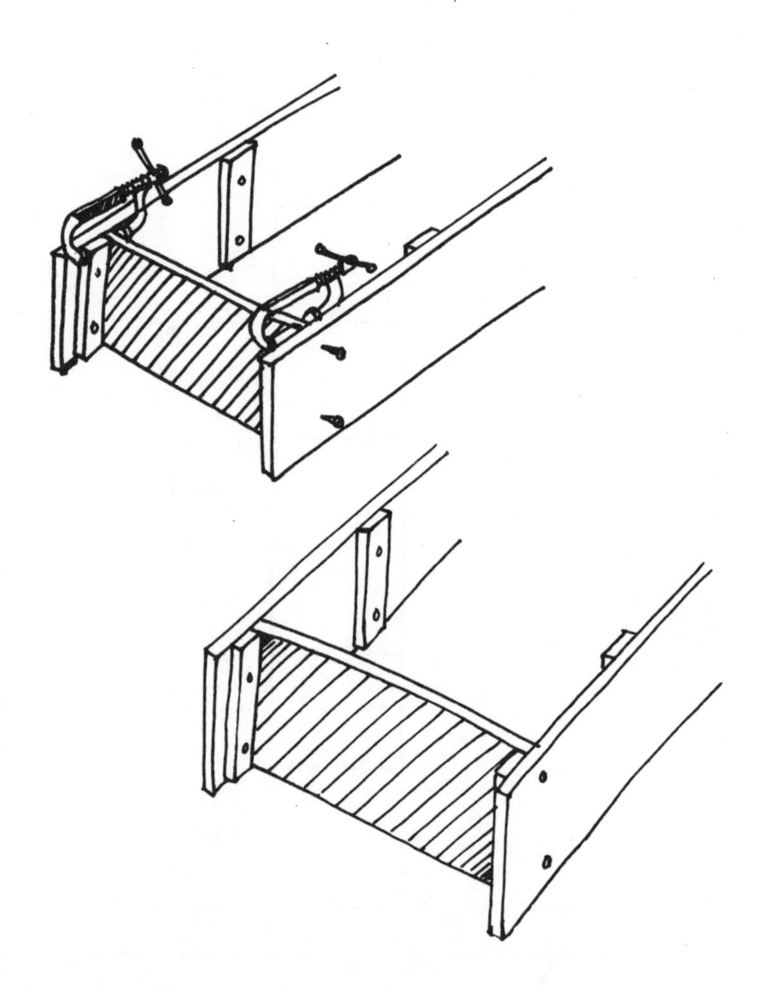

Fig. 10. Starting construction of shelves

Remember you want the screwheads countersunk flush to the surface or else counterbored below, so they're not sticking out. If counterbored, you can even putty them over later if you like. *Don't* counterbore more than 1/4", however, or the 1 1/4" screws will break through the outside of the bookcase.

Have a wet rag or roll of strong paper towels handy to wipe off the excess glue. Ideally, you use just enough glue to cover the joined area but not so much that it will squeeze out as the screws are tightened.

After the screws are in, either remove the tack nails or drive them in all the way. Use a nail set to sink them below the surface.

Assembling the frame and shelves

Usually try to assemble the basic structural frame of a piece first. In this case, that means the two sides, the top, and the bottom shelf. When this frame is set, the basic shape of the piece is set and the remaining pieces should fit in place easily.

The project is best assembled with the front of the bookcase on the floor. This allows you to keep an eye on the back edges, to keep the shelves flush with the sides so that the back piece will fit well against them later.

Join the top to one of the sides first. Tack some finishing nails along both ends of the top piece, their points just protruding (*9*). Spread glue on the top edge of the side piece. Position the two pieces with their front edges on the floor; press them together till the nail points catch and hold. Be sure the pieces butt correctly, flush everywhere. Tap the nails in part way and check the alignment again. You may find yourself wishing you had five hands; try kneeling on the floor instead, holding the side piece between your knees and feet. Hold the butt joint together with one hand and hammer with the other.

Next, glue and tack the other side piece to the top. It's not important to keep these joints perfectly square yet. The final adjustments can be made when the back is attached. Do try to keep them roughly square, however.

When the joints look good, drill two holes at each joint through the top into the end of the side piece, and screw them together. If the top piece is slightly cupped and not flush to the side piece in the center of the joint, put a third screw in the center. Again, either remove or set the finishing nails.

Do the bottom shelf next. Spread glue on its ends and on top of its corresponding cleats. Slide it into place, again keeping the frame roughly square. Tack it through the side pieces with finishing nails, or clamp it to the cleats (*10*). Hold the shelf tightly against both the side and cleat as you tack or clamp. Drill two holes through each side into the ends of the shelf; drive the screws. Remove the clamps and excess glue.

Repeat this process with all the remaining shelves. If a side is bowed out, pull it in to meet the shelf as you clamp it. You may have to brace one side with your foot or against a wall so you can pull on the other side.

Attach the back

The bookcase will probably still have some lateral sway in it. Play with it till the corners are square. The easiest way to check the square is to measure the diagonals: when they are equal, the corners are square *(11)*. If the diagonals are unequal, you've got a parallelogram. This is always true for any four-sided figure whose opposite sides are of equal length.

Now measure the dimensions for the back piece. Take the measurements along the edges, from corner to corner, in case the middle of one side is bowed a little. Cut the plywood to a hair less than this size. (If your floor is at all irregular, you should make the back an additional ¼″ short.) Keep and mark one square factory corner on the piece.

Glue the back edges of the bookcase and lay the back piece on. Align that square corner you marked with one upper corner of the frame *(12)*. Start tacking finishing nails through the plywood into the frame along one of the sides that forms that square corner. Don't drive them all the way yet. Keep that side of the plywood flush to the bookcase edge; if the frame member is a bit bowed, try to pull it in to meet the plywood flush. Check the diagonals again; adjust the frame if necessary. Tack down the other side of that square corner the same way. Check the diagonals of the frame again. They should be equal or fairly close (within ⅛″); if not, you'll have to pull the nails from one side of the back and readjust the frame. Once the frame is set, tack down the other two sides of the back. Check the diagonals once more to be sure, then drive the nails in all the way.

The construction is now complete. If you want to putty the screwholes and nailholes, do so now. Do a final sanding if you want, and then apply the finish of your choice.

Variations

Inserting the back

If you don't want the side edges of the back to show, you have to make room for the ¼″ back within the frame. This means cutting all the shelves between the top and bottom ¼″ narrower. You must also run cleats around the inside of the frame for the back to sit against and be screwed to. One method is to insert small cleats between the shelves *(13)*; the other is to run full-length cleats around the inside of the frame, run the shelf cleats up to them, and notch the back corners of the shelves to fit around the vertical cleats.

Inserting a back is much easier with a router. Just rabbet the back edges of the frame. But the shelves still must be made narrower. (See Cabinet project for method.)

Making all the shelves equal depth

If your 1x10s are of unequal widths, there are several methods to get the shelves and frame flush and equal widths.

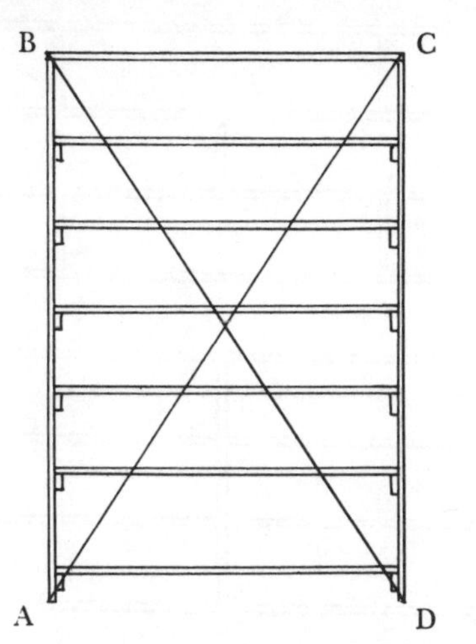

Fig. 11. AC should equal BD

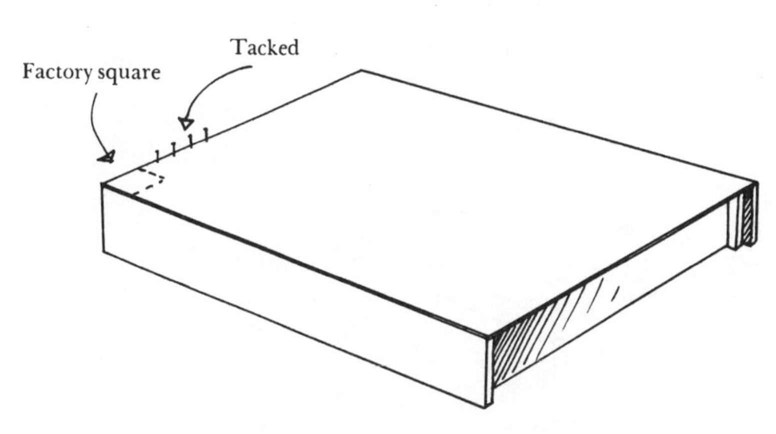

Fig. 12. Starting back

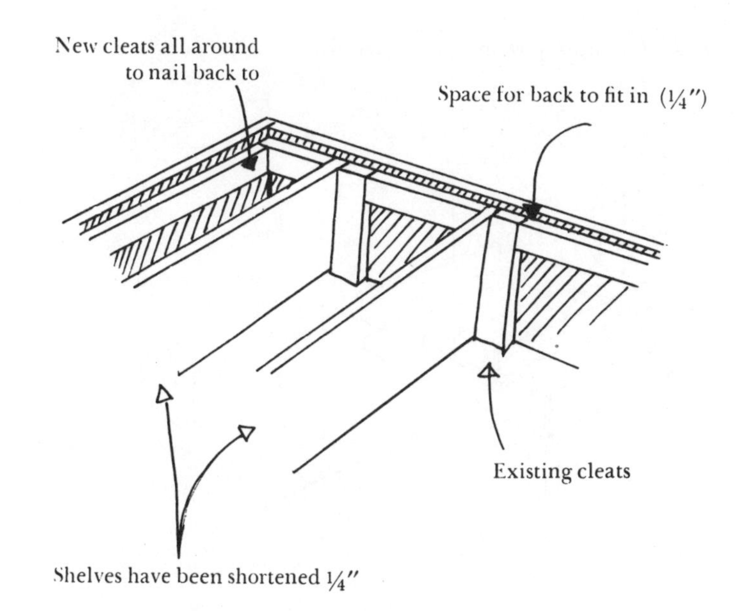

Fig. 13. Cleat Set-up for Back

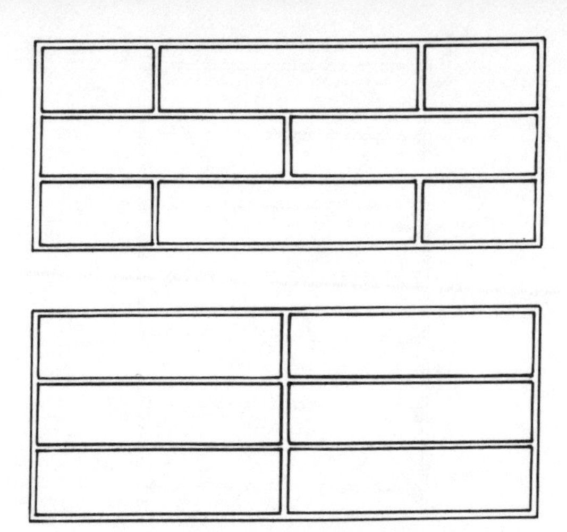

Fig. 14. Bookcase Shelf-Support Variations

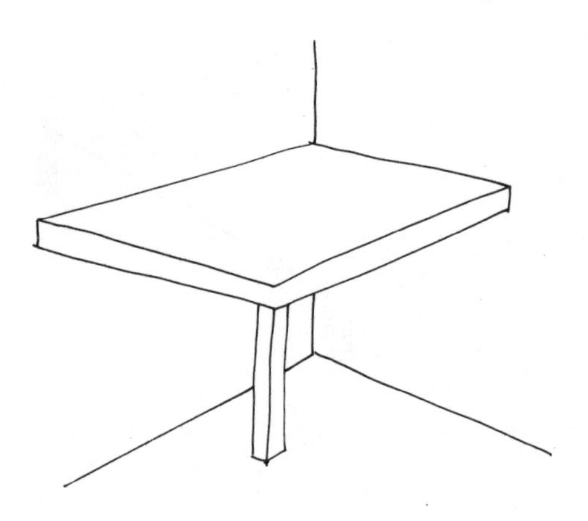

Fig. 1. Rough Drawing of Loft

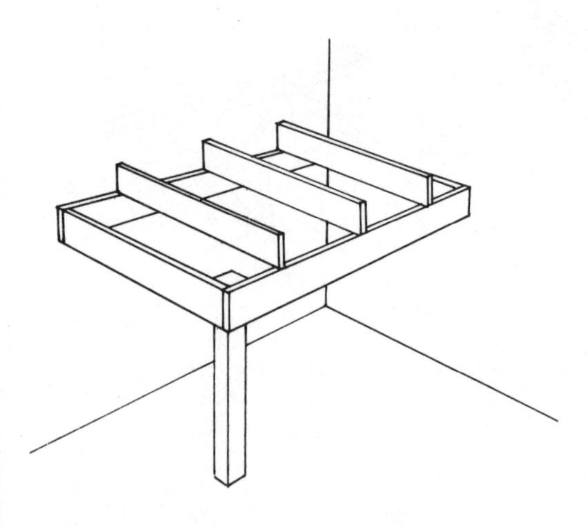

Fig. 2. Incorrect Loft Structure

First, of course, you can simply rip all the pieces to the same width with your power saw and straightedge guide. Chances are, however, you may still be a little bit off.

If you are good with a plane, plane the edges down. This takes a lot of time and skill. Try clamping all the shelves together and planing or sanding them even.

The other method is to sand the edges flush after the bookcase is assembled. Use the narrowest pieces for the sides, then assemble the shelves with their back edges flush to the back of the sides, so the back piece will fit well. Then sand down the shelf overhangs at the front of the bookcase. A belt sander is best; a finishing sander does the job but takes a long time. Hand sanding will take forever. Actually you can cheat and just sand the joints flush, tapering the rest of the shelf edge. This actually results in a slightly convex front edge of the shelf, but it's barely noticeable.

Dealing with baseboard molding

It's quite possible there is a bothersome strip of molding on your wall along the floor that prevents your placing the finished bookcase flush up against the wall. If the floor is flat so the bookcase is stable and the gap does not bother you visually, everything's fine. Otherwise there are several possibilities. You can cut away the molding right behind the bookcase; the disadvantage is that the gap shows if you ever want to move the bookcase.

You could also notch the sides of the bookcase to fit around the molding, in which case you would also have to make the bottom shelf shorter in the back, and stop the back piece above the molding height.

If you don't mind the gap, or if the bookcase is between two other pieces that hide the gap, but the bookcase is unsteady, attach a cleat to the wall behind the bookcase and then attach the back piece to the cleat.

Shelf supports for long spans

As mentioned earlier, ¾″ shelves under heavy loads should not span more than 32″ without extra support. Figure 14 shows some examples of how to support long shelves. All joints could be dadoed, or cleats could be used at the ends, with the support pieces simply butted, glued, and screwed in place.

Other bookcase designs

See the Variations at the end of the Cabinet project. Although they are assembled with dadoes, every one could easily be done with cleats instead.

LOFT BED

A loft bed magically creates extra space in cramped apartments. Bed above, space below. No rule says the sleeping area has to be on the same level as the kitchen and living room. In fact, a loft bed can give a limited amount of privacy to one person while another is still up and about in the same room. Kids, particularly, like sleeping high in a loft, whether they need the extra space or not.

If your ceiling is high enough—say 10′—you can build the loft high enough for a six-footer to walk under. Even with an 8′ ceiling, the under-space can be used for storage or a sitting area. By building vertically you get twice as much use out of one area.

Planning

Designing for your home

Why do you need a 10′ height? Assume for the nonce that a loft platform and mattress will take up approximately 1′ of vertical space. If you've seen any lofts, you could estimate—about 6″ to 8″ for the structure plus the thickness of your mattress. You need about 3′ clearance above the mattress to be able to sit up comfortably, and at least 6′ clearance under the loft. Kids need less, basketball players more.

Let's see how you would go about planning and building a loft bed for your home. We will examine all the steps from the first thought through planning, ordering materials, construction, and finishing.

Your first consideration is the height of your ceiling. Do you have 10′ of space? If not, you will have to sacrifice some head room above or below the platform. Decide what's more important to you. You'll be using the top every night; maybe you'll be using the bottom area only sporadically, to get a book from the shelves or as a sitting area.

After you determine the structure and its size, you can figure your loft height and the clearance exactly. Your particular loft might easily take up less than a foot of vertical space.

How big will the platform area be? Single or double mattress? Will you want extra space around the mattress—on the loft floor, that is—for a clock, radio, Kleenex, books? Slippers? TV? Any shelves, cubbyholes?

Look around the room and pick a place to put the loft. The more walls you can attach it to, the better. With one wall, you need two posts. Two adjacent walls means just one post, and three walls means you don't need *any* posts. A free-standing no-wall, four-post loft is the most prone to swaying, since it isn't tied in to the house structure.

Pick a wall that will be easy to work with and yet provide good support. Wood studs are best; brick or masonry are fine but mean more work. Avoid walls where the main supporting structure is more than 2½″ behind the surface, such as plaster on lath with brick 3″ back.

Consider the ventilation situation. It's always much warmer nearer the ceiling. That's fine in the winter, but will you be near a window in the summer? Can you get a fan or air conditioner near the loft? Will you even be there in the summer? Will you have any light up there? Or do you need to bring a new outlet up? Will an extension cord do? How do you get up there? A ladder? Which side can you put the ladder on so it will be least in the way? Do you want a railing? On how many sides?

These are all considerations to deal with. Once the loft is up, it may be too late. Draw a rough sketch of your final design, even though you still don't know how it will be put together. Let's say it's for a single mattress (39″x75″), with some floor area around it—3′10″x7′10″, with the platform 7′ off the ground. It's to go in a corner against two walls, both ¾″ plaster on brick. So you only need one post at the open corner. Your rough sketch might look like Figure 1.

Structure

Now you have to figure out the structure, how the pieces support each other and combine to make one unit. A number of different structures will work, but each is based on the simple structural principles discussed in "Structure."

As with any unknown structure, try to break it down to its basic components. Trace the stress between the load and the main structural resisting force. The load here is you and your mattress. The main resisting force is your house—the walls and the floor. If you want to put yourself on a mattress 6′ in the air, how are you going to transmit your load to the house?

We know that the mattress has to rest on a platform, some sort of floor. Your floor can be a sheet of plywood. How do you hold the floor up? Like any floor, with joists. Nail the ply to joists. Find the right size joist for the span (see Span charts in Reference) and all the load will be supported.

Now how do the joists transfer the load to the building? How do you attach them to the walls and the real floor? One logical solution would be to put them on top of a frame. Attach the frame to the walls and put a post at the outer corner. The frame idea is good. Putting the joists on *top*, however, is awkward to build, an inefficient use of vertical space, and it looks bad (*2*). More important, it is an inefficient structure, because there is no tight, solid joint between the joists and the framing beams. The structure would stand, but with time it might sway, and start to pull out from the walls.

Think of a better way to join the joists to the frame. The joists could be notched and nailed into the beams, or they could rest on cleats attached to the inside of the beams, and be fastened through the beams with nails or screws (as in the Captain's Bed project). Or you could even use joist hangers. All of these methods use the vertical space efficiently and unify the entire structure. You would not (we hope) slip the joists between the beams and just nail or screw them in. That would mean that most of the load would be on the nails. The stress dispersion from joist to beam is only as strong as its joint. Remember you want to make the whole piece act as one structure. The joints must be as strong and tight as possible.

Our own preference is to notch (*3*). It looks better from underneath and is more fun to build.

The frame corners—the joints between the framing beams—should also be joined strongly to rigidify the structure. Whether or not you notch for the joists, notching these beam joints is wise. The beams fastened to the *walls* should be notched so as to support the others (not as in Figure 4). This gives added strength and also simplifies the assembly and fitting of pieces. With one end of a beam in its notch, you're free to work at the other end. You don't need an assistant to help hold the piece up.

All the corners should be bolted. If you want the loft permanent, you can also glue these joints. It's not necessary, though. The joint at the inside corner of the two walls can be notched and nailed, since you can't get a bolt in there.

If for some ungodly reason you don't like the look and ease of notches, you can get away with simple bolted and glued butt joints. But you would be well advised to stick a 4x4 block up inside the corner and glue and bolt to that also.

Now we know the simple structure: plywood nailed on joists notched into a frame attached to the wall and a post. This is actually similar to the basic structure of a wood frame house.

Wood sizes

What you need now is the proper size wood for each element of the structure. A regular floor can be ¾″ ply or composition board on joists spaced 16″ apart, so you know this is more than adequate for a loft floor. Actually a loft floor won't undergo as much stress as a regular floor, particularly if you can't stand on it and the mattress is dispersing your weight. From experience we know you can get away with ¾″ ply or composition board on *24″* spaced joists, or ½″ ply or ⅝″ composition board on 16″ spaced joists.

What size joists? That depends on the size of the platform, of course, and how much weight you expect to have up there. In general, for lofts you can use 2x4 joists (spaced 16″ on center) for spans up to 6′ wide. If the loft is fastened to and rigidified by only one wall, 2x4s might creak a little; you might use 2x6s instead for spans 5′ to 6′. For larger spans up to 9′, 2x6s will be good joists. If you have plans for supporting five or six people up top, or anything larger than 9′ wide, you'd better check the Span charts in Reference for the proper joists.

The proper size framing beam for your loft is dictated by the joist size and the construction method. Cleats require a beam the next size bigger than the joist, so you can fit the strip and the joist within the face of the beam—a 2x6 beam for 2x4 joist, 2x8 for 2x6. With joist hangers, you can use beams the same size as the joist, but you should use at least a 2x6 beam for lofts longer than 6′. For notching, use a 2x6 beam for 2x4 joists, and either 2x6 or 2x8 beams for 2x6 joists when the loft is up to 8′ long. Use a 2x8 beam for lengths up to 12′. You can cut our requirements a little and still have a loft that will stand, but it may creak and sway under heavy use. The loft may sleep only one or two, but you may have extra people climbing up at a party, or you may want to store heavy boxes of books up there. So build it strong.

Your post will of course be a 4x4. That's standard post material. You will have to fasten it to the floor of the room somehow. This floor joint doesn't have to be too strong, since the load of the loft will prevent it from lifting up. But you don't want the bottom to shift out of position. Toenail some nails through the post into the floor, or countersink screws or bolts through it at an angle into the floor. Or simply nail molding to the floor around the base of the post.

Final touches

Anything else need figuring out? If you've made a rough sketch, you can see that the edge of the plywood will show above the frame. If this bothers you, cut the plywood a little smaller and put some molding around it (*5*). Use ¾″ wide molding and the ply will still have ¾″ of the beam edge to be nailed to—more than enough. A 1x2 on edge would project ¾″ above the ply and keep the mattress from slipping off.

Your plan is almost complete now. One more thing before figuring the materials lists. A ladder. Stairs are unnecessary for a sleeping loft, and they would occupy floor space you're trying to liberate.

The easiest ladder to build is made of wooden dowels between two side pieces (*6*). Dowels 1¼″ in diameter between 2x4s are fine. A space of two feet between the sides should be comfortable. Drill holes throughout the 2x4s and stick and glue the dowels in. Run the ladder straight up to the loft, or at a comfortable angle.

A ladder with flat steps will be easier on bare feet. It's a bit more difficult to build, but not much. Use 2x4s for the steps and the sides. The steps have to be attached well, or the sides will eventually pull loose. The fancy way is to dado the sides and glue and screw the steps in. Rougher looking, but very strong, is the cleat method. Attach 1x2 cleats to the sides with glue and either nails or screws; then glue and lay the steps on top of the cleats, and nail or screw through the ladder sides into the steps—just like the shelves in the Bookcase project. The complication with building flat steps comes when you want to run the ladder at an angle rather than straight up; the steps have to be fastened at an angle so they are parallel to the floor and easy for you to walk up (see Variations). Round rungs are the same angle no matter which way the ladder runs.

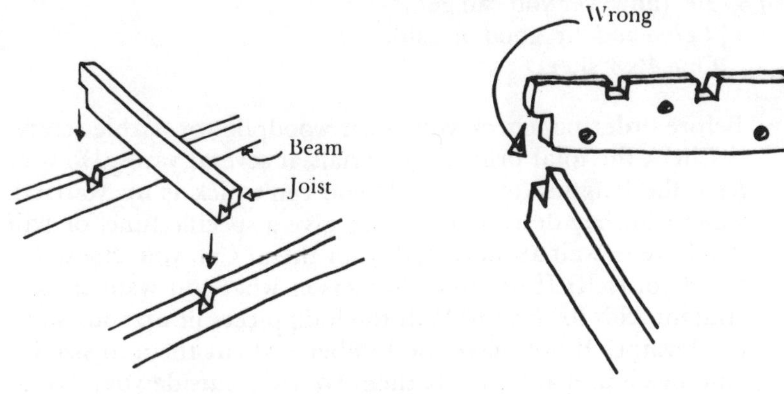

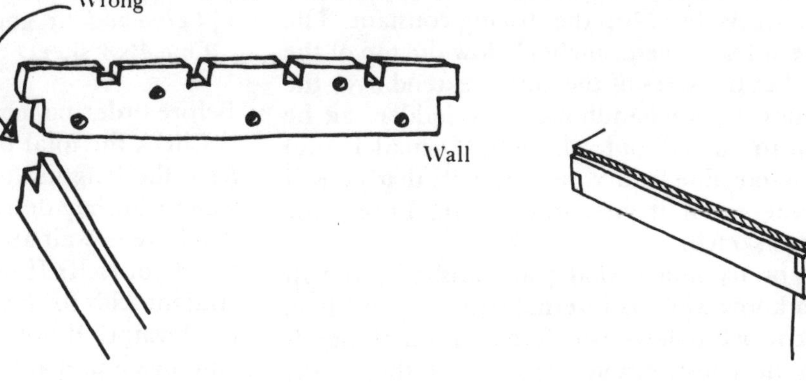

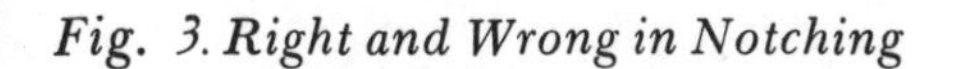

Fig. 3. Right and Wrong in Notching

Fig. 4

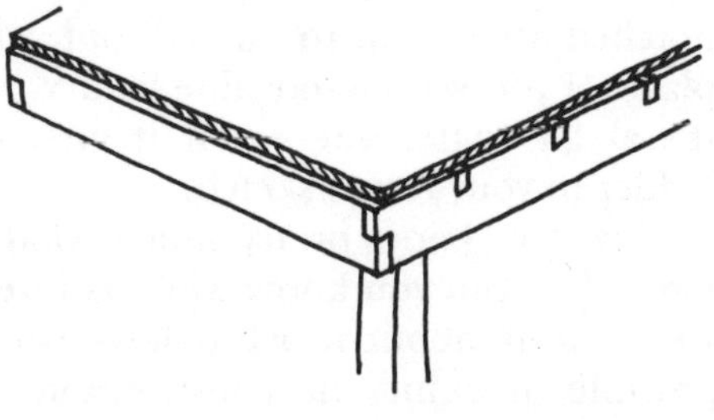

Fig. 5. Leaving space on edge for 1x2

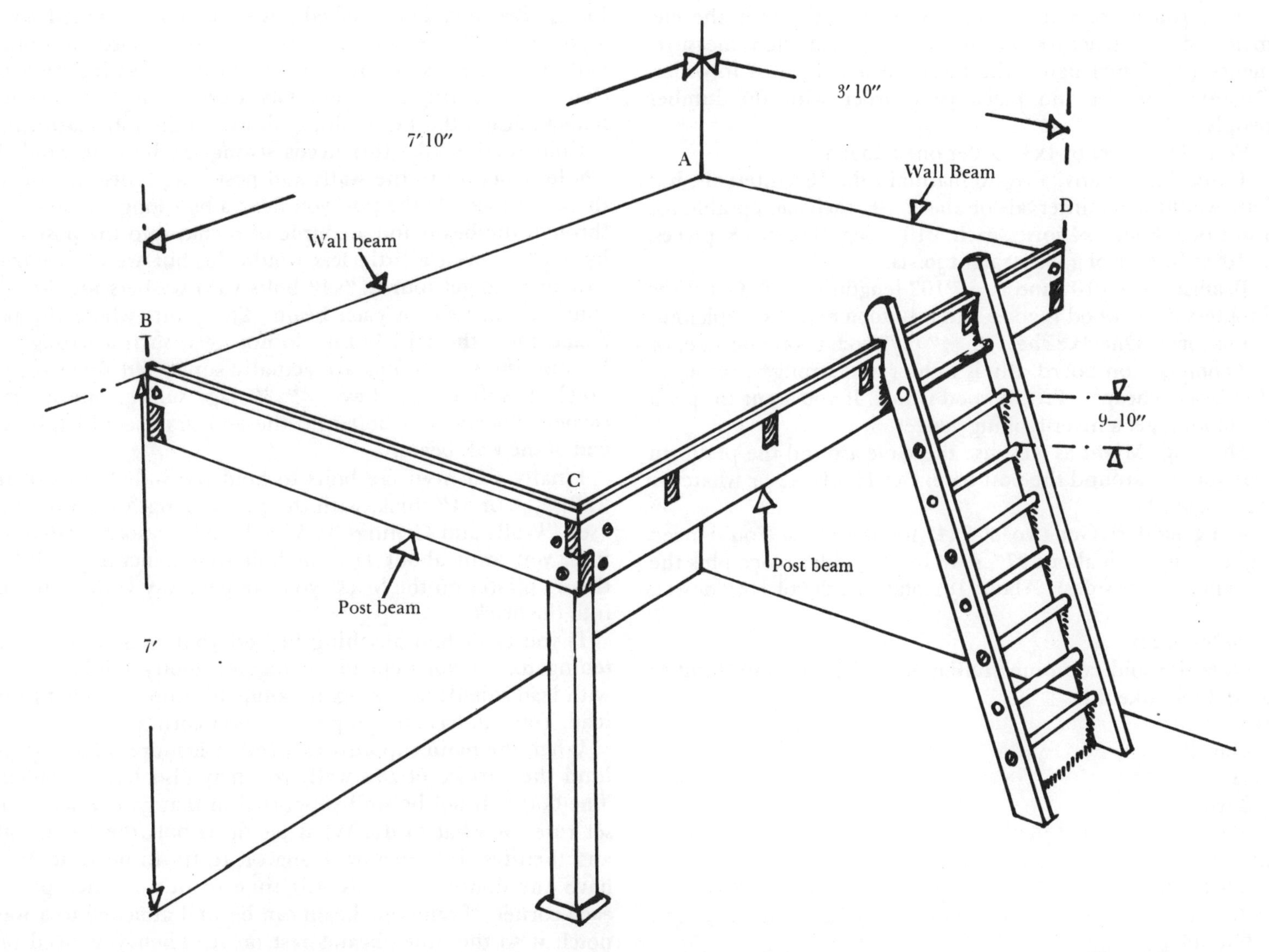

Fig. 6. Loft Layout

The steps of the ladder should be about 10″ from center to center. You can vary this, but keep the spacing constant. The top rung should be at least several inches below the top of the loft, for toe room. Let the sides of the ladder extend over the top of the loft about 6″, for handholds. The ladder can be notched at the top to "hook" onto the loft. Toenail it into place. If you want a movable ladder, just leave it; that's OK if it's at an angle, but not if it goes straight up. Locate the ladder in your rough sketch.

Now you know pretty much what your finished loft will look like. And you know what its internal structure will look like. Think about it a bit—have you forgotten anything? If possible, imagine the construction steps; see if they make sense. Do you have all the tools you'll need? If not, can you get them?

List of pieces

Now you make a list of your materials. Go over the elements of the structure one by one, figuring their measurements (*6*). Then figure the most efficient lengths to order. Organize the list and place your order with the lumber people.

Post: Seven feet of 4x4. Order one 8-footer.

Joists: How many? Five to maintain the 16″ intervals, but four would make intervals of about 19″ each—acceptable for a loft bed. Four 2x4 joists, each 3′10″ long. Get two 8′ pieces, or 10′ to be sure of good straight joists.

Beams: Two 3′10″ and two 7′10″ lengths of 2x6. Get three 8-footers if the wood is good, three 10-footers if it's suspicious.

Platform: One 4x8 sheet of 3/4″ plywood, good one side; or 3/4″ composition board. Ply is lighter and stronger, composition board cheaper. Fir plywood is fine. If you want to spend the money, get a nicer-looking veneer.

Molding: Might as well use the same around the platform that you use around the post base. Get 14′ of 1x2 or whatever shape you like.

Rung ladder: Get two 8′ 2x4s for the sides. You'll need eight rungs, each about 27″ (24″ inside ladder space plus the thickness of the sides). About 18′ total. Get 20′ of 1 1/4″ dowel.

Lumber order

Organize and combine similar materials so your lumber order looks like this:

4x4s
One 8′
2x4s
Two 8′
Two 10′
2x6s
Three 10′
1x2s
One 14′
1 1/4″ dowel
20′ (however you can get it)
3/4″ plywood, fir, good one side
One 4′x8′ sheet

Before ordering, check your own woodpile for usable scraps.

Check the total price of materials at several yards. Do they have the lengths you want? If you can't pick it up yourself, when can they deliver? Can they give a specific time, or will you have to wait around all day for them? Can you choose the wood yourself? If not, tell the person what you want it for—straight 2x6s for beams. Will the long pieces fit up your stairs or elevator? If not, have the lumberyard cut them to size for you, or be prepared to cut them yourself outside your home.

Hardware

You need nails, bolts, and glue to put this together. The plywood becomes a floor, so get a pound of 1 1/4″ flooring nails, for 3/4″ flooring. The notched joists are to be toenailed, so get a pound of 3″ common; 2 1/2″ or 3 1/2″ would also do. Longer nails are unnecessary, and their extra thickness increases the chance of splitting the wood. Get a pound of 2 1/2″ finishing nails to fasten the 1x2 molding down on the loft platform.

Construction like this needs strong lag bolts to hold the whole structure to the walls and post. Lag bolts 5/16″ or 3/8″ thick are good. At the post you need a bolt long enough to go through the beam and a couple of inches into the post—5/16″ by 3 1/2″ or even a little less would do, but we always overstructure; so get four 3/8″x4″ bolts with washers for the post connections, two for each beam. The joints where the post beams meet the wall beams do not need such a strong bolt because the post beams are actually supported there on the notched wall beam. Two 1/4″x3″ lags suffice, one at each corner. Too thick a bolt into the end grain could split the end of the wall beam.

Finally you need lag bolts to hold the wall beams to the wall: 5/16″ or 3/8″ thick, with the proper length for your wall (see "Walls and Ceilings"). Wood stud, masonry—whatever it is, you want about 1 1/2″ of bolt into something solid. If there's plaster on the brick, you can get away with 1″ of bolt into the brick.

If you can't find anything behind your plaster worth fastening to, do *not* depend on toggles, molly bolts, or bolts with lead shields in the plaster alone to support such a heavy load. You will have to put posts at every corner.

When the main supporting member is more than 2 1/2″ behind the surface of the wall, you may also have problems. The bolt will not be well supported in that gap. There is no set rule for what to do. What *we* do is bolt the beam tight and then test it somehow. Hang on it, try to move it. If we have any doubts, there is still time to put another post at each corner. If only one beam can be well attached to a wall, notch it so the other beams rest on it. Then you need one more post at the other end of the second wall beam.

For this project we arbitrarily assume you have about 3/4″ of plaster on top of brick. Get 5/8″ lead shields 1 3/4″ long, which puts an inch of shield in the brick. Get 3/8″ lags, 3″ long.
1 lb. 1 1/4″ flooring nails
1 lb. 10d (3″) common nails
1 lb. 2 1/2″ finishing nails
Four 3/8″x4″ lag bolts and washers
Two 1/4″x3″ lag bolts
Nine 3/8″x3 1/2″ lags
Nine 5/8″x1 3/4″ shields
Don't forget the washers. The hardware dealer probably will. Any other materials? A good wood glue if you intend this as a permanent piece. Wood putty if you want to fill any holes. Stain, paint, polyurethane, or whatever you wish to finish the piece with.

Tools

Before any project, figure what tools and bits and blades you need and gather them together. You'll always forget something, but try to minimize the extra trips to the hardware store. They interrupt the exquisite flow of your work. This planning also helps you visualize the actual steps of construction. Good planning as you relax on a cushioned chair makes for less sweat later.

For layout: Tape measure, rafter square, level, long straightedge or chalk line. An assistant.

For construction: A saw, of course. Preferably a portable circular saw to speed things up, particularly the ply cutting. It also cuts notches accurately, although a saber saw would be quicker here. Goggles. Blades; crosscut or combination plus a plywood blade. For composition board, a plywood blade is fine, but use an old one if you have it. Two clamps might help when you cut the ply or composition board. Hammer. Nail set for the finishing nails. Chisel to finish the notches.

An electric drill is essential for the brick; a 1/2″ variable-speed drill is best. Bits for the bolt holes: 3/8″ wood bit for the wall beams, so the 3/8″ bolt can slide through to the shields inside the wall; 5/16″ or 11/32″ wood bit for the 3/8″ post bolts for a tight fit between wood and wood; 7/32″ for the 1/4″ bolts. A 5/8″ masonry bit to drill holes in the brick for the lead shields—get a good one with as thick a shank as your drill will take. Good ones are expensive but last much longer, and work faster. Test the brick for drilling difficulty. If it's very tough, you might also get a smaller masonry bit, 1/4″ or 5/16″, to make pilot holes for the bigger bit. Get a 1 1/4″ wood bit for the dowel holes in the ladder. You should also have a 1/8″ wood bit handy, to predrill for the toenailing. If you're going to countersink the bolts, you'll need a bit larger than the washers. The 1 1/4″ bit for the dowels will probably do it.

A heavy-duty extension cord, with a three-prong adaptor if you need it. A rasp and/or sander to smooth out the notches and dowel ends. You need some kind of wrench to drive the bolts. If they are countersunk, you might be able to drive them all the way with an adjustable wrench, but a ratchet with the proper sleeves would be best.

A pencil or two. A nail apron or tool belt to keep things organized is optional. Anything else you might particularly want? A breathing mask against the sawdust. A saber saw will do the end notches faster than a circular saw.

Construction

Now you should be ready. You have the tools and the hardware, the room is cleared for construction, and you are waiting for the lumber delivery. They said it would be there by 10 A.M. and it's now a quarter past two. Call them and complain. They'll say, "Gee it's on the truck," and an hour later it'll show up. *Check it out before accepting delivery*. You told the man you wanted good pieces when you ordered it. If you can't get 7′10″ of relatively straight beam out of their 10′ length, send the offending member back. Have the driver write it all on the receipt. Other people usually load the lumber, so don't get too mad at him. He or you will probably have to call the yard to get things straight. You will have to wait another day to get good pieces, but it's usually worth it. One bad piece can cause you hours of extra labor, not to mention screwing up the entire structure.

Finally you are ready to work. Pile the lumber neatly on the floor next to the loft location.

Your plan of attack is well crystallized at this point. You know how the pieces are to fit together. Go over your plans. Figure the order of steps from start to finish, so you don't box yourself into any corners. Which pieces go up first, and what has to be done to them before they go up? The more carpentry experience you gain, the easier such organizing becomes. At first it seems mind-boggling. How can you see what's going to happen if you've never done it before? Do what you can. With common sense, you will avoid most errors, and mistakes can be remedied. Here are the steps we would follow, followed by a detailed explanation. Start with the pieces that support all the others—the beams and the posts. The wall beams first, since they can stay up on their own, and the back wall beam first of all, because it can partially support the side wall beam. Be careful not to rush the beams up without first notching them. Once up, they're almost impossible to notch.

Loft construction outline

1. Lay out lines on wall for beams
2. Prepare back wall beam
   A. Cut to length and notch both long beams. Lay front one aside
   B. Drill for bolts and shields
   C. Sink shields, insert bolts and washers in beam, and attach to wall
3. Prepare second wall beam same way; put up

4. Attach the post beams
   A. Cut and notch smaller post beams
   B. Tack beams and uncut 4x4 in place, level and mark 4x4 height and base
   C. Cut 4x4
   D. Tack beams and 4x4 in place again
   E. Set post in floor
   F. Bolt in beams (and glue)
5. Cut, notch, and install joists, one by one
6. Put plywood in place, mark for fit, cut, and install
7. Ladder
   A. Find length
   B. Cut sides, drill for rungs
   C. Glue and insert all rungs in one side
   D. Fit second side over ends of rungs (with glue)
   E. Attach ladder
8. Finish
   A. Cut and attach molding
   B. Sand (if you want)
   C. Stain or paint or otherwise finish or not

1. Layout

If the plywood is to be 7′ off the floor, then the beams will be 6′11¼″ above the floor. Mark this height at the inside corner of the walls, corner A. Hold a long straightedge, or the actual 2x6 beam, against one wall at that height. You could use an assistant here. Level the piece and draw a line on the wall along its top edge. A line on the top won't show once you put the plywood up. It may be hard to see up there, but you're going to have to stand on something anyway to get any drilling leverage up there. Or make several marks and chalk a line across. Do the same on the other wall. Check the height along the lines. If floor undulations cause the height to vary so much that you might have trouble with head room (on top or bottom of the loft), raise or lower the lines accordingly.

Unsquare Corner Confusion

One other problem to watch out for is a grossly unsquare corner at *A* (7). This makes a difference only if the platform has to be an exact size—if it's being built to the exact dimensions of the mattress, for instance. The ramifications are illustrated, with the solid lines showing what the loft will look like if the *post beams* (*BC* and *CD*) are cut to the planned size, and the dotted lines showing the adjusted lengths if you want a square corner at the post and enough space to fit the mattress in. In this situation, the best thing to do is to lay out the platform area on the *floor* first, making beam *BC* the correct length, and corner *C* a right angle. Then cut the other beams to whatever length this layout demands. You will end up with an extra triangle of space along *AD*, but you can't do anything about it unless you want to rebuild your house.

2. Prepare wall beam AB

Cut both long beams to their 7′10″ length (or their adjusted lengths, in the above situation). Lay them on edge, flush at the left end, and mark the centers of the joist locations across both top edges (*8*). The joists should be evenly spaced, of course; 18¾″ on center is close enough, making the last interval 19″. (If the beams are different lengths, due to an odd angle *A*, the last interval on the right won't match. That's all right. The joists will be parallel to beam *BC*, not to the side wall at *AD*.)

The exact width of each pair of notches depends on the thickness of the joist that goes in them. Your 2x4 joists, unfortunately, may vary in thickness by as much as ⅛″. To cut your notches to fit rightly, you will have to number each pair to correspond to a numbered 2x4; then measure the thickness of the 2x4 and transfer that measurement to its corresponding notches. Don't cut the 8-foot 2x4s to length yet; once the beams are set in place, your planned measurements may be slightly changed. Just assume that each 8′ length will become two different joists.

Also, measure the thickness of the short 2x6 beams and transfer the measurements to their respective notch locations on the long beams. Note that the ends of the front beam *CD* are notched *on the bottom* (*9*). Draw the notch outline with a sharp pencil and a square. Accurate cutting on these lines will guarantee tight-fitting notches. These are lap joints, really, and are cut as described in "Working Techniques."

Decide how deep to cut the notches. The 2x4s will be notched about half their depth, or 1¾″; therefore you'll cut 1¾″ out of the 2x6s. Brace each 2x6 well against something as you cut—and wear your goggles, of course. Set your saw blade to 1¾″ depth, test it on a scrap of wood, then cut out the joist notches *within* your notch marks on both long beams. Chisel and rasp out the pieces. Test all the crosspieces for a thickness fit. Rasp out the notch sides if the fit is too tight. Do this now—it's too difficult to do after the beam is up and you're actually installing the joists.

At the ends of the beams, where the 2x6s are notched into each other, the depth of the notches is 2¾″, half the width of a 2x6. Your circular saw probably can't cut this deep. Make the notch with a handsaw or a saber saw. Outline the notch on the face of the beam and cut it away. (If you have a bad corner as in step 1a, you'll have to duplicate that angle in notches at *A* and *D*.) Now lay beam *CD* aside for a while.

Drill the boltholes in the beam *AB*. Stagger the holes approximately as shown, so as not to weaken the grain. Make the countersink holes first with the 1¼″ bit deep enough for the washer and bolt head, and then drill the boltholes through the centers with the ⅜″ wood bit. Drill straight. Using the ⅜″ bit first would make it hard to center the 1¼″ bit, and you would end up with a sloppy-looking cutout.

Let your assistant hold the beam in place on the layout line, while you slip a pencil through each hole and mark its

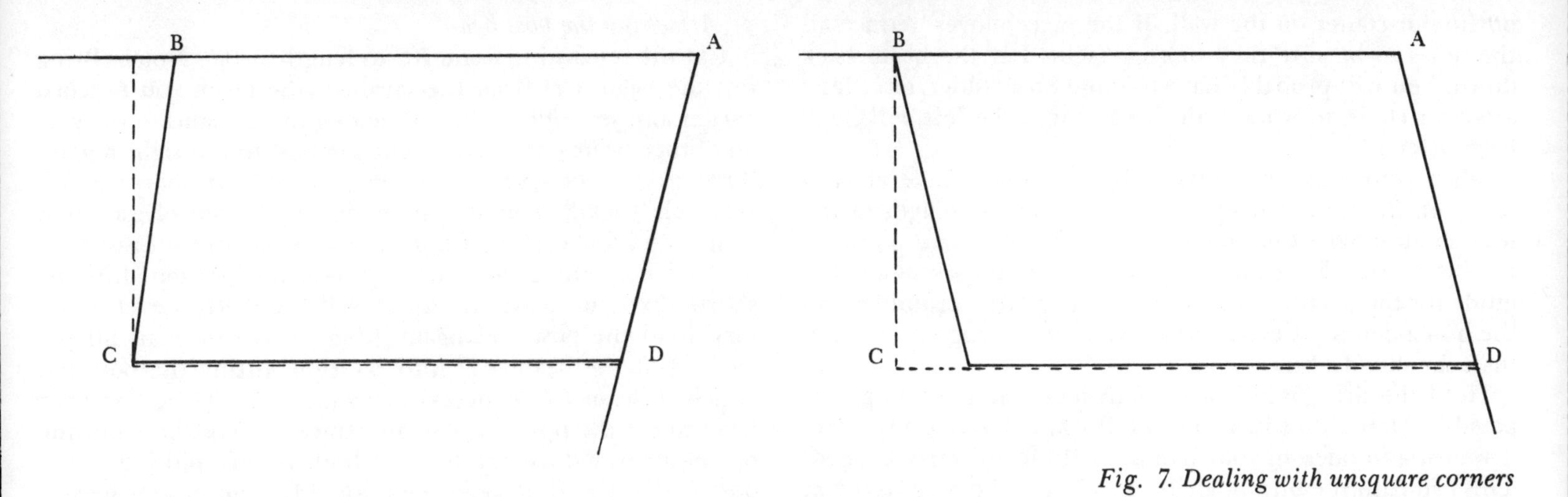

Fig. 7. Dealing with unsquare corners

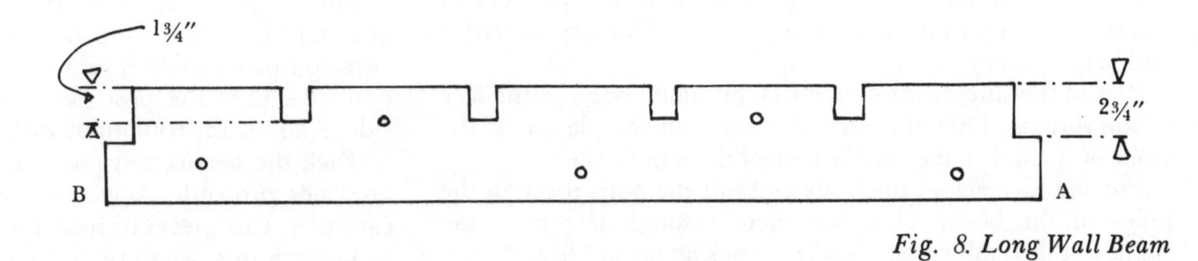

Fig. 8. Long Wall Beam

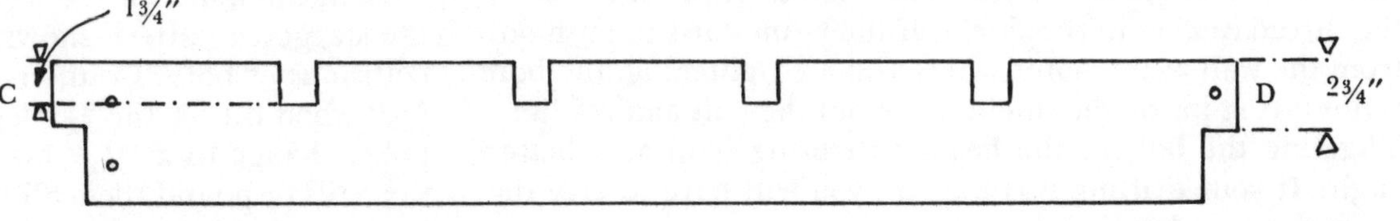

Fig. 9. Front Beam

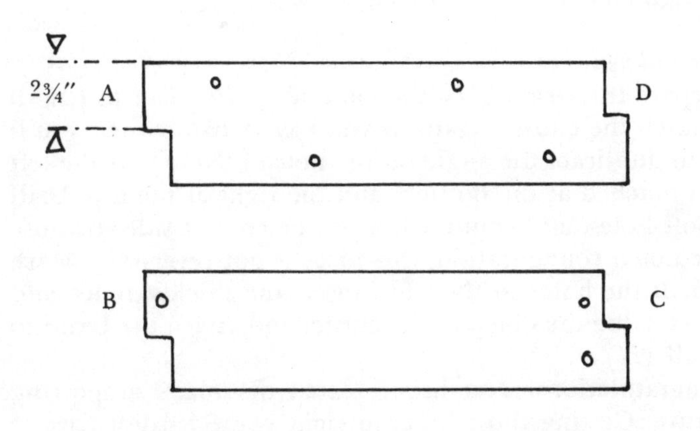

Fig. 10. End Beams

outline or center on the wall. If the piece moves, retrace all the holes to be sure they line up right. Lay the beam back down. You will probably have to stand on a ladder, or at least a box or chair, to work with these beams; the loft will be 7′ high, after all.

With your masonry bit(s) drill a 5/8″ thick hole at each point on the wall as deep as you need for the length of the lead shields. Start each hole with an awl, or a nail point, so the bit doesn't dance off. Put some tape on the bit as a depth guide for the length of the shield. If the plaster crumbles and the hole opening is extra large, drill deep enough so you can inset the shield a bit.

Hold the drill tightly with both hands and as straight as possible. Don't force it. You can tell you're forcing if you feel it wanting to buck in your hands, or if the bit starts to bend. This can ruin the bit, the drill, the plaster, the hole, and you won't get finished any faster. Drilling a pilot hole with a smaller carbide bit can save a lot of time on really tough brick. You can also dip the bit tip in lubricant (preferably cutting oil) every minute or two (don't use water) to cool it and help it to cut.

Pound the shields into the holes, but don't damage the face of the shields. They should end flush with the plaster; if the front of the hole has crumbled, push them in farther.

Put the washers on the bolts and put the bolts through the holes in the beam. Hammer them through if they're too tight. Let the ends come out of the back about 1/4″ to 1/2″.

Lift the beam into place carefully, trying to set the tip of each bolt in its shield. Tighten each bolt a couple of turns till the thread catches in the shield. If the beam starts to push out from the wall at the point where you are tightening, the bolt is not fitting inside the shield. Back out the bolt and readjust. Wiggling the bolt in the beam will bring it in at a better angle. If your drilling was way off, you will have to take the beam down and redrill the hole.

After all the threads catch, tighten each bolt *almost* all the way in. As you do the first ones, watch that the others don't pull out from the wall. Tighten them again if they do. Finally tighten each one *all* the way.

3. Second wall beam

Prepare this beam, *AD,* the same way (*10*). Cut to length and notch the ends. (Again, if you have a bad corner, you'll have to duplicate the angle on the notch.) Note that the left end is notched at the bottom, and the right at the top. Drill the bolt holes, and countersink on the correct side—because of the notch configuration, this piece is not reversible. Mark and drill the holes in the wall, insert the shields, bolts, and washers. Glue the notch at the corner and attach the beam to the wall (*11*).

Congratulations. You have created the main supporting structure. Getting those bolts in right is the hardest part of the project.

4. Attaching the post beam

Cut the remaining beam *BC* to length, 3′10″. Notch. Bring out the beam *CD* from the sawdust—the beam you notched earlier and set aside. Before attaching these beams to the post and other beams, we have to cut the post to the right height. This may not be seven feet exactly, since your floor may not be level. Tack 3″ common nails into both ends of each post beam, after first drilling for the nails. Wake up your assistant. Set and hold the post in its approximate location. Lift the shorter 2x6 into position on the wall beam *AB*. Level it and tack it to the post and beam (don't drive the nail all the way). Lift the beam *CD* into position, lifting the post and attached beam *BC* as necessary to meet *CD* (*12*). Use your level to get the post vertical, and trace the post base outline on the floor. Adjust the level of both beams, pulling a nail out if necessary, but keep their top edges flush—you need a flat nailing surface for the plywood. Mark the post where the *tops* of the beams meet it. Take the pieces down.

Cut the post just below this line. Your circular saw needs one cut from each side to cut through a 4x4. Square the cutting line around all four sides of the post. Cut carefully on one side, turn the post over, and cut through the opposite side. Smooth any roughness with your rasp or file.

Tack the beams and post back up again. Be sure the post does not protrude above the beams. Measure the post base carefully. Cut pieces of molding to fit tightly around it—butt or miter joints, whatever your eye desires. Nail the pieces to the floor with finishing nails (*13*).

Drill the holes for the lags at the post connections. Follow the staggered pattern shown in Figure 14, to avoid head-on collisions of bolts. Countersink first with the 1 1/4″ bit. Use a 5/16″ wood bit for the 3/8″ lags. If this is too tight to turn the bolts, change to a 11/32″ bit. When the bolts are in, remove the nails or pound them all the way in. Now fasten the beams to the wall pieces. Pull the nails first, countersink drill as before, then drill with either a 7/32″ or 3/16″ wood bit for these 1/4″ bolts. Squeeze some glue into this joint, and run the bolt in.

5. Preparing and installing joists

An easy way to mark the length of the joists, and to lay out the notches at the same time (*15*), is to lay each 2x4 upside down into its notches in the beams. With a pencil, mark the length at the outside of beam *CD;* and trace line on joist at top of beam and, with a square, *up* along the faces of the joist for the notch outline. Thus, the part that is now in the notch is not cut away. Be sure to notch so that any crooked 2x4s will (when they are put in right-side up) crook *up* in the middle.

Lift each joist into place and check the fit. Rasp away if it's too tight, get some shims ready if it's too loose. One by one, toenail each joist in (*16*). Predrill holes for the nails; shim when necessary; glue if you wish. The basic structure is now complete (*17*).

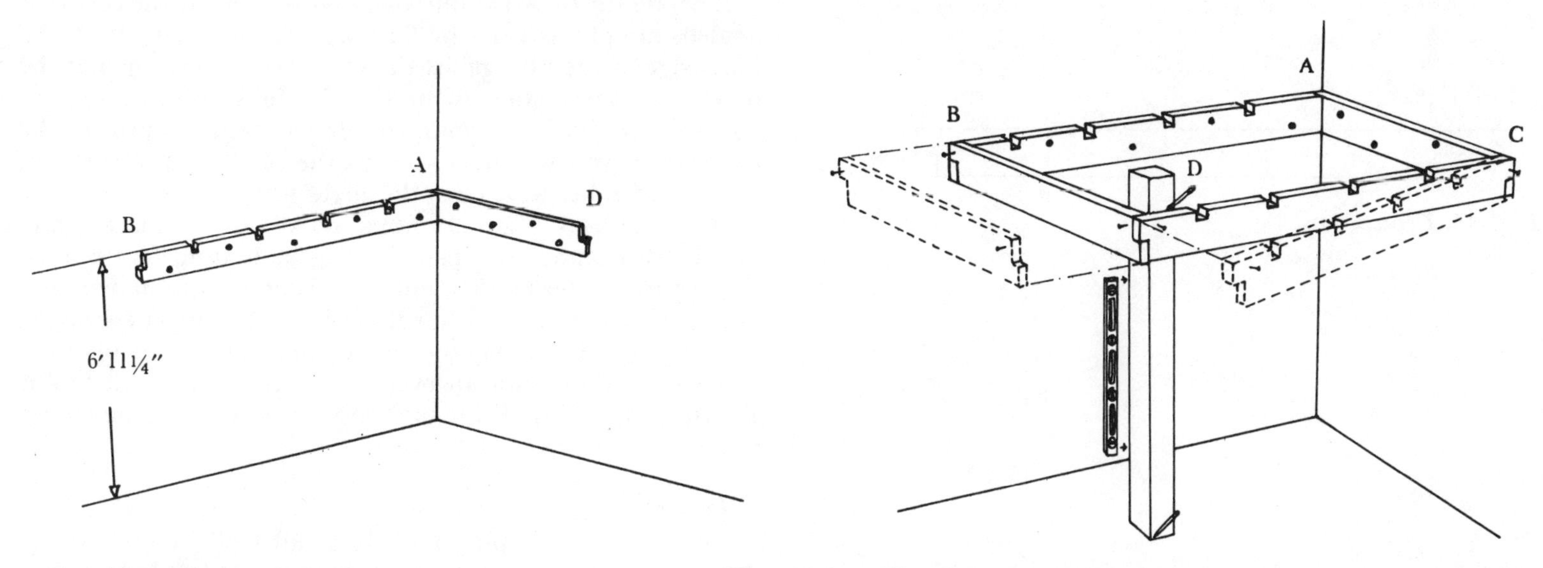

Fig. 11. Wall Beams Installed

Fig. 12. Installing side and end beams and finding correct post height

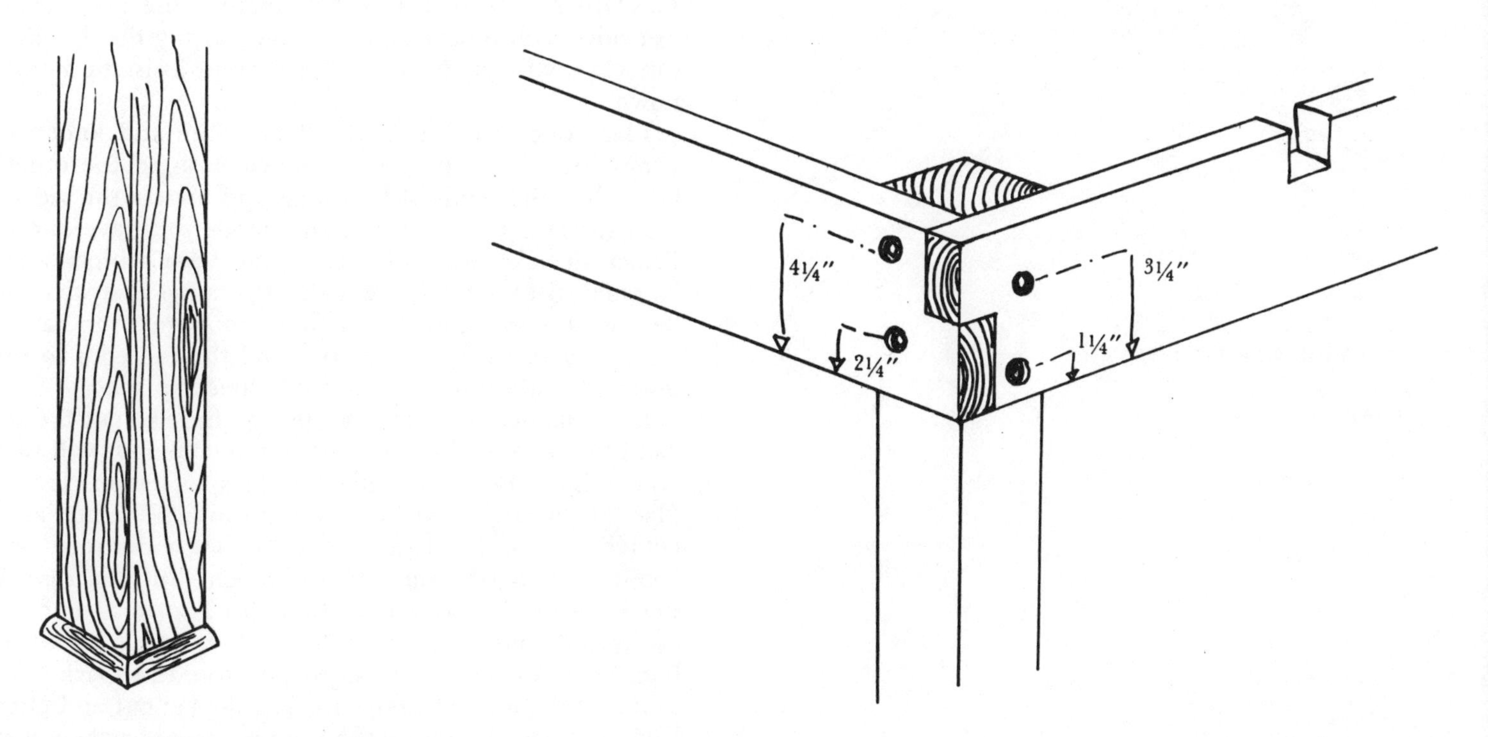

Fig. 13. Molding on post at floor

Fig. 14. Bolt positioning

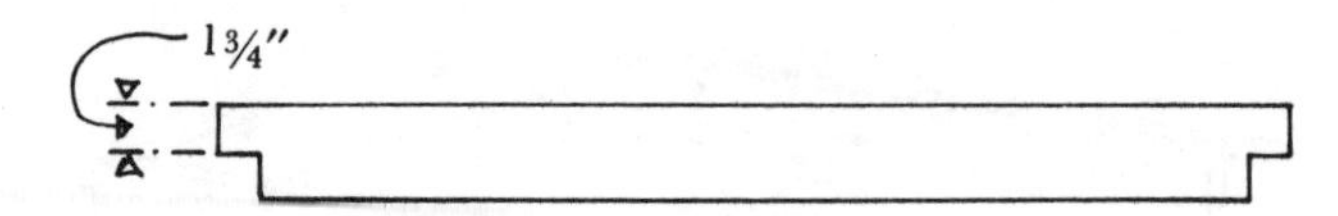

Fig. 15. Typical Joist

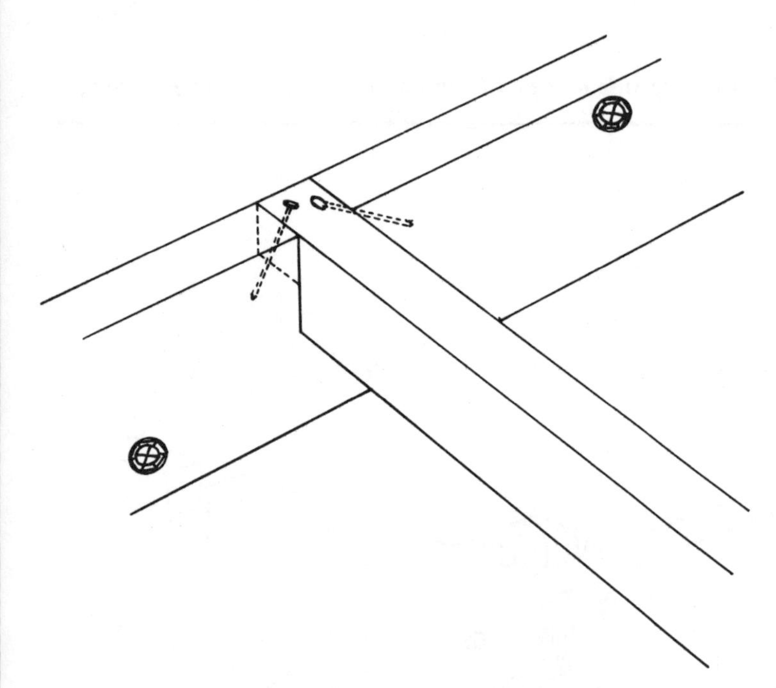

Fig. 16. Nailing joist in place

6. Installing plywood

Levitate the plywood into place on the loft. If the corner is angled, the ply will not be flush against the walls. Push the long edge of the ply against the wall. Measure the gap at the other wall, mark, and cut to fit (*18*). Be sure you switch to the plywood blade on your saw. If you want to protect the top veneer, you will have to turn the ply over to mark and cut it; remember which way the angle goes.

Once the back edges are fitted, set the ply in place again. Stand below and run a pencil along its bottom face against the outside of beams *BC* and *CD*. Take the ply down, and draw new lines back within these lines the thickness of the molding, ¾″. Cut along these new lines and put the ply back up. It should fit with an even ¾″ strip of the post beams showing. Don't install the molding yet; wait until the ladder is up.

7. Ladder

The ladder is simple. If it's to stand straight up, cut the 2x4 sides to about 7′6″—6″ more than the 7′ loft height. Figure where you want the top rung. Say 8″ below the top of the loft (giving toe room between the rung and the bottom of the beam), or 76″ from the floor. Since the rungs should be about 10″ apart, you can space eight rungs in this 76″ distance at 9½″ intervals. Clamp the 2x4s together, mark the intervals on one, and drill 1¼″ holes through both at once for the dowels (*19*). Put some backing under each hole to prevent splintering as you drill through. If the rungs are to be a standard 2′, cut them to 27″ to include the thickness of the 2x4 sides. You'll have an easier time sawing the dowels if you can clamp them down, or have your assistant hold them down.

Place one 2x4 flat on the floor. Glue and knock all the rungs into it, running them down flush to the other side. Glue the other ends of the rungs and finagle the second 2x4 onto the rungs (*20*). The more hands you have, the better. Knock the 2x4 down over the rungs with a block and your hammer, till the rungs are flush. Any rungs that stick out can be rasped, sanded, filed, or left as a monument to hand-hewn work. It's also a good idea to round the top ends to prevent midnight stabbings. Toenail the ladder in place.

If the ladder rises at an angle, you have to cut the bottom (and top, if you wish) parallel to the floor so it will sit right. Here's how. Before cutting the 2x4s to length, lay one in place at the angle that looks good to you. (The best angle for ladders is usually when the base is out a distance equal to about one-quarter the vertical height—in this case 1¾′.) Place your rafter square as shown in Figure 21 and mark a horizontal line along the tongue of the square. Cut at this line; replace 2x4 in its angled position and mark the horizontal at the top (6″ above the loft floor); cut and put piece back in place. To locate the rung positions, make a mark 8″ vertically below the platform level on the angled 2x4; divide

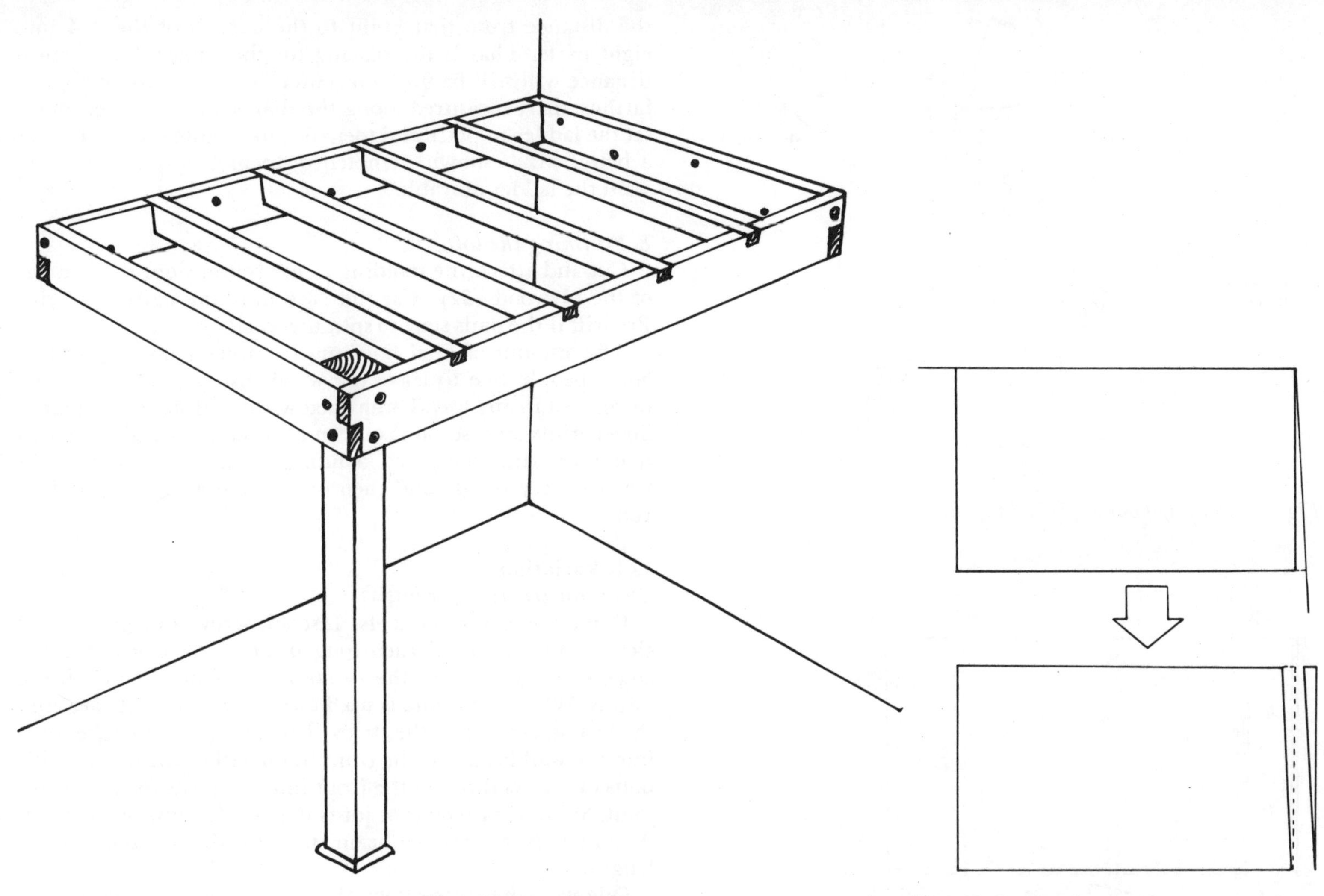

Fig. 17. Completed Structure

Fig. 18. Cutting floor to fit unsquare shape

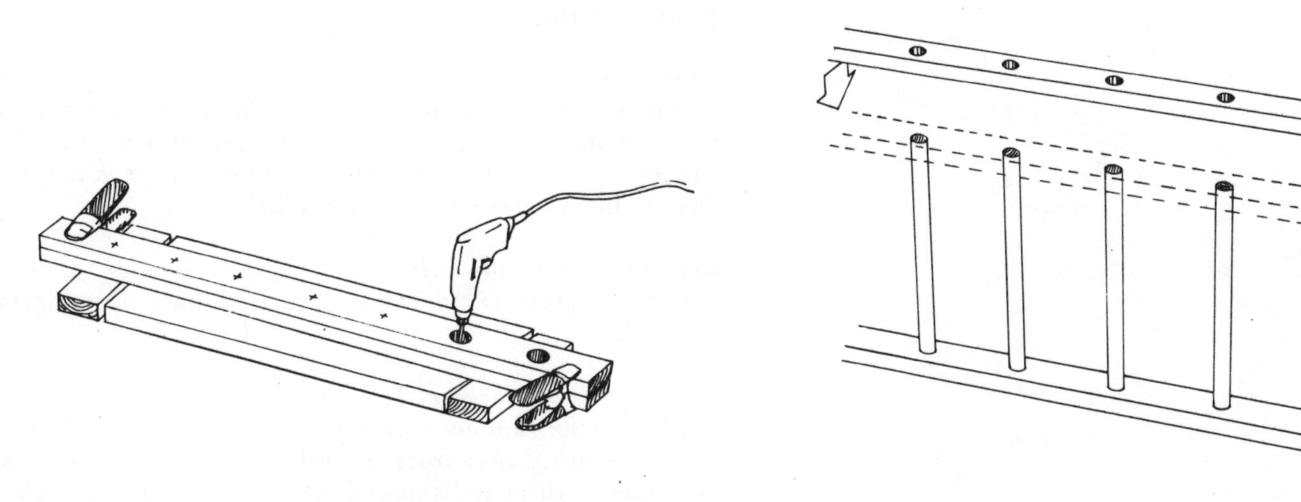

Fig. 19. Drilling for Ladder Rungs

Fig. 20. Assembly of Ladder

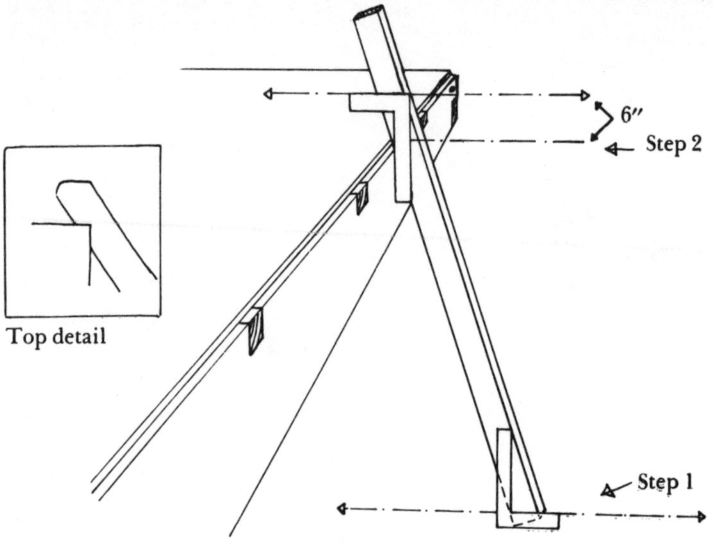

Top detail

6″

Step 2

Step 1

Fig. 21. Finding floor angle and top angle

the distance from that point to the bottom of the 2x4 into eight parts. That is the spacing for the rungs. The vertical distance will still be 9½″ on center, but they will be slightly farther apart measured along the diagonal 2x4. When done, set the ladder in place and toenail it in or notch out at top for a better fit, as shown. If notched, toenail only if you don't want the ladder movable.

8. Finishing the loft

Cut and attach the molding to fit around sides *BC* and *CD* of the plywood (*22*). Use the 2½″ finishing nails, carefully. Predrill if the nails start to split the wood.

The amount of final finishing is purely a matter of taste. Some people like to leave the wood just as it is, rough and uncolored in any way. Unfinished wood will show dirt marks, fingerprints, and so on. You can also stain, varnish, polyurethane, or paint the piece. Sanding is important only if the wood is very rough, and then only if the roughness bothers you.

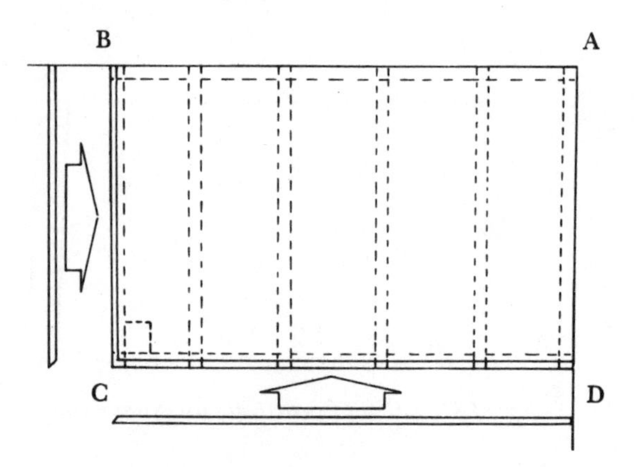

Fig. 22. Applying edge strips

Loft Variations

Cleat construction for joists

Don't notch for the 2x4s. Instead, screw and glue a 1x2 cleat to the inside of each long beam, 3½″ below the top edge. Cut the joists to the *inside* measurement between the beams. When the frame is up, fit *and glue* the joists between the beams, resting on the strips. Toenail the ends of the joists into the wall beam. At the front beam, either toenail or drive bolts or screws through the front into the joists for a stronger joint. Shim or rasp ends of joists that are too low or too high. Also stick shims between beam and joists that are too short in length.

One variation is to attach the cleats about 1¾″ below the top edge, and then notch the joists to fit inside the beams and on the strips (*23*). This variation is strong and doesn't let the ends of the notches show on the outside of the beams. Some people like that.

Joist hangers

Again, don't notch for the 2x4s. Cut them to the inside measurements between the beams and attach with joist hangers. Easy, strong. The advantage over cleats is that each 2x4 can be raised or lowered individually.

Lofts fastened to one wall

Use two posts (*24*). Try to run the longer beam against the wall.

Lofts fastened to three walls

This is the famous look-Ma-no-posts loft (*25*). It looks beautiful and is no harder to build than any other loft. Requirements: three walls spaced the distance you want. A window in the far wall helps ventilation.

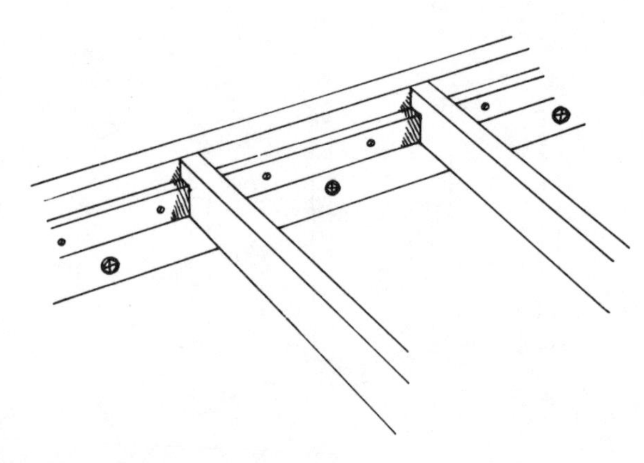

Fig. 23. Cleat Support for Joists

Railing

Build a simple 2x3 or 2x4 wall-type structure around the sides. Bolt vertical 2-bys flat to the side of the loft, and nail a 2-by on top of them. The type of loft we just built needs a railing only a foot or so high, as in Figure A showing diagonal bracing (Three-Step Approach). A loft with standing room on top needs a rail 3′ to 4′ high. Such a railing needs more support. Fasten it at the ends to the walls, or run an end up to the ceiling and fasten it, or use a diagonal brace onto the platform if you have the room. Or you could plan the 4x4s to continue through and above the platform and tie the railing into them. The plywood would then have to be notched around the posts at the corners, with 1x2 cleats on the 4x4s under the plywood for support.

Extending the posts above the platform is a good idea if you want to build any other structure on top of the loft—bookcases, closets, headboards. The posts will support anything you want. *Plan such a move from the beginning.* Once you cut the posts to the loft height, gluing the cut-off section back on won't help you. Also remember to leave enough room *between* the posts for the mattress.

Step ladder

Cut the sides, as with the rung ladder. Lay out step intervals. If steps are 2x4 material, cleats are 1½″ below surface of steps, so lay out cleat lines at same intervals as steps, but start 1½″ closer to the bottom of ladder. In other words, the first cleat is 8½″ above floor for 10″ step intervals, next cleat is 18½″, etc. Glue and screw 1x2 cleats with 1½″ No. 10 screws, countersunk, to both sides. Cut steps to exact 24″ width. Lay one side on floor, cleats facing up. Place top and bottom steps in place against cleats. Glue top ends of steps and bottom where they meet cleats. Lay the other 2x4 in place over the steps gently, straddle it, and sit on it carefully. Your pressure and incredible balance will hold the pieces together while you drill and screw two 2½″ No. 10 or No. 12 screws into each step. Hold the step tightly against the cleat as you drill.

Turn the whole thing over and fasten from the other side the same way. Glue and slip the remaining steps into place, drill, and screw them in.

Step ladder angled

Cut the sides as with the rung ladder. Use the bevel square to lay each cleat in at the same angle. Or improvise with your rafter square in this way: Lay the 2x4 in place at desired angle. Lay rafter square against it on the floor, the bottom blade extending away from the loft. Clamp a 1x2 or scrap piece to the square as shown (*26*). As you slide this assembly along the angled 2x4, the bottom blade of your square will describe horizontals to the floor. Check back to description of Figure 21, and locate cleats the same way as dowels for angled rung ladder (*27*).

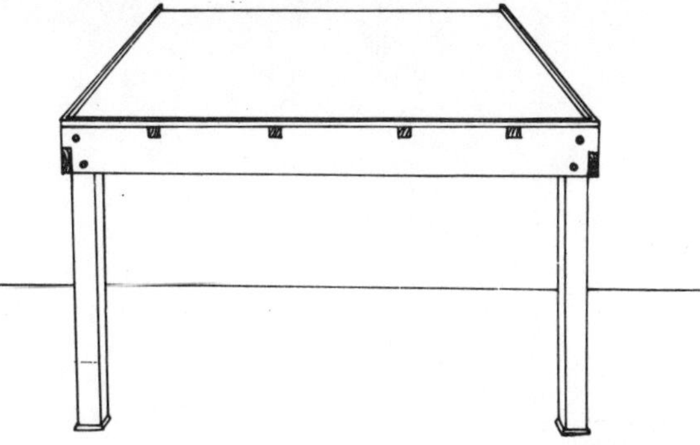

Fig. 24. Loft on One Wall

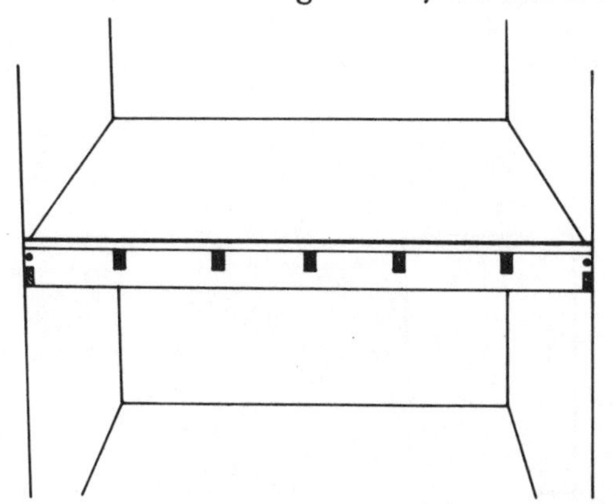

Fig. 25. Loft on Three Walls

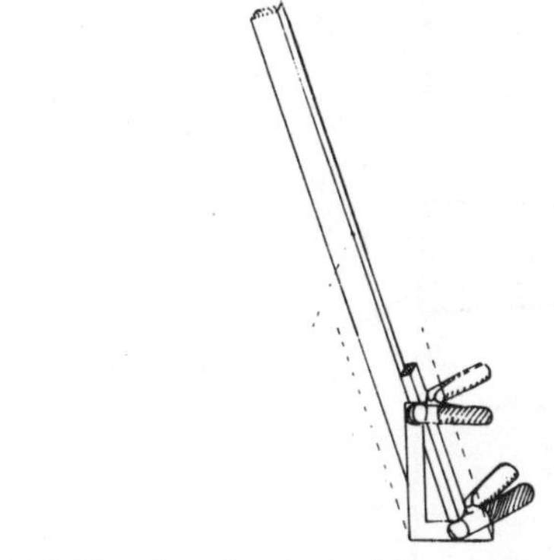

Fig. 26. Finding level positions for step ladder

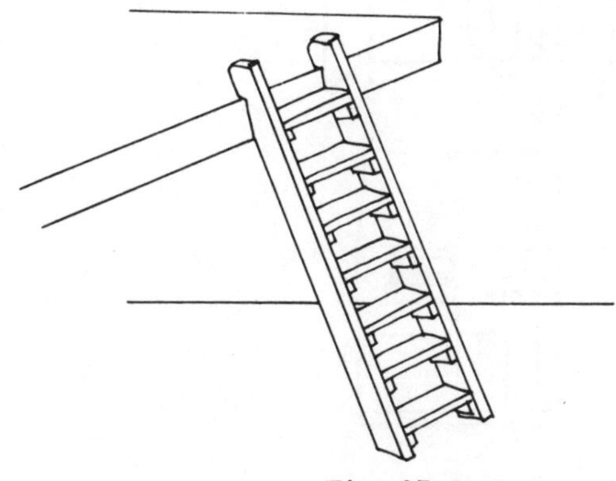

Fig. 27. Step Ladder

CABINET

Your closets are overflowing with boxes of books, your records are piled on the floor. The new hi-fi has been temporarily set up on the kitchen table for the last four months, while the couch is covered with magazines. Your liquor sits in the kitchen cabinet behind the canned corn. You need a cabinet.

A cabinet is both a display space and a readily accessible storage area. It can include drawers, doors, drop-leaf desk tops, adjustable shelves, pigeonholes, and much else. It will stand on its own or hang on the wall.

A strong, good-looking cabinet, even one with many different features, is easy to build. Just take the elements one at a time. Make a full plan for every detail and relate each to the whole structure.

Our project has a number of different elements, not all of which you may want in your own cabinet. But from seeing how we organize all the elements, you should be able to adapt what you need to make a cabinet uniquely your own.

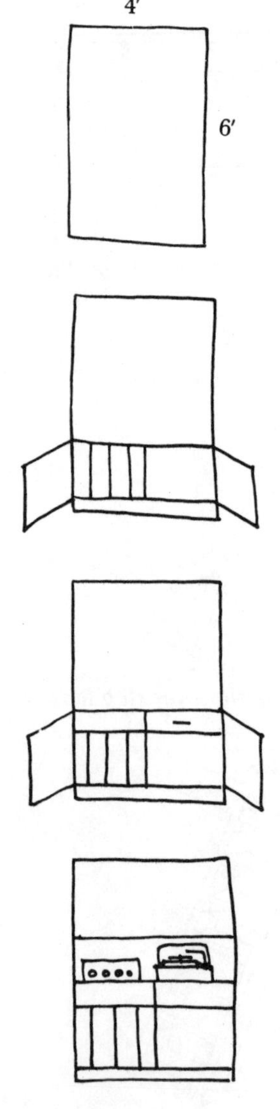

Fig. 1.

Planning

Designing your own

The first step in designing a cabinet is to find a location for it. This helps determine its shape and maximum size. A hanging cabinet is attractive and saves floor space, but it offers only a couple of shelves within easy reach. It also causes extra problems in fastening it securely to the wall. A standing cabinet offers the most accessible shelf space.

Let's say you have enough floor space along one wall to place a cabinet 4′ wide. A simple rectangular box shape is the easiest to build. You can alter it (see the Variations at the end), but the principles of design and construction remain the same. Make it 6′ high for easy access to the top shelf. Make the depth at least as much as the deepest object you'll be keeping in the cabinet. If that item is a record album, measuring about 12½″x12½″, your depth could be 13″. Make rough sketches as you go.

How do you divide up this 4′x6′x13″ space? Look at all the things you want to store in it. Figure the *size* space you need for each, and fit that space into a simple sketch of the 4′x6′ cabinet (*1*). Think about what *kind* of space you need for each element. Open or closed; drawers or doors? Which way should the doors open? Adjustable shelves? Also *where* do you put each space? Do you want a certain thing at eye level? Stooping level? Out of children's reach? Think a lot about how you use things, how often you use them.

You've got sixty to seventy-five record albums. The shelf space should be about 14″ high, leaving the standard 1½″ finger space on top; and about 2′ wide, to hold these and future discs.

A helpful addition can be several dividers within the space. Also a door to keep out dust. You could locate this section at eye level, or down at the bottom if you like to kneel or sit as you search for the proper sounds. For our project we'll place it at the bottom, leaving 3″ below for toe space. We can plan on a toe plate under that shelf, 3″ high and located 3″ back.

You have a half to a dozen bottles of liquor and wine—to be used only occasionally. An area the size of the record space will do. You could even fit in a good-sized hookah. This space too should be covered with a door. The logical spot is right next to the record area.

Shorter compartments just above, 5″ to 6″ high, would be handy for accessories like phono cables and jacks, a record cloth, cleaning brush, cocktail napkins, jiggers, mixers, plastic pink elephants. In fact, at least one drawer would be perfect for the very small items—they might get lost in the back of such a deep low area. The doors could be extended up to close over the drawer and small shelf.

On the next shelf up you can place your stereo receiver and turntable, with room to spare. No reason to hide these electronic marvels behind doors. Make the shelf height equal to the clearance needed to remove your turntable dust protector—we'll say about 15″. (If your turntable or receiver is deeper than the 13″ shelf, you can either make the whole cabinet deeper, or make that one shelf deeper.)

Still to go are the dozens of books, magazines, knickknacks, *objets d'art.* The art pieces you want at eye level; the magazines and books should be handy for easy reference and browsing. Now, there is no structural reason these shelves have to be permanently fixed in place. You'll be buying new knickknacks, larger books, or taller stereo components—adjustable shelves allow you to change the design to suit the contents. There's room for two or three more shelves, depending on the height of your objects. If you're not sure, plan on three. You may not need the third one right away, but you won't feel like making an extra shelf a year from now.

If you've been drawing these spaces into the outline of the cabinet, you now have a simple basic design for your needs.

Structure—wood choice

How do you make a strong structure out of this design? First choose your wood. Two-bys make an extremely strong cabinet, but there's no need to go to the expense. One-bys are standard for shelves and cabinets. *Common grade* 1-bys wider than 10″ unfortunately tend to be warped; *clear* 1x14s, if you can find them, are good for this project. For the same price, however, you could get a beautiful birch-veneer ¾″ plywood and put veneer tape on the edges to look like solid wood. Taping takes time, but if you decide to insert the shelves in dadoes you'll have an easier time with plywood. Even clear 1-bys (particularly 1x14s) may have a slight warp which plywood, because of its laminated construction, is not prone to; warps make it very hard to slip a piece into a straight dado.

Fir plywood is cheaper and will do the job but won't look as elegant. It has plug marks and a less attractive grain. It's all

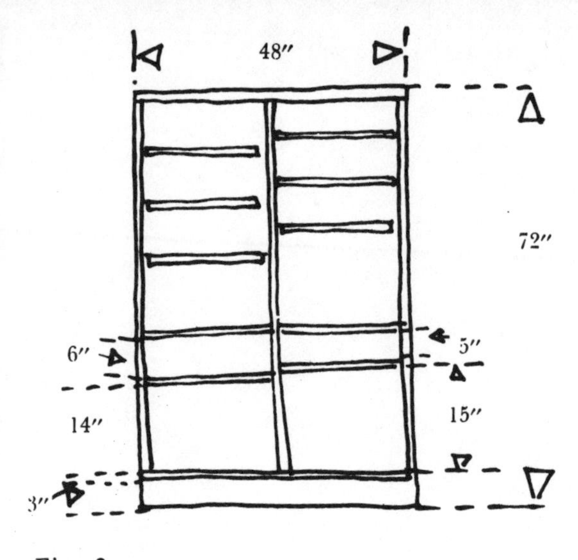

Fig. 2

right if you paint it, but we're praying that you will *not* be painting this beautiful wood cabinet.

With heavy loads like books, ¾″ shelves should not span more than 32″ without support. This is a standard rule of thumb. You can extend the span a few inches for lighter loads, but then you never know what loads you'll be supporting next year. Too long a span causes the shelf to sag, weakening the whole structure. Eventually the shelf can pull out from its sides.

This cabinet is 48″ wide; the shelves need support. One possibility is to stick vertical supports between them at the center, or in some interesting staggered pattern. This is impractical for adjustable shelves, however. Instead, run one long vertical piece up the middle from bottom to top and put half-length shelves between it and the sides.

This arrangement also allows more latitude in shaping each space to its contents. For instance, the turntable needs a 15″ height, but the adjacent amplifier only needs 7″ or 8″. The liquor cabinet can be raised an inch more than the record cabinet to give more space for tall bottles.

Structure—analyzing the structure

To help with the remaining structure decisions, make another rough sketch with these new adjustments (*2*). Draw the thickness of the wood; problems at the joints and intersections, particularly, are obscured in a single-line sketch.

The fundamental structure of the cabinet consists of the three verticals, the top and bottom, and the four permanent shelves above the bottom. The adjustable shelves *rest on* or *in* the structure, but they are not structurally joined to it. Basically we have an open box frame with some irregularly placed cross-bracing. As we learned in "Structure," an open box is liable to fail under lateral stress. To rigidify this structure (or any open box) attach a back to the frame. The back ties everything together, helps make the whole structure work as one unit. Thin plywood—⅜″ or even ¼″—will do: the stress is not *on* its surface as with a shelf, but *across* it.

A back is also a good design factor. The cabinet will seem a more self-contained piece. Thin objects can't fall behind the cabinet.

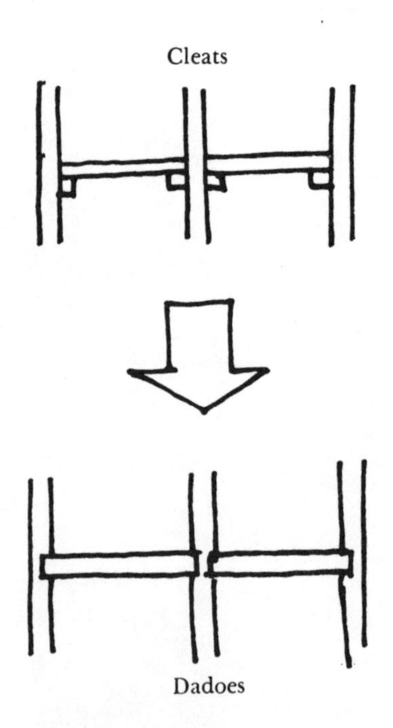

Fig. 3

Structure—Joining the elements together

Now you know *what* parts to join for a strong structure. *How* do you join them together? What kinds of joints will be strong enough? The piece will not be rigid if you make weak joints. Consider the joining techniques we're familiar with.

At the shelves, simple nailed or screwed butt joints do not disperse the stress efficiently. Most of the load would be right *on* the nails or screws, under Joints and Joining in "Working Techniques." It's much better to use cleats as supports for the fixed shelves as in the Bookcase project; or to rout dadoes in the verticals to accept the shelves (*3*).

Cleats are easy to build with. The disadvantages are that

cleats look bulky and destroy the lines of the piece; they require screwholes that will either show or have to be filled; and they take up space within the shelf area.

Dadoed joints make the structure even stronger than cleats. They are also cleaner looking, and simple to make with a router or a dado attachment for your circular saw. The standard dado depth for ¾″ shelves is ¼″. The shelf end is glued and fitted into the dado, and then fastened through the vertical with 2″ finishing nails. The fitted wood joint makes screws unnecessary. This also eliminates the possibility of ugly gaps between the shelf ends and the verticals.

If you don't have a router or dado attachment, you will probably want to use cleats. Consider investing in a router at this point, however. A cabinet like this, built as well as you are going to build it, costs several hundred dollars at a store—even more if it's custom-made. Since you will be saving all but a hundred dollars or so (for materials) of that total cost by building it yourself, you can perhaps afford to invest fifty to seventy dollars in a router, one of the most versatile and enjoyable tools you can own. The dado attachment for a saw is cheaper—fifteen to twenty-five dollars—but it's limited to cutting straight grooves. Either tool, however, will increase the value and looks of this cabinet by at least its own cost. All in all, it's a good way to get a useful tool without feeling the bite too much.

Our own preference obviously is to rout for the shelves. We'll follow that method for this project and later, under Variations, explain the construction differences for the cleat method.

The very top shelf of the cabinet, the top of the outer frame, can be joined to the sides with miter joints, glued and nailed. Mainly a design refinement, this common cabinet joint creates a neater seam between the pieces, and is adequate in strength.

The center vertical support transmits most of the stress on it straight down, through the wood itself; a simple butt joint at top and bottom, screwed and glued, would be sufficient. Routed joints would look better, however, and also enhance the vertical's bracing effect on the frame.

Now what about the back? Gluing and nailing it right to the back of the frame is strong enough. Unlike the shelves, it undergoes only relatively light lateral stress. If you draw a three-dimensional sketch, however, you'll notice a design problem with this arrangement (*4*). The edges of the tacked-on ⅜″ ply are visible from the sides. It's preferable to slip the back inside the frame, rabbeting a lip for it all along the inside back edge down to the floor, with a ⅜″ straight router bit; glue and nail it with 1¼″ finishing nails (*5*). You could also use the ¾″ bit used for the shelf dadoes—set your guide so half (⅜″) of the bit is on the wood, half off.

Note that this insertion decreases the depth of the shelves and the middle support. To keep the necessary 13″ shelf depth, increase the two cabinet sides and top piece to 13⅜″.

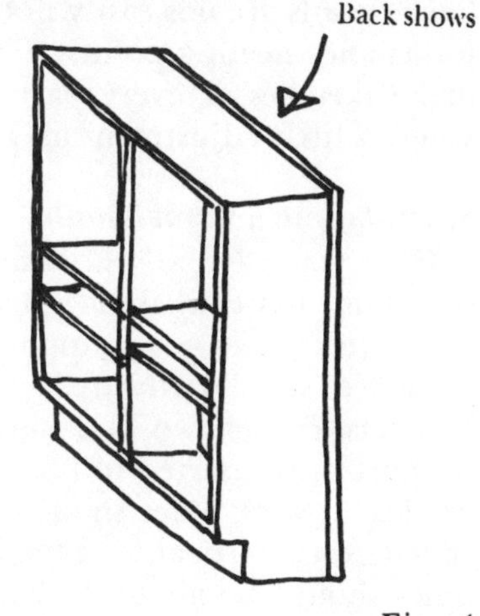

Fig. 4

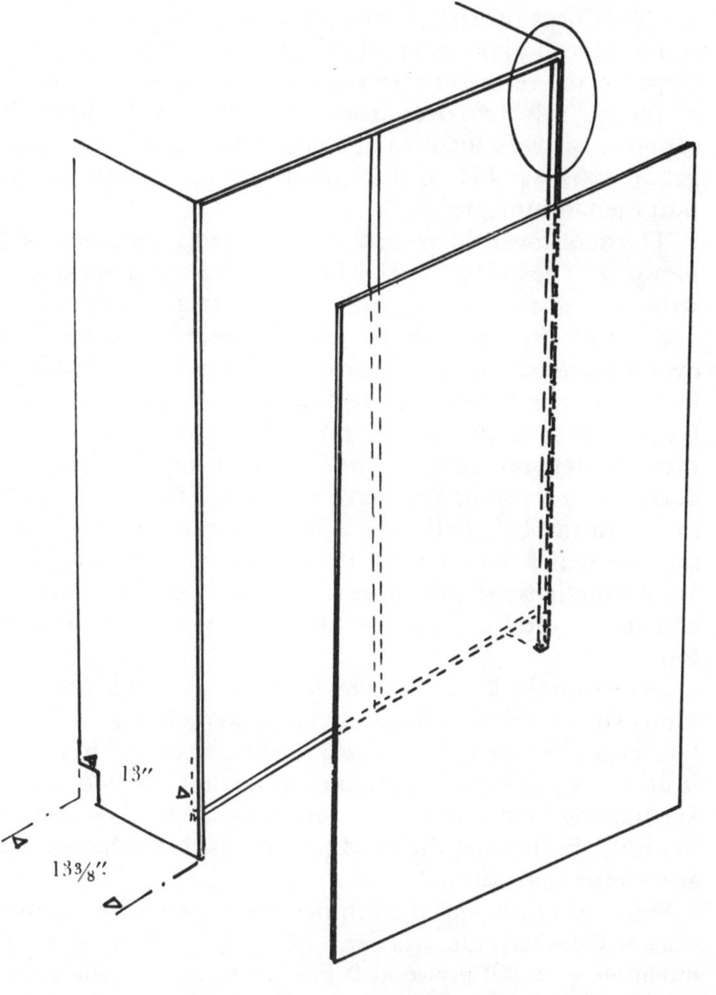

Fig. 5. Back Joint

Such details are not easily noticed without sketches of all the joints and meeting points. That's why it's very important to make sketches of every part of a project. You never know where a little adjustment may be necessary.

Secondary structural details

Now you have all the main structural problems worked out. Time to work on the remaining details.

There are two good support systems for adjustable shelves, discussed in "Hardware." One uses shelf standards and brackets (which we find ugly); the other consists of shelf-support pins inserted in holes drilled into the verticals. One vertical row of holes should be placed about 2″ from each edge of each vertical, so four pins can support each shelf. The holes should be about 1″ apart in each vertical row. Since we'll obviously need holes on both faces of the center vertical, those holes might as well be drilled all the way through; if the pins have 3/8″ shanks, it will be possible to put one in from each side of one hole. The adjustable shelves should be cut to 1/8″ to 1/4″ less than the span from side to side, for clearance when slipping them in and out. This time we'll say 1/4″. Check the size of your shelf pins to be sure.

The record dividers are under no great stress—they're just there for the records to lean against. Might as well use some of the 3/8″ ply left over from the sheet for the back. The supports can be routed in at top and bottom, if you want to get an extra 3/8″ bit; or just butted, nailed, and glued. We'll butt them for the project.

The doors over the records and bar area should be made of plywood at least 3/4″ thick. They are not held rigid by the structure, so that 3/8″ or even 1/2″ ply might warp over the years. The simplest cabinet door is hinged at the side and overlaid against the cabinet front. The flush and lipped doors look neater, but cause more trouble in fitting. (See under Cabinet Door Construction in "The Ubiquitous Box.") Also they shorten the shelves (just as inserting the back piece does), which means you have to enlarge the sides again to maintain the 13″ shelf depth, plus make tricky blind dadoes for the shelf fronts. Unless the overlay door grossly offends your esthetic sense, we suggest you use it. Make it just high enough to totally cover the shelf edges at its bottom and top.

Decide on the kind of hinge for the doors—there are several kinds you could pick for an overlay door. If it's the regular butt type, be sure it's narrow—no wider than 3/4″ per leaf, so it fits the side edge of the frame. Also select the door catches. We like the touch latch a lot. The door opens when you press in on it, eliminating the need for door handles that stick out and scrape your shins.

Since we're routing the cabinet, we might as well rout the drawer together. The drawer bottom will not be supporting much weight; 3/8″ plywood is fine for it, 3/4″ for the sides. If you don't want to invest in a 3/8″ router bit for this bottom piece and the record dividers, you can use cleats to lay in the bottom. The cleats won't show unless you lie on the floor and look up. Use the 3/4″ bit to cut dadoes to join the drawer sides together.

We'll put the drawer in the 5″ high space on the right side of the cabinet. The front should fit with about 1/16″ clearance on top, bottom, and each side—so our front would be 22 3/4″x4 7/8″. Make the sides and back of the drawer 1/4″ narrower than the front or 4 5/8″.

The drawer can slide either on special roller hardware, or on built-in cleats. The rollers work beautifully but cost a few dollars. Simple wooden cleats work well enough. Waxing them helps. Also get an inset or flush-type drawer handle, so it doesn't block the closing of the door.

This seems to complete all the details. All this heavy brainwork helps unify the structure. Later you won't have to tack on pieces that you've forgotten about. For instance, if you didn't plan ahead to inset the back, you would likely build the shelves the same width as the sides.

Working plans

A working plan of some sort simplifies construction and is the first step toward an economical ordering of lumber. If you haven't done it already, make a three-dimensional sketch or orthographic projection of the cabinet with all its details (*6*). It doesn't have to be in scale unless you'd like to see the exact proportions. Write in all the measurements—total distances, dimensions between shelves, door and drawer sizes, etc. (*7*). Do isolated sketches of any area that is not perfectly clear to you (*8*). Check and double-check your figures. Patience pays.

Now make a list of the measurements of every piece of wood. Note that pieces fitting into routs, such as the lower shelves, are a bit longer than the open space they span. For instance, the total width of the cabinet is 48″; the middle and two side vertical supports are each 3/4″ thick; that leaves 45 3/4″ for the two shelves. The usable or open shelf space for each is half that, or 22 7/8″; the actual piece of wood for the shelf is 1/2″ longer, or 23 3/8″, because it has to fill the two 1/4″ routs at either end.

Double-check your figures by adding up all the individual measurements across; we hope you end up with a grand total of 48″.

In your list, put all the across-the-grain measurements first, to save time and confusion when you cut the plywood and want to know how to orient the piece. The grain runs left and right on the shelves, and up and down on the verticals. The doors can be oriented either way. On the drawer sides, the grain runs the long dimension; on the bottom it doesn't matter. With solid wood the grain direction is important structurally. With plywood it's a design factor, meant to give the appearance of solid wood. You can change the orientation if you want, but keep your list consistent.

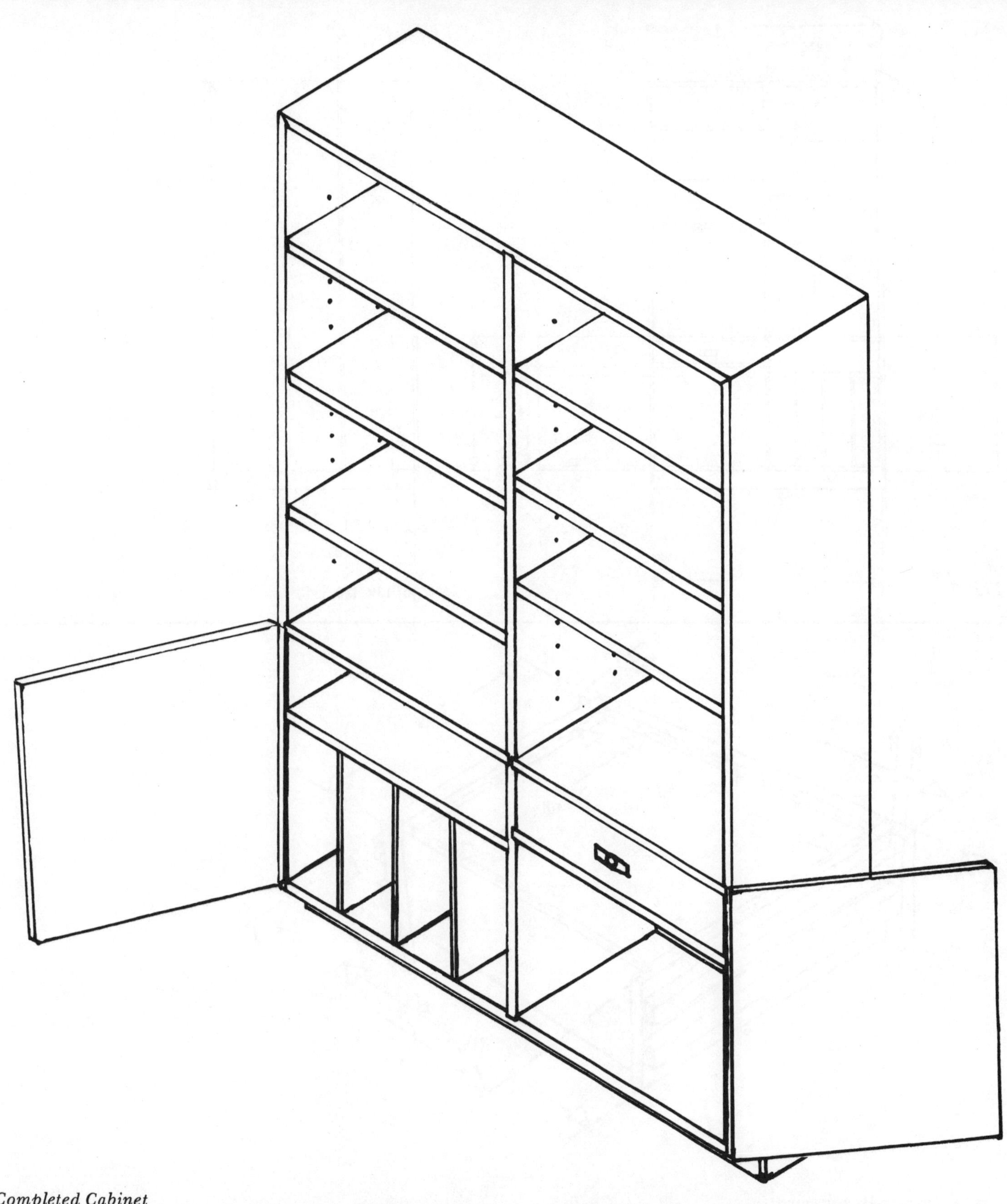

Fig. 6. Completed Cabinet

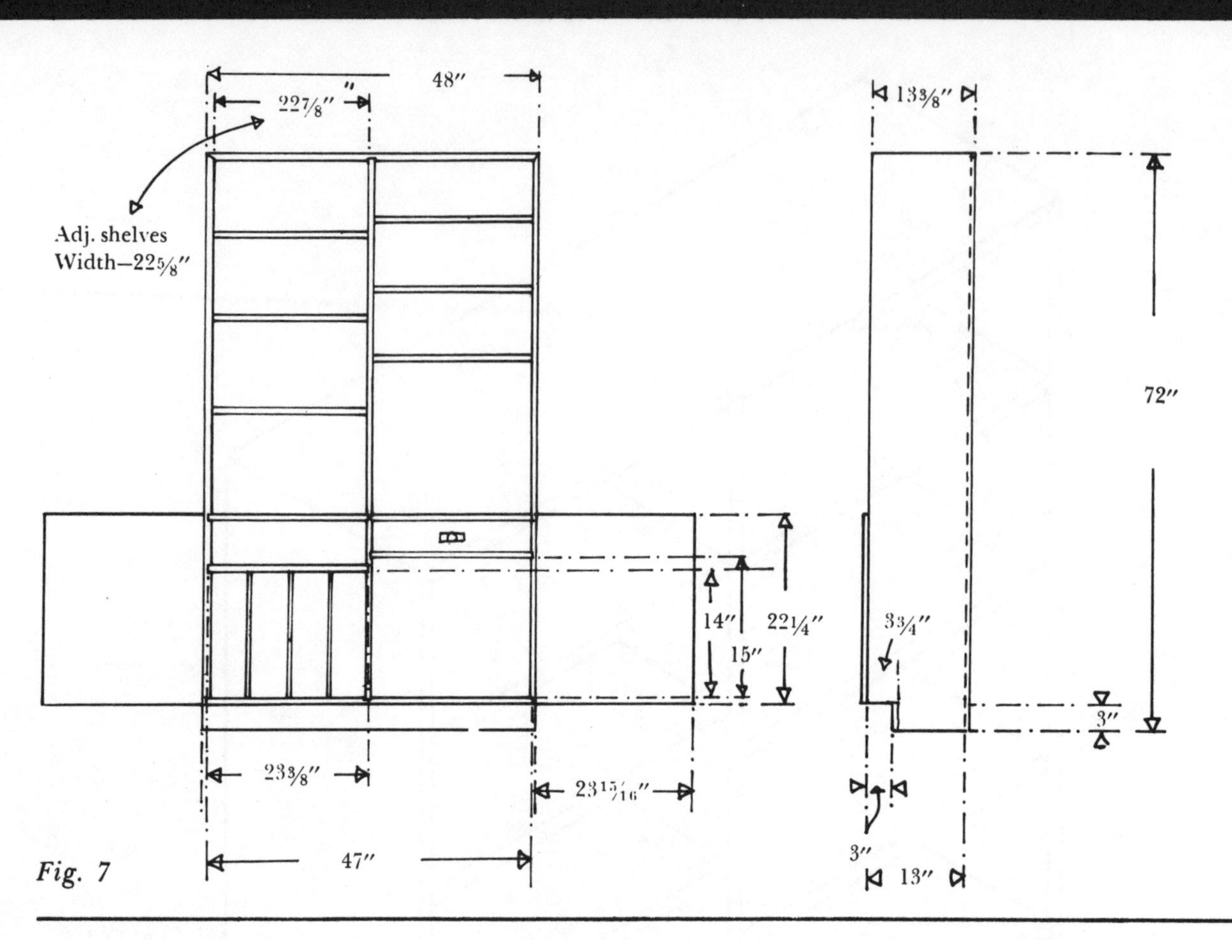

Fig. 7

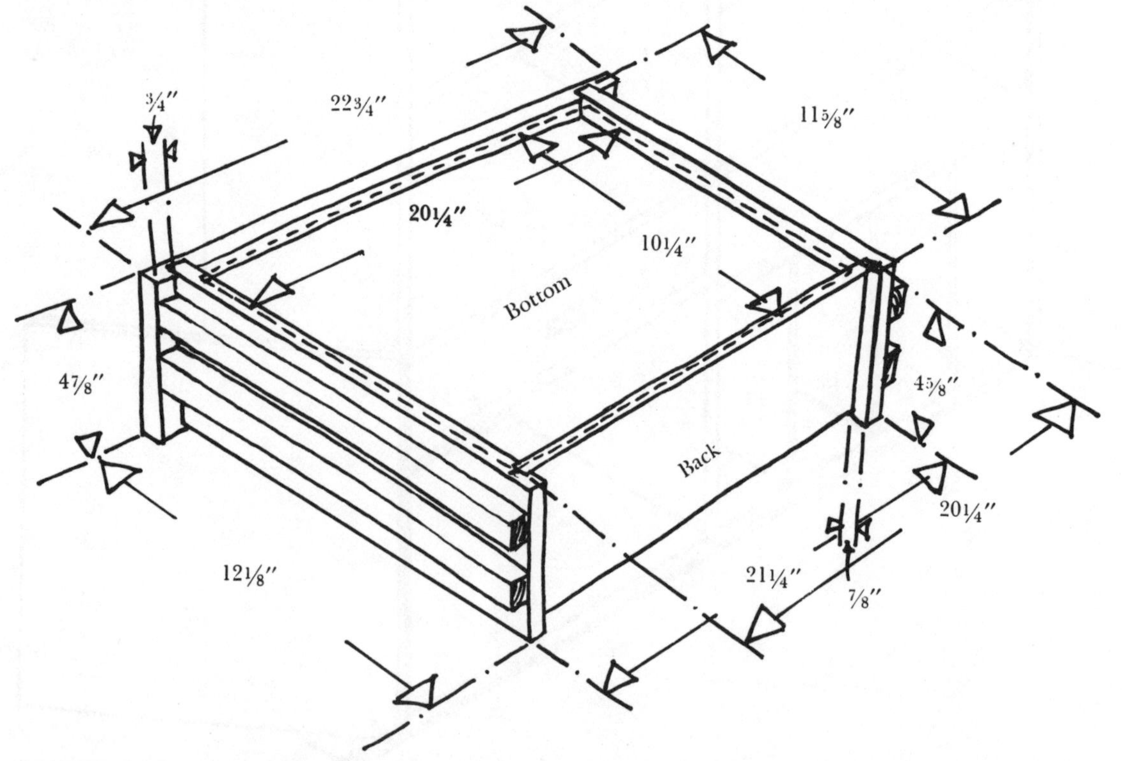

Fig. 8. Drawer Measurements

List of pieces

¾″ plywood

Left and right verticals—two at 13⅜″x6′, with 3¾″x3″ notch in bottom front of each

Bottom permanent shelf—one at 13″x47″

Permanent shelves—four at 13″x23⅜″

Adjustable shelves—seven at 13″x22⅝″

Top—one at 13⅜″x48″

Center vertical—one at 13″x68″

Toe plate—one at 3″x48″

Doors—two at 22¼″ high by 23¹⁵⁄₁₆″ wide, grain running either way

Drawer front—one at 4⅞″x22¾″

Drawer sides—two at 4⅝″x11⅝″

Drawer back—one at 4⅝″x20¼″

Drawer runners—six at 1″x10″

⅜″ plywood

Back—47½″x71½″ (possibly cut a shade less both ways)

Record dividers—three at 13″x14″

Drawer bottom—10¼″x20¼″

Cutting plan

Next step is to arrange the pieces for a cutting plan (*9*). This tells you how many plywood sheets you should order, and also serves as a guide for cutting the sheets. (See Layout in "Working Techniques.") Here is one layout that uses three sheets of ¾″ and one of ⅜″. There is a lot of waste on the third piece, but there's no way to fit those 13″-wide pieces on the first two sheets. If the price of the third sheet is too much to bear, you could get all the pieces from two sheets by making the adjustable shelves shallower (10″ is enough for books), eliminating the extra shelf, cutting the top piece from the space saved, and juggling the little pieces.

Your lumber order

Get your ¾″ birch plywood with a plywood core—a composition-board core would be super heavy, and a lumber core expensive. For the ⅜″ ply for the back, you can probably only get fir or Philippine mahogany, but ask for birch.

If you add up the front cabinet edges, you'll see you need about 80′ of veneer tape. Some lumber yards carry 50′ or 100′ rolls of tape, the bulk price being cheaper than ten 8′ standard rolls from the hardware stores. Your list will be:

¾″ plywood (birch veneer, plywood core)

Three 4′x8′ sheets

⅜″ plywood (birch veneer if possible)

One 4′x8′ sheet

1″ birch veneer tape

One 100′ roll (or two 50′ rolls)

Hardware

Four cabinet hinges for the door, narrow enough to fit on the ¾″ thick cabinet edge

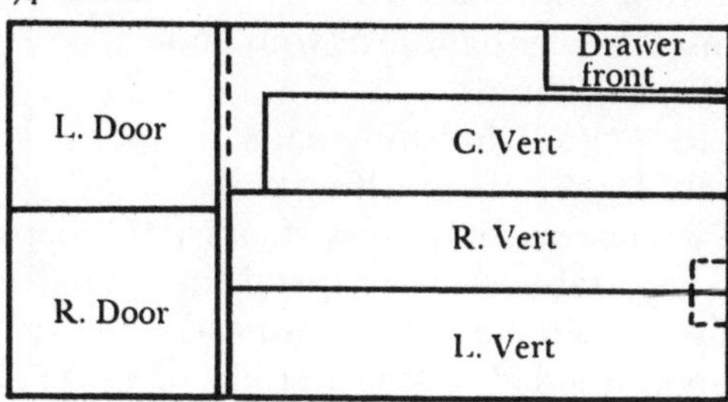

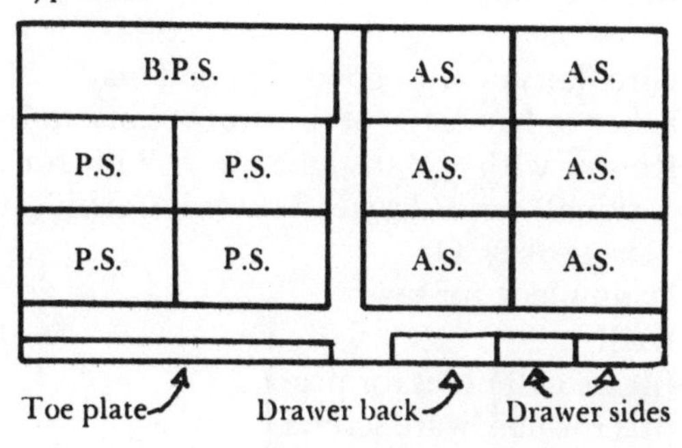

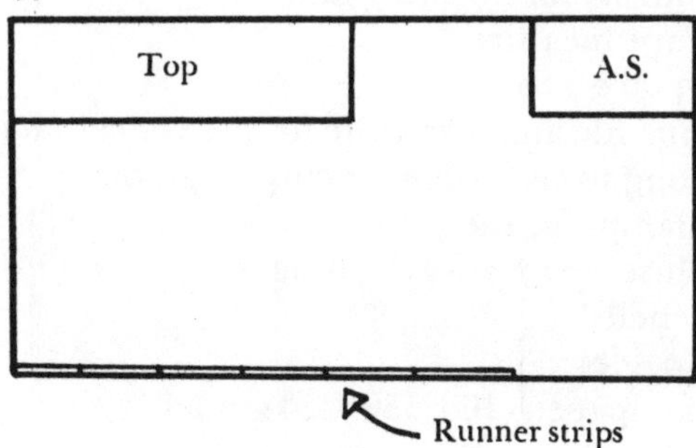

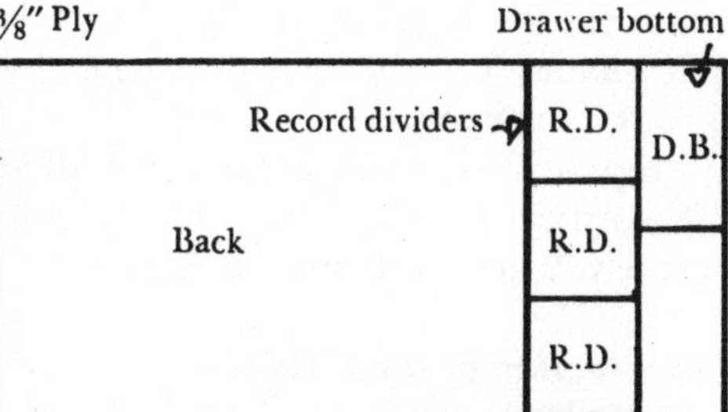

Fig. 9. Cutting Plan

Two pressure catches
Inset drawer handle of your choice
1 lb. 1¼″ brads
1 lb. 2″ (6d) finishing nails
1 lb. 1¼″ (3d) finishing nails
Two dozen shelf-support pins, ⅜″ depth, ¼″ diameter, for adjustable shelves. Spend time searching for the metal ones
Two dozen No. 8 1¼″ flathead wood screws for the runners
Good wood glue—liquid hide, or white glue
Contact cement for the veneer tape

Tools needed

All of this planning should help you see pretty clearly what tools and accessories you will need. Gather them all together beforehand.

Circular saw with good plywood blade
Saber or handsaw for notches at bottom front of sides
Router with ¾″ straight bit. ⅜″ bit for record dividers (optional) and drawer bottom. Carbide if you can afford it, steel okay. Or,
Dado blade for saw
Drill
Bit to drill holes for pegs
Bits for hardware screws
Screw bit for 1¼″ screws
Hammer
Small screwdriver for hardware
Nail set for finishing nails
Tape measure
Square
Fine file and mat knife for the veneer tape
Long straightedge for cutting guide
Clamps for the guide
Chisel to mortise the hinges
Pencil
Goggles
Sandpaper—100, 150, 220 grades

Birch veneer ply will come finely sanded. You can sand it further with a finishing sander if you have one, or by hand. You won't need a belt sander unless you make a sloppy saw cut and have to trim an edge down. Don't use the powerful belt sander on the veneer surface—you might sand right through the veneer.

Prepare your work area. Have all the tools ready, get all the hardware. When the lumber comes, check it carefully before you accept it. Store it neatly.

Construction of the Cabinet

Cutting preparations

Your first order of business is to ready all the pieces to be cut to size. Tack your cutting plan up in view where it won't be blown away by the saw or covered by sawdust. If you don't have a sizable workbench or sawhorses, set blocks on the floor to support the plywood as you cut.

Look at each piece of plywood. Pick the side you want to show. Plan for flaws or bad marks to end up on a waste area, or where they will show least in the cabinet; line up interesting grain designs on pieces that will show most (*10*). The selection and positioning of grain is an art in itself, one you'll really appreciate when you've lived with a piece of furniture for years and noticed its grain swirls every day.

Follow the usual preparations for sawing (see Saws in "Power Tools"). *Wear your goggles whenever you run the saw.* Install the plywood blade, set it no more than a sawtooth below the wood. Find the guide measurements if you don't know them (plywood blades are slightly thinner than crosscut blades). Clamp the guide to a scrap of plywood (or designated waste area on the ply) and with the saw base against the guide, cut in a few inches. Remove the saw and measure the distance from the guide to the left and to the right of the kerf. Write these measurements down and keep them handy. Every time you cut a piece, consider which side of the blade the edge of the finished piece will be on, and set the guide accordingly; place the saw kerf on the waste part.

If you haven't already, you might look now at the Cutting Plywood section in "Working Techniques." It offers a number of suggestions to save you trouble and to improve your cutting accuracy.

Miter cuts

The tops of the sides and the ends of the top piece of the frame are mitered (*11*). This involves very careful cutting and measuring. First adjust your saw blade at a 45° angle to the baseplate, checking it with a bevel square or 45° triangle. This new blade angle changes the distance from guide to saw kerf; make some practice cuts to find the new measurements, and to get the feel of the miter cut. Hold the saw very steady and firmly; don't let it tilt or sway.

Decide which side of the ply is to be the outside of the cabinet. Lay that side face down for the cutting. Be sure the ply is fully supported on either side of the cutting line—if the ply bends under the saw weight, the accuracy of the cut will suffer. The miter cut is made *before* the doors are cut out of the sheet. Without the door pieces, you would have little support for the saw on the miter cut.

The piece is to be cut 6′ high on the *outside* surface; but you are sawing and setting up your guide on the *inside* surface, whose height is a bit lower due to the miter. How much lower? How do you know where the cutting line is? The simple rule is that the inside measurement is shorter than the outside measurement by the thickness of the wood: here 6′—¾′ = 5′11¼″. This is clear when you realize the end section of a miter is a 45°–45°–90° triangle. Both legs (the shorter sides) of such a triangle are always equal (*11*).

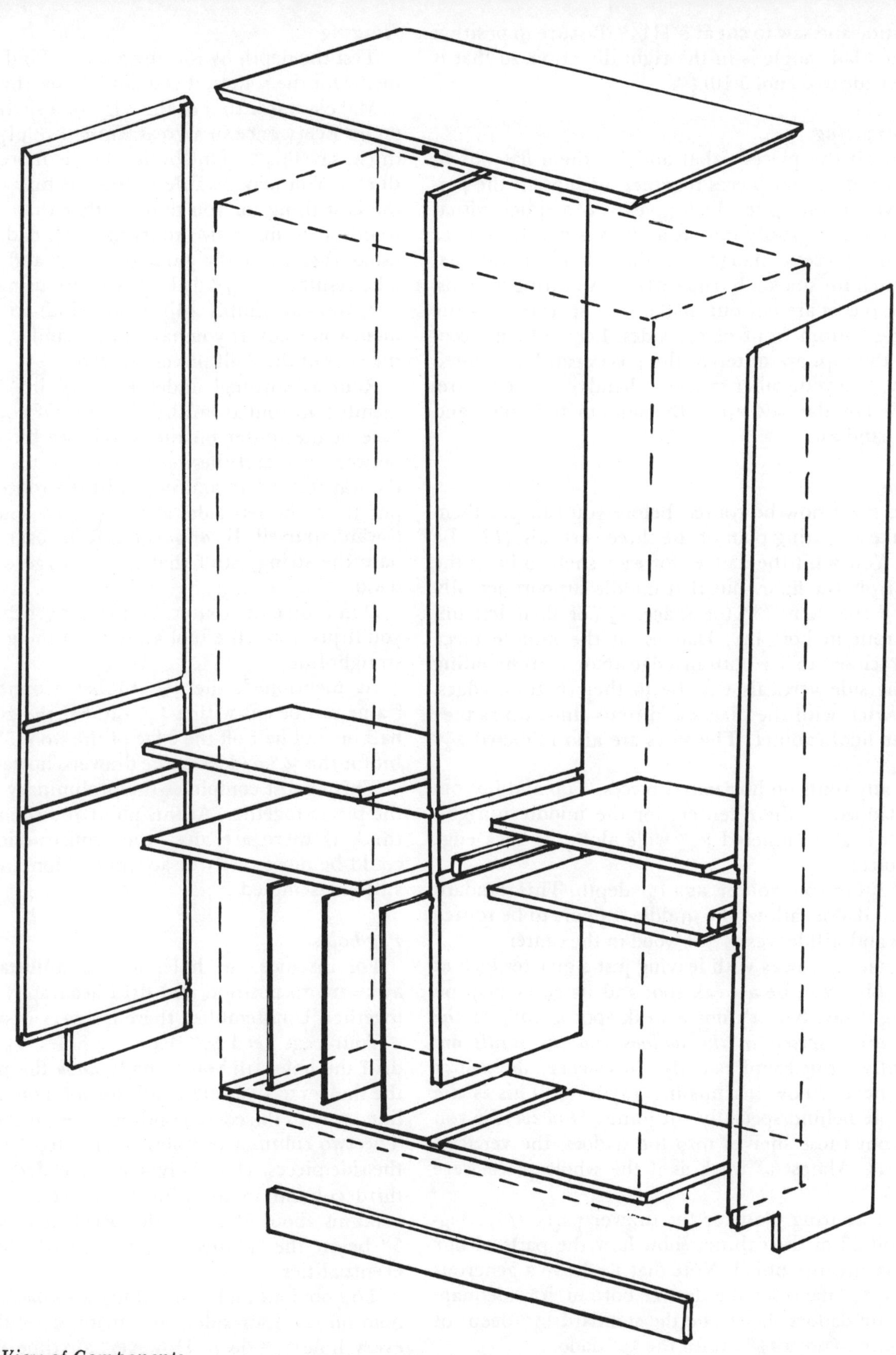

Fig. 10. Exploded View of Components

Set your guide and saw to cut at 5′11¼″. Be sure to position the saw so the blade angle is in the right direction, so that it cuts the underside to 6′, not 5′10½″.

Finishing the cutting

As you cut all the pieces, label and lay them flat out of harm's way. Keep similar pieces together, such as all the permanent shelves in one pile. This gives you a quick visual check if any piece is grossly miscut and saves you time later picking out the pieces. Change to a sharp blade if the first one dulls—watch for smoke, burnt kerfs, slow cutting as signs. When all the pieces are cut out, make your special cuts—the notches at the bottom front of the sides. Locate them accurately. Since the tops are mitered, the pieces can be oriented only one way. Use your saber saw, or a handsaw; or cut carefully with the circular saw up to the sides of the notch and finish with a handsaw.

Routing plans

The pieces must now be routed before you can put them together. Make a routing plan of the three verticals *(12)*. Be careful now. You want the dadoes for each shelf to be at the same height from the floor. But that middle support actually starts 3½″ *off* the floor (3″ toe space, ½″ of shelf left under it after rout in bottom). Dadoes on the middle piece will be 3½″ closer to its bottom edge than corresponding dadoes on the side verticals will be to their bottom edges. Draw your sketch with the pieces and routs lined up as they will be in the final cabinet. The sides are also rabbeted ⅜″ for the back.

Now plan any routs on horizontal pieces. Top and bottom shelves get dadoed in their centers for the middle upright. The top shelf is also rabbeted ⅜″ wide along its back edge for the back piece.

Set the ¾″ bit in your router, at a ¼″ depth. This standard depth for shelf dadoes allows the middle support to be routed on both sides and still leaves ¼″ of wood in the center.

"But how can I get away with leaving just a quarter-inch of wood there? That will be a weak spot and its gotta snap on me," you might say. No, it's not a weak spot *as long as the other pieces are inserted in the dadoes and the joints are fastened tightly* (glued and nailed). In essence, the center vertical will never know it's missing anything. This is the whole principle behind specially cut joints. *However,* if you were to *not* put those shelves into the dadoes, the verticals *would* be weak. Almost as weak as if the whole piece were only ¼″ thick.

Also make a routing plan for the drawer parts *(13)*. Figures 8, 21, and 22 of the cabinet show how the parts of our dadoed drawer are assembled. Note that we leave a generous ½″ below the ⅜″ dado for the drawer bottom, for adequate support. All the dadoes should be the standard ¼″ deep; of course, you have to use a ⅜″ bit for the ⅜″ dado.

Routing

Test the depth by routing a scrap. Find the guide measurements for the router, if you don't know them.

Make a test fitting of the edge of a shelf into a dado in the scrap. Every once in a great while, ¾″ plywood will be extra thick, say 13/16″. (One-by lumber is more likely to be extra thick.) You may be able to force it into the dado; otherwise the best thing for you to do (other than using different plywood) is to make two routs for each dado. Make the first as usual, then move the guide over 1/16″ and make another pass. The result is a 13/16″ dado. If you are using a dado blade on a saw, you can simply adjust the width to the exact measurement you want. If you have a belt sander, you can sand down the ends of the shelf pieces, instead.

Rout as outlined under Routers, in "Power Tools." Remember to rout from the ends in. Of course, start with the base of the router on the wood, the bit off. If the guide is between you and the router, start on the right, rout most of the way to the left, and stop. Lift the router out *after* it stops, put it on the left side, and cut until complete. Never rout toward yourself. *Wear goggles.* Rout back and forth to eliminate the stringy stuff that may emerge on the edges of the dado.

Don't force the router. You won't get the dado done faster; you'll just force the tool away from the guide and ruin your straight line.

As mentioned, the ⅜″ rabbet around the back of the frame can be cut with a ¾″ bit. Set the guide so the bit cuts half on and half off the edge of the wood. You must use a ⅜″ bit for the ⅜″ dadoes in the drawers, however.

This almost completes the preliminary work needed to fit the pieces together. At this point you should always stop and think if there are any other construction operations that could be more easily or accurately done while the pieces are still unassembled.

Peg-holes

For instance, the holes for the adjustable shelf pegs are awkward to position and drill accurately when the piece is together. Unassembled, there is a very easy way to do it.

Your pegs need ¼″ diameter holes, ⅜″ deep. Don't ever drill the holes till you actually *have* the pegs. Of course, for the shelves to sit straight, all the holes on one vertical have to line up with the corresponding holes in the opposite vertical. The two columns of holes are located 2″ from the edges of the side pieces. (For shelves over 16″ deep, you would need a third column in the center for extra support.) Begin the columns about 5″ above the permanent shelf and end them 5″ below the cabinet top; this should leave room for any eventualities.

The obvious and painstaking approach is to lay out every hole on the four sides, measuring carefully, and then drill every hole ⅜″ deep. However, anytime you have to repeat

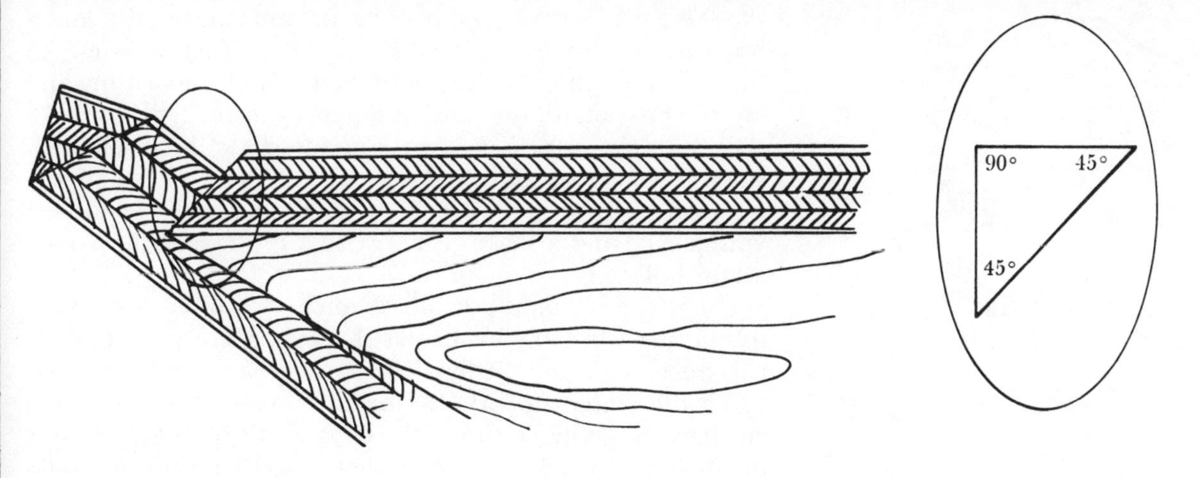

Fig. 11. End Miter Cut

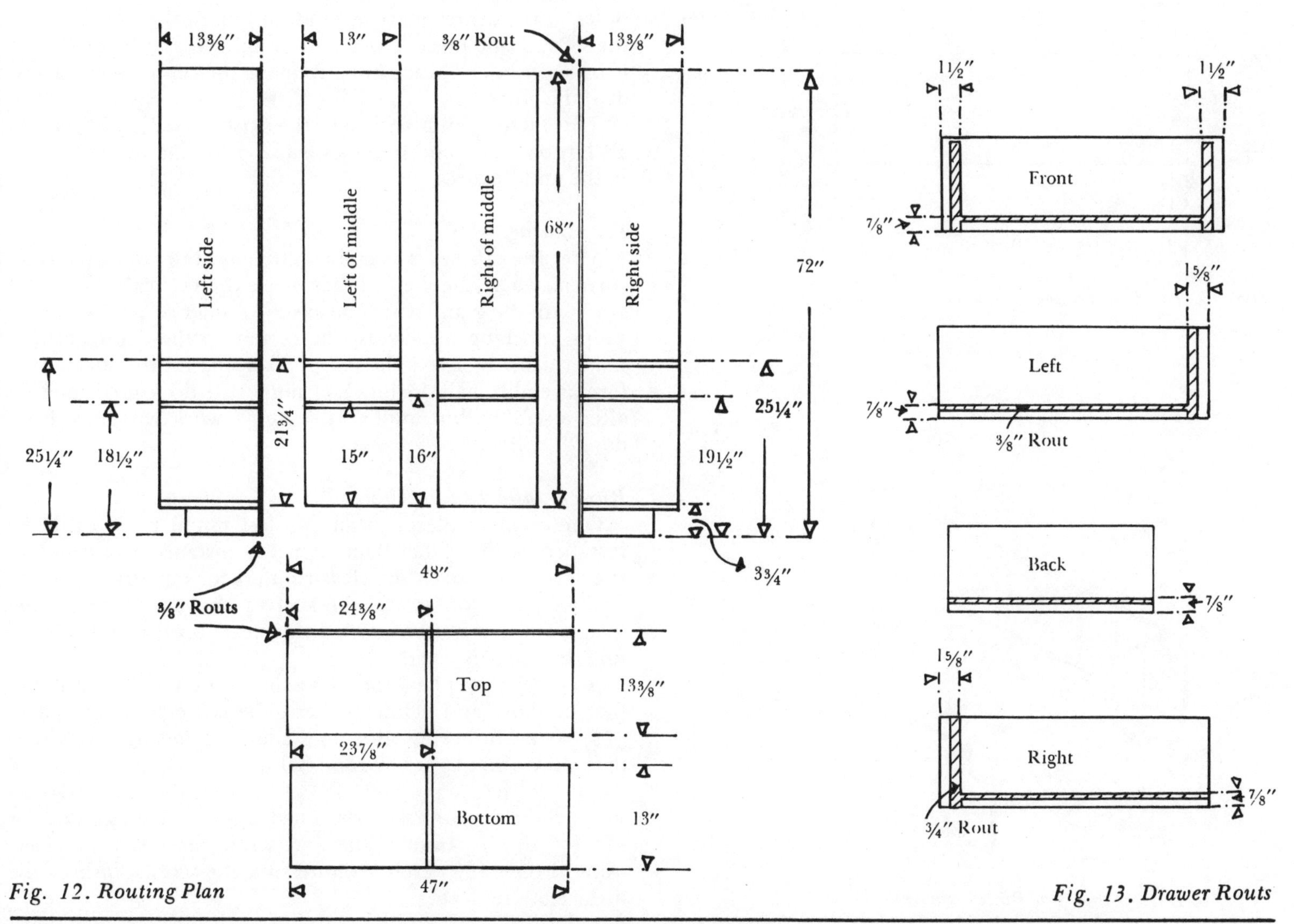

Fig. 12. Routing Plan

Fig. 13. Drawer Routs

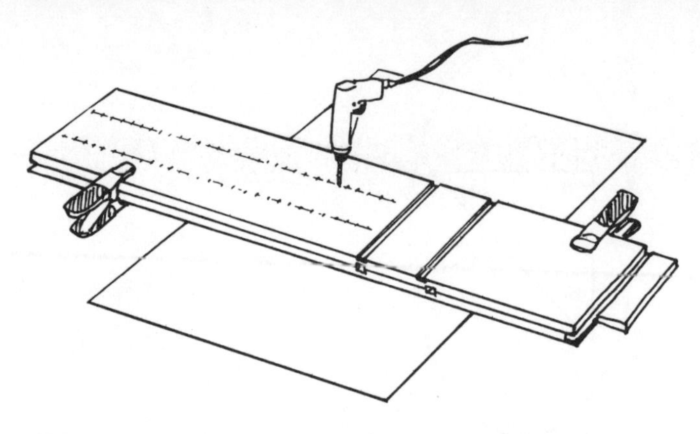

Fig. 14. Drilling shelf pin holes

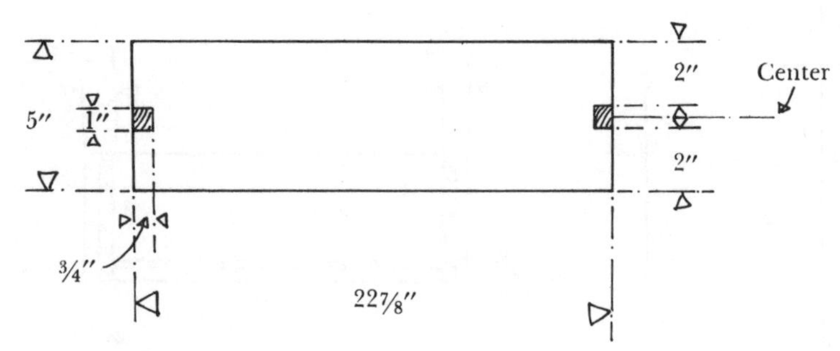

Fig. 15. Inside of Drawer Space

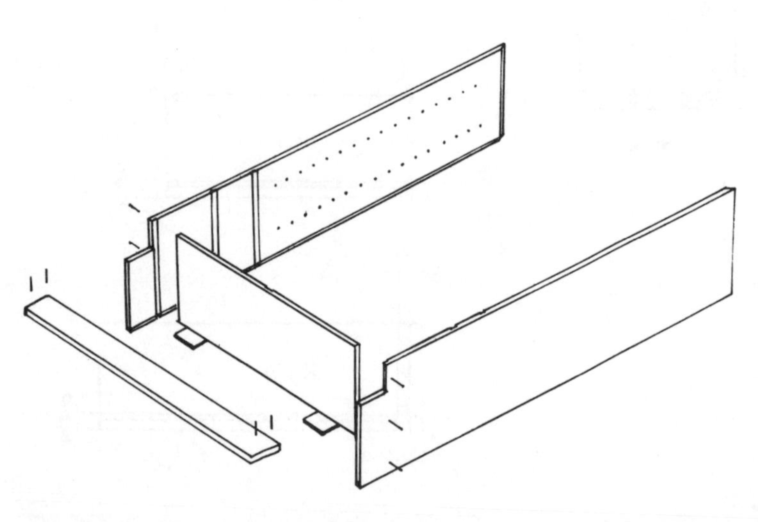

Fig. 16. Bottom and Toe Plate Assembly

the same process over and over again, you can usually find a way to speed things up—even if it's only a quicker layout. In this case you can eliminate more than half of the drilling and layout. First of all, the middle support can be drilled all the way through, since it has corresponding 3/8″ holes on both sides. That means only one side of the piece needs layout. Very carefully measure and mark all the points to be drilled with a tape and a sharp pencil. Use a center punch to start all the holes.

Usually you would put blocks under the piece to prevent splintering when the bit breaks through. Here you can drill two holes with one bit by aligning one of the side verticals under the center support. (Remember to allow for the different lengths of the verticals—the ends won't be lined up, but the dado grooves will.) Clamp them together with the routs lined up (*14*). Make a drill depth guide by putting some tape on the bit at the total depth you want to drill—3/4″ + 3/8″ = 1 1/8″. Then drill all the holes, replace the side vertical with the other side vertical, and use the through holes in the center piece as guides to mark the new holes in the second side piece. Use a nail or an awl, and be careful to mark the *center* of each hole. Remove the center vertical and drill the holes.

This method eliminates half the drilling, tedious layout on all but one face, and provides backing for the through holes in the middle piece.

Pre-sanding

After the cabinet is assembled, it is hard to sand the inside corners, where the shelf surface meets the verticals. You must sand with the grain, which means the sander or sanding block keeps knocking up against the corner (although an orbital sander can run across the grain). Sanding all the surface before assembly helps a lot. You must still do a final sanding after assembly, but most of the rough work will have been done.

Runners attached to cabinet

One or more pieces in any project might more easily be installed earlier rather than later. For instance, you may find it awkward to attach the cleat runners for our drawer to the cabinet *after* you attach the shelves above and below the space. To save trouble, install the cleats on the verticals before the frame is together.

According to plan, the cleats are centered between the shelves above and below or, here, the corresponding dadoes (*15*). Draw this center line on the middle and right verticals, then draw lines 1/2″ above and below to mark the outlines of the 1″ runner. Make a mark 3/4″ in from the front edge to show where the runner starts. The front of it acts as a stop for the 3/4″ drawer front. Glue and screw the cleats in place, using three screws each. Countersink the screws flush to the surface of the cleats.

Thinking out the assembly

Use your brain before your back. Devise a reasonable, efficient order in which to put the pieces together, one that produces the strongest structure and joints. Since the outer frame of the cabinet is what all the other pieces are fastened to, plan to square and assemble it first.

This frame consists of the two vertical sides, the bottom shelf that fits into dadoes on the sides, and the mitered top. Join the bottom to the sides first—the miter joint at the top will be weak till the glue sets and might loosen while you work on the bottom shelf. The easiest way to assemble this project is with the pieces up on edge on the floor, as if the cabinet were lying on its back. Once the frame is together, all the innards slip easily down into their dadoes, and you can nail from the sides.

Gather your materials together. For the frame assembly, you'll need a hammer, 2″ finishing nails, a nail set, glue, a damp sponge or paper towels to wipe off glue, some blocks of wood (2-bys) , and a rafter square. Remember at all times to protect the ply surface—from hammer marks, excess glue, block marks, whatever.

Fitting the frame together

First, as decided, we join the bottom shelf to the sides. Usually, a plywood shelf fits easily into its dadoes, which makes the joining pretty clean-cut. Start three nails in each side piece *(16)* . Try to get the point of the front nail to just barely come through on the inside of the dado. Spread glue in a zigzag pattern inside the dado on one piece. Prop the side up on edge and bring the shelf into its dado. Get the pieces flush at the *front*. Remember that the shelves all are 3/8″ less deep than the sides, so that the back can fit inside the frame. You can make that alignment easier by setting the shelf on bits of 3/8″ ply. If the shelf is cut too wide put the extra at the front, not the back. There has to be room for the back piece to fit into its rabbet, and it's easier to sand down the front of the shelf than the back, which is slightly inset.

If that front nail point is sticking through, you should be able to press the shelf against it, which will help hold everything in place while you hammer. Try to hold the shelf tightly between your knees, with your head and hammer hand over the outside of the side piece; hold the juncture of the two pieces with your other hand. A co-worker is obviously very helpful here. She can offer resistance at the other end of the shelf as you hammer. In fact, she can put the other side piece in place and support against that. If the pieces slip, readjust.

Once a nail catches into the shelf, hammer slowly. Feel the veneer on both sides of the shelf with your fingers for suspicious bulges, indicating that the nail is off-course and about to pierce the veneer surface. In this case, stop immediately. Drive another nail near it, then remove the first nail with your hammer claw and a block of wood to protect the veneer.

Don't hammer the nail in all the way yet. Leave enough showing so you can remove it if you have to. Glue the dado in the other side piece and hammer its front nail in the same way. Hammer in the rest of the nails and then sink them all the way, using the nail set to get them below the surface. Don't damage the wood with the hammer or the set.

Wipe off excess glue immediately with the damp sponge. This is important, since stain doesn't penetrate evenly over dried glue film.

With your rafter square, check that the shelf is approximately perpendicular to the sides—you can still adjust the pieces by hand. But this U or H shape will stay unstable till some more framing is joined in. Just be sure the shelf is tight against the inside of the dado. If the nails are set and there is still a gap, put some support against the one side and, using a block to protect the wood, hammer the other side.

In rare cases the shelf will not fit so neatly into its dadoes. You must then support the two sides very firmly—one side up against a wall for instance, or against an assistant, or both sides against blocks nailed to the floor—and lower the shelf into its dadoes. Force it down evenly with a hammer and a block of wood. Keep moving the block toward the high end of the shelf, the end most reluctant to go down. This movement also keeps the shelf edge from being marked up too much by the block. Constantly check that the sides are not pulling away—hammer them back in, if so.

Be patient. The pieces may fall apart on you once or twice. Start over, support them better. Wipe off the smeared glue.

The next piece to be permanently joined to the cabinet is the toe plate. Glue and fit it in place, making sure the joints are square and that it does not protrude below the side supports. Drive two finishing nails into it at each end, and four or five down through the bottom shelf, measuring back on the shelf the corresponding 3″ to 3¾″.

Mitered joints

Before gluing, check the fit of the top shelf against the side pieces. This shelf extends all the way to the back. It should be flush front and back to the side pieces, and the mitered edges should form perfect right-angle corners. Adjust the frame till the diagonal measurements are about equal; this squares the frame. The rafter square is of little help here, since the pieces may be slightly bowed.

Because there is no dado or notched joint to hold the pieces in relative position, your main job is just that—to keep the miter flush, to prevent the pieces from sliding off each other as you nail. Predrilling the nailholes helps. Clamping blocks in place, as in Figure 17, helps immensely. Your aim is to first tack two nails through the top at one end, then two at the other end; then two nails into the miters through each side piece. Finally drive all the nails in all the way.

Insert a bit into your drill that is slightly thinner than your 2″ finishing nails, and lay the drill within reach. Glue

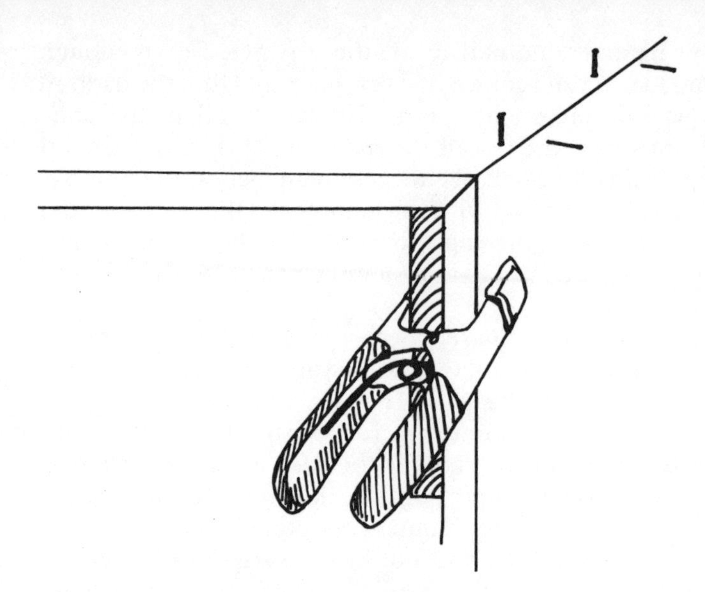

Fig. 17. *Holding miter for joining*

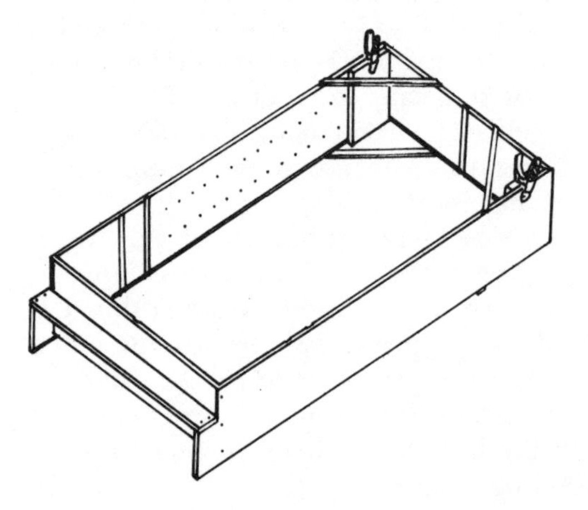

Fig. 18. *Top added and braced*

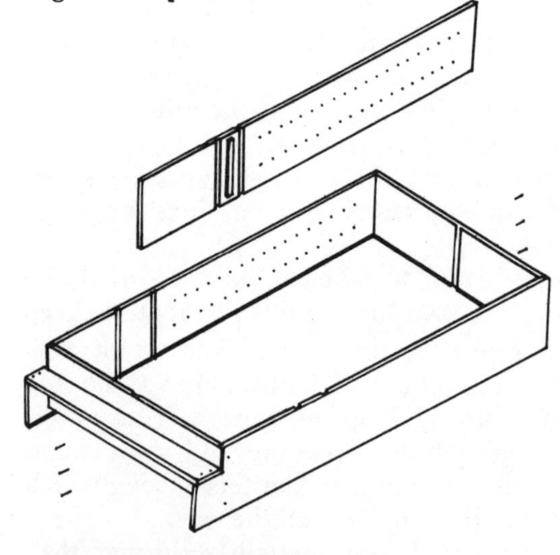

Fig. 19. *Center Support*

the edges of the side pieces, but not the top shelf. Less glue means less accidental slippage, yet the joint will still dry strong if you slide the pieces together to spread the glue around. Hold the pieces in place at one end while trying valiantly to keep the other end as flush as possible (an assistant is a great help here again), drill two nail holes through the mitered joint and tack the nails in part way. It's fairly easy to drill without jarring the pieces apart, and you'll then need less hammering. You might even start the holes through the top piece before you set it in place, to improve your odds.

Work slowly and carefully, constantly checking the miter and wiping off the glue. If one nailhole causes the pieces to be off, drill another. Adjust the joint at the other end of the shelf and square the cabinet again. Drill and nail that second miter together. If the miter is nailed crooked, pull the nail.

Go back to drill and tack in the other two nails at each joint. Hammer the eight nails in all the way with a nail set. Check the diagonals again, and square the frame.

If your miters are not perfectly cut and do not make a perfect corner, use common sense to adjust them. For instance, any gap between the pieces should be adjusted to show only on the top, not on the more visible side. Gaps can also be puttied. An overlapping edge can be sanded.

Bracing the frame for further work on it

Now you have a choice to make. The glue in the miter hasn't set yet, so the joint might be jarred apart by other work performed on the frame. You should either let the frame set overnight and work for now on separate elements such as the drawer, or taping the doors and adjustable shelves; or you could tack some braces to the front and back of the mitered corners with finishing nails and continue working. Your decision may depend on what time it is. If it's late, let it set overnight. If you still have a lot of time to work, nail in braces (keeping the frame square) as in Figure 18 and set the whole frame up on blocks to protect the bottom edges from the floor. Position the braces with room to slide in the center support and permanent shelves. The nailholes you make in the ply will be covered when you put on veneer tape.

Center vertical

Once the frame is set or braced, join the center vertical to it. This piece is really part of the frame. Again, square the frame. Glue the dadoes at top and bottom and slip in the support (*19*). If you like, prenail this support and the remaining shelves. Shims under the shelves may not be necessary, since the pieces may stay put by the pressure of the assembled frame. Be sure the support is facing the right way—peg holes on the top, dadoes for two shelves on the bottom, dadoes matching the ones opposite them on the side pieces. It should be flush with the bottom shelf front and back, while reaching only to the back 3/8″ rabbet on the top piece. Nail it in, wipe off the glue.

Permanent shelves

Glue, insert and nail the left and right bottom permanent shelves next (*20*). Then install the 3/8″ record supports and nail them with 1 1/4″ brads. (If you had installed all the shelves first, you wouldn't be able to get your hammer in to nail these supports.)

Install the upper left permanent shelf. Use your nail set so you don't smash the dado with your hammer. Then put in the other permanent shelf on the right. Toenail through the center vertical, or leave it unnailed if the frame pressure seems to be holding it tight. The nails are mostly just to hold it in place till the glue sets.

Constructing the drawer

First consider the handle for the drawer. A flush handle requires mortising—easier to do on the front piece before it is attached to the drawer. Center the handle along the drawer at the height you want, and carefully trace the outline to be mortised. Chisel down into the entire outline first, to the required depth, and then chisel out the interior. Install the handle.

The drawer assembly is a snap. First, slip all the pieces together to see if they fit properly (*21*). Take them apart and squeeze glue into all the 3/4″ routs, but not the 3/8″ ones for the drawer bottom. Slip the drawer bottom into its dado on the back piece. Fit the side pieces in place and nail them to the back piece. You can rest the drawer on one side as you nail through the other. Glue the dadoes in the front piece, slip it into place and nail it to the sides (*22*). No need to glue or nail the frame to the bottom piece—it will stay in. The drawer automatically squares itself about the bottom piece, but check it anyway.

Attaching the drawer runners

Measure the gap between the runners on the cabinet verticals (the cabinet runners) and the shelf above (see Figure 15). The *real* gap, as opposed to your planned gap of 2″, is what decides the location of the cleats on the drawer sides (the drawer runners). Remember you want the drawer front, in the closed position, to be 1/16″ below the shelf; and note that the drawer front is 1/4″ higher than the sides (*23*).

A 2″ gap above means the upper drawer runner is to be 1 11/16″ below the top of the drawer *side*. (That's the 2″ gap minus the total of the 1/16″ and 1/4″.) If the gap is 2 1/8″, the upper cleat is lowered 1/8″ to 1 13/16″. And so on. (See Drawer Construction in "The Ubiquitous Box.")

Put the front ends of the runners 3/4″ back from the front edge of the verticals. This will let the front of the drawer slide in flush with the shelves, and keep it from going in too far.

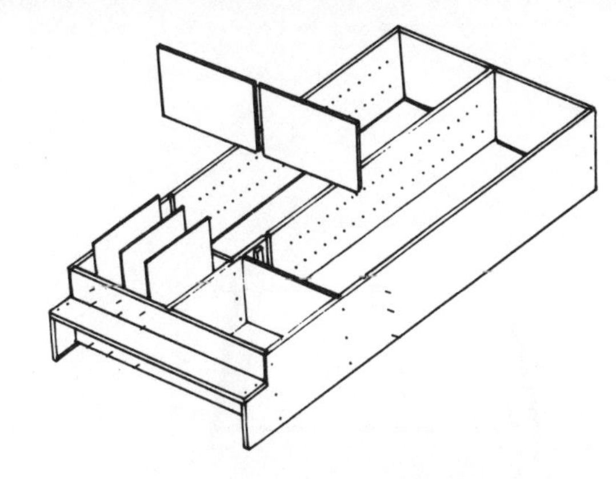

Fig. 20. Shelves and Dividers

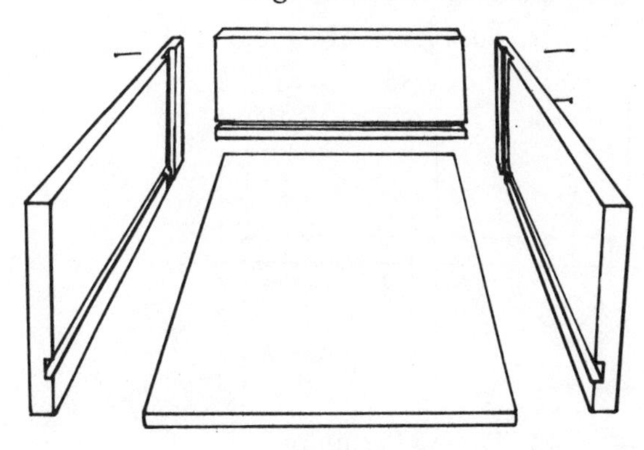

Fig. 21. Drawer Sides and Back Assembly

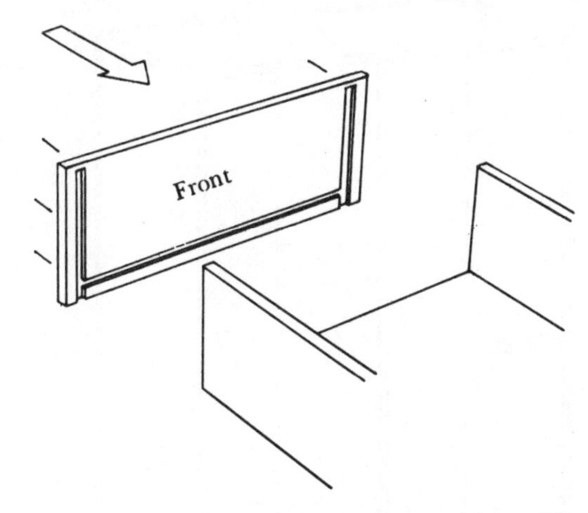

Fig. 22. Drawer Front Assembly

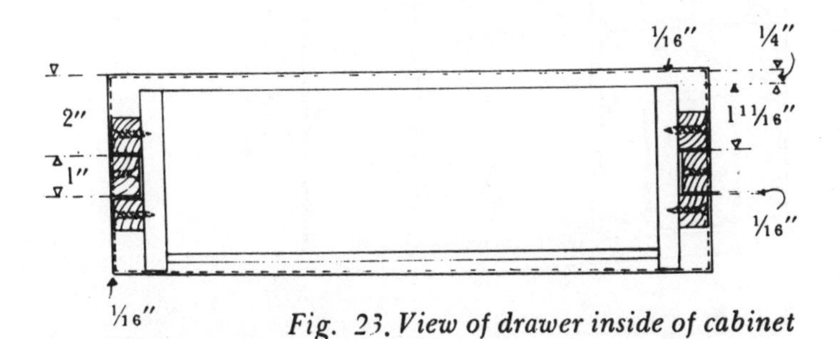

Fig. 23. View of drawer inside of cabinet

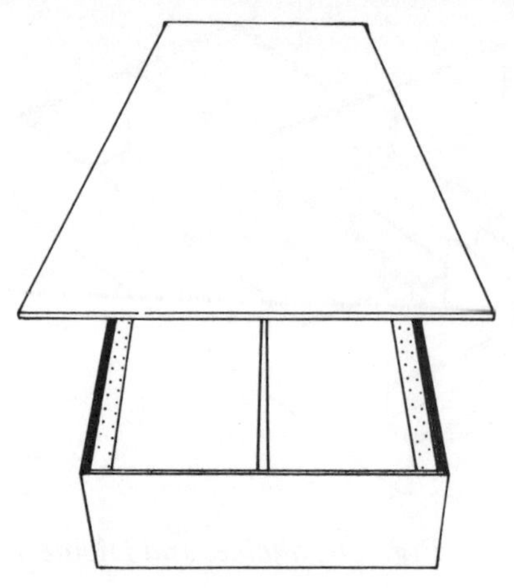
Fig. 24. Back Assembly

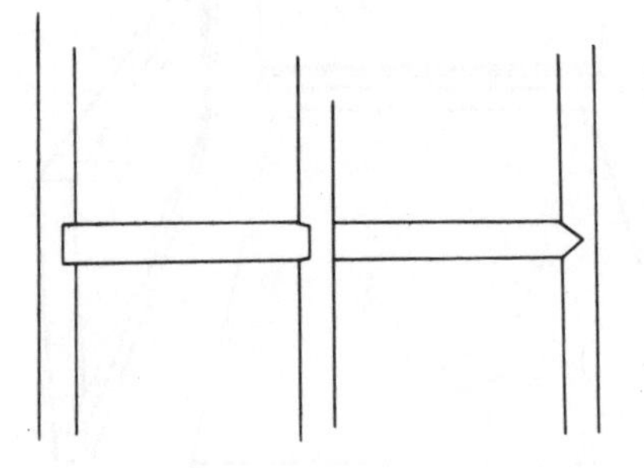
Fig. 25. Taping Variations

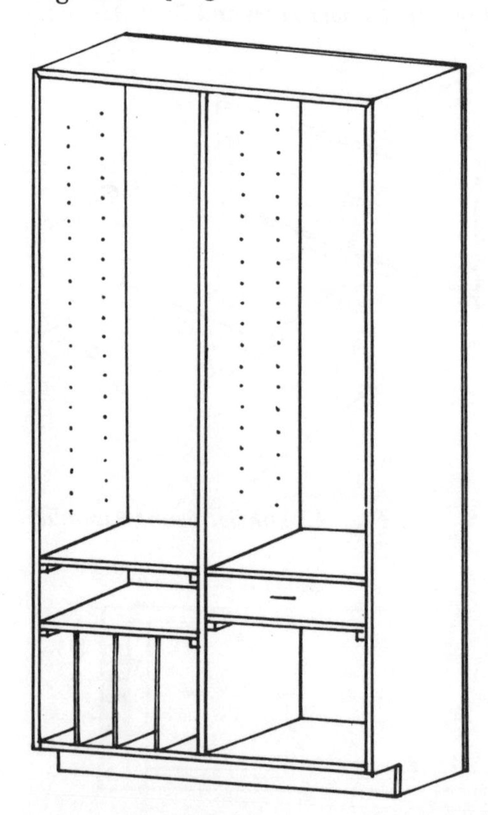
Fig. 26. Cleat Structure

Tack the upper runners in place with a couple of 1¼″ finishing nails, driven in all the way.

Test the drawer, see if it slides easily and closes in the right position. If not, pry off the drawer runners carefully, so as not to break them, and reposition them. When you get them right, trace the outlines, pry them off again, and apply glue. Leave the nails in this last time, the points just sticking out. When you glue it back in place, the points will guide the piece into the proper position. Sink the nails, and attach the runners permanently with three countersunk screws each, as with the cabinet runners.

The bottom drawer runners are located below these at a distance equal to the height of the cabinet runners (supposedly 1″), plus 1⁄16″ for leeway. They prevent the drawer from rattling and jumping as it moves.

Waiting for the glue to set, and the braces to come off

All that remains to complete the cabinet is attaching the back, the doors, and the push latches, veneer-taping, and finishing. You can't attach the doors till the cabinet edges are taped, and you can't do that or attach the back till the braces are off. You *can* tape the doors, the drawer (if you want), and the adjustable shelves.

The taping is simple, as explained under Preparing Wood in "Working Techniques." Save time by cutting and tacking down a few strips; gluing them and their corresponding plywood edges; and then preparing a few more strips while the first batch dries. Be sure to wait till the cement on the tape and the ply is dry to the touch before putting on the tape.

Get the tape flat to the surface. Puncture it with a pin to get rid of obstinate air bubbles.

Tape *opposite* sides first on the door. When the glue has set, in about an hour, file the ends and attach the tape to the other two sides.

Removing the braces

After the miters have set overnight, remove the drawer and turn the cabinet over onto its face. Set it on blocks to protect the edges. You'll probably need some help to turn it over. Remove the back braces by pounding them up from underneath, keeping them level.

Installing the back

While the back is facing up, glue the ⅜″ rabbet around the back of the frame, and put a thin bead of glue along the vertical support and the permanent shelves. Insert the back piece (*24*). If it's a little small, that's all right, as long as it doesn't fall through. From the front it will look all right. Nail into it through the frame every 6″ to 9″ with 1¼″ finishing nails. Mark on the back the location of the center vertical and the permanent shelves. Nail into these pieces through the back every 9″. Now is the time to drill any holes through the back for your stereo wires. Also, when you turn the cabinet over, drill holes through the shelves and/or center vertical, so you can connect the hi-fi components to each other.

Final veneer tape and doors

Turn the cabinet over again. Remove the front braces. Tape all the front edges. Usually you do the longest lengths first, butting strips between them, but fancier patterns are possible (*25*).

Mortise for the hinges and attach the doors, according to the kind of hinge you've chosen.

Mount the touch latches. They should come with complete instructions. One part goes on the inside of the door, and one part on the side of the center support. The latter piece is screwed in through extra-wide holes which allow you to adjust the piece back and forth for maximum efficiency.

Finishing

The construction is done. Fill the small nailholes with a wood putty. We prefer powdered wood putty mixed with water to plastic woods. If you're going to stain the piece, mix the putty with any water-based pigment colored to match the stain. Clean off the putty around the nail holes. Putty any other holes or imperfections you don't care for. When the putty is dry, sand the whole piece. You sanded the unassembled pieces earlier, so there should be very little sanding to do now. Pay particular attention to any pencil marks or dirty spots you may have made, any dried glue, and, of course, puttied areas.

Birch-veneer plywood can be sanded almost as smooth as glass. If that's what you like. Always sand *with* the grain. Be careful at inside corners not to bump up against the side, which leaves rough marks.

Remember to sand the adjustable shelves, which are probably lying over in the corner.

There are many different finishes you could put on a piece like this. The Finishing chapter will explain all the varieties and how to apply them. This piece will be getting a lot of use, a lot of fingermarks, so we suggest putting some kind of protective finish on it—such as butcher's wax or Minwax—three coats at least. Stain it first if you like. Get a really fine look, with a lot of work, with several coats of varnish or polyurethane. Sand each coat lightly before applying the next coat. Fine cabinetmakers will put on a dozen or more coats of varnish, each coat very thin. Three coats should be good enough, however. Again, stain it first, if you like.

As with all decisions on finish, test some possibilities out on scrap pieces of the material first.

You can save yourself some work by not finishing the back of the back, assuming it won't show.

You should take the doors off, the drawer out, and remove any other hardware to make the finishing a little easier and neater.

When you are done, put the cabinet in place. Put the adjustable shelves in where you want, and admire your beautiful work.

Variations

Cleats

If you have no router or dado blade, cleats join the cabinet strongly (*26*). The lines of the cabinet won't look as clean, (although the cleats will be hidden by the doors), and a lot of screwheads will need covering.

Use 1x1s or 1x2s as cleats to support the permanent shelves and to join the outer frame together, as in the Bookcase project. Indent them a bit from the front to give a better look. The top can still be mitered.

Screw and glue the cleats to the verticals, glue the horizontal pieces, lay them on top of the cleats, and screw *into* them through the verticals (*27*). Countersink all screws; you can putty or plug the screwheads, or leave them exposed.

Cleats require some changes in the overall plans. All pieces that previously fit into routs—the permanent shelves and the center vertical—are now ½″ shorter, the total depth of two dadoes.

The back is a special problem. You can attach it right onto the back of the frame (making it a full 4′x6′), but it will show from the sides. Or you can run cleats all along the back inside of the frame and fit the back inside against them. The first method allows you to cut all the shelves and frame of the cabinet to the same depth, 14″. Glue the back edges of the frame, center vertical and permanent shelves, and nail or screw the back piece to them.

The second method looks better, but the extra frame of cleats means that *all* the shelves have to be notched around them. Also the top and two side verticals have to be cut ⅜″ deeper than everything else (as in the routed cabinet).

Construct the cabinet in the same order as in the routing method. Draw a sketch of how the verticals look with the cleats attached—the lengths and locations of the cleats will be less confusing. Make a cutting plan, including the length of all the cleats and the notches in the shelves. Cut all the pieces except for the drawer bottom and then attach *all* the shelf cleats (and the double drawer cleats; see later) to the unassembled pieces. Use 1¼″, No. 10 flathead screws.

Attach the long bottom shelf to the two sides first. Glue all the areas to be joined. Use the 1¼″ screws, three to a side. Hold the shelf tight against the cleat and the side as you drill and screw.

Attach the mitered top as usual. The center vertical does not need cleats if you cut it accurately. Glue and three screws at each end will be fine. Attach the permanent shelves in the same order as in the routing method.

The third shelf cannot be screwed through the center vertical because the shelf next to it is in the way. Instead, toenail some finishing nails through the side or through the top of the shelf into the cleat.

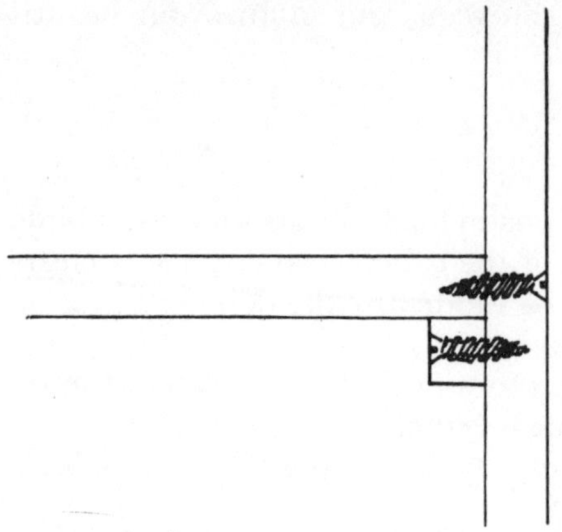

Fig. 27. Joining cleat and shelf

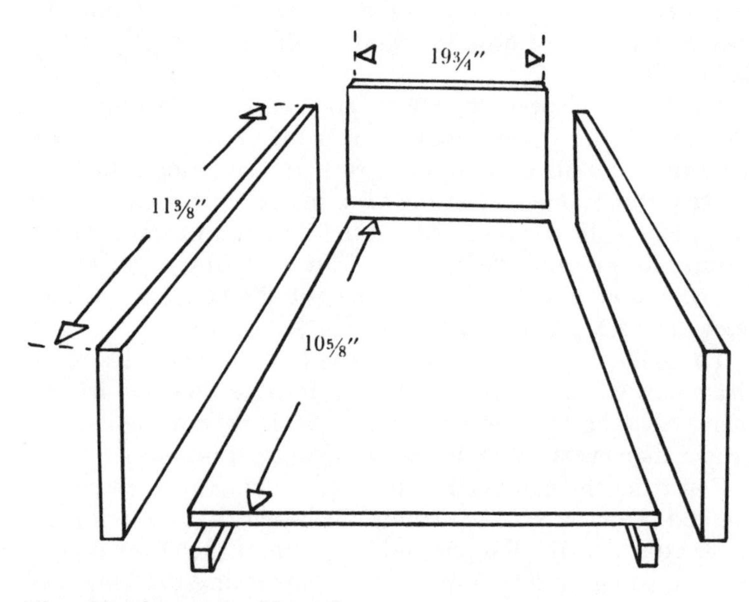

Fig. 28. Cleat Assembly of Drawer

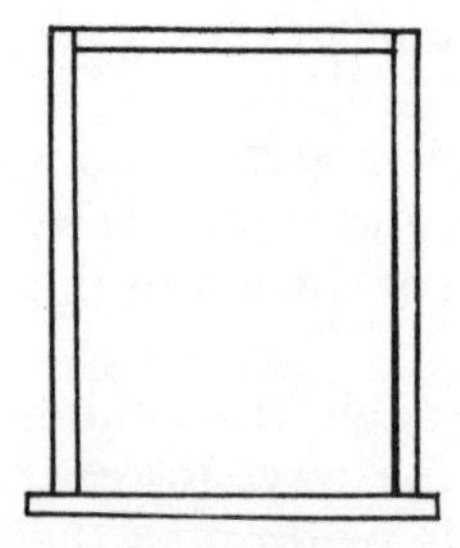

Fig. 29. Butting Pattern for Drawer

The record supports do not need cleats either. Glue the supports in and fasten with finishing nails.

The drawer is built with butt-jointed sides and a bottom resting on cleats (*28, 29*). Cut the front, sides, and back first. Attach the handle to the front. Nail and glue the four sides together, then attach the bottom cleats with glue and either ¼″ finishing nails or 1¼″ screws. The cleats must all be flush on the top surface to seat the drawer bottom neatly. They can be off at the bottom, as long as they don't protrude below the drawer sides. Perfectly butted joints between cleats are unnecessary.

You didn't cut out the drawer bottom earlier, because you want a perfect fit. A neat way to do this is to lay the assembled drawer upside down on the ⅜″ ply and trace the inside dimensions onto the ply with a *sharp* pencil. Angle the point so you draw the line right on the edge. Then cut this piece out and glue and nail it to the cleats.

Since you already have one cleat in the drawer space (supporting the shelf above), you might as well use that as part of the runner assembly, putting two runners on the cabinet walls and only one on each side of the drawer.

Door variations

If you are mounting the doors *flush* within the openings, leave 1/16″ space all around the sides (*30*). Similarly, with *lipped* doors, leave at least 1/16″ space between the opening and the inner section of the door, plus room for the hinge.

All shelves and record supports within the opening have to be shortened in depth by the thickness of the flush door, or the part of the lipped door that is inset. Similarly, the drawer has to be less deep.

Other door constructions include a material such as straw webbing, canvas or other materials stretched between a frame. These also can be flush, lipped, or overlaid doors. See Cabinet Door Construction in "The Ubiquitous Box."

Using shelf standards for adjustable shelves

If you plan to lay these in dadoes, flush to the surface, save money and trouble by buying ones the exact width of the router bit you have. Why buy an extra bit before you really need it? Standard standard sizes are ¾″ and ⅝″. Standards meant for surface mounting also come in ⅞″ widths.

The standards do not have to be glued in place; screws are enough. Most standards are 3/16″ deep, but check the ones you get. Make some practice routs till you get the right depth.

Lattice stripping instead of veneer tape

Some people prefer the look of solid wood lattice strips over the plywood edges. Lattice is ¼″ thick, which means the doors have to be cut ½″ smaller in both dimensions. Lattice on top of the drawer frame means the drawer should be built ¼″ shorter.

Lattice comes in ¾″ widths, among others, so you don't have to rip it. But cutting to length requires care. Support the pieces well as you cut. Using a miter box is ideal. If you have trouble, try cutting them a shade long and sanding the ends down to size.

Measure and cut the strips to length, glue and nail them in place with 1″ finishing nails.

Dividing a large cabinet into two parts

Perhaps you want as much cabinet space as we just designed, but you want it to be more movable. This cabinet could easily have been built as two separate units *(31)*. This method requires a little extra wood, for one more vertical support, and slight changes here and there in construction—e.g., extra drilling of peg holes. To make the two cabinets look more like one unit when they are set in place, join them through the center supports with carriage bolts or countersunk screws. Bolts show more, but they are more easily removed when you want to separate and move the units.

Some units, including this one, can be divided into two parts along the horizontal. This is particularly useful for tall units when one 8′ sheet of plywood cannot span the entire height.

Use special construction at the juncture points so the top part stays put. (See Figure 40, in "Working Techniques.")

Using 1-by's for cabinets

Solid-lumber eliminates the work of attaching veneer tape or lattice strips. It saves cutting only if you make your depth the exact width of the 1-bys. Chances are, however, that the width of the pieces will vary as much as ⅛″. This is all right for simple bookcases, particularly those without backs where the depth variations won't be noticed. But it causes problems in projects like our large cabinet, where the shelves and sides have to be sized accurately. Further ripping, planing or sanding would probably be necessary for 1-bys in such projects.

Backs must still be sheets of plywood. Doors can be either plywood or laminated 1-bys.

If you are assembling with dadoes, we suggest getting clear 1-bys. Any warps in the wood will give you a lot of problems when you try to slip the shelves into their dadoes. Common-grade lumber will do for cleat construction. Order about 20 percent more than you need, however, to be sure you get enough good wood.

Fastening a standing cabinet in place

If the floor under your cabinet is not flat, the cabinet may be dangerously precarious, liable to attacks by leaping cats and children. Stick some shims under it first—and we don't mean matchbooks. Use real shims or scraps of wood. It is also wise to put one or two screws through the back into the wall, in a high but inconspicuous spot. If there is no back, mount a cleat under a high shelf and screw through that.

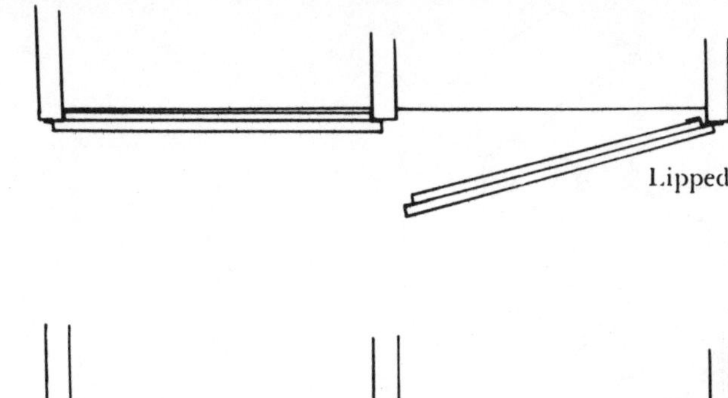

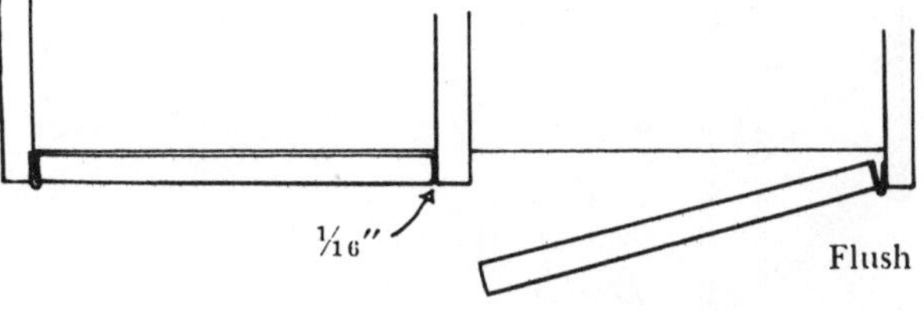

Fig. 30. Door Variations

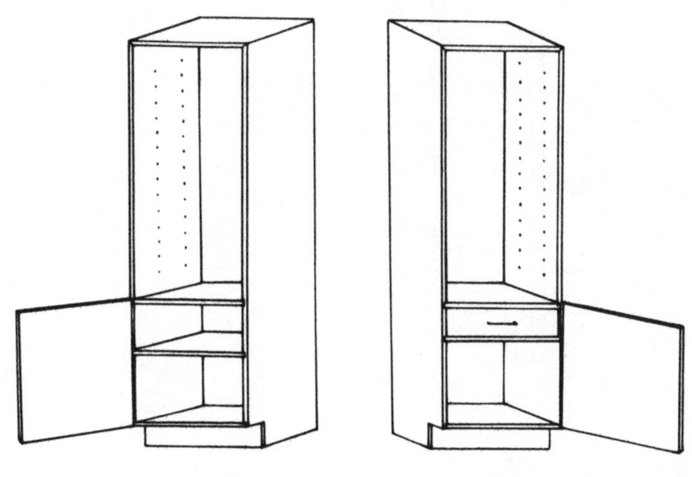

Fig. 31. Two-Section Variation

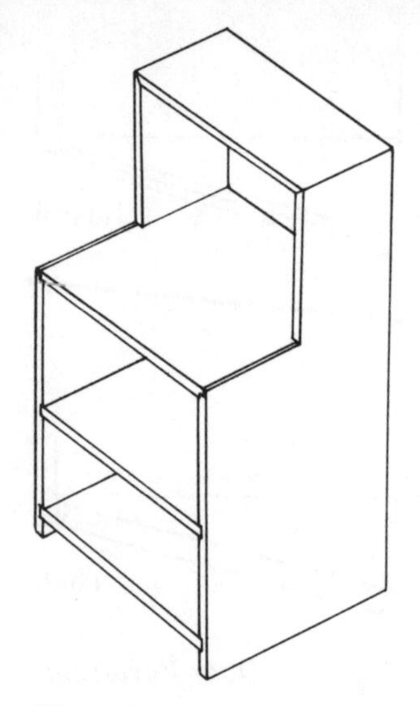

Fig. 32

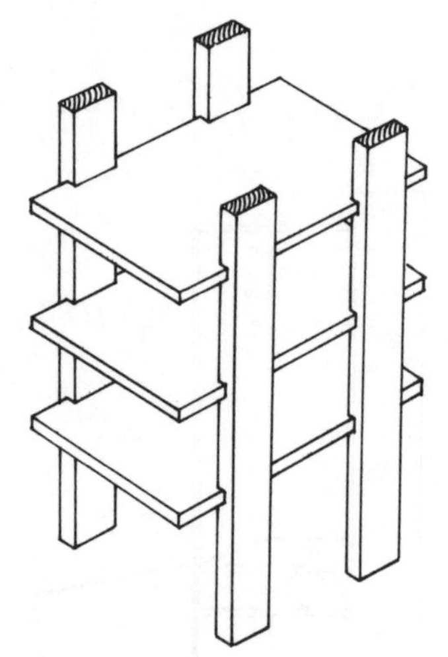

Fig. 33

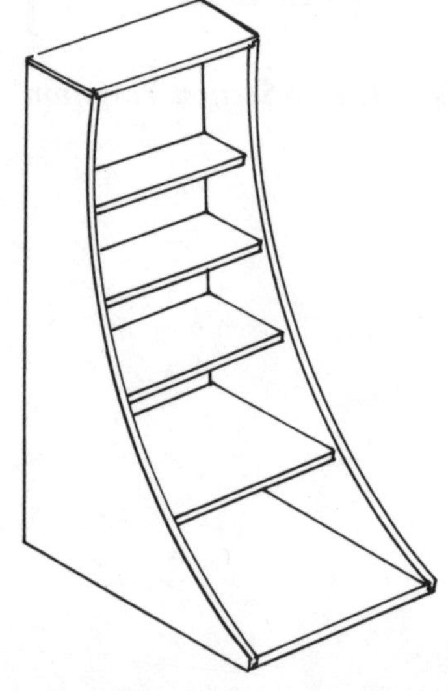

Fig. 34. Other Cabinet Variations

Hanging cabinets on the wall

A cabinet is mounted to the wall through its back. Use whatever kind of bolts are necessary to hang a heavy load to your particular wall (see "Walls and Ceilings"). Obviously, the back must be strong and well joined to the rest of the cabinet; the whole cabinet must be one unified structure. We like to use at least 3/8" plywood for the back, 1/2" or more for very heavy loads like books and records. Attach it *inside* the frame into rabbets (with glue and 1" or 1 1/4" finishing nails) or to cleats (with glue and screws).

Be sure to use washers with the bolts into the wall and to tighten them well against the wood. Use your own judgment as to how many bolts to use. As a rough guide, one 3/8"-diameter lag near each upper corner, and one near the center of the bottom, would be sufficient for a cabinet 4' wide by 2 1/2' high, assuming the lags reach well into the strong supporting structure of the wall (again, see "Walls and Ceilings").

Other cabinet variations

The possibilities are infinite, limited only by your imagination. Figures 32 to 34 (and the Variations in the Bookcase project) should help your imagination to get going; you'll see other ideas wherever you look. These are all done with dadoes and rabbets, but they could as well be made with cleats.

BUILT-IN KITCHEN CABINET

A built-in is a piece of furniture that uses the floor, walls, or ceiling—or some combination of the three—as part of its structure. Several framing pieces are rooted solidly to the house, and the skin or outer structure is joined right to these pieces. A minimum of inner framing is needed because the tie-in to the house structure makes the piece stable and sturdy.

This in-place, piece-by-piece assembly is ideal for kitchens and bathrooms, where immovable pipes resist a neat installation of prefabricated furniture. Built-ins also fit better in irregular areas such as unsquare corners and bowed or sloping floors. Between two other objects that form an irregular, unsquare opening, a squarely built prefabricated piece will leave ugly gaps, whereas a built-in can be fitted and shaped, piece by piece, to fit the space exactly.

A well-fitted built-in looks like it was born there. Which it was. It can lend unity to the design of a room. Rather than a number of kitchen cabinets in a row—some sticking out more than others, some tilted a bit—you can have what looks like one long countertop, with lots of cabinet space underneath.

The disadvantage of a built-in is that it cannot be moved, not without taking it apart. And it is fitted so well to its birthplace that it probably won't fit in very well anywhere else.

There is no strict order of steps to follow in constructing a built-in—it depends on the situation. But there are principles and elements common to all built-ins. We'll show you how to build a basic kitchen cabinet, with no obstructions, as an illustration of one way to put these elements together. The Variations section following shows, among other things, how to adapt for and build around obstructions.

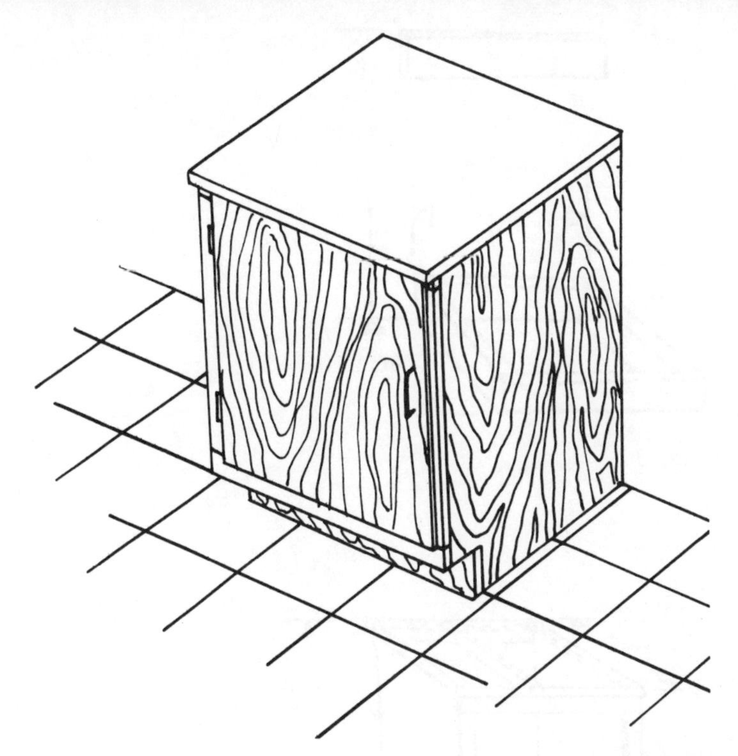

Fig. 1. Finished Cabinet

Planning

Basic features

You have a space in your kitchen that you want to fill with a built-in base cabinet. What features do you want? This is a rather simple piece that you need for performing only a couple of functions. You want a good countertop that you can cut on, work at, and clean easily, and that is waterproof. And you want storage underneath—say two shelves, depending on the available space—for cans, dishes, extra utensils. And you want a door to close things off (*1*).

These are your general needs. Now plan the dimensions. If this built-in is to be placed right next to another cabinet or counter area, you may want to keep the same height and depth for uniformity. The width can be determined by the space available. Without such guidelines, work from the fact that a standard counter height is about 36″. Take this as an approximation, not as gospel. You may be taller or shorter than the norm. Research indicates that most kitchen work is most comfortably done on a surface 3″ below elbow height,

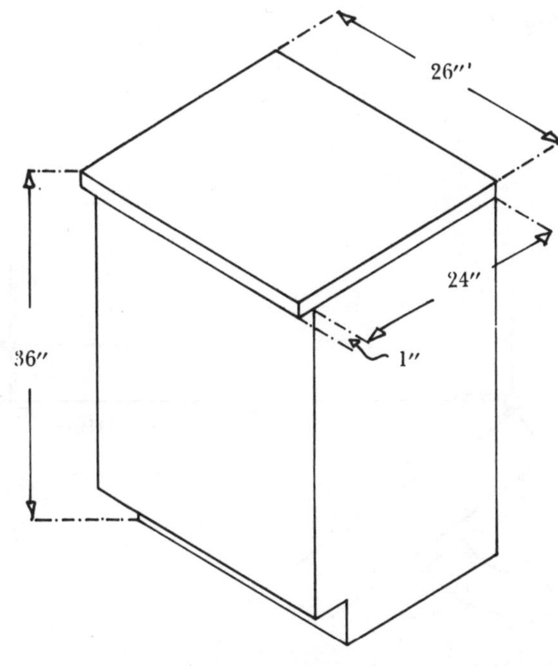

Fig. 2. Measurement Plan—Rough

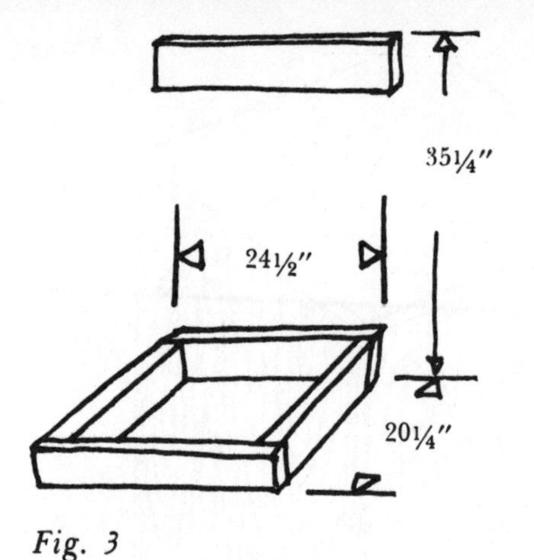

Fig. 3

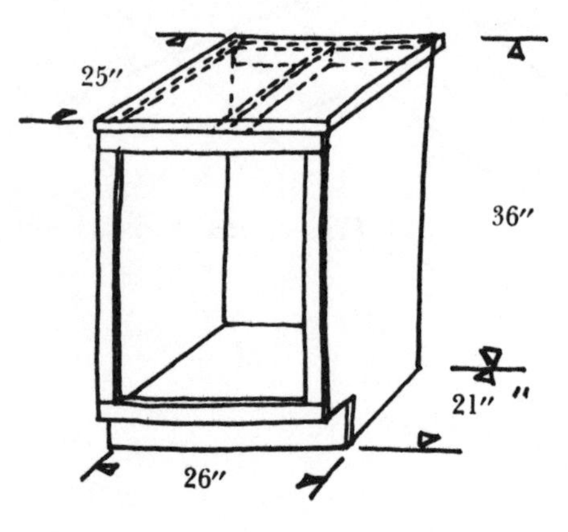

Fig. 4.

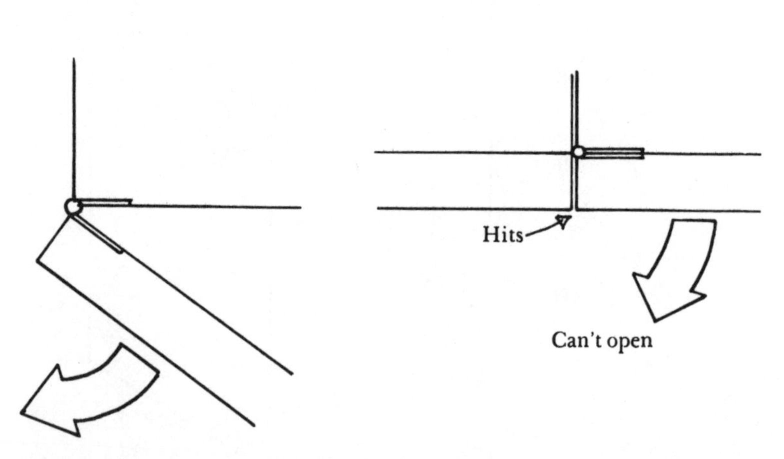

Fig. 5. Why flush door will not work against another cabinet

but mixing is easier on a lower surface, about 7″ below the elbow. Practice slicing a tomato or something about 36″ off the floor, and see how it feels. Pile books on a table to the 36″ height. Adjust the height as you feel necessary. Standard cabinet depth is 24″. Arbitrarily, let's say the width of your space is 26″.

You must have toe space to stand comfortably at the counter—about 3″ high, 3″ deep. It's also standard to extend the counter top one inch over the cabinet front, partly as extra toe or knee space, and also to be sure that the top extends beyond the door.

Make a sketch of the cabinet's plan so far *(2)*. We know we want a door and a couple of shelves, but first let's design the inner structure of this cabinet. Then we'll know what we have to work with, and we can more clearly relate the door and shelf design to the total structure.

Structure

Common to all built-ins are the framing pieces attached to the room. This kitchen cabinet has to support only a static load; no extra frame is needed to supplement the tie-in to the house.

Common floor framing for most built-ins is a leveled frame of 2x4s, with the 1½″ edge on the floor *(3)*. Another framing piece, a 1x2, is fastened to the wall. This piece is as wide as the inner cabinet dimension and is placed just below where the countertop will be.

The cabinet skin can be ¾″ plywood *(4)*. The sides are fastened to the 2x4 frame at the floor and to the 1x2 at their upper rear corners. The top is fastened on top of the sides and the 1x2, and holds them together. And that is the basic structure of the built-in. The rest depends on the variations you want. Structurally, no back is needed. The joint to the 1x2 makes the wall the back, making the whole piece very rigid.

Shelving

The bottom shelf can sit on the 2x4 floor frame. Adjust the height of the toe space to the same height as that of the 2x4 frame, or 3½″, for an efficient use of space. You could put a second shelf about halfway up, supported on cleats, in dadoes, or on adjustable shelf pins. We'll use the adjustable method here. You may have a few very tall things, like a plumber's helper; you can fit such things in the bottom front if you indent the second shelf about 6″.

Door

How do we put the door on? Consider what will be next to the cabinet. An overlay door, flush to the edge, can't swing open if its hinged edge is up against another cabinet *(5)*. You could flush-mount the door. A fancy solution is the lipped door. Use a router or a dado blade to make the lip. Or make a number of passes with a regular plywood blade, chiseling out the remains (not as neat). We'll use a lipped door on this cabinet, with special hinges designed for it. (See Cabinet Door Construction in "The Ubiquitous Box.")

Most factory-made cabinets with lipped doors add a facing frame to the front for the door to close against, rather than just leaving the edges of the side pieces (*6*). It has no structural necessity, but will use it, since many kitchen include such cabinets and you may want a uniform design.

Use 1x2s for the facing. The top piece, fastened only at the ends, is weak in the center. Join another 1x2 between it and the framing piece on the wall.

A pressure catch for this door is *not* a good idea. As you work at the counter, you might lean your knees against the door and accidentally release the catch. This can get to be a nuisance. Use a simple roller or magnetic catch, and a flush handle that doesn't poke out at you.

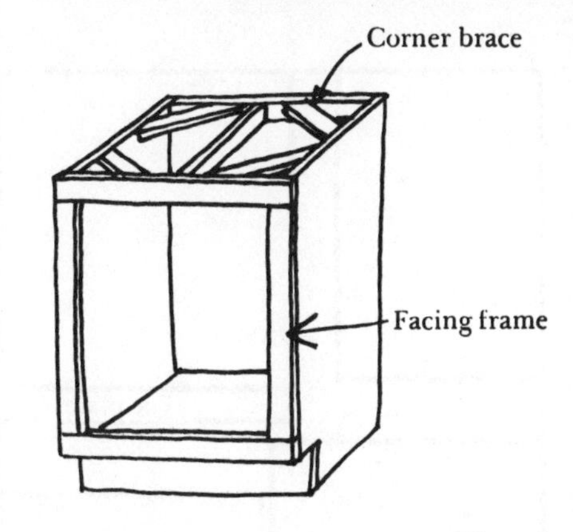

Fig. 6

Counter top

As mentioned, the top is made of ¾″ ply with plastic laminate glued on top. The laminate is glued on oversized, and trimmed with the router and a laminate-trimmer bit. You do this before the top is attached to the sides, or you won't be able to fit the router along the back edge. This causes a slight joining problem, since you don't want to nail or screw through the laminate into the sides. How then can we fasten it down? Add a little bit of structure to the frame—bracing pieces at the upper corners, flush to the top of the sides (7). Screw *up* through these pieces into the counter. Make the braces out of ply, say 45° triangles with a 5″ hypotenuse, or out of 1x2s about 5″ long, mitered at both ends. (For applying laminate tops without a router, see Variations.)

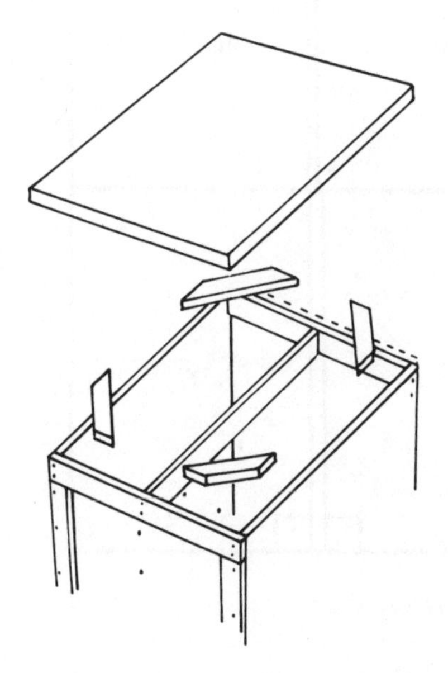

Fig. 7. Exploded view of top and its supports for joining

Joining the pieces together

No special joints or notching are necessary for this built-in. Except for the countertop bracing, all the pieces are simply butted together, glued, and nailed. The tie-in to the house structure makes butt joints adequate for this basic box structure. Any kind of static load is dispersed from the countertop to the sides and down to the floor, or to the wall. Nothing can really move or sway because of the joints to the solid framing.

Finishing touches

You can cover the plywood edges on the door that show when the door is closed and the front edge of the second shelf with veneer tape. The inner edges of the door lip are more trouble to tape neatly, and aren't necessary to do; but do them if you wish. The front edge of the countertop can be covered with a narrow strip of plastic laminate. If the side edges will show, put laminate on them also.

Final plan and design

You must figure the measurements of all the pieces so you can order the lumber and make your cutting plans (*8*). As-

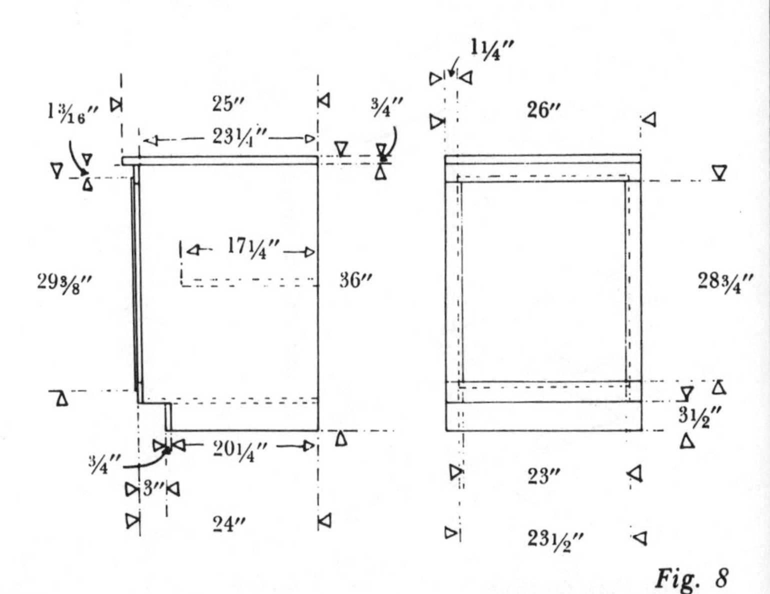

Fig. 8

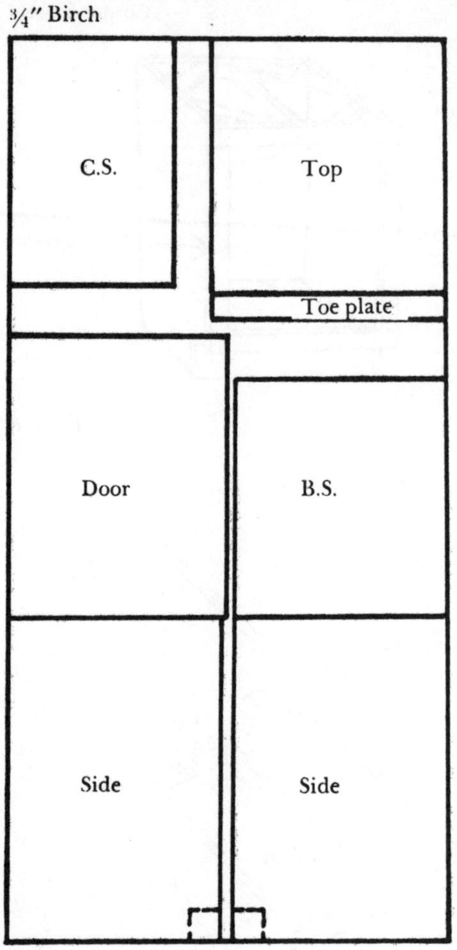

¾″ Birch

C.S. | Top | Toe plate | Door | B.S. | Side | Side

Fig. 9. Cutting Plan

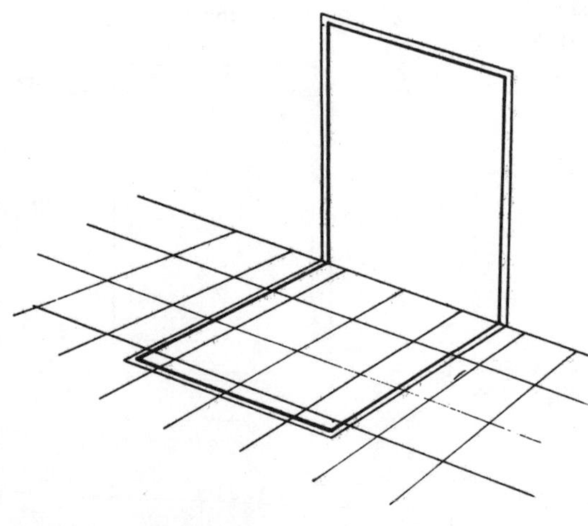

Fig. 10. Plan Outline

sume, for the ease of figuring, that the actual size of a 2x4 will be 1½″x3½″, and the size of a 1x2 will be ¾″x1½″. (Measure the wood when you get it, however, since real sizes sometimes vary.) Work from the outside, whose measurements you know, to the inside. If the width of the whole piece is 26″, then the width of the 2x4 frame and also the back 1x2 is 24½″. The total depth is 24″ (the countertop is 25″), so the plywood sides are 23¼″ wide, with the thickness of the 1x2 facing making up the other ¾″. Thus the bottom shelf is 24½″ wide and 23¼″ deep. The toe-space notch is to be 3″ deep; subtracting that and the thickness of the toe plate from the 24″ depth leaves 20¼″ for the depth of the 2x4 frame. Also remember to make the adjustable shelf ¼″ narrower than the 24½″ inside width, leaving room for the shelf pins, and 6″ shallower than the inside depth, or 17¼″.

The door can be tricky to fit. The usual lip for a ¾″ cabinet door is ⅜″ deep and ⅜″ wide. The back section of the door must fit into an opening 23″x28¾″, with at least 1/16″ clearance on all sides, and extra clearance on the hinge side for your hinge. Measure your particular hinge to be sure, but for now we'll say you need an extra ⅛″ for it. Therefore the back of the door—the part left after the lip is routed—will be ⅛″ shorter and ¼″ narrower than the opening, or 22¾″x28⅝″. Since the rout, or rabbet, is ⅜″ wide on all sides, the dimensions of the front of the door will be ¾″ larger, or 23½″x29⅜″.

Sketches of each tricky area will illuminate all the gremlins. Be sure to draw in the thickness of the wood.

These measurements may not be the exact final measurements of your piece; you may be trimming or enlarging a piece to fit an unlevel floor, or a curve in the wall, for instance. But they will do as a working plan.

List of pieces

¾″ *plywood* (in general we want the grain to run vertically on vertical pieces, and right to left on the shelves—but it's not essential if it means having an inefficient cutting plan)

- Side verticals—two at 23¼″x35¼″
- Door—one at 23½″x29⅜″
- Countertop—one at 25″x26″
- Center shelf—one at 17¼″x24¼″
- Bottom shelf—23¼″x24½″
- Toe plate—3½″x26″

2x4s

- Floor frame sides—two at 17¼″
- Floor frame front and back—two at 24½″

1x2s

- Facing, top and bottom—two at 26″
- Facing, sides—two at about 28¾″
- Framing piece on wall—one at 24½″
- Center support—one at 22½″
- Braces for counter—four at about 5″ each

Plastic laminate

One piece at least 30″x30″. (Wear gloves when carrying, and handle with care—the edges are very sharp.)

Cutting plan for plywood

We can get all the pieces out of one sheet, with little waste (*9*).

Ordinarily we prefer to use 5/8″ or 3/4″ composition board as a base for plastic laminate. It's a little smoother than plywood, and cheaper. If you have a scrap around, use it. Otherwise, since one plywood sheet can be used so efficiently, use the plywood.

Lumber order

Make a neat list that will please your lumberyard. You can get all the 1x2 pieces out of two 8-footers, by cutting three of the long pieces and two of the short 5″ pieces out of each length.

3/4″ plywood (Birch, if the wood will show. Fir, good one side, is rougher but cheaper, if you're going to paint it.)
One 4′x8′ sheet

2x4s
One 8′

1x2s (clear)
Two 8′

Plastic laminate
One piece, at least 30″x30″. Look for remnant pieces, since a full 4′x8′ sheet is expensive and much more than you need.

Other materials

1 lb. of 3″ (10d) common nails
1 lb. of 2″ (6d) finishing nails
Dozen 1¼″ flathead screws, No. 10—for braces
Fasteners for attaching framing piece to wall
Cabinet hinges for lipped door
Door catch
Flush-type door handle
Wood glue
Veneer tape and contact cement
Wood putty
Shelf support pins
Finish of your choice

Tools needed

Try to imagine what you will need and gather them all together beforehand.

Hammer
Nail set
Drill and standard drill bit set
Proper bit to get into wall for framing piece
Level
Screw bits for 1¼″ No. 10 screws
Straightedge
Router
Laminate trimmer router bit
Saber saw
Circular saw with plywood and standard blades
Tape measure
Screwdriver
Rafter square
Fine file
Mat knife
Finishing sander, or sandpaper and sanding block
Putty knife
Goggles

Construction

Important: With built-ins, cut the pieces as you work, not all at once. Dimensions may have to be altered if the floor is not level, for instance.

Clear area

First, move the adjoining furniture or appliances out of the way. Before moving them, mark where they go on the floor; measure their heights and levels if you want your new cabinet flush to them. Pieces that can't be moved, like other built-ins, pose special problems; see Variations.

Layout

Draw the outline of your cabinet base on the floor, showing the thickness of the plywood (*10*). Using the level, continue the lines vertically along the wall. The inner lines show the thickness of the plywood and the exact location of the framing pieces. Don't forget these lines—they eliminate a lot of confusion. Make sure all the corners are square.

Floor frame

Cut the 2x4 pieces to size. Glue and nail them together with simple butt joints. Lay this frame in place and level it with shims if need be. Nail the back in first—3″ nails into the wall or toenailed into the floor. Then toenail the rest from the inside of the frame (*11*).

Cabinet sides

If the floor and wall are straight and level, you can cut the plywood sides to the plan and notch them for toe space. If the cabinet sides will show, keep the good surface on the outside; otherwise put it on the inside. If the floor or wall is not straight, you will have to adjust the size and shape of the sides. The important things to remember are that you want the front of the ply to be 35¼″ high; the front top corners to be right angles; and the top to be level. For instance, if the floor is higher in the back, then the plywood is cut that much shorter in the back. If the floor level is off from left to right,

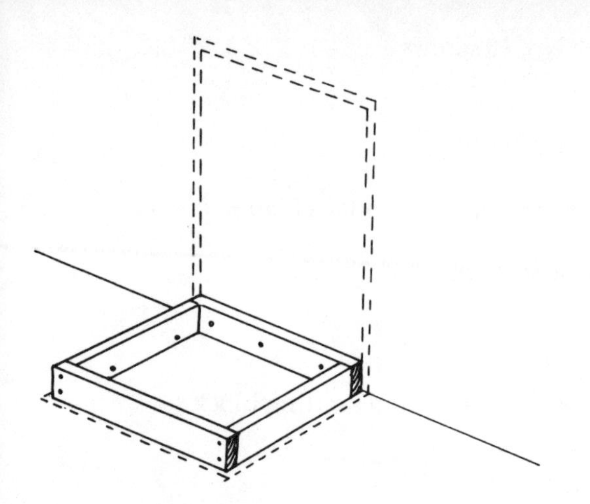

Fig. 11. Base Frame

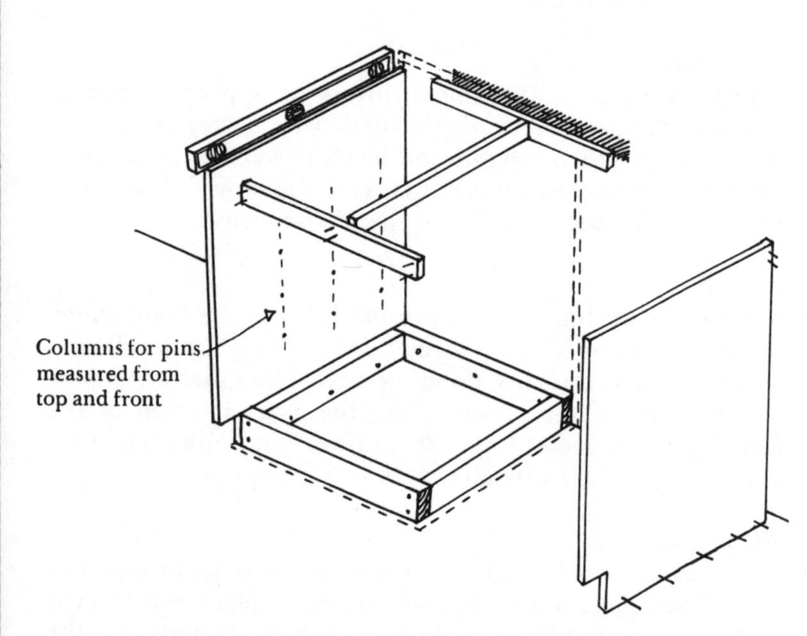

Fig. 12. Leveling sides and adding center top supports

one piece may have to be cut higher than the other. Similarly, if the wall has any bumps or curves in it, you may have to contour the backs of the plywood so the fronts will line up with the floor frame.

Holes for shelf pins

Drill the holes for the shelf pins *before* installing the sides. Drill three columns of blind holes in each piece, as deep and as thick as the pins you got. Lay out the hole centers for drilling by measuring from top and front of the ply (since the other sides may be angled or contoured to fit). Measure carefully, so the holes on one piece line up with the corresponding holes on the other. Indent all the marks with a center punch or nail point, and then drill. (See Cabinet project.)

Prepare top facing piece

The sides are now ready to be attached to the frame. To help hold the pieces up, use the top facing piece; cut it out, start two finishing nails in each end, and keep it handy.

Fastening sides to frame

Start five or six finishing nails at the bottom of each side, and two at the top back corner (*12*). Put the pieces in place and check each one's level from front to back and the level across the tops. If your level isn't long enough, lay a straight piece of wood across and put the level on it. An assistant can be a golden blessing to help hold things. If your cutting and measuring was accurate, the pieces should be level. If they're still off, find where, and either trim some off, or, if the sides won't show, shim. Pull the sides away and spread glue along the side of the 2x4 frame. Put one side back and start one or two nails. Do the same to the other piece. Check levels again. If off, pull nails and adjust. Finish nailing. Glue the ends of the facing strip, place in position, and start a nail at one end. Align other end with outside edge of plywood and tack in place. Check both ends for alignment and drive the nails all the way.

Wall framing

Cut the 1x2 framing piece for the wall. Hold it in place by hand and measure the distance to the front facing strip. Cut the center support to that length; glue and nail it to the center of the back piece. No special joint or screws are necessary because there won't be much downward stress on the center support. Its purpose is just to keep the top front facing piece from bending or warping inward. Place this T-structure in position, flush at top with the rest of the structure, and screw or bolt the back piece firmly to the wall. (Consult the "Walls and Ceilings" and "Hardware" chapters to determine the proper fastener depending on your type of wall.) Glue and nail it to the front facing, and at the back corners of the plywood.

Toe plate and shelves

Measure, cut, glue, and nail the toe plate in place to the sides and 2x4 frame (*13*). Measure, cut, glue, and nail the bottom shelf to the 2x4 frame with 2″ finishing nails. Be sure the shelf is flush at front to the sides. Also, cut the adjustable middle shelf to size—1/4″ less than the total inside cabinet dimension, to provide clearance. Veneer-tape the front edge and rest the piece on the bottom of the cabinet for now. If you waited to install it *after* the front facing is up, you might have trouble slipping it in and straightening it out.

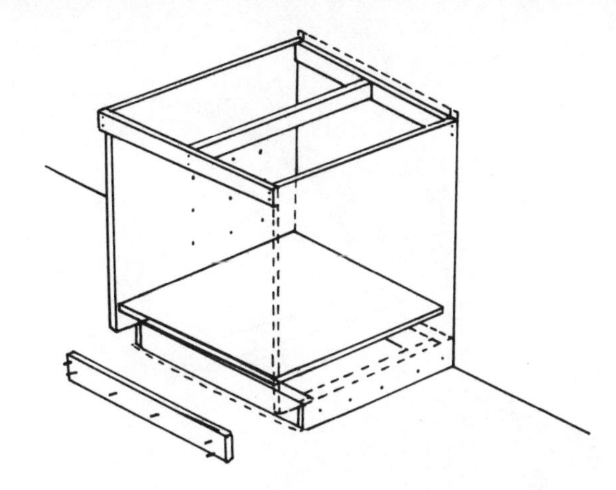

Fig. 13. Adding bottom and toe plate

Rest of facing and top bracing

Measure, cut, glue, and nail the bottom facing piece, then the vertical ones (*14*). Be sure they overhang on the inside, not on the outside of the cabinet. Cut the verticals accurately so there are no gaps at the joints.

The last bit of interior construction is the countertop bracing. Cut the 1x2 pieces, with mitered ends, about 5″ long. Check the square of the cabinet again. Glue the 1x2 ends and screw them into the corners as shown, with 1 1/4″ flathead wood screws, countersunk. Keep the braces flush to top of cabinet.

The glue on these pieces and the facing pieces should set overnight before you attach the door and the countertop. You can, however, prepare the door and countertop.

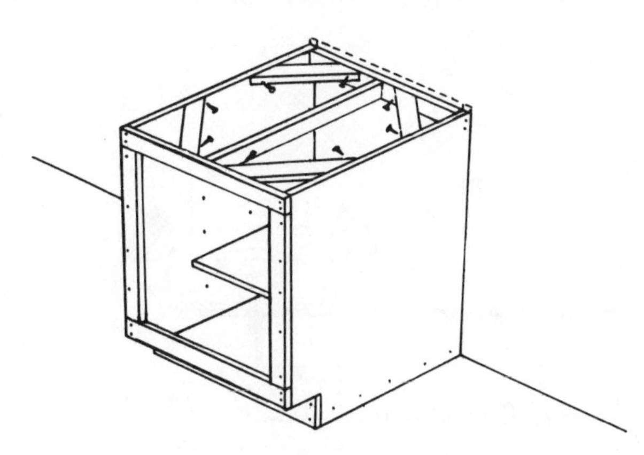

Fig. 14. Face 1-Bys added and top support diagonals in place

Prepare door

Cut the door to size. Plan the best surface to show. Put on your goggles and rout a lip 3/8″ thick and 3/8″ deep around the edge of the door (*15*). Use a guide, of course. Since you are routing along the edge, any straight bit 3/8″ or larger will do. If you have only a 1/2″ bit, for instance, set your guide so 3/8″ of the bit is on the wood, 1/8″ off. Make some practice cuts, to check accuracy (see Routers, in "Power Tools"). Attach veneer tape around the outer edge of the door. Tape the inside lip, if you want.

A flush door handle requires mortising. Best to do that now before hanging the door. Position the handle in place and trace the outline to be mortised. Chisel it out carefully to the proper depth. Attach and insert the handle as required.

Prepare countertop

Measure the top of the cabinet carefully for the exact size of the countertop. Add one inch to the depth for the overlap at the front. You could even lay the remaining ply on top and trace the outline from underneath (particularly helpful if you built off square). Cut the top to this size. Set it in place to check. If the wall is uneven, either try to contour the back of the plywood, plan to put a piece of molding at the back to cover the gap, or ignore the gap.

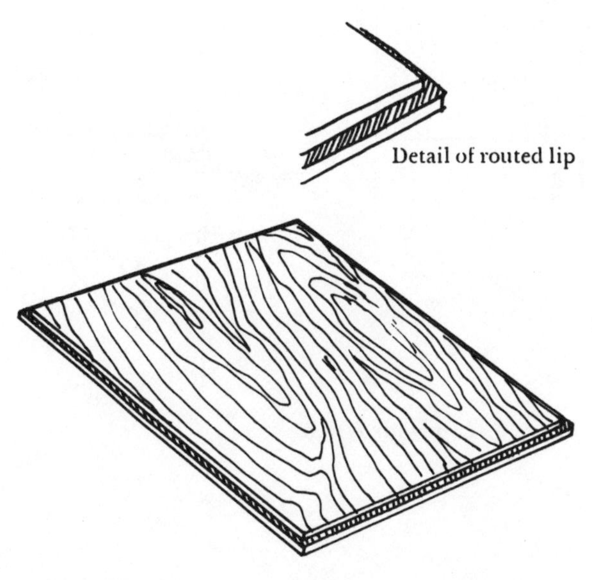

Fig. 15. Lipped Door

Installing the plastic laminate

Laminate is always cut an inch or so larger on all sides than the piece it is being laminated to. Once it is glued and set,

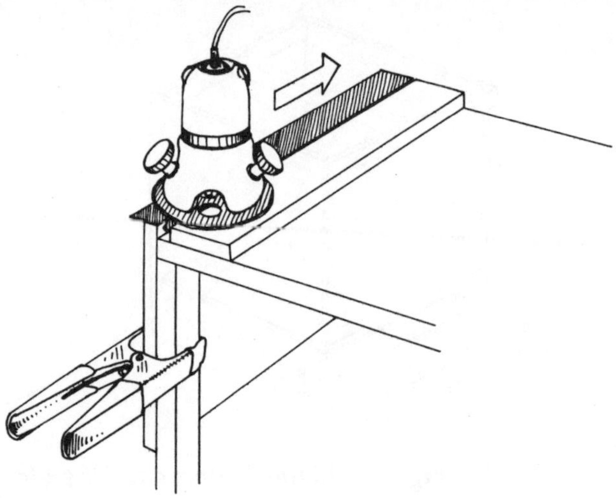

Fig. *16. A way to support the router for edge trimming*

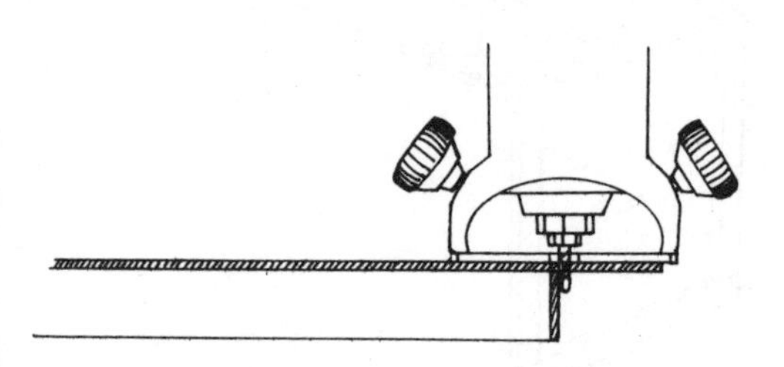

Fig. *17. Trimming the laminate*

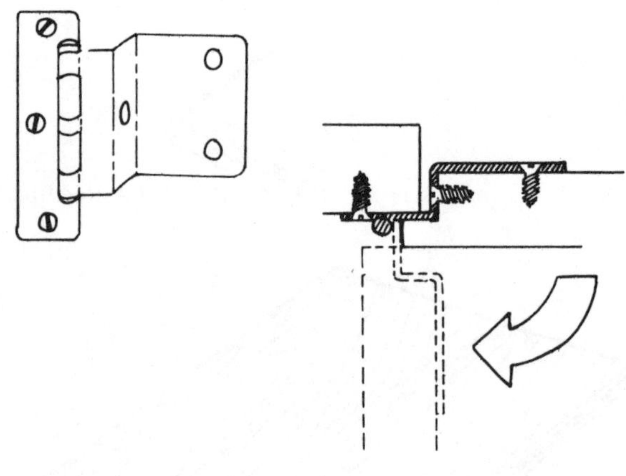

Fig. *18. Lip Door Hinge*

the excess is trimmed off with a router and laminate-trimmer bit. Trying to cut to the exact size is risky—a saw blade leaves a ragged edge, and can sometimes chip little pieces away. And, if you cut it too small, there's nothing you can do. Trimming off the excess, on the other hand, is quick, easy, and perfectly exact. A trimmer bit has a pilot head at the bottom that follows the edge of the wood. Thus the laminate can be cut to the exact contours of the wood—whether it's perfectly straight, curved, or angled.

You need laminate pieces for the front edge and top surface of the countertop. Do the front strip first, and lay the top piece over it, so no seam will be visible on the finished top surface. The front edge is ¾″x26″; turn the laminate upside down on a clean, smooth surface and outline a strip about an inch bigger in each dimension—1¾″x27″. Whenever possible, cut laminate with the good side down. Cut the strip off with your saber saw; don't worry about a good cut, so long as you stay around the line. Wear your goggles and handle the laminate with extreme care; the edges are cutting sharp.

Spread contact cement along the plywood edge and the back of the strip. Let them dry to the touch (usually about fifteen minutes) and stick the strip on carefully, overlapping on all sides. Press the strip down everywhere, again being careful of the sharp edges. You can tap it lightly with a block and hammer.

Now clamp the piece so the laminate strip faces up. Put the trimmer bit in your router, put your goggles on again. Start by resting the router base flat on the surface, the bit *not* touching the laminate. Start the router. When it reaches full speed, bring the bit into the laminate as far as it will go. The pilot guide will stop it flush to the wood. Trim all around the perimeter. The only trick is keeping the router base flat on the surface. This calls for care on a thin edge like this, or else a setup as in Figure 16, where a wider surface area is provided for the router base.

Now outline on the bottom of the remaining sheet a shape about an inch bigger than the countertop. Cut it out. If the sheet is close to the desired size, don't even bother cutting it. Attach the piece to the countertop the same way you did the strip. Spread glue, let dry, attach, tap all over, trim (*17*). No special trimming setup is needed here.

A fancy finishing touch is to bevel the front edge with a smooth mill file. It looks a little better, and is smoother to the touch. File on the downstroke. The standard bevel angle for a front edge is 45°; for the inside edges (if you had laminate on the sides) 22½° to the vertical. Obviously, you don't have to be exact. There are trimmer bits that bevel as they cut, if you want to spend some more money.

Install door

The next day, after the glue on the triangle braces and 1x2 facing strips has set, attach the lipped hinges to the door and hang the door in the cabinet (*18*). Attach the catch. Fastening the door before the countertop gives you access to the inside of the door, even when closed. This facilitates hinge and catch installation.

Attach countertop

Now spread glue (wood glue, of course, not contact cement) all along the top of the cabinet and bracing. Lay the countertop in place (*19*). Clamp it at the front, lay a heavy object on top, or have someone hold it in place while you drill a hole through each triangular brace. Be ever so careful not to drill through to the laminate top. Put some tape on your bit as a depth guide to be sure. Countersink the screws, being sure the countertop is well held down as you screw. Add a piece of molding on the back edge to cover any bad gaps between counter and wall, and to keep your knives and garlic from falling behind the cabinet.

Finish

The construction is finished. The adjustable shelf can be lifted onto its pins now. Check that all nails are sunk. Putty the nailholes and any gaps such as between the facing strips. Sand the wood areas and apply the finish of your choice.

Variations

Middle shelf on cleats

Attach the cleats to the sides before nailing the sides in place. Position them from the top and front edges of the plywood. Install shelf before facing.

Middle shelf in dadoes

Rout blind dadoes before installing the sides. The half-shelf we used in this project cannot be slipped into the dadoes once the sides are up since the dadoes do not extend to the front.

Here's what you do. First see if you can get a helper to hold things. Start some 2″ finishing nails on the outside of the dadoes. Fasten the top 1x2 facing piece to the side pieces. Stand this assembly up and slide it several inches back along the 2x4 frame—just to get the correct distance between the bottom edges. Glue the edges of the shelf and slide it into its dadoes from the back. Slide the whole assembly back into place. Drive a 2″ nail on each side near the back, part way into the shelf. Then proceed as before to check the levels and fasten the whole assembly in.

Building the cabinet without a router

You can still make a lipped door using a dado blade, or even by making a number of saw cuts ⅜″ deep and chiseling the excess away.

Plastic laminate can be cut and fitted with nothing but hand tools. It takes much longer, however. Use the plastic laminate cutter (see "Hand Tools") to cut the pieces. Cut them closer to size than you do with a power saw, because

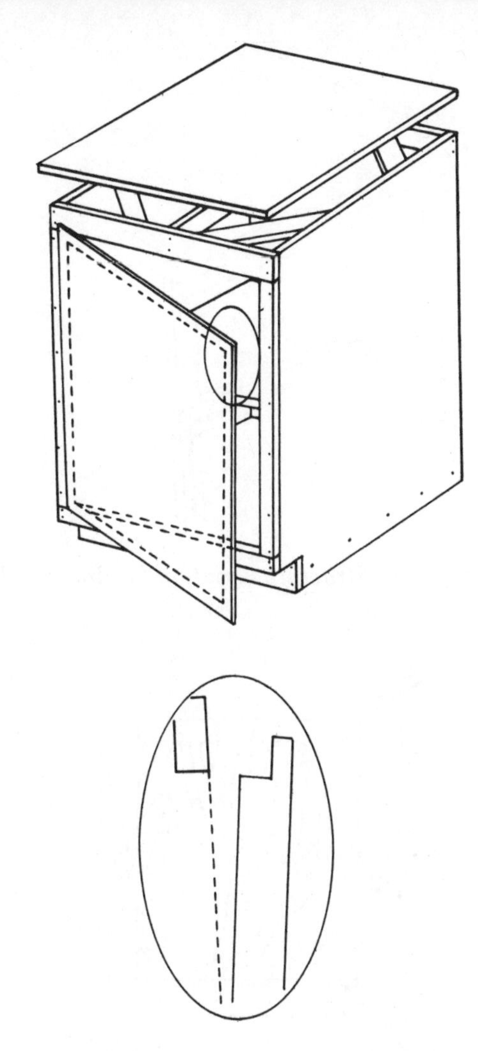

Fig. 19. Door in place and top being applied

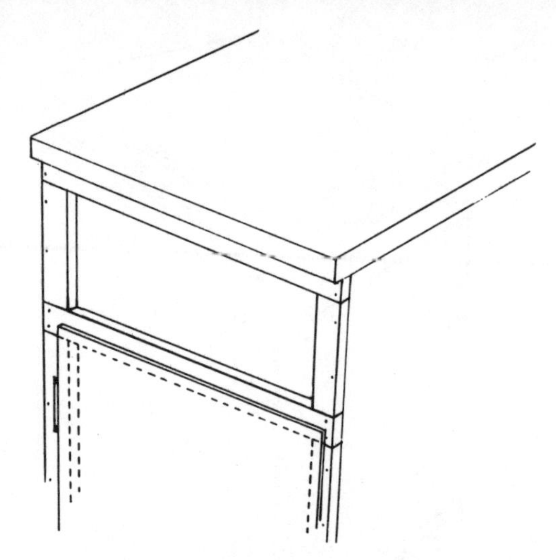

Fig. 20. *Structure variation for drawer addition*

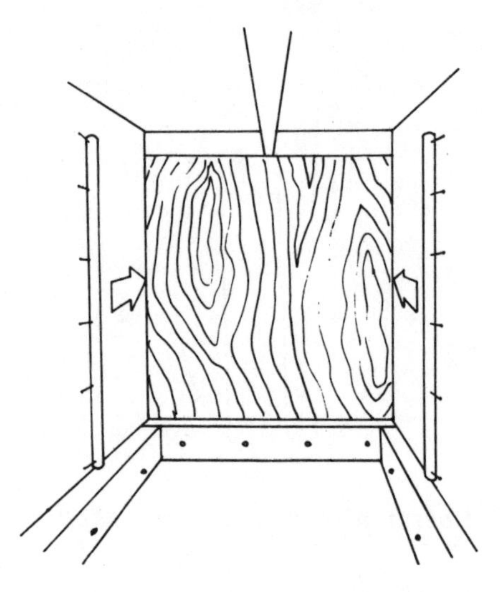

Fig. 21. *Applying back support molding from inside*

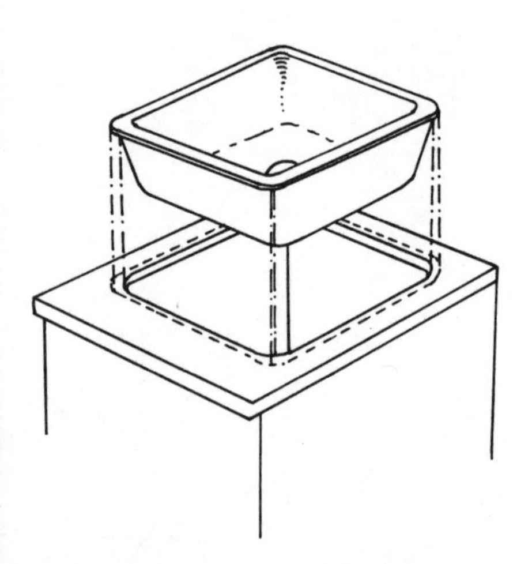

Fig. 22. *Sink Insertion*

once you glue the piece on you will have to trim it to size by hand. Use a block plane, with a very fine cutting opening, to trim the piece close to the edge, but not all the way. Plane along the edge, but angle the blade down from the finished surface. Do the final trimming with a file, as in the router method. File downward or at an angle, but away from the finished surface.

If you don't want to do this, you could hunt up a stock top at the lumberyard or hardware store. This is usually a piece of ¾" composition board with the plastic laminate already on and trimmed. You'll have much less choice as to color and size. Build the cabinet very carefully to the width and depth of the stock top. You can, with a decent chance of success, trim a stock top with your saber saw, but we don't recommend it unless there's no other way. Remember to measure and cut from the bottom. Cover the edges with veneer tape.

Adding a drawer

A drawer is very useful in a kitchen, and very easy to build. Follow any of the drawer designs in "The Ubiquitous Box." Make the door shorter, and add another facing strip between door and drawer (*20*). You can make the drawer front lipped like the door, flush, or overlaid.

Drawer dividers of thin plywood can be routed or butted in, fastened with glue and finishing nails.

Add a back

No structural need, but it looks better when you open the door. It also makes it a little harder for mice and roaches to get in (or out). The back can be made of thinner ¼" ply.

After the sides are up, measure and cut the back to fit between the 2x4 frame and the bottom of the back 1x2. Slip it in. Fasten it in place by pressing and gluing quarter-round molding strips against the perimeter and nailing them to the plywood sides (*21*).

Adapting cabinet to take a sink

The idea is to build the whole cabinet and then cut the hole for the sink out of the top. The plumbing must be completed up to the drain trap before you start building, so you can plan the center of the sink over the center of the drain. Build the frame and sides as usual, this time fitting the piece around the plumbing and eliminating the middle shelf and the top 1x2 center support. Run your level (or a plumb line) up from the center of the trap and check the distances from the sides, front, and back of the cabinet. Attach the countertop as before, and mark the center of the trap on the top.

Center the outline of the sink opening around this center point. Most sinks come with templates, so this should be easy. Otherwise, measure the sink's dimensions carefully, from *inside* the lip. The sink is supposed to sit *in* the hole, and the lip *on the edge* (*22*).

The hole is cut with your saber saw. Here you *have* to cut with the good side up. Unfortunately, you can't mark the laminate surface with a pencil—it just won't take. And a chalk line will just blow away as you saw. What we do is make our chalk lines, then apply masking tape to the outside of the lines. Then if the chalk blows away, the edge of the tape acts as a guideline. The tape also protects the surface from scratches from the saw base.

Start with a plunge cut, safely *within* the outline. Or drill a pilot hole near, but inside, a corner. Wear your goggles. Cut all around the outline. Try to fit the sink in. If it's too tight, trim the opening some more. A perfectly neat cut is unnecessary, since the sink lip covers it.

Building next to an immovable piece

This situation makes it impossible to nail the plywood sides to the set floor frame. One possibility, if the immovable object is wood and you don't mind nailing into it, is to nail the cabinet side to it, with the framing wedged up against the side piece. Or you might eliminate the side piece, and nail cleats to the immovable object.

The other alternative is to partially prefabricate the cabinet. Build the 2x4 frame, slide it in place, shim, but don't fasten it. Measure, cut, and level the side pieces in place. Mark the level and positioning of the 2x4 frame on the sides. Bring all the pieces out from the wall. Glue and nail the sides to the floor frame, lining up the marks you made. Slide the top framing pieces into this assembly, glue and nail them. Slide the whole structure in again and fasten in place. Proceed with the rest of the construction as usual.

Building between tilted or unlevel objects

The problem here is that if you build square, gaps will show between the cabinets. There are several solutions, none of them ideal. You can try to match the tilt, which may cause problems at the joints if the angle is large. Or you can fill in the gaps with "dummy" pieces as in Figure 23.

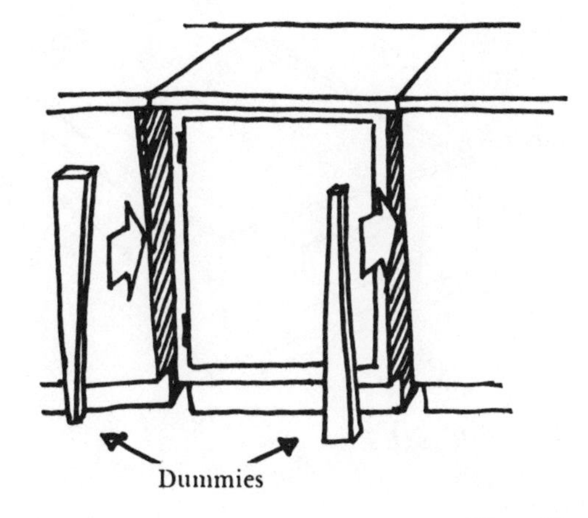

Fig. 23

Building around obstructions

This is mainly a matter of common sense. Notch or trim each piece around the obstacle as you put it in. Make sure what is left is not too flexible or weak. If you have to cut away part of a 2x4, be sure to nail both remaining parts down. If a pipe is running through the middle of a side, measure its position from the floor and wall carefully, mark on the plywood, and cut a slot slightly larger *(24)*.

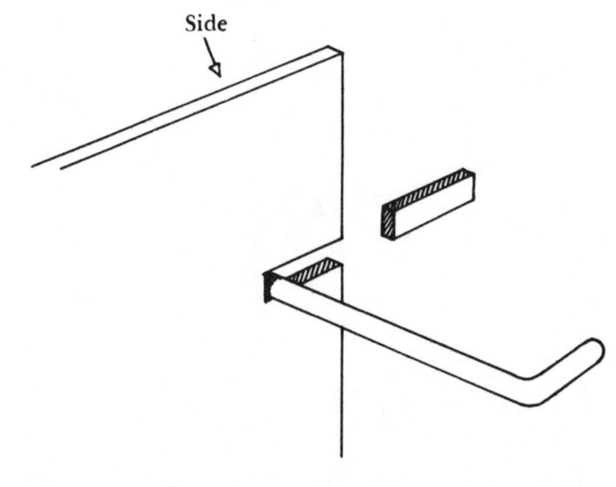

Fig. 24. Cutting around obstructions

Wall cabinet

Constructing a wall cabinet is more like building a freestanding cabinet, which is then attached to the wall. Build a simple cabinet with doors and a firmly attached back. Consult the Useful Dimensions table in the Reference section if you aren't building to match existing units.

CAPTAIN'S BED

A captain's bed uses the space underneath the mattress as storage for boxes, sextants, starboards, etc. The raised platform gives you toe room as you tuck in the sheets, or heel room when you sit on the bed with your legs over the side. Easy to build, it's essentially a box, sitting on a frame.

Storage access is by drawers, or simply doors that open out. We'll use doors for this project, since they're easier to build. If you plan on using the storage space often for extra sheets, everyday clothing, etc., drawers are worth the extra construction effort (see Variations).

Planning

Design

First we want the mattress to fit down into the frame an inch or so, so it won't slide over the edge. Let's say an inch.

The height of the bed is whatever's comfortable. Do you sit on the bed like a chair? What's comfortable for you to fall into late at night, or crawl out of in the morning? We'll say 24″ for this project. That also allows ample height for storage. The toe space is a standard 3″ deep and 3½″ high.

The width and length of the bed are determined by the size of your mattress. Let's say you have a 39″x75″ twin mattress. The *inside* of the frame into which the mattress fits should be slightly larger than the mattress—say 40″x76″. This allows space for the bed coverings and your fingers as you tuck them in.

For easiest access, hinge the doors on the side or bottom; opening out, of course, for inside clearance. Thus the bed you design might look like Figure 1.

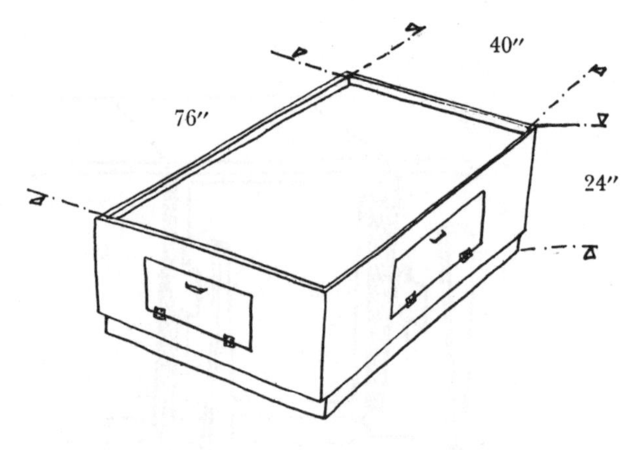

Fig. 1. Rough Plan

Structure

The bed is really a box sitting on a smaller platform. The platform is just a frame of 2x4s on edge—standard toe-space construction for all sorts of furniture. Miter the corners for looks; glue and nail them together. Screws are unnecessary since the box bottom will hold the frame together.

The box is also simple. Use ¾″ plywood, standard thickness for structural furniture members. Since the structure is under a dynamic load, you must join the box strongly, so it won't separate or collapse. Such a load transmits stress not only downward, but sideways, putting the corner edge joints under great stress. The sides should be mitered for the seamless look of our design; then glued and screwed (*2*).

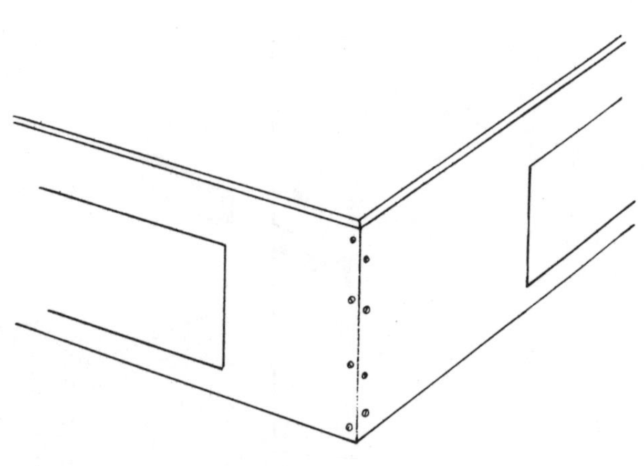

Fig. 2. Miter Joined Corner

The box bottom sits on and is nailed to the 2x4 frame. The inside bottom edges of the plywood box sides are rabbeted ⅜″ deep and ¾″ wide, so the bottom sheet of ply slips in like a cabinet back. Glue and screw it through the sides. The rabbet is ⅜″ deep, instead of the usual ¼″, to give the sides a little more area to sit on.

Finally the mattress needs something to rest on, a "floor." This can be a sheet of ¾″ or ½″ ply, or ¾″ composition board, which in turn needs to be nailed to and supported by joists. Five 2x4s are fine for this 40″ span and would be good even for a double-mattress span of 55″. Five joists can be spaced 19″ on center, adequate for a bed. The joists, of course, should be joined to the box to rigidify the structure as much as possible to keep the tops of the sides from bending out under the lateral stress. You also want to hide the supports from the outside, for design, and you want the plywood floor level to be 1″ below the top. Therefore, the problem is how to support the supports on the inside; the solution is to glue and screw 1x2 cleats to the insides of the box, and rest the joists on them.

There are three ways to join the joists to the cleats. One, just set the 2x4s on the cleats and screw into the 2x4s through the sides. Also screw into the end joists through the head and foot of the bed, for extra rigidity of the frame. Nails do not grip well enough to resist the lateral stresses working in direct opposition. Countersink the screws. Either putty or plug the holes, or let the screwheads show. Brass screws, particularly, look very attractive when lined up neatly. This is the method we'll use.

A second way is to set the joists in place, without joining them. Instead, screw a couple of 1x2 crosspieces to the cleats to combat the lateral stress (*3*). When the plywood floor is nailed down to the joists, they will be steady. This is not a super method, definitely not suggested for actual floors; but it's fine for a bed. And no fasteners show.

Third is to notch the cleats and glue and toenail the joists into the notches (*4*). In this case the cleats must be 2x4s, glued and screwed from the outside. The nails are at angles to each other and to the lateral stress, and won't pull out easily; the glued notch forms a strong joint in itself.

Secondary structure

If you plunge-cut the door openings neatly, the pieces you cut out can be used as the corresponding doors, thereby keeping the grain pattern continuous.

How big can the opening be without weakening the structure too much? No exact answer to that, but you can't make it much wider than two feet. Leave enough ply above the opening so the cleats and joists don't show. *Structurally,* the ply below can be cut out entirely, though you'd have trouble hinging the door. Thus the size partly depends on how you are attaching the joists—notching takes up the least vertical space and allows the highest door. Also try to position the doors along the center of each piece—certainly no closer than 4″ to a corner.

You could also make two smaller doors to a side; but one door in the center gives almost complete access to the space underneath.

In fact, if you cut the door openings flush to the bottom of the cleats, you will have the maximum height for the doors, and also be able to attach the door catches to the bottom of

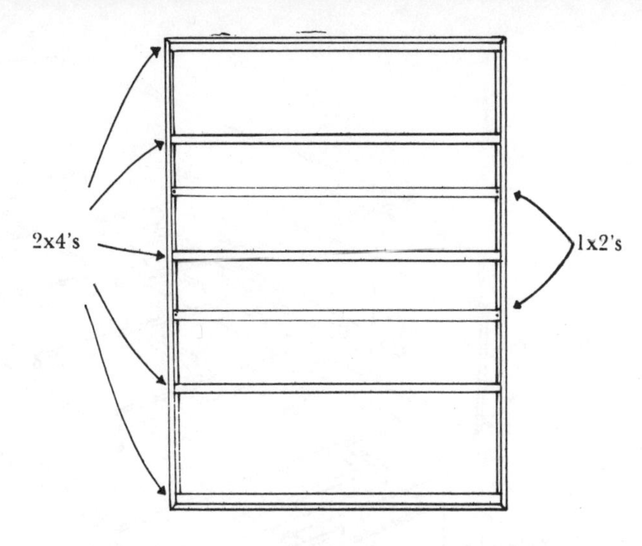

Fig. 3. 1x2 variation where joists rest on supports and are not joined to sides

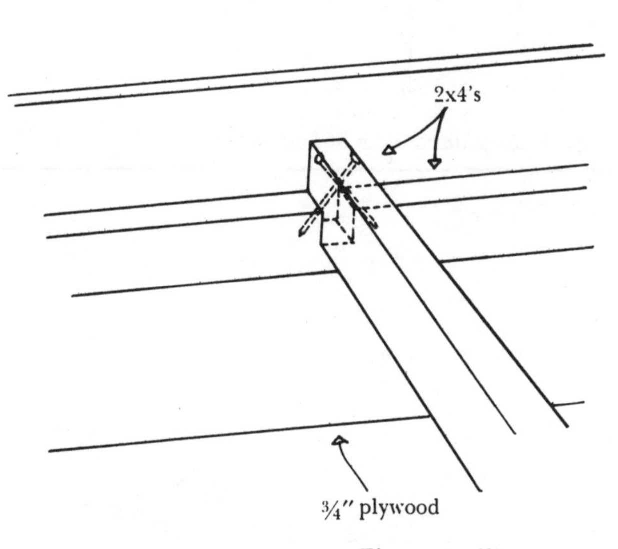

Fig. 4. Nailing of joist

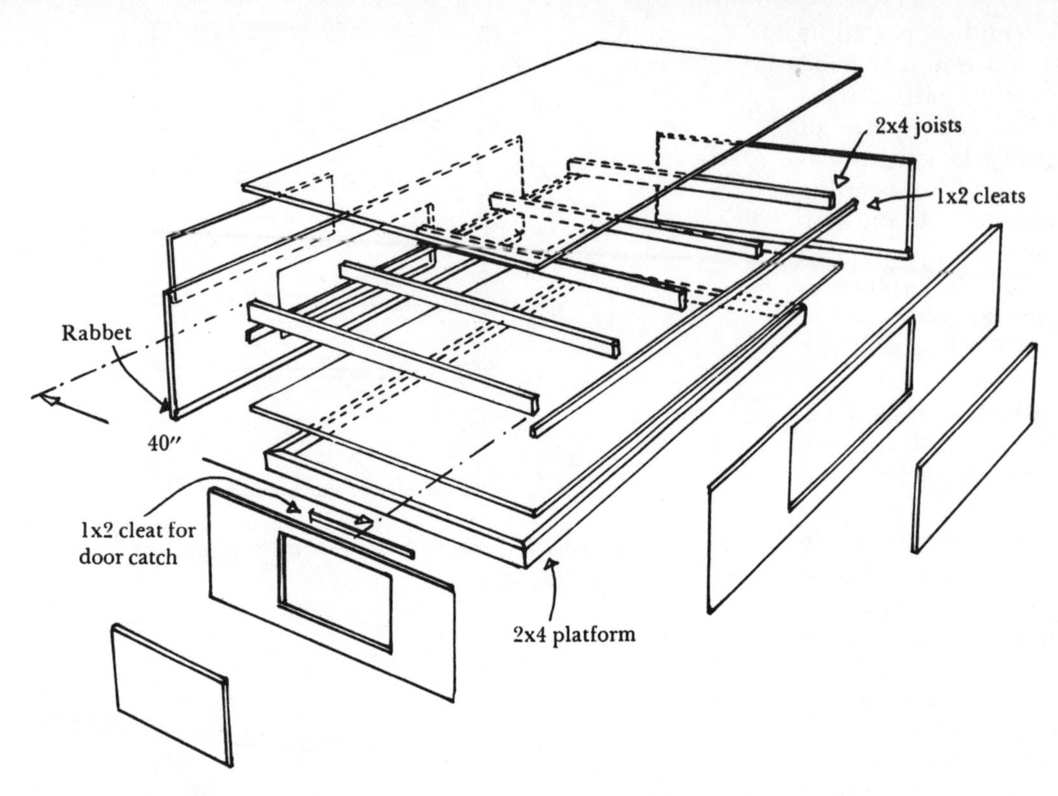

Fig. 5. Exploded view of bed

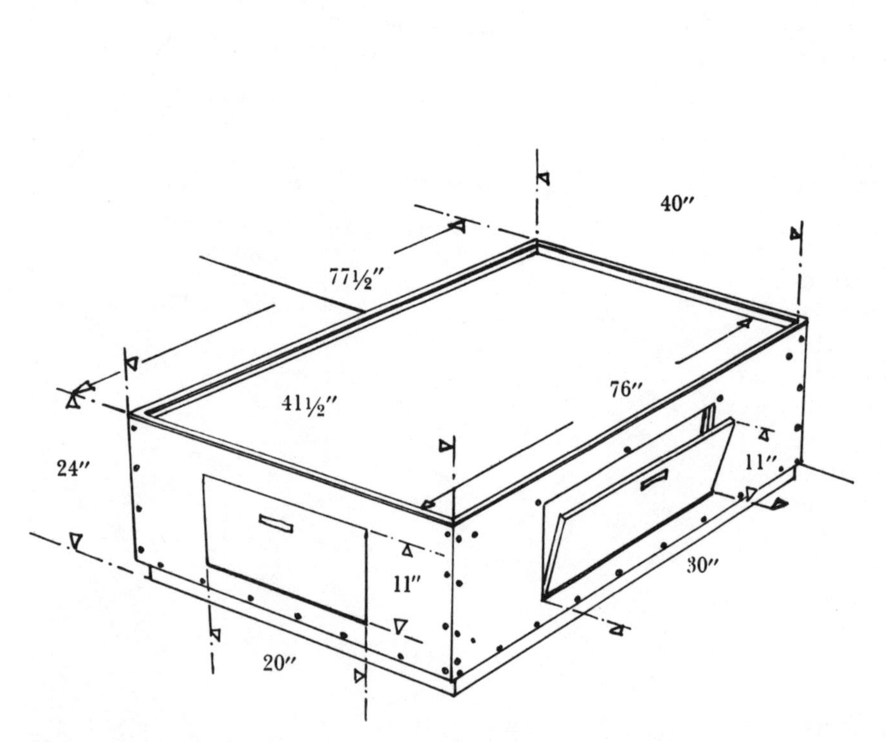

Fig. 6. Plan of Measurements

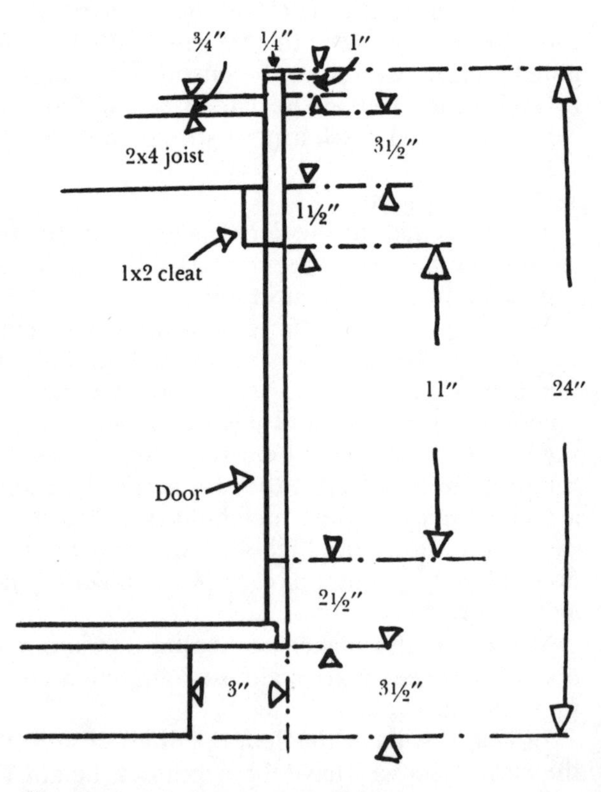

Fig. 7. Measurements from Side

the cleats. Note that there is no cleat for the joists at the foot of the bed, so you will have to put one in just for the door catch.

Touch latches are no good here because you might kick them open accidentally. Also, cats have been known to figure out touch latches; and they love those dark recesses under beds where you'll never get them out. Use piano hinges for a neat appearance. Attach lattice strips as door stops on the sides of the inside face of the opening so that the door closes flush to the front surface.

The top edges of the plywood sides should be covered with ¼″x¾″ lattice stripping. These edges take a lot of abuse, and solid wood lattice protects the edges better than veneer tape would. Tape the edges of the door and openings only if you want—they show only when the doors are open.

To finish we suggest a stain, and several coats of polyurethane, varnish, or wax. You can, of course, paint the bed (see Finishing).

So here's your final design (*5* to 7).

List of pieces

¾″ plywood
- Sides—two at 20¼″x77½″
- Head and foot—two at 20¼″x41½″
- Bottom—one at 40¾″x76¾″
- Top—one at 40″x76″
- Side doors—two at 11″x30″
- Door at foot of bed—one at 11″x20″ (all three doors plunge-cut from the foot and two sides, if possible)

2x4s
- Floor frame sides—two at 71½″
- Floor frame head and foot—two at 35½″
- Joists—five at 40″

1x2s
- Cleats for joists—two at 76″
- Cleat for door catch at foot of bed—one at about 30″

¼″x¾″ lattice
- Top side edges—two at 77½″
- Top edges of head and foot—two at 41½″
- Door stops (for the vertical edges)—six at 12″

Plywood cutting plan

There is a lot of plywood left over on the second sheet; this may come in handy if you can't plunge-cut the doors neatly out of their sides (*8*).

Lumber list

¾″ plywood, birch veneer
- Two 4′x8′ sheets

¾″ plywood, fir veneer
- One 4′x8′ sheet (for the bottom)

½″ plywood, fir veneer, or ¾″ composition board
- One 4′x8′ sheet (for the top)

¾″ Birch

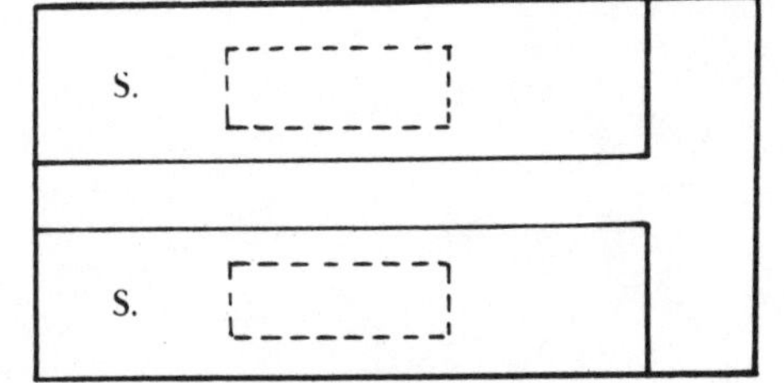

¾″ Birch

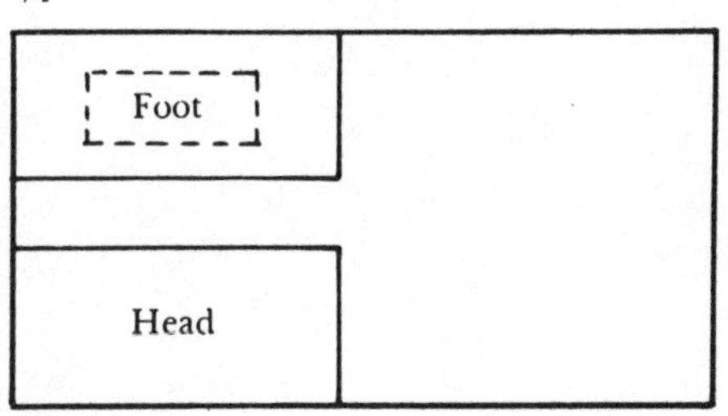

½″ Fir or ¾″ Comp.

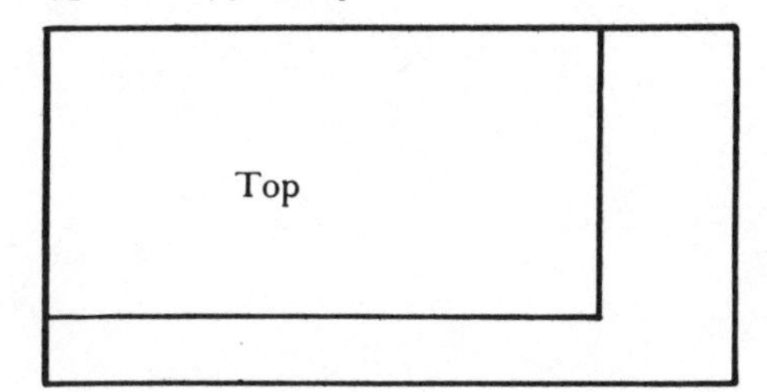

¾″ Fir

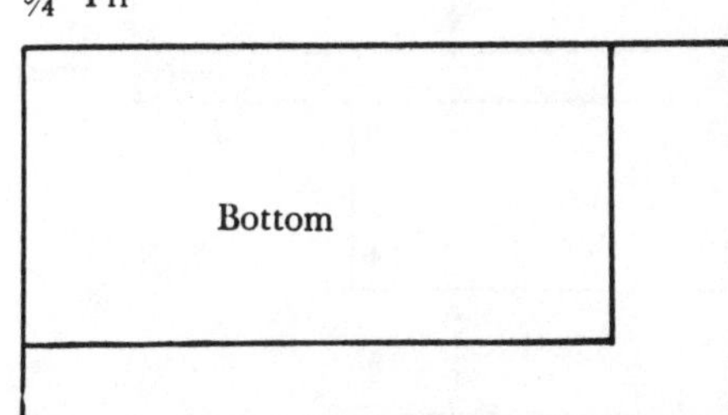

Fig. 8. Cutting Plan

2x4s
Two 10′
Three 8′
1x2s (clear)
One 8′
One 10′
¼″x¾″ lattice
Four 8′

Materials
1 lb. 1½″ (4d) finishing nails (for top and bottom sheets)
1 lb. 1″ (2d) finishing nails (for lattice)
1 lb. 3″ (10d) common nails (for frame)
Box of 100 1¼″ No. 10 flathead wood screws (for miters, cleats, rabbet); brass if desired
1 dozen 1½″ No. 10 flathead wood screws (for joists); brass if desired
Wood glue
One 6′ length brass piano hinge (you cut it to two 30″ lengths)
One 2′ length brass piano hinge (to be cut to 20″)
Screws for hinges
3 catches
Veneer tape and contact cement, if you want to tape doors
Putty
Finish of your choice

Tools needed
Tape
Adjustable square
Long straightedge
Clamps
Hammer
Nail set
Circular saw with plywood and crosscut blades
Hacksaw
Router with ¾″ straight bit (or a dado blade for your power saw)
Drill with screwsink-type bits for 1¼″ and 1½″ No. 10 screws
Bits for hardware screws
Screwdriver or screwdriver bit for drill
Goggles

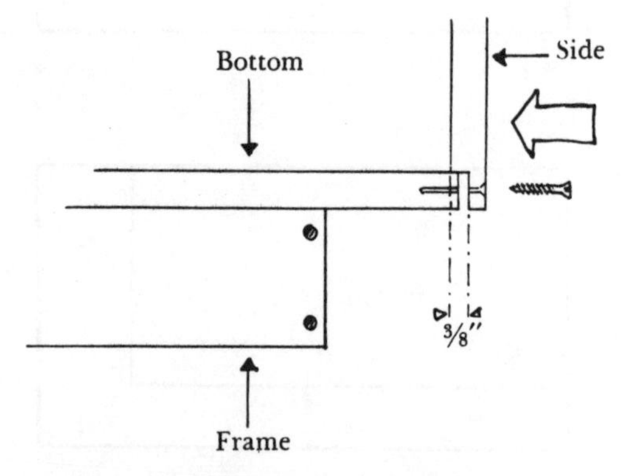

Fig. 9. Assembly of Side to Bottom

Construction

1. Cut the 2x4 pieces for frame, mitered. Glue and nail together. Set in place. Toenail to floor if you want.

2. Cut out the four plywood sides of the bed, mitering the ends. (See Miter Cuts, in Cabinet project.)

3. Before assembling, rabbet the bottom edges of the plywood pieces for the ¾″ bottom piece. Make the rout ¾″ wide and ⅜″ deep (*9*).

4. To cut out the doors, outline the opening on the inside, set up a guide, and start each cut with a plunge cut. Use your circular saw for most of the line, and finish the corners with the saber saw, or do the whole thing with the saber saw. Mark the doors to correspond to their openings—you want the grains to line up later. If the cut-out piece is too ragged to use as the door, you have enough extra plywood on sheet number two to cut several doors.

5. Glue and screw the sides together with 1¼″ screws. Tack or clamp temporary blocks at the inside corners to help position the miters (as in the Cabinet project). Countersink all screws. Use brass ones if you plan them to show.

6. Take measurements for the bottom piece across the frame from inside one rabbet to the other. Take the measurements near the corners, since the sides may be bowing out a little. They will straighten out when joined to a square bottom piece. Cut out bottom piece. Check fit by lifting plywood frame onto it. Take frame off.

7. Place bottom on 2x4 frame, with 3″ overlap on all sides, and nail in place.

8. Spread glue in dadoes or on edges of bottom sheet; lift ply frame onto bottom *(10)*. Countersink 1¼″ screws through the ply into bottom edge.

9. Cut and attach 1x2 cleats to inside of sides with glue and 1¼″ screws *(11)*. Position them with the bottom edges flush to the top of the door openings.

10. Cut lattice strips as stops for sides of door openings. Glue and nail them to the insides, letting about ¼″ overlap into the opening *(12)*. We want to attach the doors before the joists and top, so we have access to the inside.

11. Measure closed hinge thickness, add it to height of each cutout door. The total may be too big for the opening. You want about ⅛″ clearance at top. Retrim bottom edge of doors if necessary.

12. Mortise and attach flush door handles.

13. Veneer-tape doors and openings if so desired.

14. Cut hinges to exact length with hacksaw. Cut slowly and carefully, to get a straight edge. Screw hinge to door *(13)*. Put one end screw in first, then a middle screw, then the other end—to insure straight fit. Then fill in the other screws. Bring door to corresponding opening, lay open end of hinge in place and screw in. Do the end and middle screws first again, check if door swings freely, then add the other screws.

15. Install catches. Inside part of catch should fit bottom of 1x2 cleat. (If you use a shorter door than our design, you'll have to add a cleat on inside for the catch.)

16. Joists *(14)*. Cut the joists to proper length. Watch out for slight bows in the top of the plywood. These can be straightened when the joists are fastened. Place one at each end, space the others evenly at 19″ on center *(15)*. Glue the joist ends. Put one screw into each end through the sides of the bed *(16)*.

17. Take inside measurements for the top piece of ply. Cut, and nail in place. No fastening is needed through the sides; the joists are already holding the sides tightly.

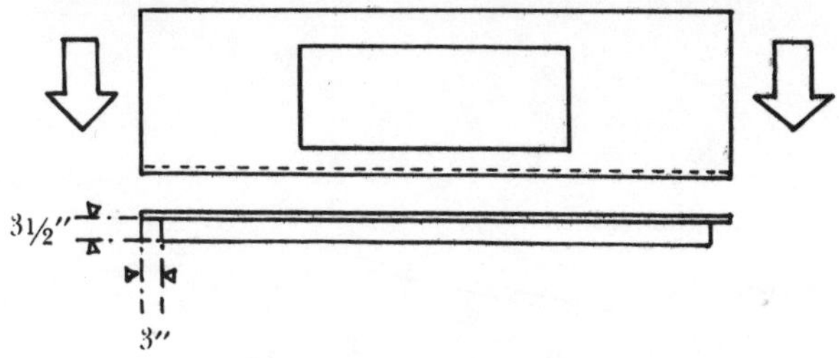

Fig. 10. How top joins and sits on bottom

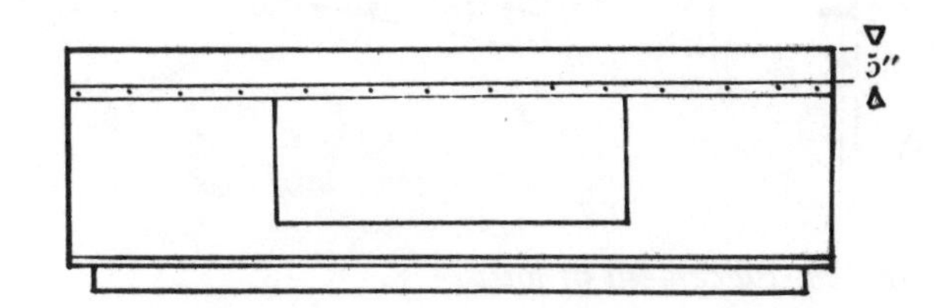

Fig. 11. Position of joist support viewed from inside

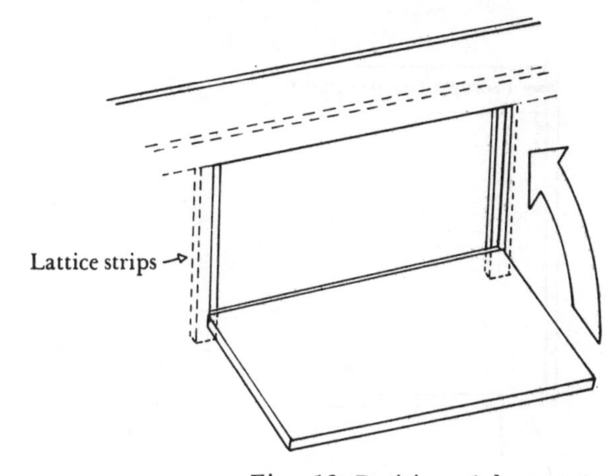

Fig. 12. Position of door stops

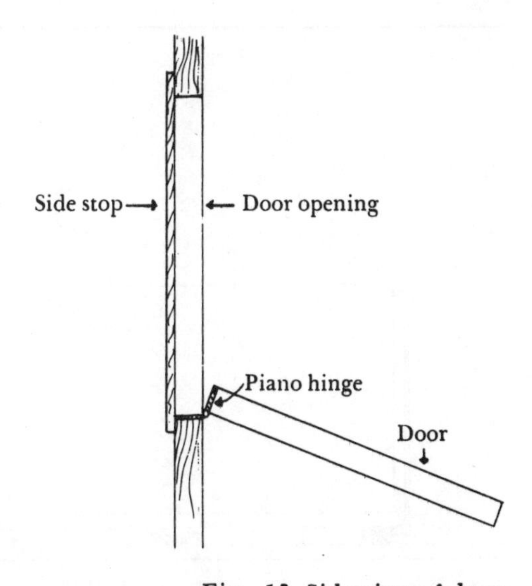

Fig. 13. Side view of door

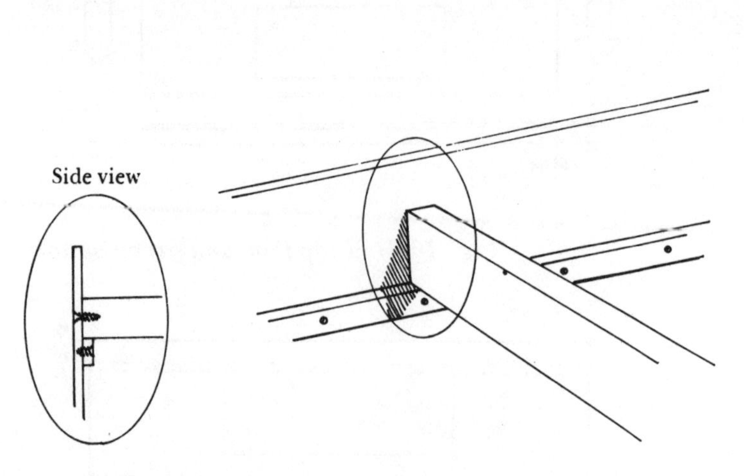

Fig. 14. Placement of joist

Fig. 15. Spacing of joists

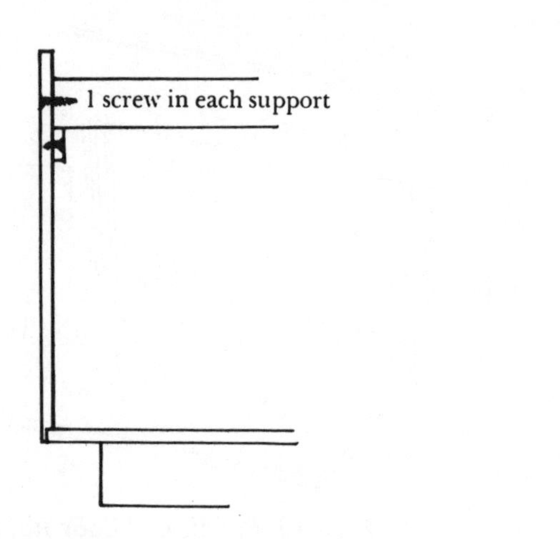

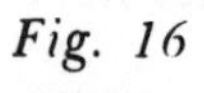

Fig. 16

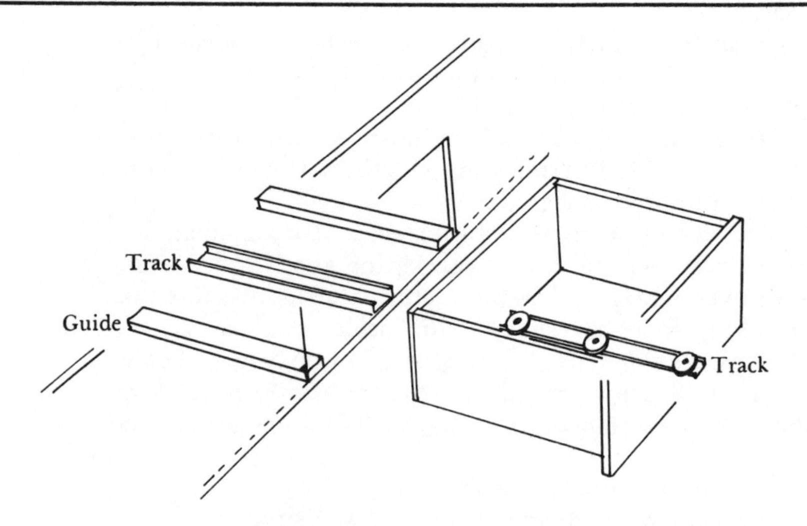

Fig. 17. Whole view of drawer

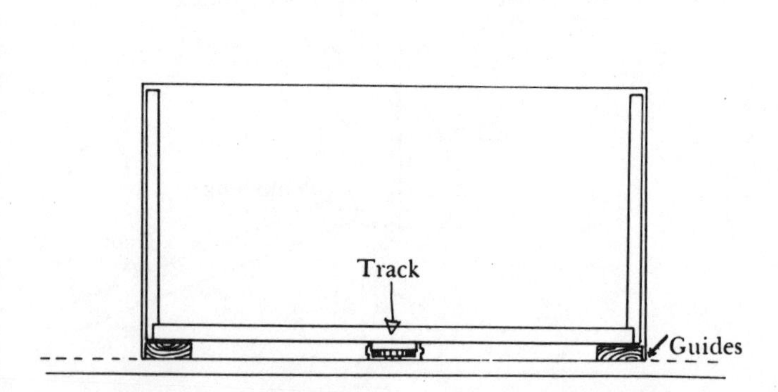

Fig. 18. Drawer variation

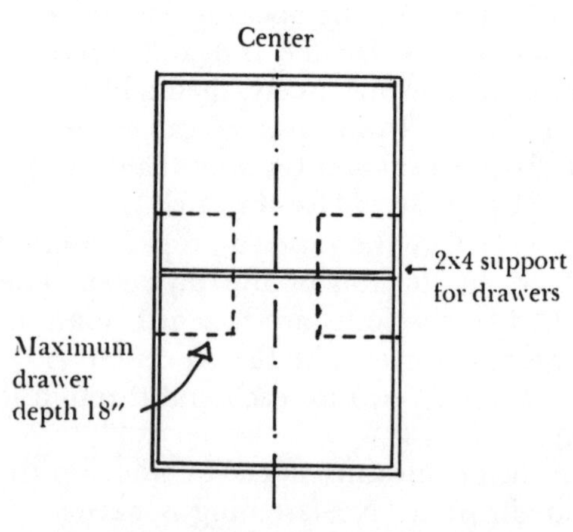

Fig. 19. Specifics of adding a drawer

18. Cut the lattice strips for top edges of box—miter the joints. Glue and nail with 1″ finishing nails.

19. Fill in screwholes, if you wish, with putty or plugs. Sand. Finish. You could stain the piece, and then wax, varnish, or polyurethane it.

Variations

Drawers

If you use the storage area every day, drawers are preferable to doors. Doors require you to kneel down and reach in; drawers bring everything out to you. Cut bottom of the opening level with the bottom piece. Use rollers or wooden cleat runners on the bottom sheet of ply (*17, 18*).

Putting two drawers on a side is not a good idea, since any good-sized drawer leaves little of the plywood side left. You would have to reinforce the inside sections of the plywood side with 2x4 framing. Try one drawer on each side (being sure they both fit in the 40″ space when they are closed, of course). Add a 2x4 support under the plywood base located under the center of the drawer paths (*19*). Otherwise the dynamic load of the drawers moving in and out would put too much stress on the base.

Building without a router

Use 1x2 cleats to attach the bottom sheet to the sides. Assemble the sides first, *upside down.* Screw and glue 1x2 cleats all around the sides, 3/4″ from the edge which will be the bottom (when it's turned upright again). Spread glue along the edge of the cleats and drop in the bottom sheet of ply. Screw it to the cleats. Turn the whole box over onto the 2x4 frame, attach and proceed as before.

Bed wider than 48″

This means you need more than one sheet of ply for the bottom, and also for the top. Add a 2x4 under and along the seam of the two bottom sheets (*20*). And be sure to cut the two bottom sheets to approximately the same size, to equalize the stress on them. At the top either run the pieces perpendicular to the path of the joists; or cut them so the seam is on the center of one joist.

Also, if the bed is over 60″ wide, use 2x6s for the joists, or 2x4 T beams.

No platform

Quicker, cheaper to build. The plywood sides rest directly on the floor. No 2x4 floor frame, no ply sheet as a bottom.

This method gives you about four extra inches of storage height. Make the doors that much bigger if you like.

If you don't want to use the floor as the surface to store things on, you can cover it with hardboard, or old carpet. Put it in after the sides are up and in place. If you really want a new plywood base, you can screw 1x2 cleats all around the bottom of the frame and insert a sheet of ply on top.

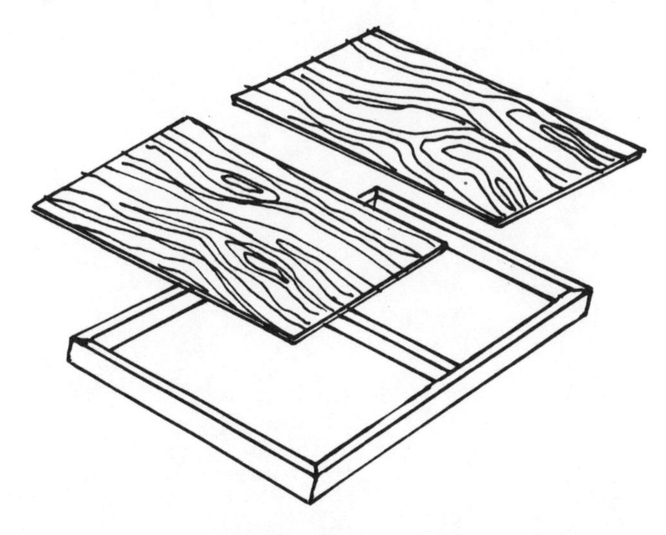

Fig. 20. Base Set-up for Larger Bed

POTTING TABLE

The need for such a table as this, plus several of its features, was suggested to us by two serious houseplant enthusiasts. Potted plants need a lot of care—repotting as they grow larger, fertilizing, leaf and root pruning, and so on. The phyllophile needs a variety of materials at hand for these jobs—bags of soil, peat moss, pebbles, assorted sizes of pots, trowels, pruners, knives, fertilizers, sprays.

This indoor potting table is designed to organize all these materials in one place, plus provide a work surface that can tolerate dirt, water, smashing pots. We've also designed it to be constructed quickly and simply, so you can get back to your plants (*1*).

Features

Our roll-out soil bin will be appreciated by anyone who's had to lug a hundred-pound bag of soil in and out of the hall closet for every repotting of a grape ivy. Buy the hundred-pound bag (much cheaper in the long run), dump it into the bin, and scoop out what you need when you need it. The bin is divided into two compartments, one for soil and one for peat, pebbles, or what have you.

A shelf at the back, above the work surface, keeps handy whatever you use most often. The shelves underneath are large enough to hold most pots and other accessories. Pots, of course, can be piled one within the other to conserve space. Hooks for tools, gloves, rags, etc. can be put just about anywhere—into the edge of the shelves, the sides of the counter, the legs.

The work surface is a simple plank top. It doesn't have to be perfectly flat, square, or well-finished like a dining room table—your pots won't mind. Several quick coats of linseed oil protect the surface and bring out the grain, adding to the overall rustic look.

Structure

The main structure consists of 4x4 legs at the corners, connected front to back by 2x4 crosspieces; the 2x8 planks are fastened across the 2x4s. The legs are notched, and the crosspieces are glued and bolted into the legs; these joints provide a good solid base for the rest of the structure (*2, 3*). The 2x8 planks (7½″ wide actually) can be glued and just nailed to the crosspieces. Bolts are stronger, but unnecessary for a potting table. The 30″x72″ countertop overhangs 3″ at the front, to give you toe space as you stand at the table.

The back legs rise through the countertop and a foot above, to simplify the fastening of the upper shelf. The middle 2x3 legs, nailed to the top, give the table extra support, but they are mainly intended to support the cleats for the undershelves. The 1x2 cleats are glued and nailed to the legs, and the ¾″ ply shelves are nailed down to the cleats. In our sturdier projects we advise screwing cleats in; nails are acceptable for this project. The very top shelf can be nailed and glued directly to the top of the back 4x4s.

The bin is simply a box on casters, not attached to the table at all (*4*). It's designed, of course, to fit easily under the left of the table, and 2x4 guides are toenailed between the front and back legs on either side of the bin, to keep it pretty much in line as it slides in and out. Another 2x4 is nailed in between the two left rear legs, to stop the bin rolling back. You could notch these 2x4s into the legs for strength.

The bin is made of 3/4″ ply with a double-thick bottom for strength. A divider in the middle enables you to separate different materials, while adding structural strength. The joints are under a great stress from the weight of the soil, so nails won't do here. Glue and screw. Butt joints will be adequate, but putty any cracks at the seams to keep soil from trickling out. Also, use two non-swivel type casters for the back of the bin, for easier steering in and out.

The table structure is actually very strong, even though nails are used in many places rather than screws. The leg assemblies are stable themselves, and the 2x8s and shelves tie them together.

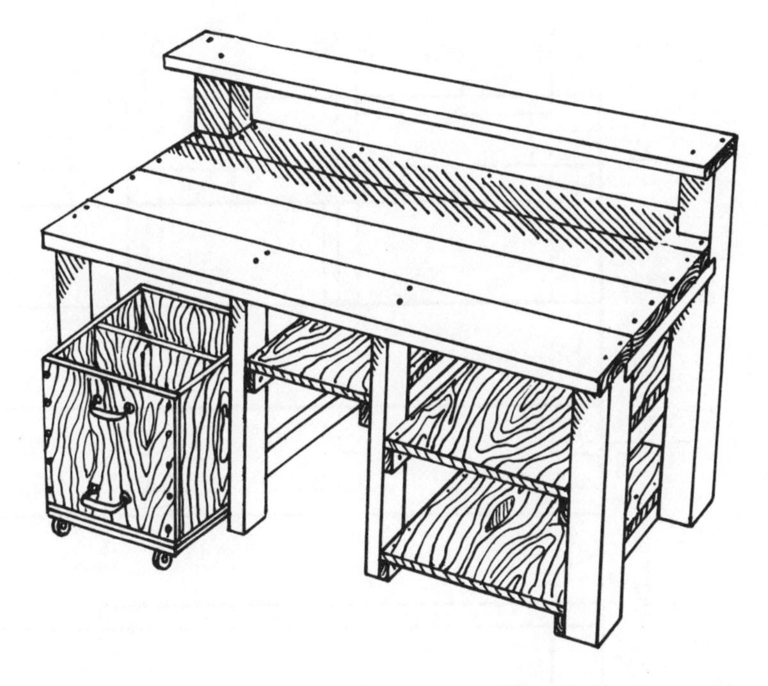

Fig. 1. Potting Table

List of pieces

3/4″ plywood
- Right shelves—two at 23 7/8″x27″
- Central shelf—one at 27″x20″
- Bin front and back—two at 17″x23 1/2″
- Bin sides—two at 23 1/2″x23 3/4″
- Double bottom bin—two at 17″x25 1/4″
- Bin divider—one at 23 1/2″x15 1/2″

2x8s
- Tabletop—four at 6′

2x6s
- Upper shelf—one at 6′

2x4s
- Top crosspieces for legs—two at 22″
- Guides for bin—one at 22″ (right) one at 20″ (left)
- Stop for bin—one at 18 1/8″

2x3s
- Middle legs—four at 34 1/2″
- Cleats on back legs for notched 2x8—two at 4″

4x4s
- Front legs—two at 34 1/2″
- Back legs—two at 48″

1x2s
- Shelf supports—six at 27″

The bin is designed to be set in 1/4″ when it's up against the back stop. There's about 1/2″ clearance on either side of it so you can fit it back in easily whenever you use it. Our plan is based on the assumption that a 2-by is 1 1/2″ thick. If the ones you get are a little thicker, adjust the 20″ space and shelf width between the middle 2x3 legs accordingly.

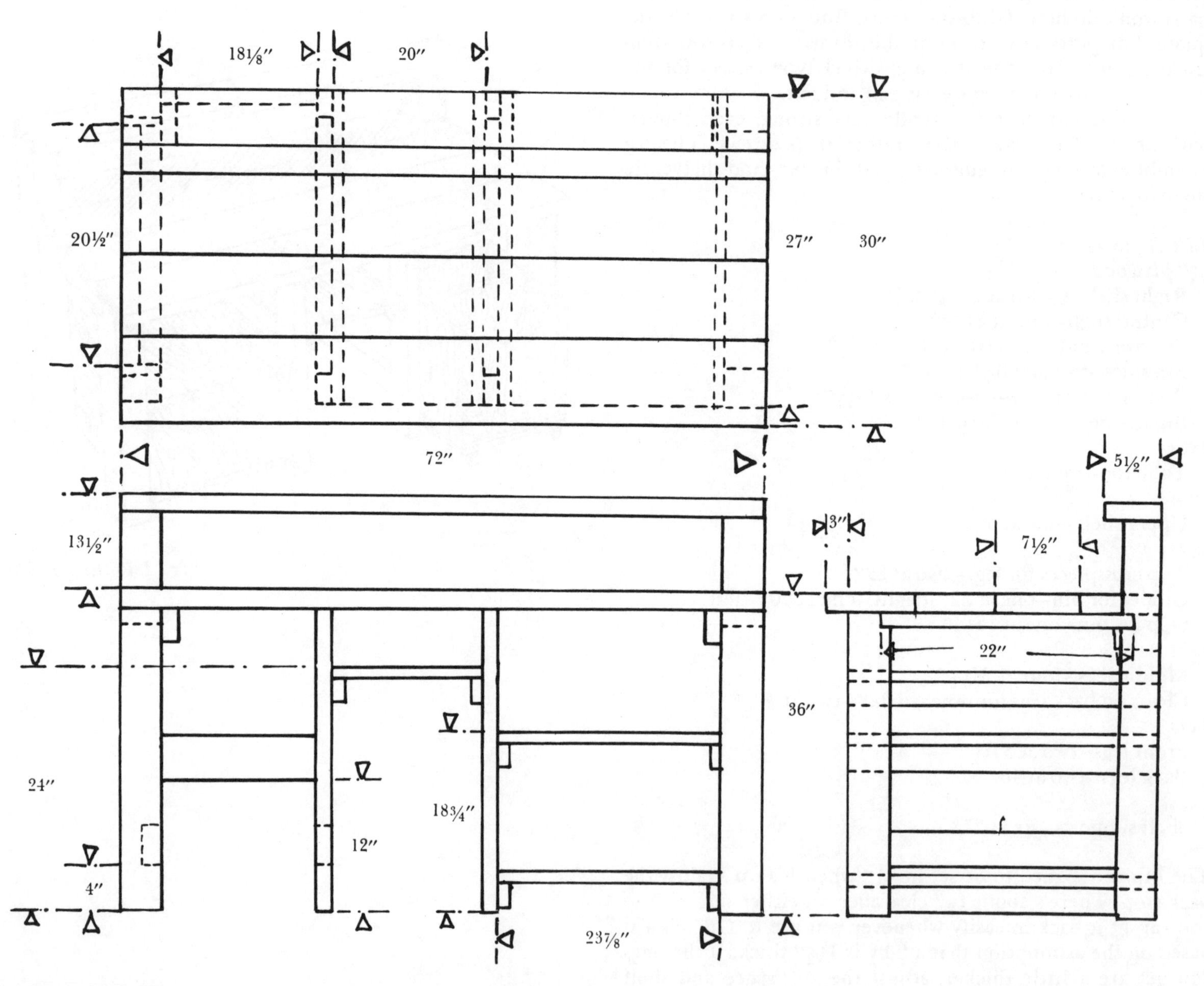

Fig. 2

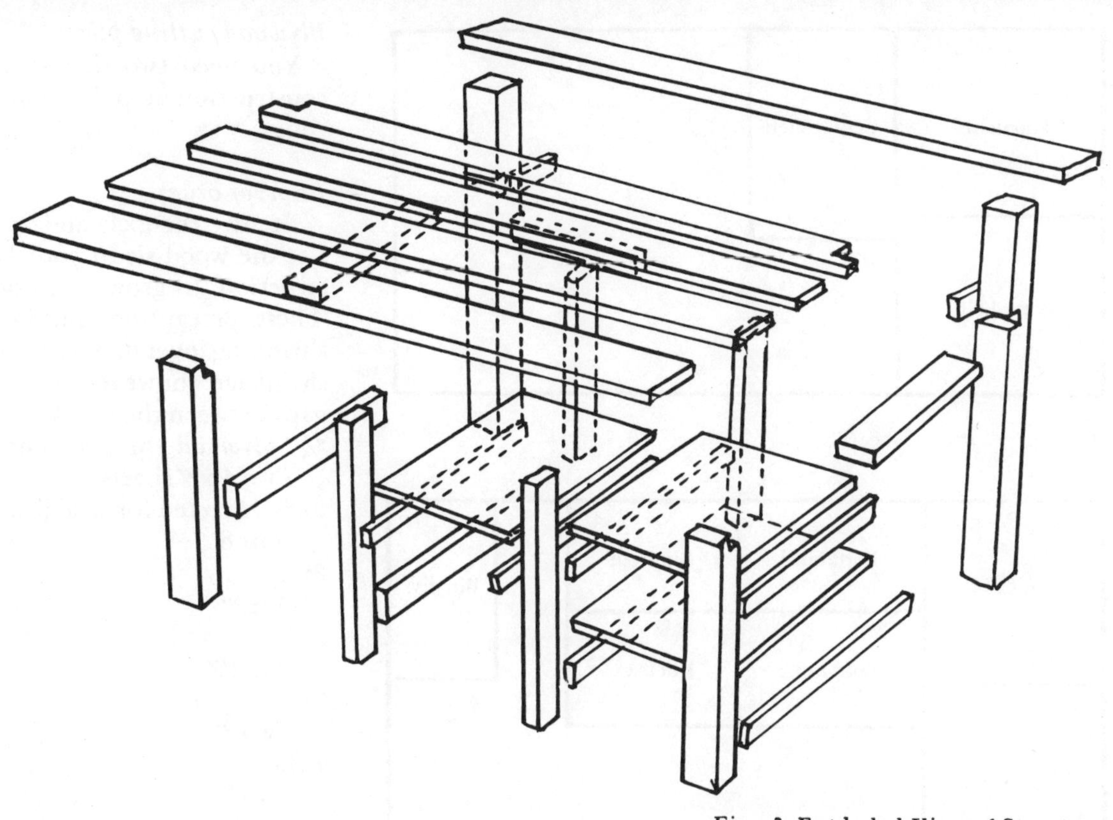

Fig. 3. Exploded View of Structure

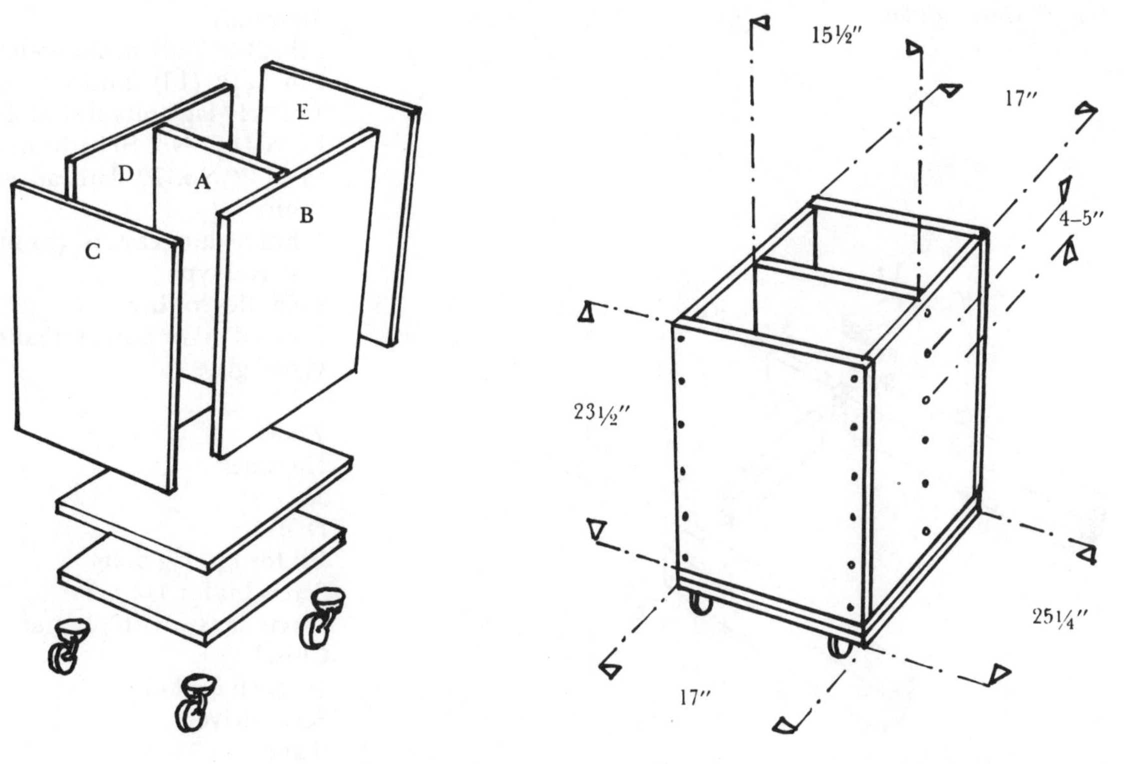

Fig. 4. Plan of Bin

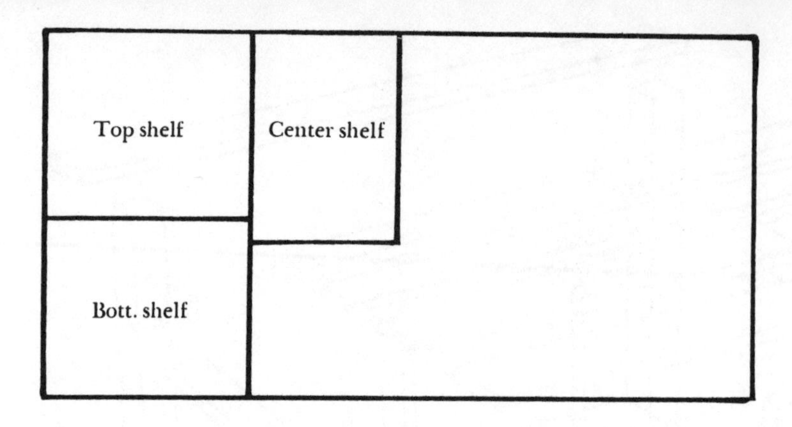

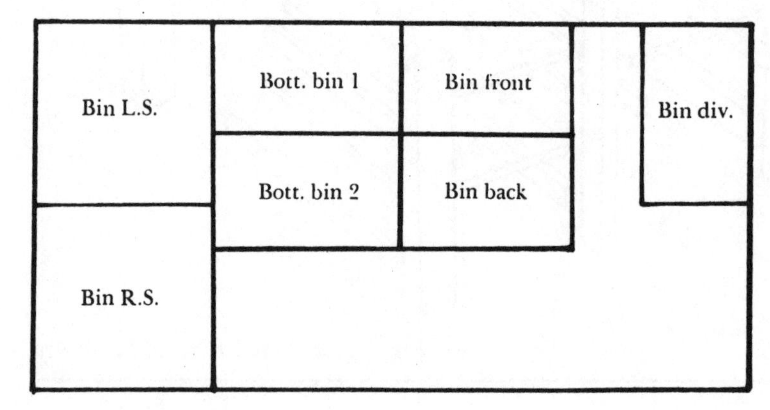

Fig. 5. Cutting Plan

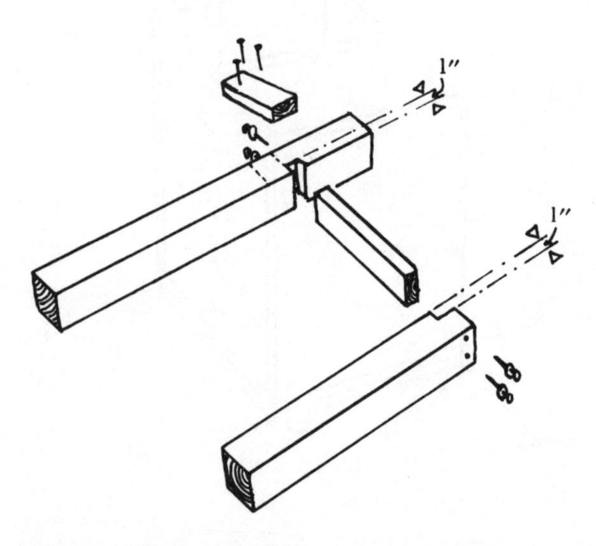

Fig. 6. Side Support Assembly

Plywood cutting plan

You need two sheets (*5*). Don't cut the shelves out till construction step 10. The bin parts can be cut out at any time.

Lumber order

Be sure the 2x4s and 2x8s, particularly, are all kiln-dried. Let the wood sit in your house forty-eight hours before construction, to grow accustomed to your humidity conditions. These precautions minimize the chances of your tabletop shrinking over the years from the wood drying out. Some shrinkage, however, is probably inevitable—causing slight gaps between the planks.

¾″ plywood (fir, good one side)
Two 4′x8′ sheets
2x8s (selected for straightness)
Four 8′
2x6s
One 8′
2x4s
One 10′
2x3s
Two 8′
4x4s
Two 8′
1x2s (clear)
Two 8′

Materials

1 lb. 2½″ (8d) common nails
1 lb. 1½″ (4d) finishing nails for cleats and shelves
8 ¼″x4″ lag bolts and washers
1 box 1¼″ No. 10 flathead wood screws for bin
35 2½″ No. 10 flathead wood screws for double bottom of bin
4 heavy-duty casters (to support 200 lb. plus), 2 of the non-swivel type
2 handles for bin
Linseed oil or polyurethane for finish
Wood glue

Tools

Hammer
Nail set
Drill
Bit for ¼″ lag bolts
Screw bit for 1¼″ and 2½″ No. 10 screws
Circular saw with ply and crosscut blades
Chisel
Wrench for bolts
Screwdriver
Tape
Goggles

Square
Straightedge
Clamps

Construction

1. Cut the 4x4s, 2x8s, and 2x4 leg crosspieces to length. Use the straightest parts of the 8′ long 2x4s for the top crosspieces.

2. With your circular saw and chisel, notch the 4x4s to take the top 2x4 crosspieces (*6*). Notch 1″ deep and as thick as the corresponding 2x4 (approximately 1⅝″, but may vary).

3. Glue and bolt the top 2x4 crosspieces to legs. It's easiest to start with the back leg. Glue and insert the 2x4 in the notch, with the pieces lying on their edges on the floor. Square the pieces and drill through the back of the leg for the two 4″ lag bolts. Counterbore for the head and washer if you want—if the head sticks out, you may not be able to push the table flush against a wall. Don't forget to put washers on the bolts. Hold the pieces together tightly and drive the bolts in all the way. Then drill and bolt at the front leg, keeping the pieces square. Repeat with the other pair of legs.

4. On the inside of the back legs, add the 4″ long 2x3 cleats meant to support the back half of the notched-out rear 2x8 plank. Glue and nail, with 2½″ commons. The cleats should be flush at the top edge with the 2x4 crosspieces.

5. Take exact measurements of the inside faces of the back legs, and notch the rear 2x8 to fit.

6. Stand the leg assemblies approximately six feet apart. They should stand on their own at this point, but an assistant would be of great help. Lay the 2x8 planks across, switching them around to find the order that provides the flattest surface and best matches the seams without any gaps. And bows should point up.

7. Tack the front piece approximately in place, one nail at each end, to help hold the front of the assemblies square and in position. Glue and nail the rear plank in place with two 2½″ common nails at each end (*7*). Fit the legs tightly within the notches of the 2x8, and keep the assembly square. Now spread glue on the front edge of this first plank and also on the 2x4 crosspieces, to receive the next plank. Nail the second plank in place, holding it tightly against the first as you hammer. Remove the temporarily nailed front 2x8, fasten the third plank in the same manner, and then the front one (*8*). The last plank should overhang about three inches. Drive the two nails at the end of each piece at opposing angles (*9*). This pattern resists any tendency for the 2x8 to loosen or lift up.

8. Measure the span for the top shelf, cut the 2x6 to length, and glue and nail it in place to the top of the back legs, flush in back.

9. Cut the four vertical 2x3 supports for the middle area. If your floor is none too flat, move the table into place before measuring for these legs. Glue and nail them in place with 2½″ common nails through the top (*10*).

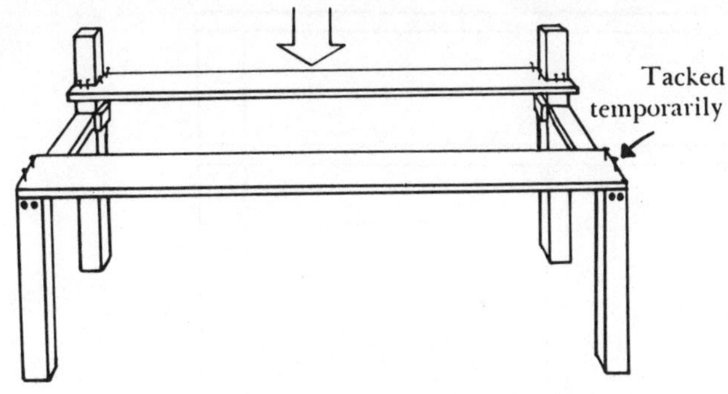

Fig. 7. Applying back plank

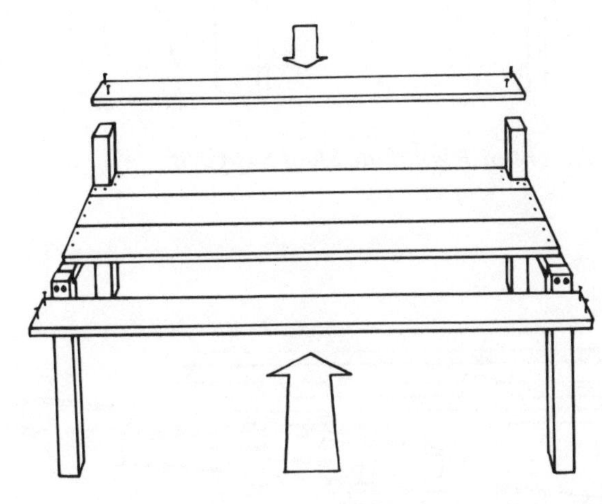

Fig. 8. Applying remaining planks and back shelf

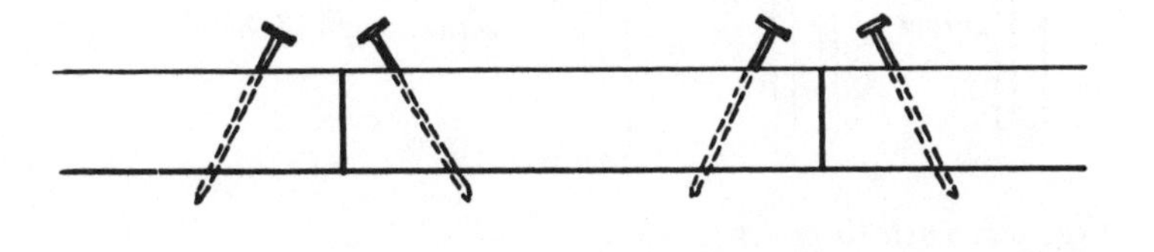

Fig. 9. Plank Nailing Pattern

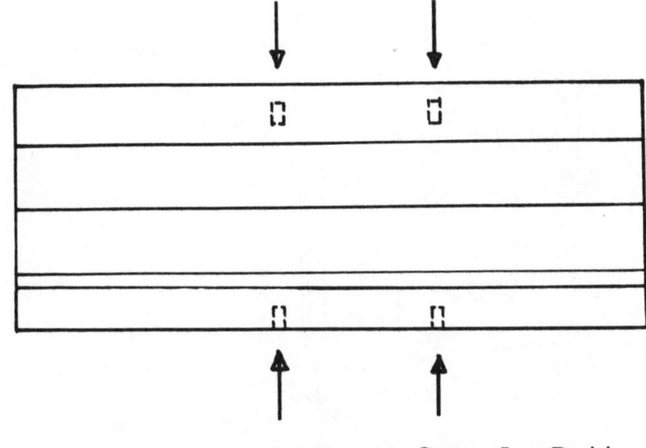

Fig. 10. Center Leg Positions

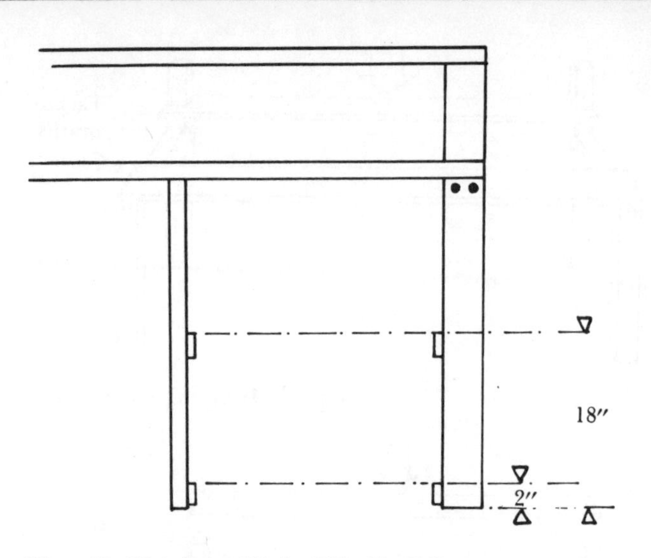

Fig. 11. Height of Right Side Shelf Supports

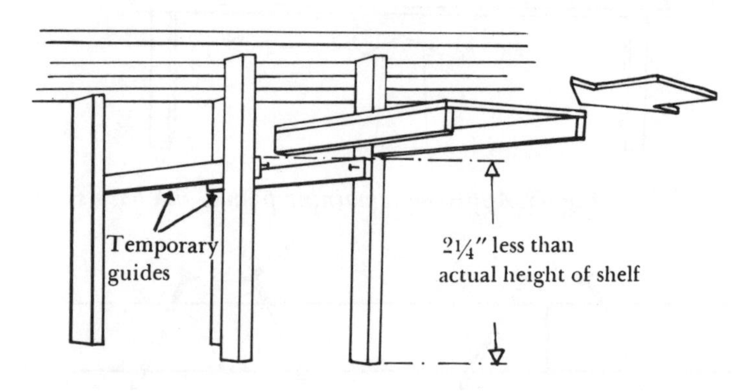

Fig. 12. Installing center shelf

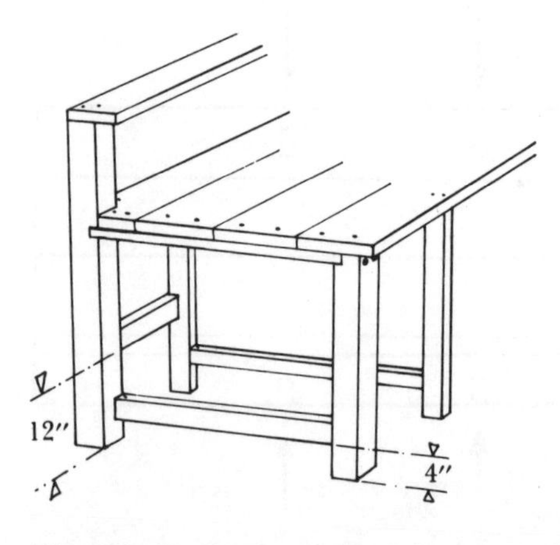

Fig. 13. Positions of Left Side Bin Guides and Stop

10. Cut all the 1x2 shelf supports. Glue and nail the four supports on the right-hand section. Use the 1½″ finishing nails (*11*). Hold each leg steady with a hand or foot as you hammer.

11. Check the plumb of the legs with your carpenter's level, measure the shelf spaces. Cut out the two shelves for the right-hand section. Prenail them with 1½″ finishing nails, glue the tops of the cleats; slip the shelves in place and nail them down.

12. Measure the space for the middle shelf. Cut it and its cleats. Glue and nail the cleats to the shelf. Spread glue on the outsides of the cleats at the ends, lift the whole assembly into place, and nail it in.

You'll find this easier if you temporarily nail some 1x2 blocks to the vertical supports, flush to where the cleat bottoms will go (*12*). You could also start some nails in the cleats before lifting the assembly into place.

You would have trouble attaching the cleats to the table legs first, and then nailing the shelf on top; the countertop is too close for hammer-swinging room.

13. Cut the 2x4 guide pieces and toenail them in place between front and back legs, both sides of the bin space. Locate them about four inches off the floor, so they will be above the bottom level of the bin (which is raised on casters). Locate them flush to the inside edges of the legs, as in Figure 13. Cut the 2x4 for the bin stop and locate it between the rear legs, 12″ off the floor. You have to toenail it into the 4x4 leg, but you can nail straight into the other end through the 2x3 leg. Place the stop in flush to the back edge of the legs.

14. Cut the pieces for the bin and assemble as shown in Figure 4, placing the screws about every three to four inches. Glue and screw all the joints. Join piece *B* to *A* flush. Piece *A* should be centered. Then attach *C, D* and *E* in that order. In each case, glue the joint and tack the pieces together with a few finishing nails. Set the pieces upright on the floor. The nails will hold the pieces while you drill and drive the 1¼″ screws. When *A* through *E* are assembled, turn the frame upside down. Glue and tack the two bottom pieces in place; drill and drive the 2½″ screws through.

15. Screw the casters to the bottom of the bin; attach the two handles. Putty any gaps at the bottom of the bin.

16. Finish as you like. Several coats of linseed oil mixed with turpentine (see "Finishing") is all the piece needs. Polyurethane is another good choice, offering even more protection.

Variations

Add a back

A back gives more space for hooks to hang things on, plus stops things from dropping behind when the piece is not flush to a wall. Pegboard is fine; nail it to the back of the countertop, top shelf, and back legs. If you want the back to continue above the top shelf, use something sturdier, like ¼″ plywood.

One or more planks very bowed

The solution is to clamp the pieces flush when you put them on, and hope that the glue will hold them flush when it dries (*14*). Another technique is to screw a crosspiece to the underside from front to back. As you tighten the screws, the bowed piece should pull down. This is a surer way of straightening than just the C clamps, but the 2-by will show underneath. Indent it 3″ from the front. Or you could simply screw a crosspiece on top of each pair of front and back 2x3 legs, then screw the whole assembly to the bottom of the planks.

An even better solution is to dowel the 2x8 edges together. But this calls for more work and precision than we want to do for a worktable.

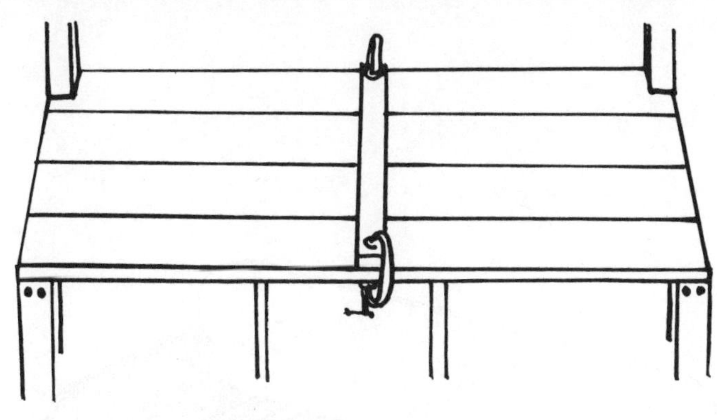

Fig. 14. Way of positioning warped planks

One or more pieces crooked—gaps between planks

Small gaps may not bother you; larger gaps may let dirt leak onto the lower shelves. This may not bother you either. At any rate, this is a problem you can avoid with careful selection at the lumberyard. Pick only straight pieces, or have them plane the pieces for you. Don't have them *rip* the pieces —this just duplicates whatever curve already exists. If there are still gaps between the pieces, you may be able to minimize them by switching the order of the planks, putting the worst edges on the front and back. Bar clamps can force smaller gaps together and hold them tight till the glue sets.

Workbench

This project can be adapted to any kind of worktable with a little imagination. Substitute drawers for shelves or the bin; add sides, back, and doors. If you want to use the table outside, make it of cedar or more expensive redwood. Close off the sides with panels and doors.

You could even make a carpentry workbench out of it. Strengthen its structure, since it will be getting harder use—hammering, sawing, etc. Screw the top down. You might even fasten it to the wall or add a floor-to-top back to rigidify it. Or add diagonal bracing to the legs, as in Figure 15. Take more time getting the top flat, or fasten a piece of plywood over it. You might also extend the top over the sides three or four inches so you can use vises and clamps, and possibly eliminate the top shelf and upper extensions of the back legs, to give yourself a larger flat work surface.

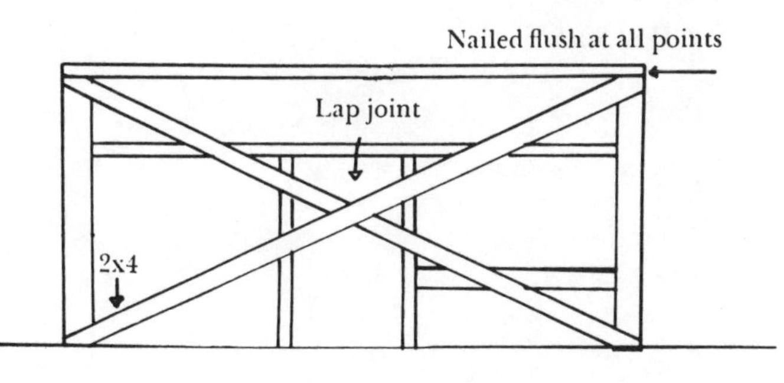

Fig. 15. Back of table showing extra bracing

Fig. 1. Wet Sink

WET SINK

This wet sink is specifically designed for a darkroom, for photographic developing, but is equally useful for any endeavor that needs a large work sink (*1*). It's much cheaper than a porcelain or metal sink. As a super economy measure, it's designed to work with no plumbing other than a drain, a short length of hose, and two pails. Photographic developing does not need vast quantities of running water; you can bring in what you need in a pail, and later empty the water and chemicals out the drain into another pail, by means of a short length of hose. There's no need for expensive runs of plumbing lines. The wet sink is mainly a basin in which you can safely use water and acids.

The bottom of the sink is slanted an inch higher to the right, so the liquid flows to the drain at the lower left end (*2*). Plan this end closest to wherever you will be emptying the pail of wastes. And if your floor isn't level, adjust the legs.

It's also useful to have a platform in the sink for trays to sit on and for liquids to flow under (*3*). Divide the length into two smaller platforms for convenience. Make them out of clear 1x2s as shown. Common grade would save a couple of dollars, but common 1x2s are usually so messed up that your trays might tend to rock and spill their contents.

The standard height for a work surface is 36″. (You might want it a little higher or lower, depending on your height—actually 3″ below elbow height is best.) Thus the height of the sink is such that the top of a common 2″ tray, sitting on top of the 1½″-thick drainage platform, on top of the sink bottom, will be 36″ high at the upper right end.

The size of the basin is determined by what usually goes in it. Our size of 25″x72″ is large enough to hold four large developing trays at once.

The height difference of the tops of the four sides is partly for design, partly for efficiency. The front should be low enough so you can work comfortably over it; the other sides are higher to protect the room from water and acid splashes. The rounded edges are a safety measure.

The sink and drainage platforms can be waterproofed and acid-proofed with two coatings of epoxy resin, brushed on like paint.

Structure

Great structural strength would be more than necessary for this project (*4*). It doesn't have to support any great loads. The basin is a simple box structure made of standard ¾″ plywood. The joints should be tightly sealed, particularly around the inside bottom edges. This is not so much for strength as for waterproofing. Dado the bottom piece into the four sides, and join with glue and screws.

The joints between the sides are not as crucial, since the sink won't be filled with water. Butt joints, glued and nailed, are fine. Epoxy resin will seal any thin cracks at the seams.

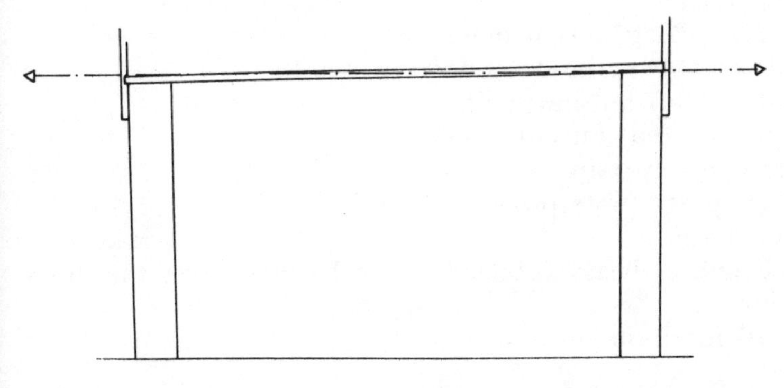
Fig. 2. Slope for Drainage

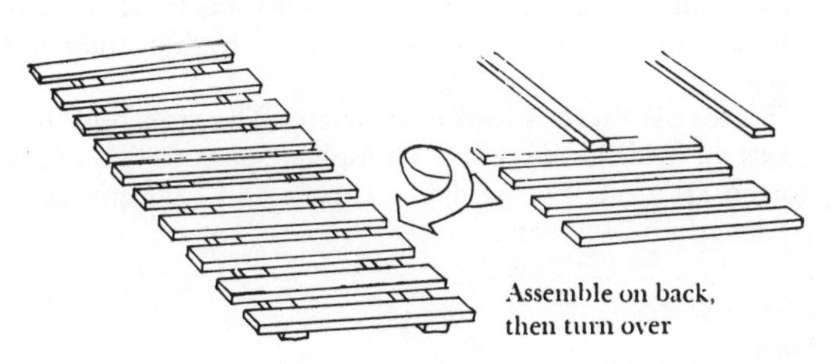

Fig. 3. Assembly of Drain Board Section

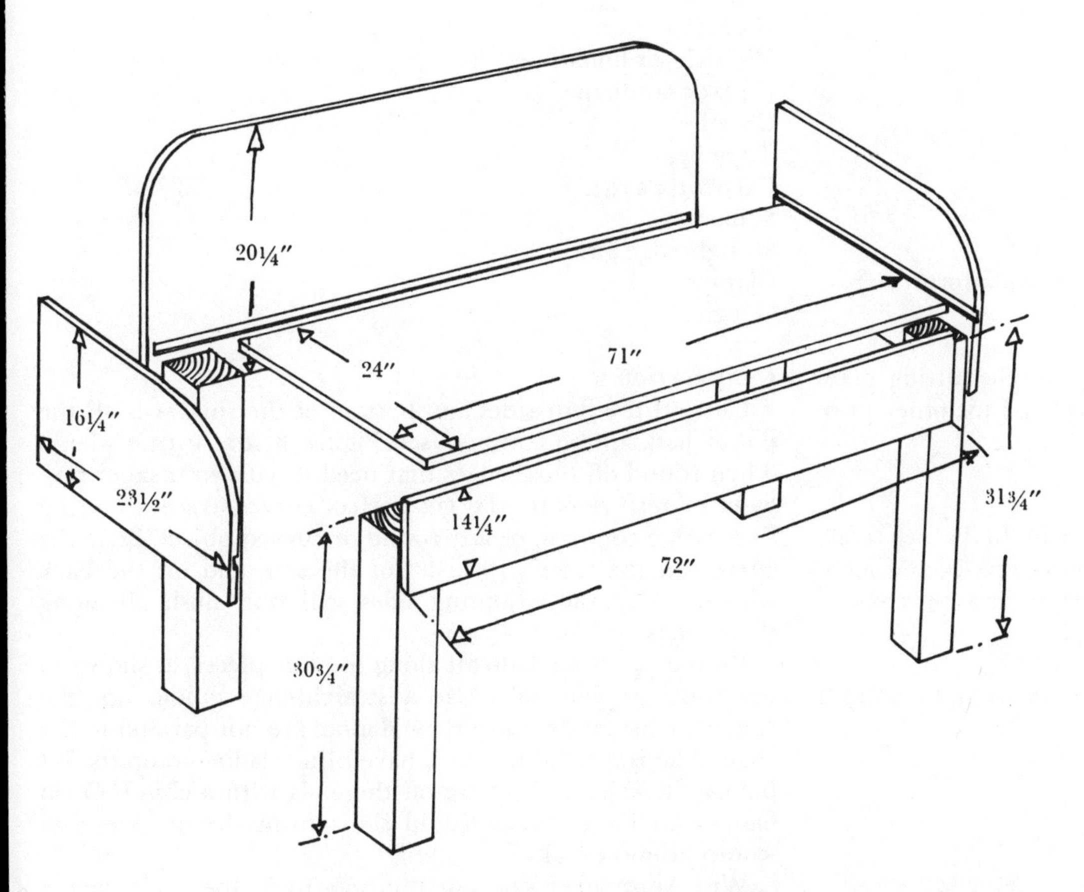

Fig. 4. Exploded View of Structure

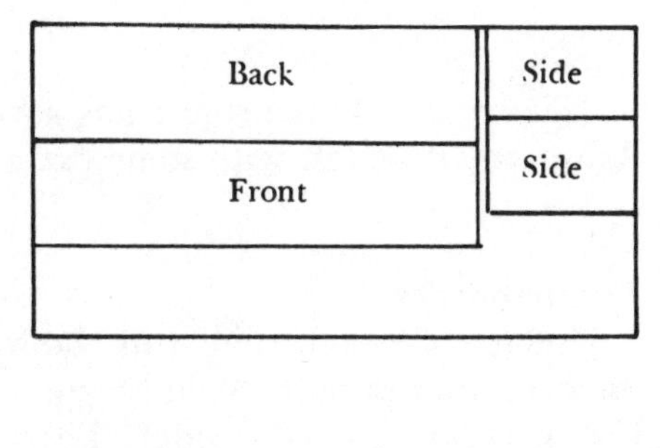

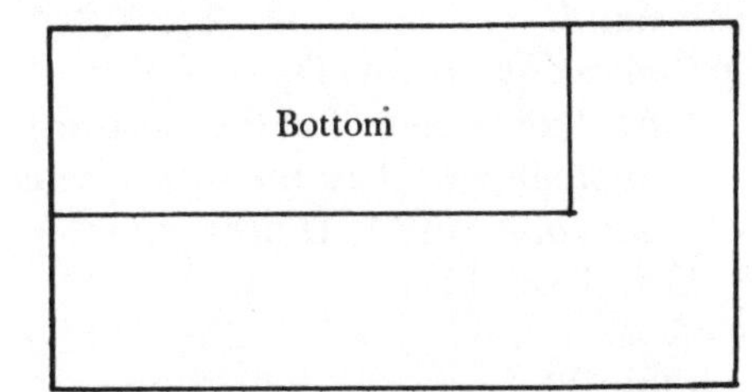

Fig. 5. Wet Sink Cutting Plan

A lattice strip on the top front edge will protect the inner plywood core from splashes as you lift liquids in and out of the sink. You could veneer-tape the other edges, but it's unnecessary; the epoxy resin could be used on them.

The legs are 4x4s for stability. The simplest joint for the legs is to extend the sides below the sink bottom about 6″ and bolt them to the legs. The legs should meet the slanted sink bottom, so the right ones are cut 1″ higher than the left ones.

That's all there is to the structure. The sink bottom needs no extra middle supports. Though it spans 6′, it's really supported all along by the dadoed front and back pieces, which are less than 24″ apart.

List of pieces
Sink
¾″ plywood
Front—one at 14¼″x72″
Back—one at 20¼″x72″
Sides—two at 16¼″x23½″
Bottom—one at 24″x71″ (The bottom is slightly more than 71″, due to the slope, but not enough to worry about.)
4x4s
Left legs—two at 30¾″
Right legs—two at 31¾″
¼″x¾″ lattice
Top of front—one at 72″

Drainage platforms
1x2s
Base pieces—four at 33″ (two per platform)
Upper crosspieces—twenty at 23″ (ten per platform)

Cutting plan
(Figure 5.) This project has a very simple cutting plan. Again you'll be left with some extra plywood for other projects.

Lumber order
Use fir plywood. The sink bottom should be relatively smooth, since puddles would form in knots, cracks, and such. Use A/D ply (good one side). The sides can be as ugly as you can stand.
¾″ plywood (fir, A/D)
One 4′x8′ sheet (for the bottom); or try to buy a scrap a little bigger than the size you need
¾″ plywood (fir, A/D or worse)
One 4′x8′ sheet
4x4s
One 12′
1x2s (clear)
One 8′
Four 12′
¼″x¾″ lattice
One 8′

Materials
16 2″x5/16″ lag bolts, with washers
3 dozen 1½″ flathead wood screws, No. 10
1 lb. 2″ (6d) finishing nails
1 lb. 1¼″ (3d) finishing nails
Can of epoxy resin
Wood putty (waterproof)
Sink drain
3′ length of heavy rubber hose, to be fitted over the drain bottom
Caulk for drain

Tools
Circular saw, with ply and crosscut blades
Saber saw
Drill with ¼″ or 9/32″ bit for bolts, 1¼″ bit for countersinking bolts, screw bits for screws
Screwdriver or screwdriver bit for drill
Router with ¾″ straight bit
Hammer
Nail set
Wrench for bolts
Rasp or sandpaper
Chisel
Tape
Carpenter's rule
Square
Straightedge guide
Clamps
Goggles

Construction

Cut out the four sides and bottom of the sink as if all the pieces had square corners (see Figure 5 for cutting plan). Then round off those edges that need it with your saber saw. Sand or rasp smooth. To get perfect curves, trace a cutting line with a compass, or any round or curved object. Start the curves on the sides within 2″ of the top, and on the back within 4″—so the adjoining sides will butt flush all along their edges.

Rout a ¼″ deep dado all along the side pieces, as shown in the routing plan (*6*). Use a straightedge guide, not the router's edge guide, since these dadoes are not parallel to the edge. The front and the back have blind dadoes—stop the bit before the edge and square off the ends with a chisel. Open dadoes on these pieces would show from the outside, and would promote leaks.

With your saber saw, cut the hole to fit the drain in the bottom piece (7). Locate it within 1½″ of the end.

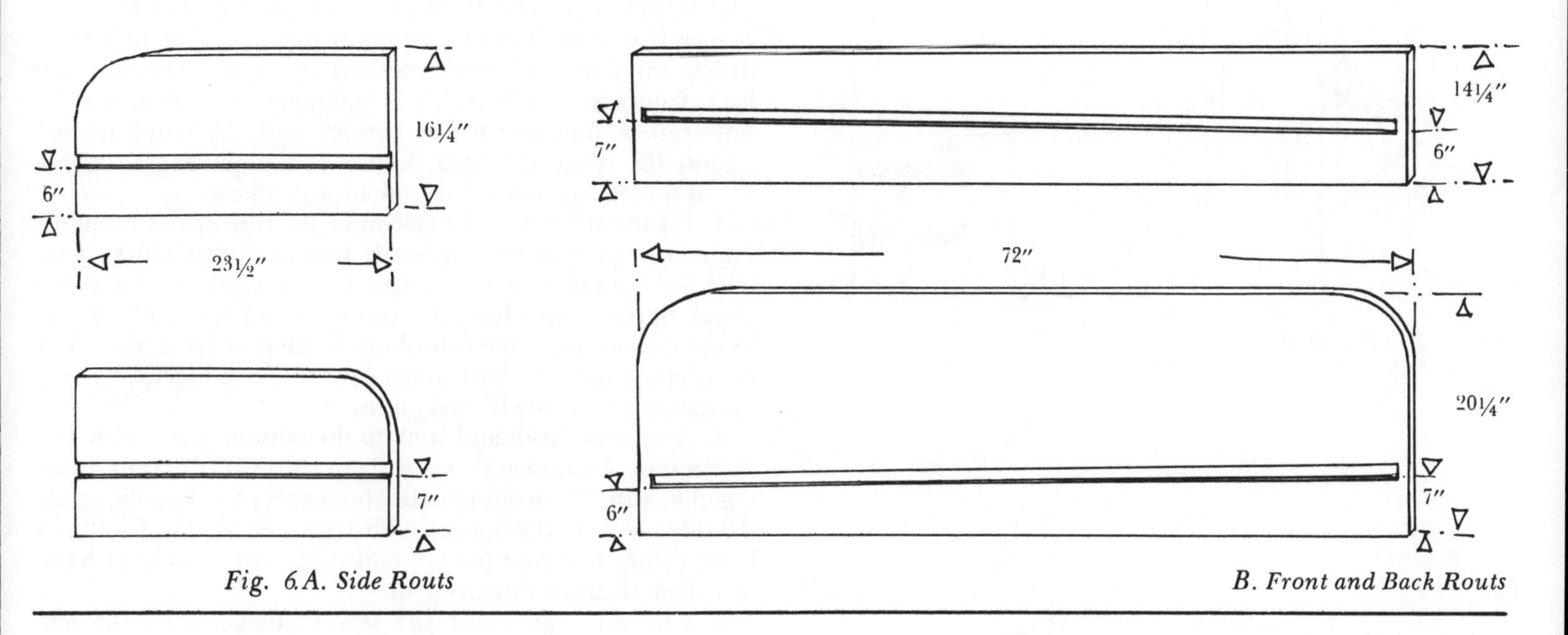

Fig. 6.A. Side Routs

B. Front and Back Routs

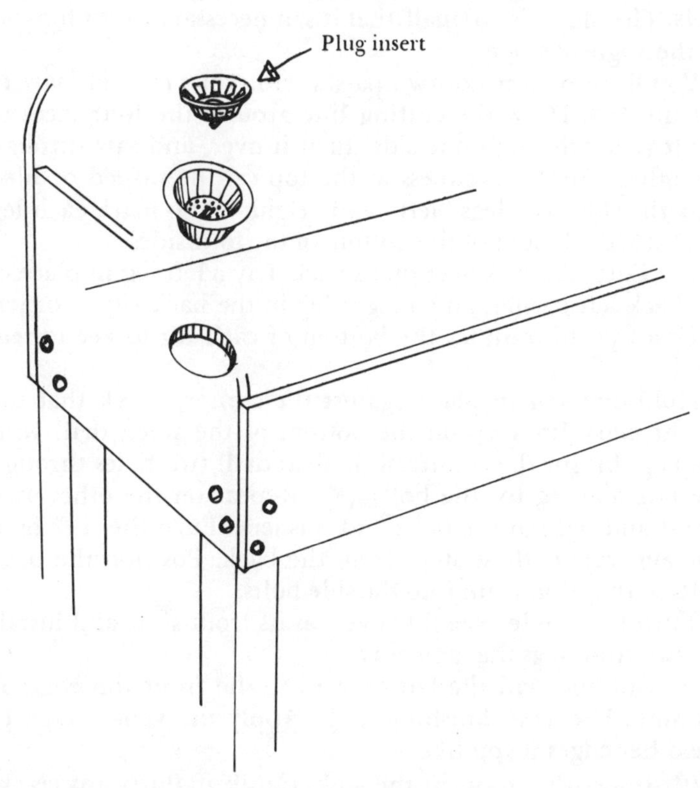

Fig. 7. Position and Cutout for Drain

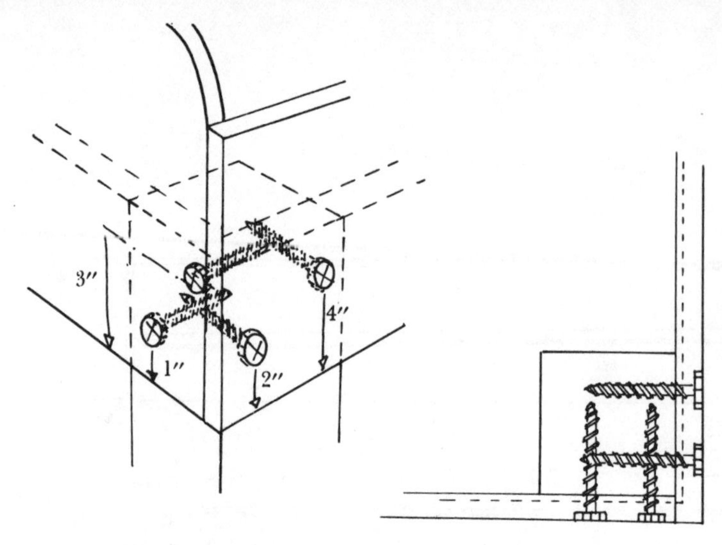

Fig. 8. Bolt Positions

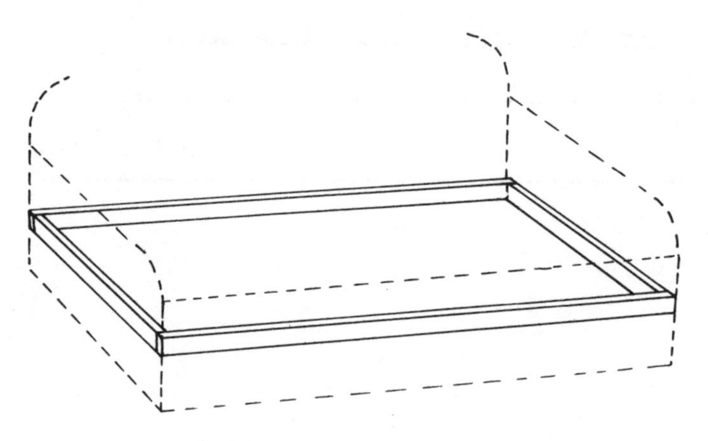

Fig. 9. Cleat Support Variation

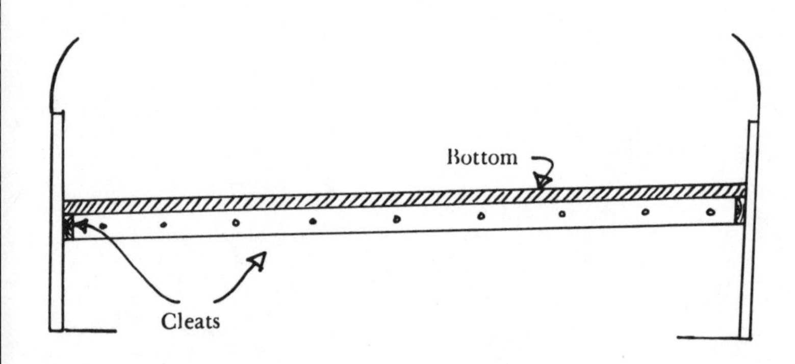

Fig. 10. Setting in bottom on cleats

3. Glue the front and back dadoes and insert the bottom, ½″ in from the ends. Tack in place with a couple of 1¼″ finishing nails. Set the assembly upright on a flat surface.

4. We didn't join the *sides* to the bottom first, because then it would be difficult to get the correct angle of slope at the side joints. The side dadoes would force the bottom in perpendicular to the sides. The front and back surfaces, however, *are* perpendicular to the sink bottom, at the same time that their dadoes carry the sink at the desired 1″ slope. Thus once the back/front/bottom assembly is complete, the left and right sides can be adjusted to the correct angle by being butted against the front and back pieces. The slight angle (about 1°) at the side dadoes will automatically follow.

5. Fit the sides into the assembly. Because of the bottom's slope, its edge may not slip easily into the side dadoes. If so, trim the bottom edge of the dadoes just a hair with a sharp chisel. Spread glue along the vertical side edges and heavily in the dadoes; push the sides into position as far as they will go, keeping the side butt joints square. Tack together along the dadoes and at the butted joints.

6. Screw the front and back to the bottom piece with the 1½″ screws, countersunk, every 8″ to 10″. Nail the butt joints together with 2″ finishing nails about every 6″. Finally, screw the side pieces to the bottom with 1½″ screws, as in front and back. Either remove the 1¼″ nails you used as tacks or hammer them in and set them.

7. Cut the right-hand 4x4 legs 1″ longer than the left 4x4s. The slope is so small that it's unnecessary to cut tops off at the angle of slope.

You'll need to make two passes with your circular saw to cut the 4x4. Draw the cutting line around the four faces of the leg, saw through one side, turn it over, and saw through the other. Any unevenness at the top can be rasped or filed smooth. Mark the legs "left" and "right." Also mark each leg at 24¾″, the height of the bottom of the four sides.

8. Turn the sink over on its back. Lay a left leg in place in the back left corner, and a right leg in the back right corner. Stick a ¾″ shim under the bottom of each leg to keep them level.

Hold one 4x4 in place against the corner, check that the height mark lines up on the bottom of the sides, drill with the 1¼″ bit for the countersink, then drill two holes through the side and leg for the bolts (*8*). Repeat on the other 4x4. Insert and tighten the bolts and washers. Turn the sink on a side and repeat these steps from the back. Position the back bolts so they don't run into the side bolts.

Turn the whole assembly over on its front side, and install the two front legs the same way.

9. Cut and nail the lattice strip to the front top edge of the sink. Use 1¼″ finishing nails. Apply the veneer tape to the other edges if you like.

10. Insert the drain in the sink. Caulk it. Putty any cracks at the joints of the sink, or deep indentations in the bottom.

11. Construct the drainage platforms (see Figure 3). This looks like and is a simple construction—though a bit tedious to make. The only trick is to assemble the platforms relatively square, so they will fit neatly in the sink. You don't need to be perfect, nor do you have to space all the long pieces exactly equally.

12. Cut all the pieces to size. Lay the 23″ strips for one platform upside down, their ends lined up flush against a block or something solid. Lay them within a 33″ width, spacing the inner ones with your eagle eye. Check with your square that the end pieces, at least, are perpendicular to the block. Lay a 33″ base piece over them, about 1″ from one end. Mark on it the approximate centers of the crosspieces, to show nailing locations; and trace the base pieces' outline on the crosspieces to show where to spread the glue.

Remove the base piece. Start 1¼″ finishing nails at the marks, and spread glue within the outlines on the crosspieces. Replace the base pieces and hammer the nails in, one for each cross strip. Pay special attention to keeping the outer pieces relatively square to the base piece. Repeat the whole process with a base piece at the other end, adjusting any ends that may have fanned out. Put a second nail in at each of the four corners of the assembly, to hold it square.

Build the second platform the same way.

Apply two coats of epoxy resin to the inside of the sink and to the drainage platforms. If you want, you can epoxy the whole sink.

Variations

Without router

Assemble the four sides. Screw and glue 1x2 cleats around the inside to support the bottom, at the same slope of 1″ over the 72″ length (*9*). Take the inside measurements with a tape or carpenter's rule; cut the bottom piece to size. Fit and glue the bottom into place, and screw it tight through the sides with 1½″ screws every 8″ to 10″ (*10*). Putty well and apply resin as above.

DINING, COFFEE, OR UTILITY TABLE

This sturdy table is very simple to build, and is almost infinitely variable (*1*). You'll recognize it as the basic table used in the Drawing chapter.

The dimensions given here would make a generous dining table for six people. You can adjust the sizes given to suit your needs; see charts in the Reference section for sizes needed to seat different numbers of people. If you make it 12″ to 18″ high, it will be a coffee table; at 28″ to 30″ high, it's a dining or kitchen table; at 30″ to 36″, it can even be used as a light utility table.

The construction is generally similar to the Loft Bed project (*2*). For larger sizes, some cross supports (like the joists in the loft bed) would be advisable. For the corner joints, lag bolts are suggested; they can be countersunk flush with the surface, or counterbored with the holes plugged for a more finished appearance. Very large screws (No. 14 or No. 16) could be used.

Clear pine or cedar should be used for the legs and frame, or even hardwood. The top can be ¾″ fir or birch plywood, or a fancier veneer, depending on your taste and budget.

Most softwood 2x4s come with slightly rounded edges. You need a flat edge so the plywood will fit neatly against the inside edge. Have the lumberyard plane one edge flat for you, or have them rip it just enough for the sharp corner you need. Hardwood usually has square edges, eliminating the need for planing; it may be difficult to find a hardwood 4x4, but you can glue two 2x4s together, clamped firmly until the glue is set.

Be extremely careful with all your measurements, particularly the notches. It would be possible to notch the legs less deeply, which would result in the legs being slightly indented from the corners, and would give more wood to bolt into.

Construction

1. Cut the 4x4 legs to length (¾″ less than full height table). Cut 2x4s slightly long.

2. Notch 4x4s, according to the thickness of your 2x4s (*3*). Remember that the *height* of the notch should be ¾″ less than the width of the 2x4s.

3. Notch 2x4s, making the notches slightly wider than needed. After the corner joints are lapped, the extra ends can be rasped flush.

4. Position the left 36″ 2x4 and the left legs carefully. Keep them at right angles, and keep the inner ends of the notches on the 2x4 flush with the edge of the notch on the 4x4s. Glue and bolt. Do the same for the right leg assembly.

5. Prop the assemblies up. Glue and bolt the 72″ pieces in place, keeping the upper surfaces of the 2x4s flush at the corners, and the corners square.

6. Measure for and cut the cleats to *exact* size, so the ends will butt tightly against the legs.

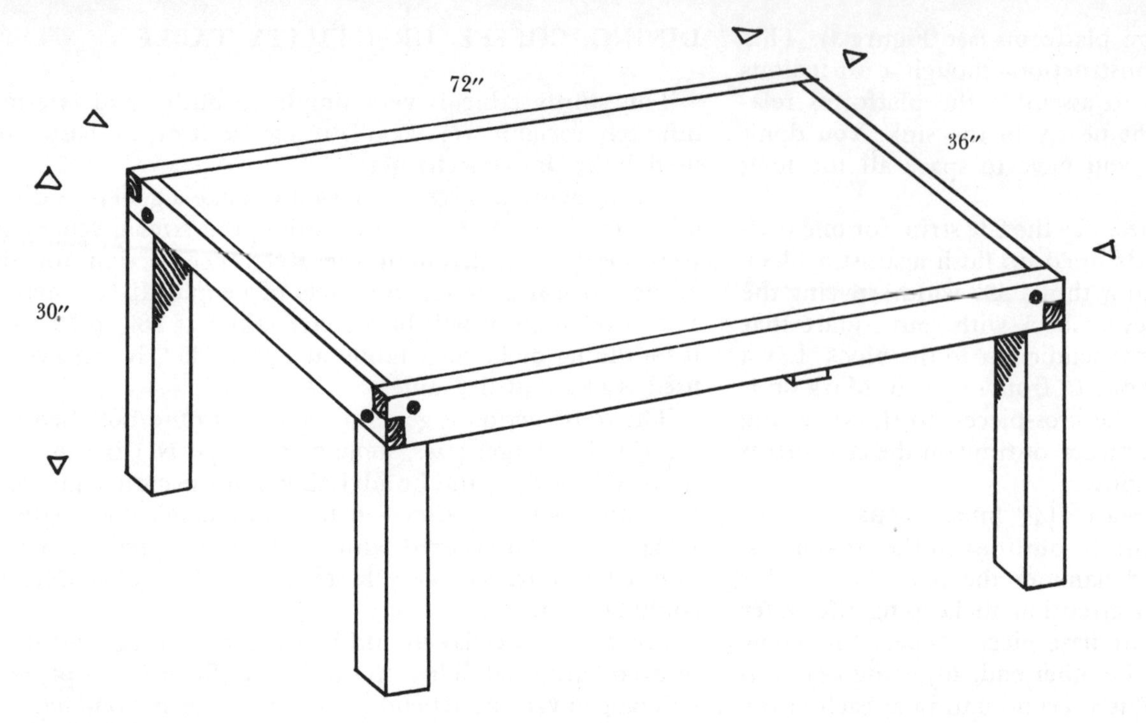

Fig. 1. Dining Table

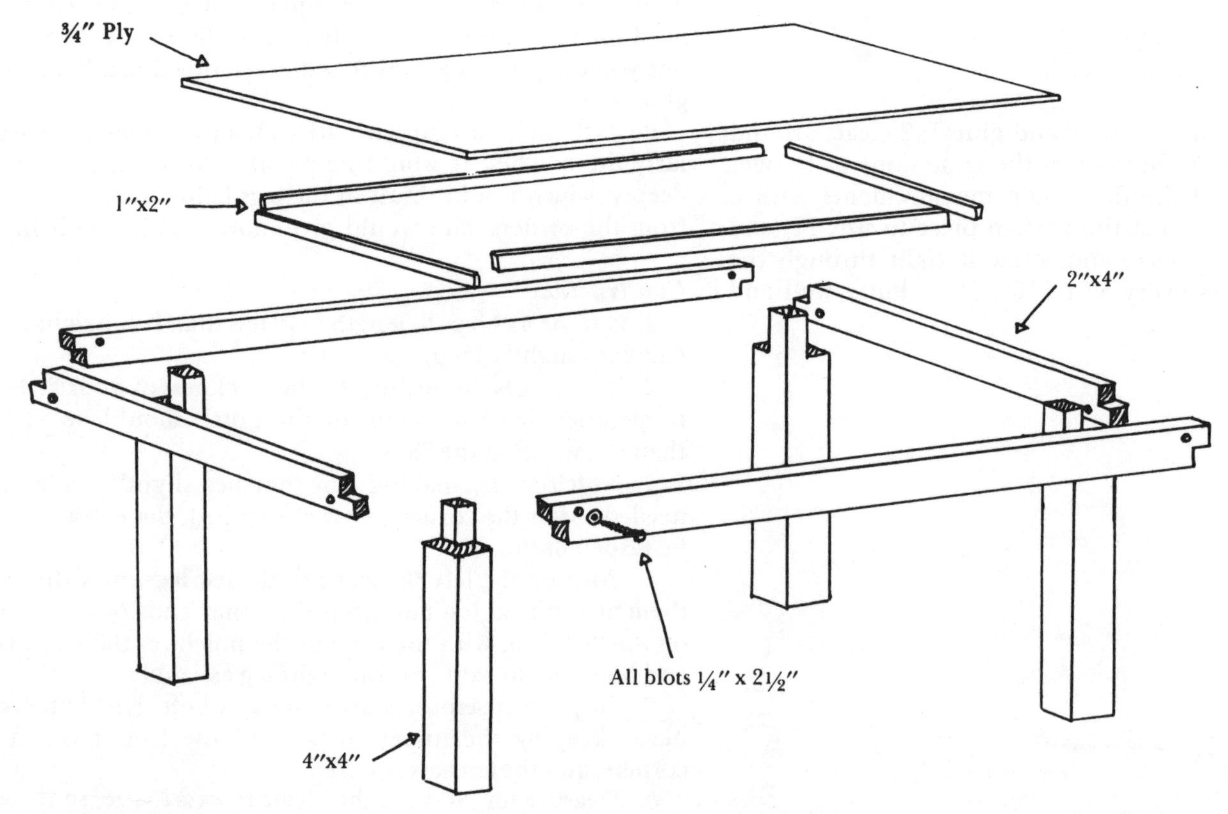

Fig. 2. Exploded View of Table

7. Glue and screw cleats in place (*4*). The butted ends help to brace the legs.

8. Measure for and cut the plywood top. Be precise.

9. Glue top of cleats and inside of 2x4s above the cleats. Lay the top in. Fasten it down to the cleats with finishing nails or screws.

10. Putty nailholes, or plug screwholes; or let countersunk screws show.

11. Rasp ends of 2x4s flush at the lap joints. Sand.

12. Stain if you like. Put on at least three coats of linseed oil, varnish, or polyurethane (see "Finishing").

Another order of construction steps, easier in some ways, harder in others, would be to cut the plywood *first*, then the 2x4s to fit around it. Cut the 4x4s to length, make all your notches as before, but assemble the tabletop first. Attach the cleats to the 2x4s and frame them around the plywood sheet. Then turn the assembled top upside down and attach the legs one by one, using a square to set each one in perpendicular. This order of steps lets you get a tight fit between the plywood and the 2x4 frame, and also makes it easier to square the table top.

You could also use 1x4s, of 3½"-wide strips of plywood, instead of the 2x4s. You would then have no problem with rounded edges. The notches in the 4x4s would, of course, have to be narrower. You should also add two equally spaced supports under the tabletop, between the longer sides.

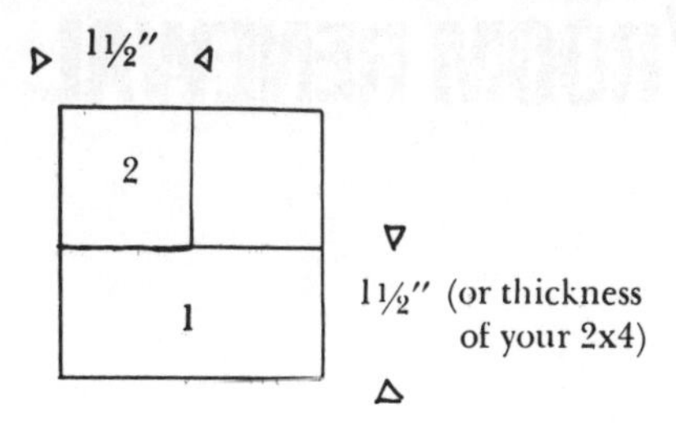

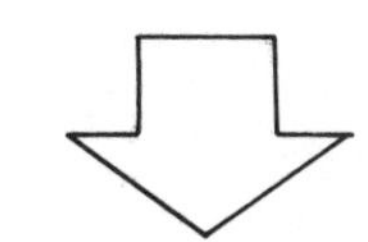

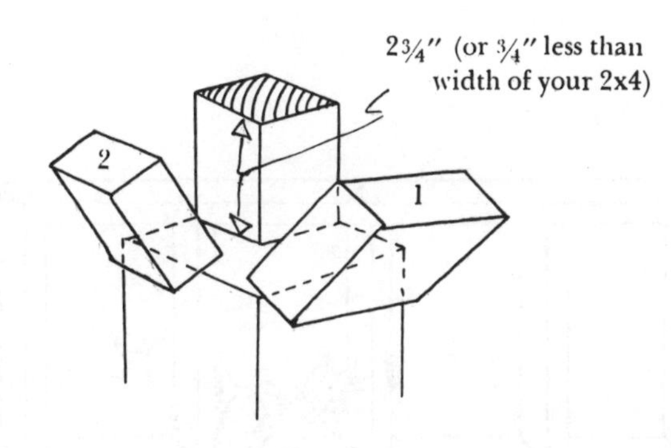

Fig. 3. Notching the legs

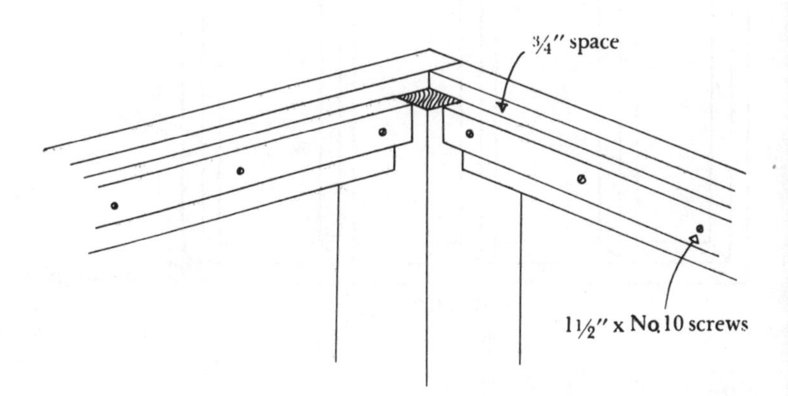

Fig. 4. Cleat Placement

11/ROOM RENEWAL

Learning How to Change Your Physical Space with New Walls, Your Visual Space with New Surfaces, and Your Aural Space with Soundproofing

You say you're tired of your living space? You want to make some changes? How about a real, honest-to-goodness, new wall to give you two rooms where you had just one? Give yourself a private study, or an extra bedroom. The first section in this chapter explains all about how to build a wall in any kind of room. In the following sections you'll also learn how to cover the wall with plasterboard, the cheapest and most common wall sheathing; how to install a door and lock in the new wall; or how to hang a new door in an old jamb.

If you don't have the space for a new wall, maybe you'd like to resurface your old walls with paneling, or tile your ceiling, or lay a fine-looking hardwood floor on your old floor. There are many ways to change your living space around, none of them really very difficult.

One of the most satisfying changes you can make is to soundproof your space. It's not always reasonable (in terms of time and money) to completely soundproof a room, but there are usually a few things you can do to quiet things down. You'll be surprised what a difference a little less noise can make in your well-being.

All of these changes are covered in the next five sections. Though they may seem like big jobs, more in the way of general construction, they are all rather simple when you know just a few things. They are all within your capabilities.

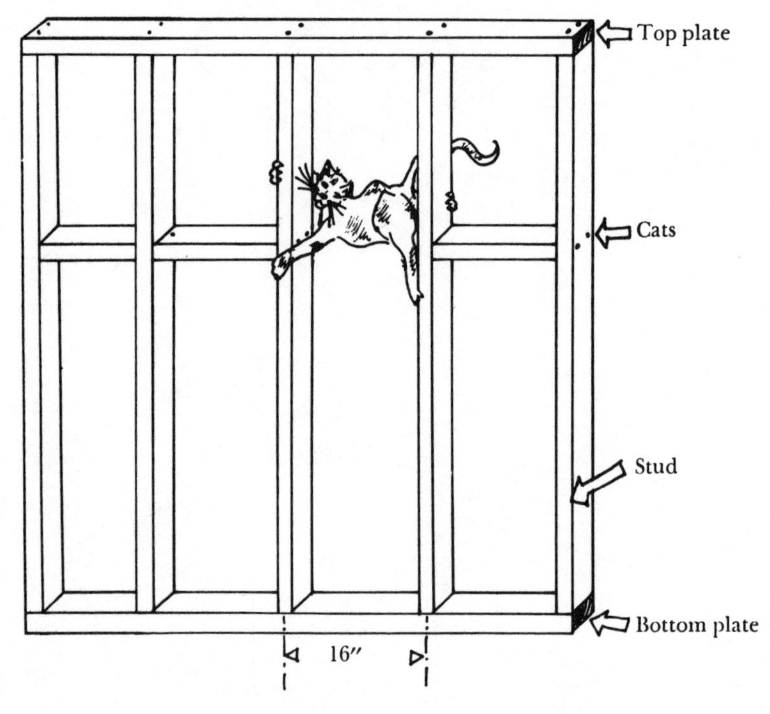

Fig. 1. Basic Wall Parts

STUD WALLS

When we were little, we both got a thrill watching carpenters erecting wood houses. It was all free-swinging hammers, 2x4s being cut in clouds of sawdust. Piles of wood were quickly transformed into beautiful, uncovered, stud walls. Walls we could run and see through.

Building a wall is one of the great pleasures and energy-releasers in carpentry. You can swing your hammer as hard as you want, you can miss the nail and smash the wood and it doesn't matter. Great accuracy with the saw is not necessary. Incredibly elephantine mistakes can be corrected by smashing the wall apart and starting over. You don't have to plane, sand, or stain. It's what kids imagine carpentry is all about when they take the hammer out of their plastic toolboxes and pound it on the nearest table. Most people seem to shy away at the thought of building a wall. But its construction is really a snap. Just hammering some 2x4s together, really.

Figure 1 shows all there is in a simple partition (a partition is just another name for an interior wall). We follow with a very detailed explanation of the construction procedure to make it easy.

The factors that mess some people up with walls are the layout problems that seem to suddenly pop up while fitting the wall in place, or those few details like how to frame out a door. All these problems disappear with proper advance planning. You don't need super skill or experience as a carpenter. You need to know what problems to look out for, so you can plan the construction accordingly. That is what you'll find soon in the section called Variations, Alterations, . . . Traumas. We devote so much space and detail to wall construction for several reasons. Building a wall, or even reading about it, is a good way to learn a lot of structural and construction techniques. The Variations section, particularly, shows ways of thinking about and solving problems that appear all through carpentry.

We want to give you enough information so that, given the space, you might actually build yourself a wall. It's as good a first project as any. Easy, quick, and gives a great sense of accomplishment. Putting up a wall can really affect the space you live in. You can have much more control over the design and use of your home. Pre-existing partitions that are nonstructural can be knocked down and new walls put up—you can even *increase* your space.

A wall can divide a large room into smaller, private rooms—very handy if you expect a baby, or need a study or work space. People with very large open areas, like living lofts, can redefine the space with new walls. You can build a closet with walls, or soundproof an area. And then some people just *like* walls: a low wall to act as a screen, a small wall to define an alcove.

The wall structure also adapts to many uses—railings, enclosures for appliances, radiators, and such things; and inside framing for built-ins.

Lumber

The required lumber for load-bearing walls in house building is 2x4s. In practice, however, we find that 2x3s are structurally sound for interior partitions that will not carry very heavy loads. And they cost 25 percent less than 2x4s. Since many building codes specify 2x4s, however, we'll use them in our discussion.

Construction-grade lumber is fine for walls. Check a few lumberyards to get an idea of how good the construction grade is. The quality seems to vary with locale and the tides.

Basic Wall Structure

A simple wood-stud wall can be built and installed in a few hours. Despite its naked beauty, it will eventually be covered with a skin, some kind of wall sheathing. *Plasterboard,* or Sheetrock, as many people refer to it, is the most common and inexpensive covering; this is what we'll refer to throughout this section. This wall structure will work for virtually all manufactured wall coverings, though, since they are all manufactured to a standard 4′x8′ size.

Each end of a stud is fastened to the top or bottom plate by two 3½″ (16d) nails. Each end of a cat is held by two nails through the stud. Screws and glue are totally unnecessary; the plasterboard skin will unify the whole structure, and the load on the wall is almost all dispersed straight down, helping the nails, not pulling them apart.

Studs

The studs are turned with their thicker dimensions (3½″) perpendicular to the surface of the wall. The thickness is needed in this direction to resist lateral stresses that might make the wall bend over—heavy objects hanging on the wall, or someone leaning against it. There is very little stress along the length of the wall, so the thin dimension is fine there.

Studs are commonly placed 16″ on center, as in Figure 1. (That phrase—16″ on center—means that the *centers* of two adjacent studs are 16″ apart; that leaves a space of about 14½″ between studs.) The first stud is flush at the end of the plate; the middle of the second stud is 16″ away from that *end.* From then on the *centers* of all the studs are 16″ apart, except of course for the last pair if the wall length is not an even multiple of 16″. This pattern is not for structural strength so much as it is for simplifying the nailing of full-size 4′x8′ pieces of plasterboard. You start the plasterboard edge at one end of the wall, and its other edge will line up exactly in the middle of the fourth stud. That leaves enough room to start the next piece, which will then end in the middle of the seventh stud, until the last sheet, which may have to be cut.

Plates

The top and bottom plates hold the studs together and give the structure unity. They also provide a nailing surface for the top and bottom edges of the sheathing material. If there is to be a ceiling or some load on top of the wall, the top plate disperses the stress to all the studs. In such cases, a second 2x4 plate is nailed to the top plate for added strength.

Cats

The cats prevent individual studs from twisting under stress and weakening the whole wall. They are usually used only when the wall is over 8′ high. The cats can be placed at the 8′ height on center to provide a nailing surface for the top edges of the 4x8 covering material (assuming you apply the plasterboard vertically).

This is the basic wall. Changes can be made for doors, windows, intersections with other new walls, obstructions, electrical and plumbing lines, and many other things. You always start with this structure, however.

Choosing and Designing the Type of Wall Construction

There are two ways to build this wall—prefabricated, or built in place. The construction methods are different, but the end result is the same as in Figure 1.

It's easiest and quickest to put up such a simple wall by *prefabricating* it in one large structure, flat on the floor, and then lifting it into place. The nailing is much easier and more accurate, and you don't have to keep climbing up and down a ladder. You need space on the floor, however—you can't build a 12′-high wall flat on a floor 10′ square. You also need room to swing the wall into place, freedom from obstructions. Too many chandeliers or an exposed electrical conduit might make the logistics incalculable. Sometimes these problems can be bypassed in a prefab wall by sneaky construction procedures, as shown in Variations.

When it's impossible to prefabricate a wall, you *build* it *in place* piece by piece. Nail the top and bottom plates in place, and one by one cut and toenail in the studs, leveling each as you go, and positioning them around the obstructions.

To help you decide which method to use, plan how your wall has to look and be built to fit the space. Consult the relevant sections in Variations. Then try to figure a way to build that wall prefab. Always choose prefab when possible.

The shapes of the new rooms will be determined by whether the wall will reach to the ceiling or to either of the existing walls, how thick it will be after adding the sheathing, whether you want it soundproof (see Soundproofing). You also should consider whether it will support heavy loads, if it will block off ventilation or heating in one of the new spaces, if it will block access to anything else, particularly doors that may seem clear but may not be able to open. Adjust its location a few inches to either side if it means avoiding existing obstructions and saving construction time and trouble. Don't center the wall on your chandelier. Plan ahead for any luxuries like doors or windows, or any other adaptations you might foresee with the aid of your crystal ball. Measure the area and make a rough sketch, particularly noting the locations and dimensions of special openings or extra studs or anything else out of the ordinary.

You should also check what's behind your old walls and ceiling. Then you can plan on how to fasten the wall in place, and whether you might need some extra wood or special construction techniques. For instance, see No Joists or . . . Studs among the Variations.

Now you should be able to tell which construction method you will be using. Check your sketch and order the materials.

A few rules will help you figure how many 2x4s you need for your wall. For the wood for top and bottom plates plus cats, figure three times the length of the wall. For the number of studs, divide the number of *inches* of wall length by 16 (the stud spacing), round any fraction upward, and add one. (For instance, for a 6′ wall, divide 72″ by 16, which gives 4½; round upward for 5; add one to get 6, the number of studs you'll need.) For the stud lengths, add at least 3″ to your ceiling height, which will leave you about 6″ waste per piece, enough to eliminate knots and splits at the ends.

Consult the Variations in this chapter that correspond to your situation, make a sketch of what you want, and add in any extra plates, studs, headers, etc., that you may need.

After your final calculations, add another 10 percent to your order, just in case.

Figure you need 8 nails per stud—4 for connecting the plates to the stud and 4 to connect the cats to it. You want 3½″ (16d) nails, which, according to the nail chart in Reference, come about 47 to the pound. To be safe, figure you can cover 5 studs with a pound of nails.

Steps for Prefabricating a Simple Wall

By a simple wall, we mean no openings; floor-to-ceiling; wood joists in the ceiling; meeting the old wall at one end; no obstructions.

1. Lay out lines on floor, ceiling, and existing walls.
2. Find and mark ceiling joists and studs in existing walls.
3. Find height of ceiling—lowest height if distance varies.
4. Choose wood for plates; cut and mark for stud locations.
5. Choose wood for studs; cut to ceiling height minus thickness of plates.
6. On floor, line up studs between plates, straightest studs at crucial spots.
7. Nail everything together.
8. Measure, cut, and nail in cats, if needed.
9. Raise the wall with helper.
10. Position, plumb, and temporarily fasten.
11. Check positioning and plumb, adjust, and fasten completely.

Prefab Steps Minutely Detailed

1. Layout

Use your chalk line to lay out a line for one edge of the floor plate. Draw some arrows inside the line to remind you later on which side of the line to place the plate. This line is *not* the line of the *finished* wall; the plasterboard still has to be added to the studs. Lay out the line accordingly, both in positioning and in length.

With the help of a friend and a plumb bob, locate three well-spaced marks on the ceiling directly above this line (*2*). Run your chalk line across these points, and again draw arrows to show the inside.

With the floor and ceiling plates laid out, you don't have to lay out the end stud lines on the existing walls. Your wall will automatically be vertical when its plates are lined up. Lay out these vertical lines only if you can't get a line on the ceiling (*3*).

2. Finding joists and studs

Find the joists in your ceiling by drilling or sinking a nail *inside* the plate line. (See "Walls and Ceilings.") Keep drilling at intervals of an inch or so till you find the joist. The holes will later be covered by the plate. Mark the locations just to the *outside* of the plate line, so they will show when you position the plate and want to nail it in quickly. Use a piece of chalk, so you can erase the marks easily when done with them.

You may be able to find a stud in your existing wall on line with the path you've laid out. If the partition layout does not meet an old stud, or if it runs between and parallel to the joists, you may be able to relocate your partition till it lines up with one or the other. For other alternatives, see No Joists Variations.

3. Finding ceiling height

The distance from floor to ceiling will probably vary along your layout lines, sometimes as much as 2″ in old buildings with sagging ceilings. You can't prefab the wall to meet the curve of the ceiling, so you find the shortest height and build to that dimension.

Measuring the height in the middle of the room is awkward with a tape or rule. Try the following method. Measure a side wall and cut a 2x4 to a length between two and three inches *less* (so you can still use it as a stud), and, to make life easier, cut it an even measurement like 10′11″ or 10′10″. Climb your ladder and, holding the 2x4 upright, run it along the bottom layout line. Find and measure the *smallest* gap between the ceiling and the top of the 2x4 for the total minimum height (2x4 length plus gap). Any wall built larger than this height will not fit in that spot.

It's a good idea, actually, to make your wall another 1/16″ to 1/8″ shorter, if you don't have much experience—even though it's best that a wall fit tight. A wall just a little too big is grief. A wall a little too short can be shimmed and fastened with no trouble, and no loss in structural strength. Be sure to write down your final height measurement somewhere.

4. Working with the plates

Look at your wood. The plates must not be crooked or twisted. Slight bows are acceptable as long as you place them so the two plates are wider apart at the middle than at the ends. That way, they will straighten out easily when nailed to the studs.

Also check the width, the 3½″ dimension of the wood. Some of the lumber might be 1/8″ or so wider than the other pieces. If you can't afford to leave these extra wide pieces out of the wall, use them as plates. It's much easier to apply the plasterboard when the plate sticks out, rather than one or two of the studs.

Cut the plates to the desired length, trying to eliminate knots or splits at bad ends. Mark them "top" and "bottom." Place them on the floor on edge, against each other in the relative positions they will have in the final wall. Keeping the ends flush all the time, hook your tape on the end of one piece and run it along the plate. If you prefer to begin the

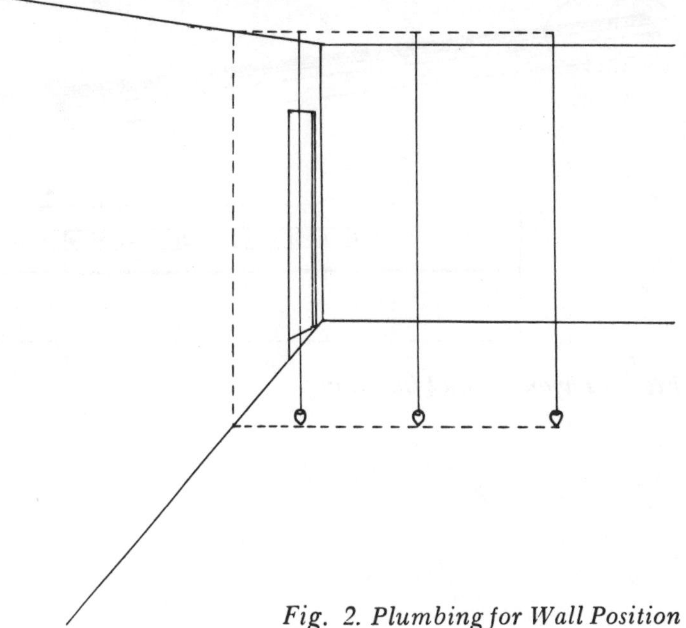

Fig. 2. Plumbing for Wall Position

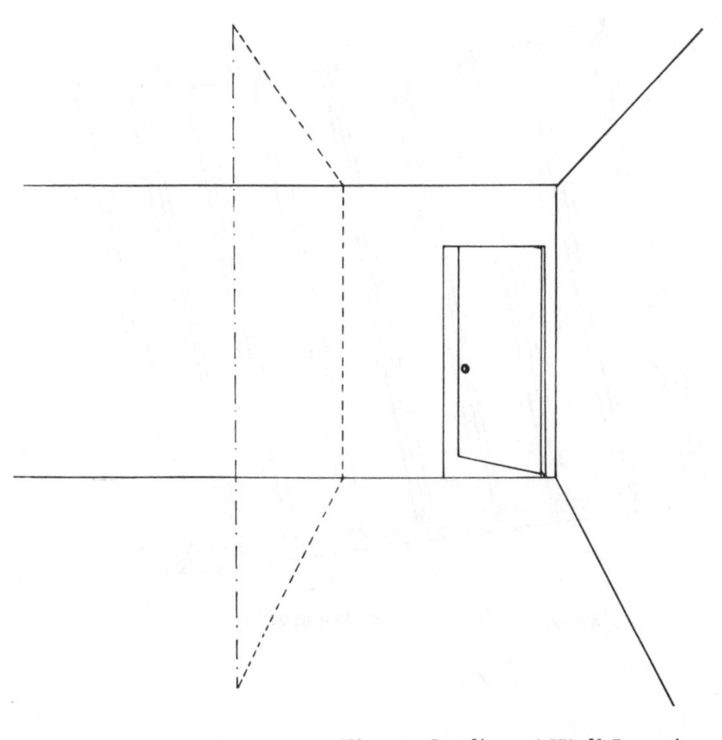

Fig. 3. Outline of Wall Location

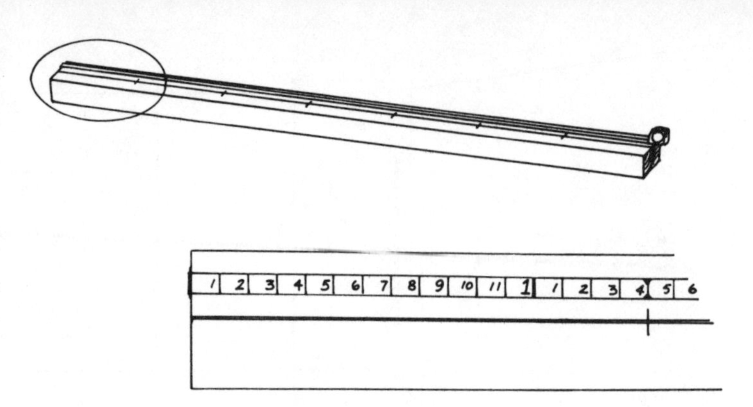

Fig. 4. Layout of Stud Location

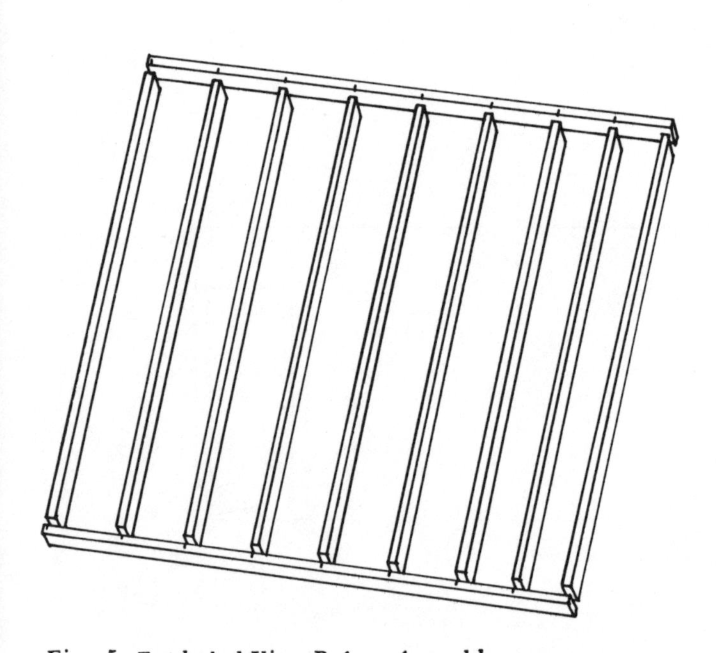

Fig. 5. Exploded View Before Assembly

sheathing at one end of the wall, start your tape at that end. Mark *both* studs at 16″ intervals (*4*). Most tapes have distinctive markings at these intervals, so you don't need a computer mind. The plate length is usually not an even multiple of 16″; if the next-to-last stud mark is at a multiple of 4′ and falls within 6″ of the end, move it to the midway point between the end and the third to last stud. It's not a good idea to put up plasterboard pieces less than 6″ wide if you have a choice.

When you are done, count the number of marks, add one for each end, and the total will be the number of studs you need to cut.

5. Working with the studs

Now choose the wood for your studs. Again, slightly bowed pieces are acceptable; their warps remain inside the wall. Slight twists and crooks are also acceptable—plasterboard and most wall coverings are flexible enough to bend over such irregularities. Structurally the wall will still be strong.

Use the straightest pieces for the end studs (and, when you have them, for the framing studs around openings). Warped studs in these spots will show and will also cause difficulty with further construction—such as fastening to the next wall, or adding a door or window. Mark the straight pieces for later recognition.

Also, if possible, try to have straighter pieces at the 48″ intervals, where the plasterboard seams will be, to make the nailing easy.

Now you are ready to cut the studs. Let's say you found the smallest height of the ceiling to be 10′7″. You lowered that to 10′ 6⅞″ to give yourself leeway. What length do you cut the studs to? Not 10′6⅞″! That's a common—indeed a notorious—carpenter's error. Look at your sketch, or ours. Remember, the studs are fitting *between* two plates. Their total thickness should be about 3″, but you should measure to be sure. The stud height is the adjusted ceiling height minus 3″—or 10′ 3⅞″.

When you cut the studs, try to cut away the bad ends. Don't leave yourself with a knot or a split at the end, which make nailing difficult.

6. Laying out pieces for construction

Spread the plates apart, still on edge, and lay the studs between them, centered on your penciled marks (*5*). The end studs line up flush with the ends of the plates, so the *edge* of the first stud is 16″ from the *center* of the second. All crooked studs should be placed with the crowns, or crooks, in the same direction—so the plasterboard will fit better. Try to place the bottom plate near and parallel to its eventual resting place—the less you move and pivot the assembled wall, the better. Plan ahead for the raising.

Lay the wall out so it *looks* square. Don't bother checking it with your rafter square. The wall will square itself when it goes up in place. (In fact, if both the floor and ceiling slope, you don't *want* your new wall to be absolutely square.) Your eagle eye is also accurate enough for lining up the studs on center with their marks. You might double-check that the outside length from plate to plate matches the wall height you want.

The main hassle in nailing the assembly together is holding the plate flush to the stud as you hammer. One trick that simplifies things is to prenail the plates before setting them on edge (*6*). Prenailing means less hammering while you hold the pieces together, and thus less movement of the pieces. Before lining up the studs, lay the plates flat and drive two 3½″ common nails just barely through the plate at each stud mark. Blunt the points of the nails with your hammer to prevent possible splitting of the wood, particularly at the ends. If the wood starts to split, stop immediately, pull the nail, and drill a hole for it.

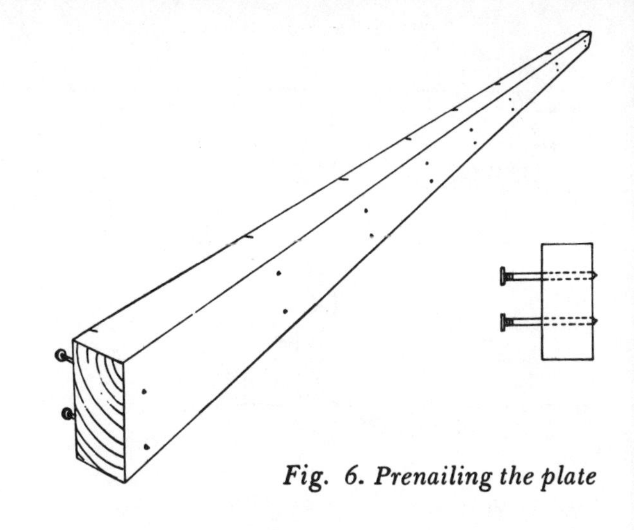

Fig. 6. Prenailing the plate

7. Nailing it all together

Once the studs are lined up and you begin to nail in one plate, you'll find the going easier if you can put some resistance against the opposite plate. Pushing it up against an existing wall (7) is a good idea (if it doesn't mean dragging the wall halfway across the room). In this case, *don't* prenail that second plate till you've finished nailing in the first plate and have moved the assembly away from the wall. Or you could nail some blocks to the floor against the opposite plate. Or have a friend wedge a foot against the opposite plate, directly behind the stud you are nailing into. Otherwise put your own considerable weight, a foot or a knee, on the plate and stud—or just on the stud, holding the plate against the stud with one hand. This is the least efficient but most common method.

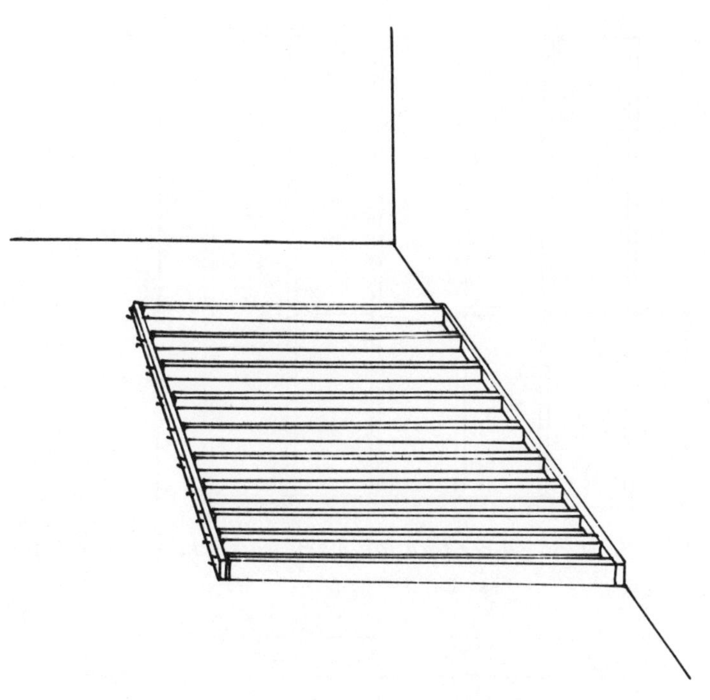

Fig. 7. Assembly

Work on one plate at a time. Start at one end, line the stud up flush with the end and the edge of the plate. If your floor isn't perfectly level, you may need to temporarily shim up either the stud or the plate to get them flush on top (*8A, B*). Anything will do for a shim here—chips of wood, cardboard, a nail. Then place your foot on the stud and plate.

Hammer the top nail in, carefully at first till it bites the stud, then hard. If you have some resistance against the other plate, everything should go swimmingly. If not, you may have to keep readjusting the pieces till the nail catches in the stud. That's why you prenail the plate—less hammering to jar the pieces until the nail reaches the stud. Once the first nail is home, check the stud's positioning again. Hammer the bottom nail in. Do the same for the other end stud.

Nail the remaining studs to this plate. It should be easier to hold the pieces together once the two end studs are in. If some of the studs are twisted pieces, try to even out the twist between the top and bottom plates. Don't line up the stud straight at top and very twisted at the bottom. Some of the studs will inevitably be a bit wider or narrower than the

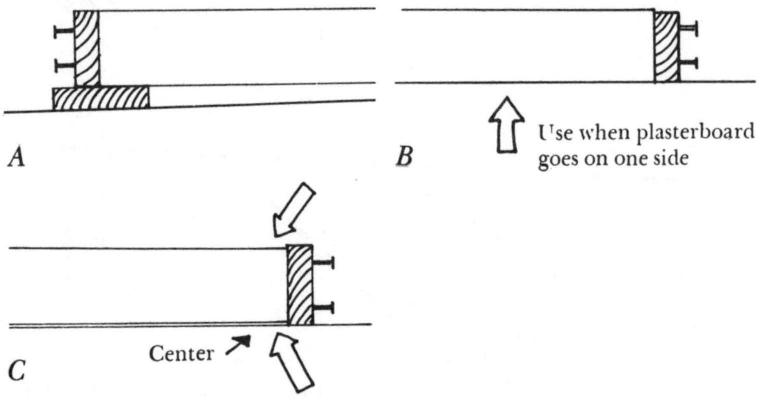

A B Use when plasterboard goes on one side

C Center

Fig. 8. Dealing with size problems

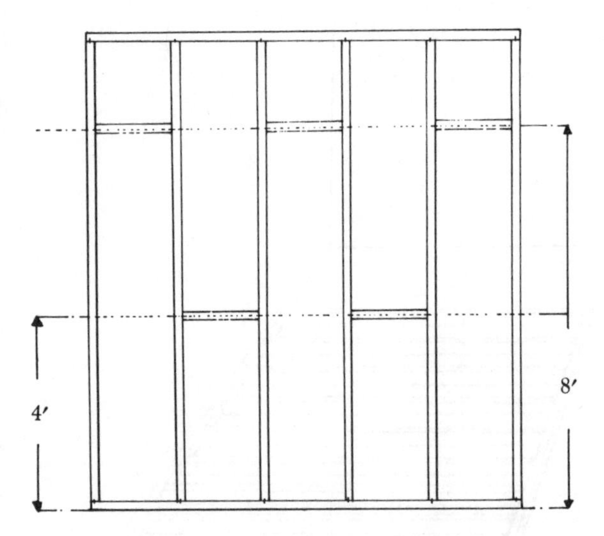

Fig. 9. Standard cat location when wall is larger than 8 feet high

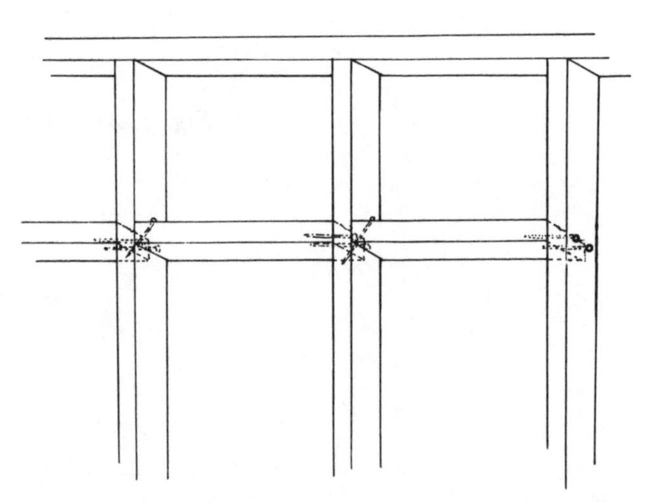

Fig. 10. Nail Positions in Continuous-Cat Construction

plate. Don't panic. Center the difference (*8C*). Most differences will be covered by the flexibility of the wall covering. If you have a choice, put such pieces in the middle somewhere, rather than at the ends or around openings.

When you've hammered all the nails in the first plate, repeat the same steps on the other plate. The hammering should be even easier—the wall is much less likely to move on you the larger it gets. You may find you cut some of the studs too short, that they don't quite meet the second plate. That's all right; nail them in anyway. If the gap is more than ¼″ when you're done, stick some shims in.

8. Working with cats

If your wall is over 8′ high, we advise putting cats in. Many large walls are built without them, and you can certainly get away without them; but they do make the wall stronger. More important, they give an even backing for the horizontal plasterboard seam at the 8′ level—and this means easier and neater taping later. There are two simple ways to place the cats. Easiest and strongest way: the top cat is *centered* at the 8′-high mark so it can receive nails from both bottom and top pieces of plasterboard; the bottom cat is centered at 4′, in the next gap, and so on (*9*). This method is good for vertical or horizontal application of plasterboard. Just measure the distances between the studs at the top plate (each distance may be a little different), cut the cats and nail them in their respective positions, with two 3½″ nails through the studs into each end.

The other method provides more nailing surface for the plasterboard at the 8′ height and thus a neater seam, but requires a lot of the more difficult toenailing. Start at one end and nail the first cat in place through the studs. Every other piece is nailed in normally from one side and toenailed from the other with 3″ nails (*10*). Toenail on the inside of the wall through the cat into the stud; brace the cat with your foot or a block of wood.

You may find a cat that doesn't fit its space, even though you cut it right. There was probably a small error in each of the previous pieces that was building up and pushed the last stud out of position. Don't force the cat in—recut it.

9. Raising the wall

The wall is built. Check that all the nails are in. You are now ready to raise the wall into position. You can do this alone with smaller walls, but in general it's easier and safer to have a helper, particularly if you're new at the game. Both of you stand along the top plate, grab it, and lift it over your heads. Then step slowly toward the other plate, sliding your hands along the studs and pushing the raised end of the wall higher as you go. (Wear gloves if you dislike splinters.) Get it as vertical as you can while still having ceiling clearance to move the wall about. The more vertical it is, the easier it will be to support. Then carry or slide the wall to its final resting place. If you slide the wall into place perfectly vertical, great. If it's a tight fit, you may have to wedge it in at an angle, locating the top plate on its layout line and then pushing or hammering the bottom plate in place. *Do not hammer the studs.* Have the ladder ready. Expect to knock loose a bit of ceiling plaster.

10. Leveling and tacking the wall in place

Once both plates are properly set on their layout lines, the wall should be perfectly vertical along its face. Check anyway with your level on a straight stud. Also check the level at the outer end—the wall may not be up square. Hammer a plate in or out to adjust. Tack one end of the floor plate to the floor, leaving part of the nail above the wood, in case you want to remove it to readjust the wood. Check the alignment at the other end of the wall, and tack it down.

Your friend is of course holding up the wall for you all this time. If you're working alone, have all the necessary tools—hammer, nails, level—nearby or in your apron, so you can grab them while you hold the wall. The few temporary nails will help steady the wall.

Look up. See if the top plate is still lined up. Tap it in if it's off, and tack a couple of nails through the end stud into the old existing wall. Check the plumb of the wall again, and the top plate alignment. If something is off, pull out some nails and readjust the wall.

11. Final fastening

When everything is copesetic, climb up and nail through the ceiling plate into the joists, whose locations you marked earlier. Hammer the nails in all the way. Make sure they are long enough to reach into the joists, of course. Finish nailing through the bottom and the end stud. On the bottom put a nail an inch or two on either side of every stud. Nail or glue some shims into any large gaps above the ceiling plate, gaps thicker than 1/4″ and wider than 16″. You're done.

Remember to let your helper know beforehand what you're going to do. You don't want to lapse into long explanations while he's supporting the whole weight of the wall.

Building in Place or BIP

If the wall you want to build is too large to prefab on the floor; if you don't want to build it in two or more parts, as in Variations; or if there are too many obstructions, such as old fixtures, or utility lines running perpendicular to the planned wall surface, then you have to build the wall in place.

Lay out your plate lines as in the prefab method. If there are no obstructions on the ceiling, mark the stud centers on both plates and nail them accurately to floor and ceiling. For thinner obstacles on the ceiling, notch out the plate (*11*). For thicker ones you will have to put the top plate up in pieces (*12*), in which case you may as well forget about pre-

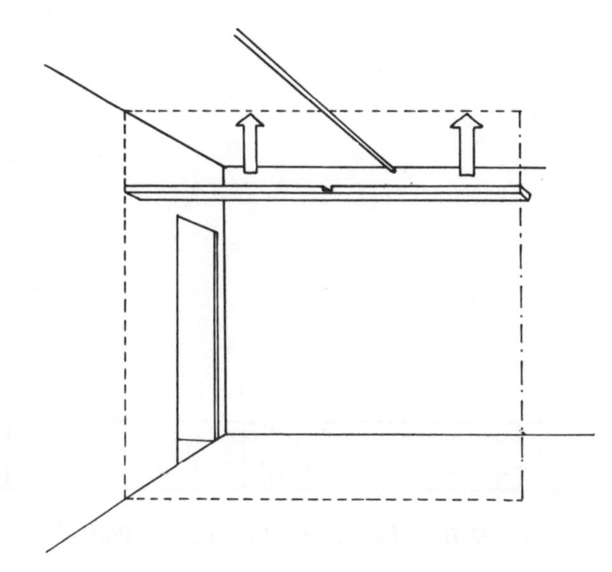

Fig. 11. BIP wall top plate allowing for obstruction

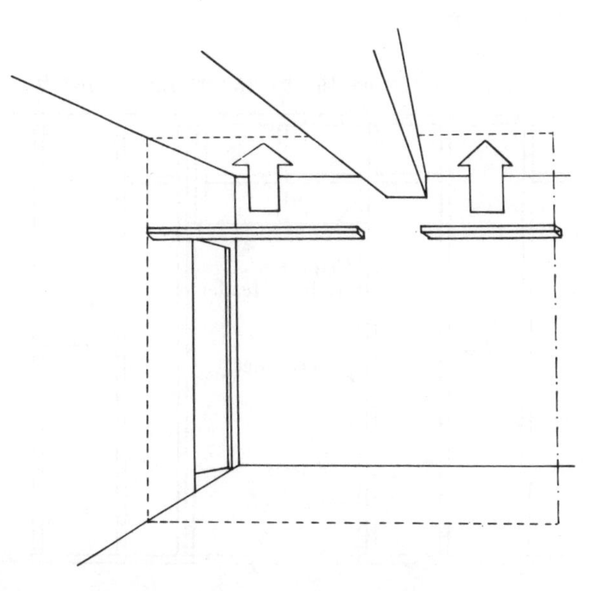

Fig. 12. Dealing with a large obstruction

marking the stud locations—just cut the plate to fit between the obstructions, and fasten each one up well, straight along the layout line. Use nails, or if there are no available joists, countersink bolts (mollies or toggles). *Then* mark the stud locations on the top plates. If obstructions occur on a 16″ interval, make the mark as close to the obstruction as you can. An interval up to 24″, if necessary, will not weaken the wall. Note the distances of any special stud intervals and then mark the stud center on the bottom plate exactly the same. All stud marks should be on the outside edge of the plates.

In BIP wall construction, each stud height has to be individually measured, since the plates follow any slope in the floor and ceiling. Find the height at each stud mark with a vertical 2x4 between the plates, as in the prefab method. Have your helper write down the heights in order as you read them off. Cut and number the pieces. Choose your wood as for the prefab with the same straightness considerations.

Slide one stud between the plates on center to its marks. Toenail the bottom in first with some 3″ nails—two nails on one side, and one on the other in the middle. The second nail is driven on the opposite side of the first; this helps push the stud back on center if the first nail pushes it off. (See Figure 8 in "Working Techniques.") Climb the ladder, holding the stud if it doesn't fit tightly at the top. Bring your hammer, nails, and a couple of small blocks of wood. Toenail the same pattern at top. If the stud is loose, hold it with your arm or hand, or hold the block of wood tightly behind it. You could also tack a nail straight into the plate to brace the stud against. If a stud is really swinging free at the top, stick a shim in the gap before you start to nail.

If the toenailing is still too much for you, give yourself a head start by pretacking just one nail into the top of the stud while it's still down on the ground. You could also drill holes for the nails, either on the ground or after the stud is up.

End studs against an old wall obviously can't be toenailed from both sides. Toenail what you can and put a few nails through the stud into the wall.

When all the studs are in, measure, cut, and install the cats, if you want them. With the wall up, you may find the cat toenailing easier through the *front* of the stud (*13*). Be sure to smash the nail heads in, so they offer a minimum obstruction for plasterboard application.

BIP walls take longer than prefab walls, because you have to walk up and down the ladder for every stud and cat, and because of all that toenailing. Otherwise, it's every bit as simple and should pose no problems if you plan carefully.

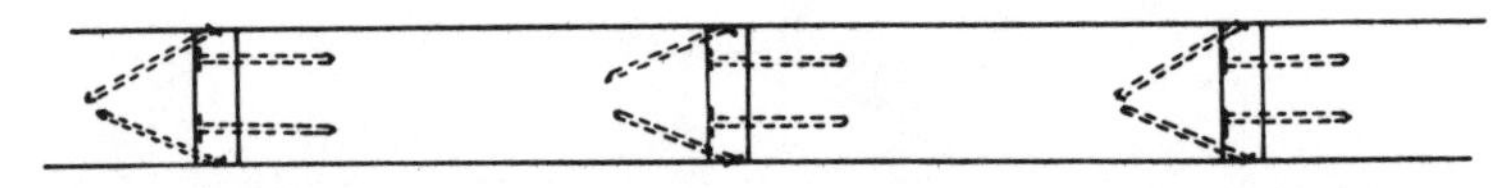

Fig. 13. Nailing Pattern for Cats in BIP Wall

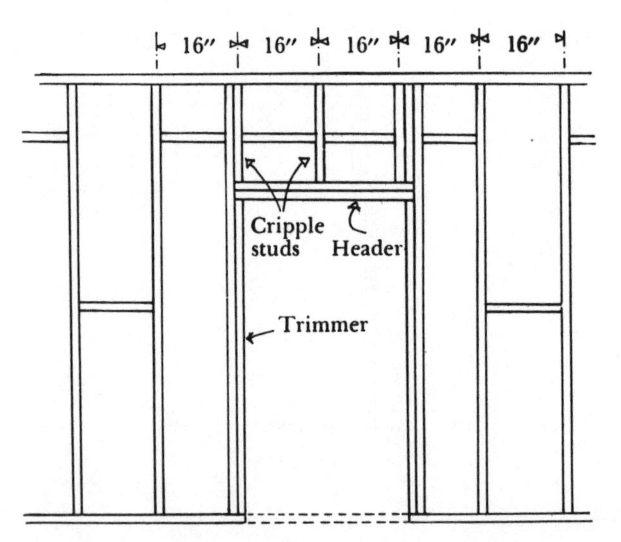

Fig. 14. Structure allowing for door

Variations, Alterations, Typical "Unforeseen" Grief-Causing Traumas Common to Prefab and BIP Walls and How to Cope with All of Them

Take a good look at your room before building the wall. See what kind of space you have for building, what obstructions there are. When you plan the wall, consider the entire system that it is a part of—other walls intersecting, openings needed for entry, the most efficient stud spacing for easy application of plasterboard. Find out what's behind the existing walls and ceiling. Then find the relevant heading(s) in this section, and the corresponding adaptation of the basic wall structures.

If the problem is slightly different from those we describe, adapt one of the solutions we offer. Double studs, double plates, blocked-stud assembly, shimming, separate plates on wall or ceiling, staggered and laminated plates, notching or boxing or building out beyond obstructions, and so on. They are all fundamental carpentry techniques.

Openings in walls

An opening in a new wall is useful. Doors are a nice touch, unless you plan on walling yourself in. An opening in a wall separating the kitchen gourmet from the dining-room gourmands makes the passing of food and dishes much easier. If you partition off a windowless study or work space out of a sunny larger room, a glass window in the partition can let in light from the larger room, let you check up on the kids, and still cut out sound. You can still draw curtains for total privacy.

As far as the wall is concerned, these niceties are just cutouts that weaken its resistance against loads. So when you make a cutout—for any reason—you have to strengthen the pieces around the opening, as in Figures 14 and 15. Add a double stud on either side, a double plate, or *header,* above, and (with windows) a single plate or *sill,* below. The shorter section of the double stud, on which the header or sill rests, is called the *trimmer.* The other shortened studs above the header and below the window sill are called *cripple studs;* they continue the 16″ spacing as if they never noticed the opening. Note how the thick header diverts the stress from above to the double studs and down, rather than through the window or door.

In nonload-bearing partitions (including those with doors), you can use double 2x4 headers for openings up to 6′ wide. For openings 6′ to 10′, the header is made of two 2x6s on edge, nailed and glued with ½″ spacers between them, to make up the wall thickness (*16, 17*). In load-bearing partitions (for instance a ceiling on the partition, or even something heavy like a loft bed mounted on the wall directly over the opening), use 2x6 headers for spaces 3½′ to 6′, 2x8 headers for spaces 6′ to 8′; 2x10s from 8′ to 10′; 2x12s from 10′ to 12′.

To construct any opening, first decide on its total width and height, and its location in the wall. If it's a door, figure the size of the door, plus jamb, clearance, and spacing blocks (see Doors and Locks section). If it's a window, figure whether it's a prehung window, or whether you have to build the jamb yourself. Include the measurements in your rough sketch of the wall. Mark *all* the 16″ stud locations on the

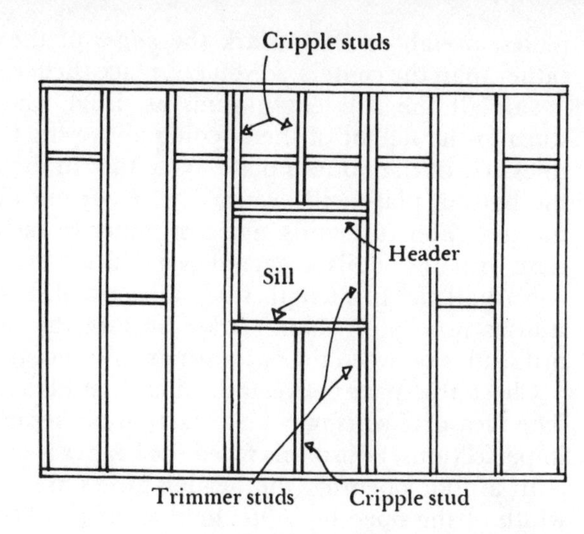

Fig. 15. Allowing for window

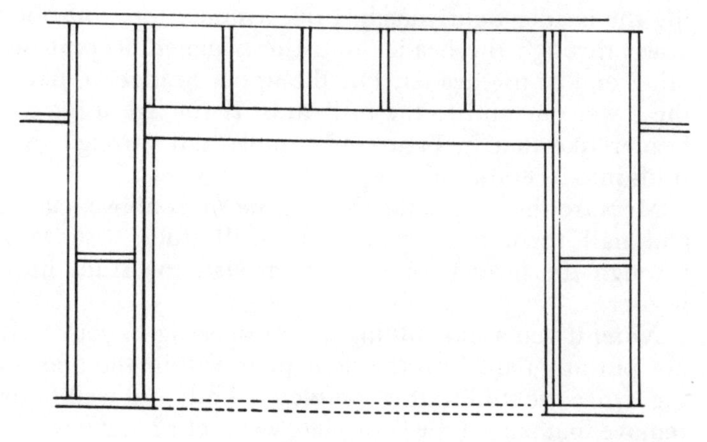

Fig. 16. Structure for large opening

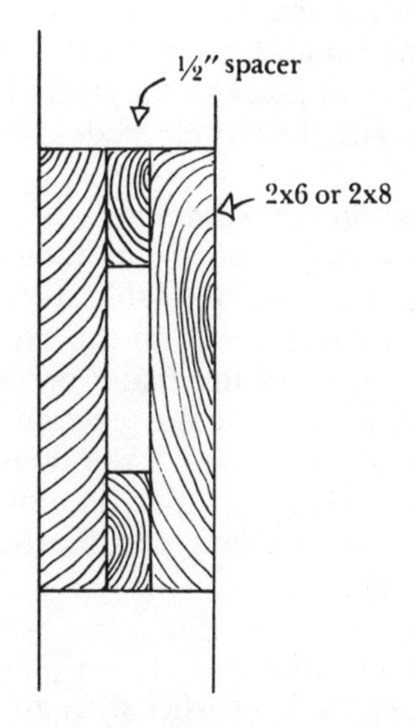

Fig. 17 Cross section of header in large opening

plates, prefab or BIP. Mark the *edges* of the double studs, rather than the centers, so you can place them exactly.

Cut all the full-length studs as usual. Cut the trimmer studs to the height of the opening above the floor minus the thickness of the bottom plate. Note that in the door opening, the bottom plate will eventually be cut out. Taking two of the straighter full studs, nail a trimmer to each, flush at bottom, with 2½″ nails staggered every 16″.

Nail all the full-length studs in as usual. Place the double studs *carefully* on their marks: be sure the *top* edge of the full stud is between the *outer* stud marks, not the inner.

Check the space between the double studs at several points. The measurements may be a fraction off because of warps or imperfections; adjust this when the header is nailed in.

In a door opening, the header pieces are next cut to the width of the opening, *plus* the thickness of the two trimmer studs. Lay the bottom header on top of the trimmers. Put two nails through the full stud into one end of the header, holding the header tightly against the trimmer top. Put one nail down through the header into the trimmer. Repeat on the other end of the header. Put the upper header in, nail it to the lower one and to the full studs. If you are using a wide header like that in Figure 17, simply nail through the full studs into its ends.

Measure the height for the cripple studs above, cut them, and nail them in as you would full studs. Use 4″ nails through the header, or toenail through the studs into the header.

Note: if you are including a door opening in your wall, do not put any nails into the floor plate within the door opening. Once the stud wall is completely built and well fastened, remove that part of the floor plate with your hand saw.

The window opening is framed in much the same way. Install the double studs with just the bottom trimmers attached. Cut and install the sill just as you did the door header. Then cut the middle trimmers to the height of the opening and nail them in place to the stud. Then install the header just as before, and the cripple studs above.

Wall ending in the middle of the room

If one end of your wall is not to be attached to another wall, as in a semi-partition, use a double stud on the end for extra strength against lateral stress on the exposed end (*18*). The inside stud fits between the plates as usual; the outer stud is 3″ longer and fits against the plate ends. If it's a BIP wall, nail the end studs together on the floor as a unit and then install. If it's a prefab wall, put the double stud in on the floor. In either case, remember to shorten the plates to compensate for the end stud.

Two walls joining at a corner

If the end is to later form a corner with another new wall, nail in a stud separated from the end stud by several 2x4 blocks (*19*). For a perpendicular wall at the middle, also use a doubled stud, but use longer pieces of 2x4 for spacers, placed with width between the studs, flush to the side where the new wall will go, since the new wall will be nailed into the spacers, not the studs (*20*). This gives the end stud of the perpendicular wall something to nail to. It also gives a nailing surface on the inside corner for the plasterboard (*21*). Be sure to check the level at the end stud. If it's off, it won't make a flush corner with the next wall.

Electrical or plumbing lines running within the wall

For flexible tubing or wiring like BX or Romex, drill holes through the studs and thread the lines through (*22*). For rigid plumbing pipe and conduits, make notches along the edges of the studs. Try not to notch in more than 2″ on a 2x4. The notches might be easier for you to make before the studs are up.

If someone else is doing your plumbing or electricity, coordinate the order of things with him before building your walls. Ask what the easiest way for him is, where he wants the studs and when. He may want to put the pipe in first and have you build around it. You tell him how that makes extra work for you, can't he work around you? He'll say sure, but it'll cost you. Then you weigh the money versus the hassle.

Obstructions on old walls

Moldings, pipes, and wires are the most common obstructions when you want to push the end stud of your new wall flush to an existing wall. You can notch the end stud to fit over the thinner pipes and such. Or mark lines on molding where the wall will intersect and either chisel that area out (*23*); or take the whole molding off, saw the area out, and put the good pieces of molding back on. If the obstruction is as thick as the stud, you might build out beyond it. Fasten a stud well to the wall in two parts, above and below the obstruction (or above it, for a baseboard). Add blocks if the obstruction is still sticking out. Then build the wall, prefab or BIP, and attach its end stud to these pieces.

Variations, Alterations and Traumas Common to the Prefab Only

More specific problems can arise in the prefab method than in BIP, because you are trying to move a complete, large, inflexible structure within very limited space. In the BIP method you are only moving individual pieces of wood.

Wall that does not go to ceiling

Use a double top plate for added rigidity. Use full lengths if possible; otherwise laminate the plates with staggered joints. This will help prevent flexing that a joint with the ceiling would have resisted. Be careful in deriving your stud measurements—remember to subtract *three* plate thicknesses from the total height. Since you can't make a layout line on

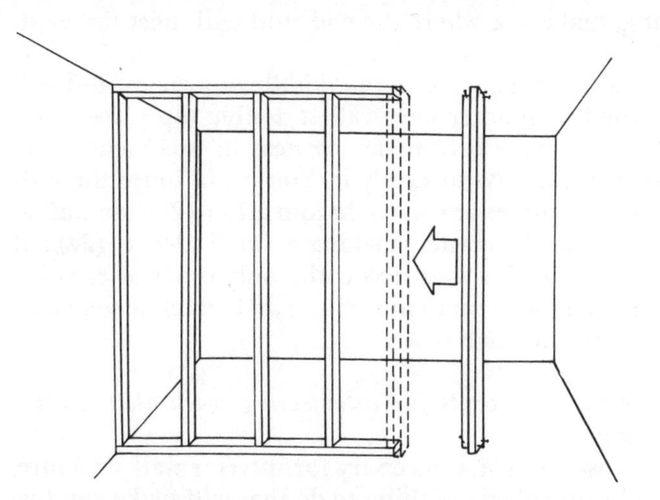
Fig. 18. Structure for wall ending in center of room

Fig. 19. Structure for wall to accept another out from the end

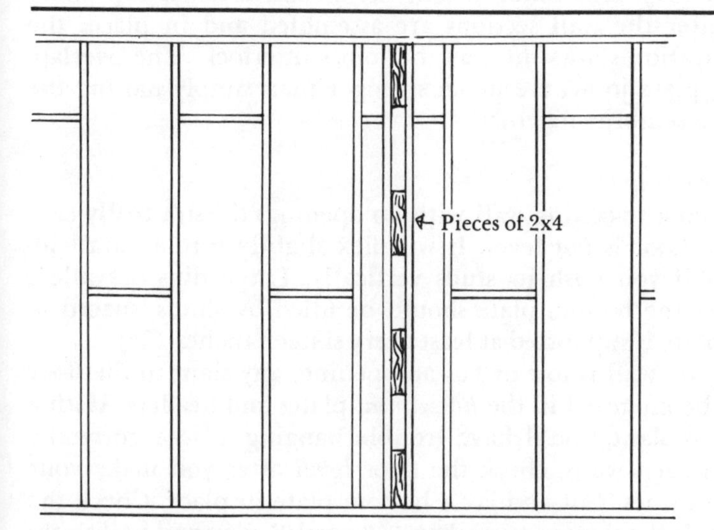

Fig. 20. Assembly for wall coming out from center

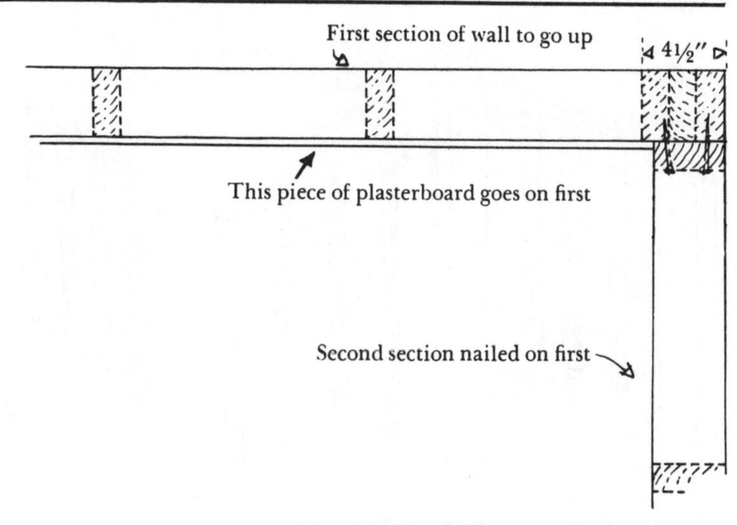

Fig. 21. Top view showing joining of walls at end

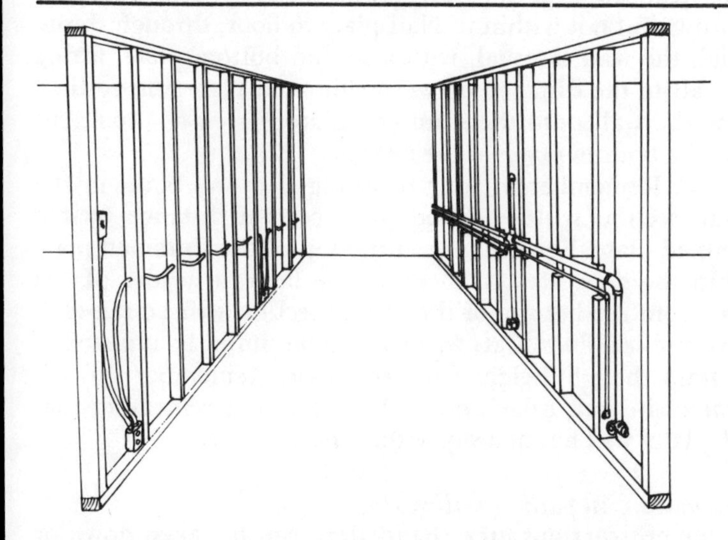
Fig. 22. How to adapt for plumbing and electricity

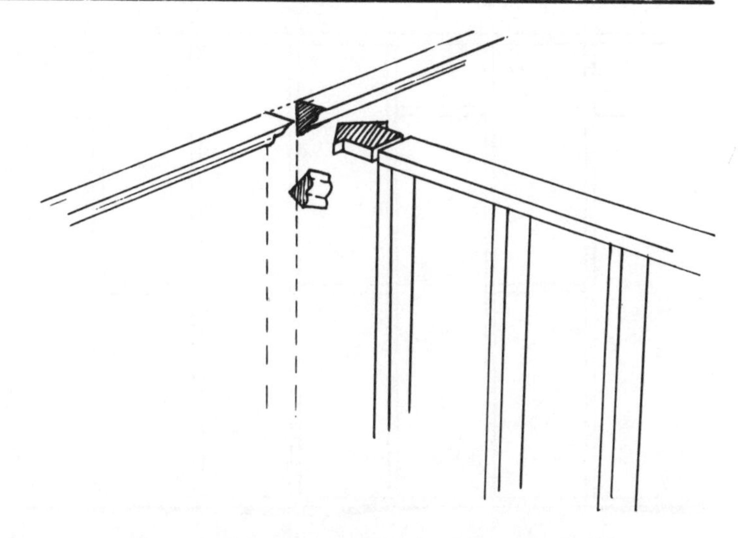
Fig. 23. Remove molding for flush mounting of wall

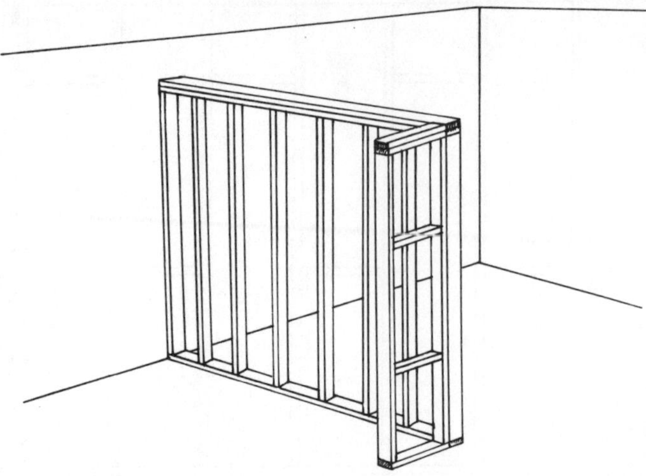

Fig. 24. *Support for short wall ending in middle of a space*

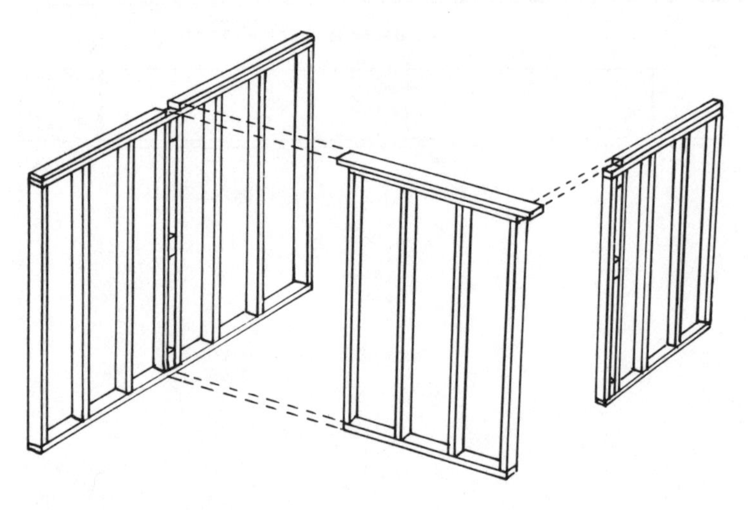

Fig. 25. *Fancy Interlocking Construction*

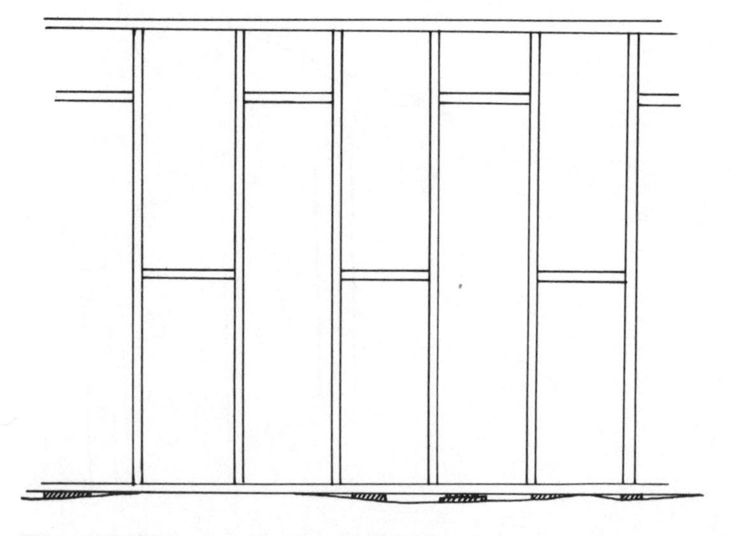

Fig. 26. *Shimming for Uneven Floor*

the ceiling, make one where the end stud will meet the existing wall.

If this wall is large, stops in the middle of a room, and will not be joined to another new wall, a double top plate won't be enough to prevent it from flexing near its unattached end. There are several ways to steady it. You could brace the wall with a diagonal piece; or you could join a 1'- to 2'-wide wall at a right angle to the corner, making a small alcove (*24*). If neither possibility fits your decor, the only other alternative is to run the end stud up to the ceiling and attach it somehow to a joist, or to one small plate.

Fancy interlocking joints for intersecting walls that do not reach ceiling

This is absolutely not necessary for interior wall structure, but it is a beautiful, classy thing to do that will make you feel good (*25*). It's easiest to nail the second layer of top plates on after the wall sections are assembled and in place; the illustration shows the way the joints interlock. The overlapping plate joints are much stronger than simply nailing the stud assemblies together.

Floor not level

A floor-to-ceiling wall with no openings doesn't really care if the floor is not level. It will flex slightly into a parallelogram if you push its studs vertically. Large dips or valleys under the bottom plate should be filled by shims, spaced so the plate is supported at least every sixteen inches (*26*).

If the wall is low or has an opening, any slant to the floor will be mirrored in the *horizontal* plates and headers. With a serious slant, you'll have trouble hanging a door correctly. With such walls, check the floor level after you make your layout lines. Cut and lay a bottom plate in place. Check the level. If the slant is more than ½" per 10', glue and nail shims underneath to level the plate. Shims on either side of a door opening, but not within it. Nail plate to floor, through shims. Prefab the wall as usual, with a second bottom plate, fitting the wall to the distance between shimmed plate and ceiling. Raise the wall onto the shimmed plate, fasten as usual, cut away the bottom plates at the opening (*27*).

Note: Remember in your planning that any point in the prefab section will be raised up a certain distance by the shimmed plate. For instance, if the top of the shimmed plate within the door area is 2" above the floor, a header placed 6'10" from the bottom of the prefab section will be raised 7' above the real floor. Cats will have to be similarly adjusted to maintain their 8' height from the floor. Remember this adjustment in your calculations. And of course remember that *both* plates will be cut away within the door area.

Obstructions in path of wallraising

Some obstructions, like chandeliers, can be taken down or moved temporarily. First, of course, try to plan so you can avoid the obstruction. Maybe you can manage by building the wall flat with the plates perpendicular to their floor layout line, then raising and angling the wall into place.

If logistics or removal are impossible, you may be able to prefab the wall in two parts and fit one in at a time (*28*). This is still easier than BIP. But where the two sections connect, be sure to compensate in the stud spacing so that the 16″ centers are maintained in the completed wall. Also, plan so that a plasterboard edge does *not* fall on the joint.

Ceiling obstructions where top plate should be

Notch the top plate to fit around thin obstacles like exposed conduit, sprinkler pipes, electronic bugs, telephone wires. For thicker obstacles—say those 1″ to 2½″ below the ceiling, build out beyond them by piecing a plate to the ceiling between the obstacles; prefab the wall to meet that ceiling plate, notching the top of the wall plate if necessary.

For a large obstruction, such as a girder running across the wall, or a low-hanging pipe, make a boxlike cutout in the wall (*29*). Prefab the wall on the floor as usual, with the usual 16″ spacing, only be sure not to locate any studs where the obstruction will be. Instead, install a stud on either side of where the obstruction will be. Nail in a double crosspiece between the two new studs, at a height that will be at least ¼″ beneath the obstacle. *After* the crosspiece is secure, cut out the top plate between those two studs. If you had to omit a regular 16″ stud, put it in now, nailing it to the new crosspiece and the bottom plate. Raise the wall carefully, since it will bend a little at the cutout until you nail the wall in place. If there is only one very low-hanging obstacle, you can also prefab the wall in two sections on either side of it. Again, keep the 16″ stud interval continuous across the sections, so the plasterboard will fit in full pieces. Add a crosspiece under the obstacle for plasterboard nailing.

If you have to make more than two cutouts, plus any other adaptation unique to a prefab wall, such as leveling a floor plate, BIP might be easier work.

Wall to wall partitions

If your room is, say, 12′ wide and you want to run a single partition from wall to wall, you will find it almost impossible to squeeze the partition into place, particularly if there's floor molding. The existing walls just won't part like the Red Sea while you raise the partition. One solution is to build the wall in two sections. The other solution is to fasten separate vertical plates to the existing walls first, and build the wall to the distance between them, in this case about 11′9″ long. Then the partition can be maneuvered in the space and slid into place.

No joists or existing wall studs to nail to

When possible, a wall *should* be attached on *all* sides to something solid—a few nails into the floor, ceiling joists, and

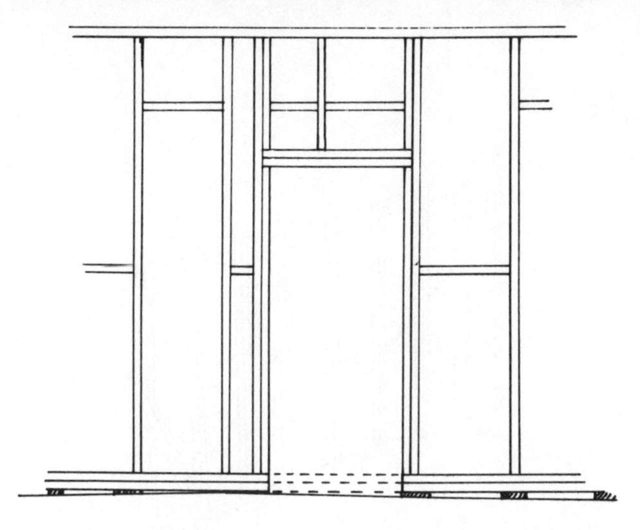

Fig. 27. Shim supports for door on uneven floor

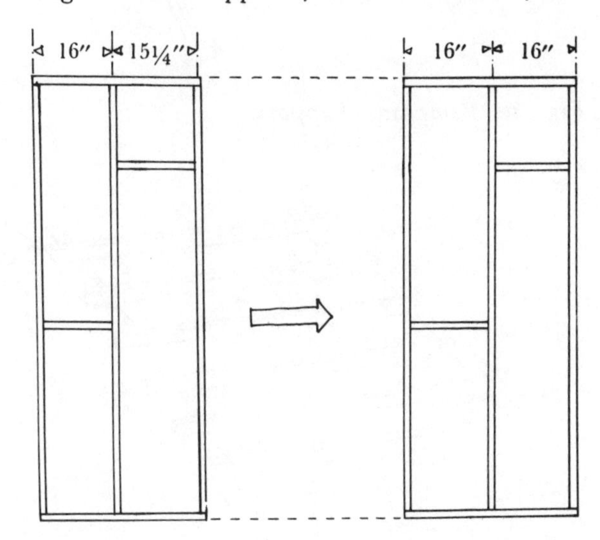

Fig. 28. Two-part wall where one prefab won't fit

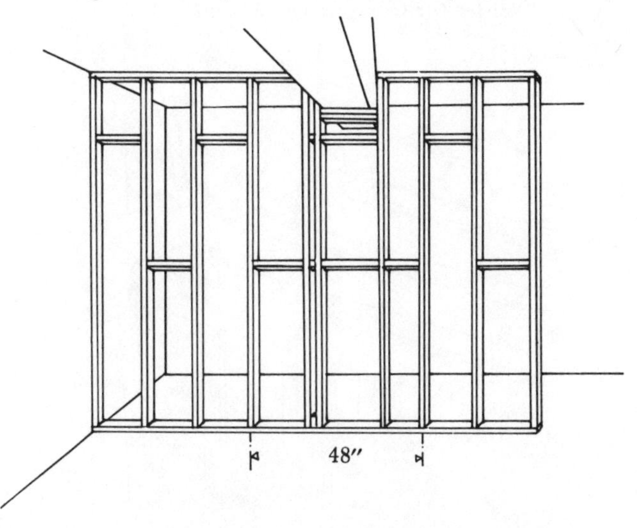

Fig. 29. Structure for Large Obstruction

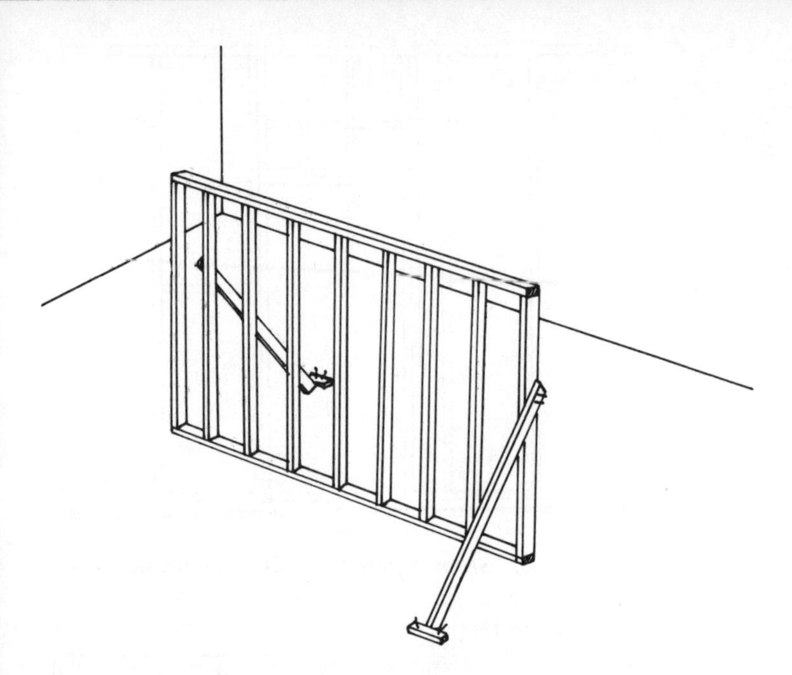

Fig. 30. Temporary Supports

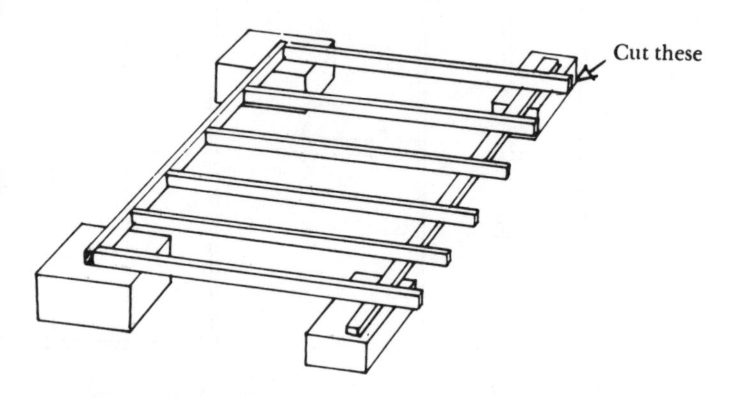

Fig. 31. Set-up for Correcting a Mistake

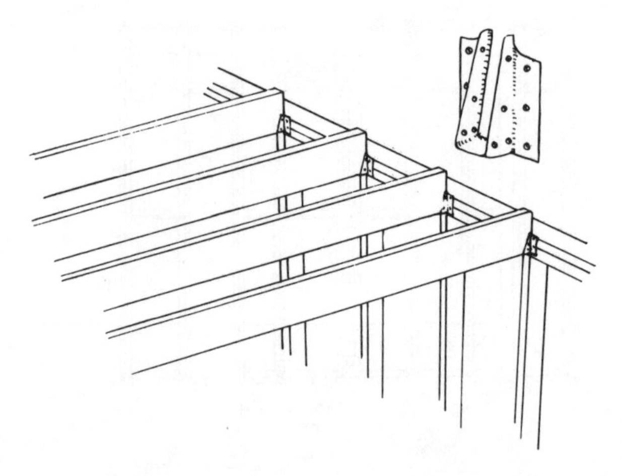

Fig. 32. Joist Hanger

old wall studs are more than sufficient. Nailing into just floor and ceiling when you can't get at the studs is acceptable. At worst, the ends of a taller wall might bend a little under stress; there is no structural danger. Nailing into just the floor and the old wall studs is all right for a partition less than 15′ long, but use a double top plate for some rigidity against lateral stress.

When both the wall and ceiling are un-nailable-to, attach *separate* wall and ceiling plates by means of toggle bolts or lead anchors, depending on the type of wall (see "Walls and Ceilings"). Then build and attach the prefab partition to these plates with nails. The partition itself is too inflexible to be attached directly with such tricky fastening devices. So always be sure to check what's behind the surface *before* prefabricating the wall. Also be sure to countersink the bolt heads so the top of the wall can fit flush against the plates.

Wall does not reach ceiling, temporarily unsupported at ends or middle

This situation occurs when you expect to add walls to either side, say to form a room. The first wall will be unsteady until you build and raise the second wall. Brace it temporarily with diagonal 2x4s nailed to studs near each end of the wall and wedged into blocks tacked to the floor (*30*). Have the braces ready before you raise the wall. Fasten them so they will not be in the way when you fasten the second wall.

Mistakes: wall built too short, wall built too high

You screwed up. You measured the height wrong. If you're only a few inches short of the ceiling, you can fix it with shims, or by putting in an extra plate or two. Glue and nail everything. If it's really short, like six inches or more, you've got problems. You could build a 6″-high wall and fasten it into the gap, but it would be much weaker than a full wall. For the extra time, you might just as well rebuild the entire wall—assuming you have more wood and can use the too-short studs for some other project.

Measure carefully in the first place.

A more common problem is building the wall too high, usually by the 1½″ size of an extra plate you forgot to adjust for. The repair is not as major as it might seem. Just frustrating. Lay the wall back on the floor and remove either plate. Do this by standing on one or two studs at a time and hammering the plate away from them. Move along to each stud, taking the plate out evenly and keeping the nails straight if possible. If the studs loosen from the other plate, hammer them back in. Hammer the nail points back into the removed plate. Now prop the wall up well off the floor, as in Figure 31, and cut the studs to their proper length. We suggest a handsaw, since you would have to use a power saw vertically—which is not the safest technique, although a skilled carpenter can and does do it.

When the studs are cut, nail the plate back on, using the old nails, and raise the wall.

Long walls

When you start to build a wall over 12′ long, it gets incredibly heavy and awkward to raise. Don't strain yourself. Build it in sections. Plan so the 16″ on-center intervals are maintained all the way, as illustrated in Figure 28.

Building a Ceiling on Your New Walls

Outside of actual house-building, this situation occurs only in serious soundproofing projects, or in renovation when the old walls and ceiling are just too messed up to bother with. Essentially you are building a room within a room. The only trick involved is in planning beforehand exactly how the ceiling and ceiling joists will be attached to the walls. Then you will know how high to build the walls to achieve a desired ceiling height.

Joists are run across the shorter dimension of the room, every sixteen inches. (You can fudge the spacing around obstructions.) For the proper size, consult the Span charts in the Reference section. The end joists should be as close as you can get them to the side walls, to give nailing surface for the ends of the plasterboard. Always use a double top plate, to transmit the ceiling load more efficiently to the studs.

One common method of attaching the joists is to hang them from joist hangers *(32)*. The bottom of a 2x6 joist would be about 3″ below the top of the plate. The other common method is to stand the joists on edge and on top of the plate and join them with framing anchors *(33)*. Here the bottom of the joist rests on top of the plate. Thus the ceiling for a given wall height is higher using framing anchors. Consider this when you plan the wall. The ceiling height will be at the bottom of the joists, not necessarily at the top of the wall plate.

Of the two methods, the joist hangers require more accurate cutting and measuring of the joists and longer wall studs. The framing anchors require longer joists, to span both plates, and they may be difficult to attach if you are very close to the existing ceiling.

You definitely need an assistant to lift and fasten the joists. Slightly bowed joists are installed with the bow pointing up. Crooked pieces can be used, but not where a plasterboard seam will fall. Check the level of each joist before fastening. Chances are the tops of the walls won't be perfectly level. Adjust by simply raising or lowering the joist hangers, or shimming under the framing anchors.

Watch your head.

Building a Curved Wall

The construction is actually simple, as shown in Figures 34 and 35. The amount you can curve the wall depends on the sheathing you will use to cover the studs.

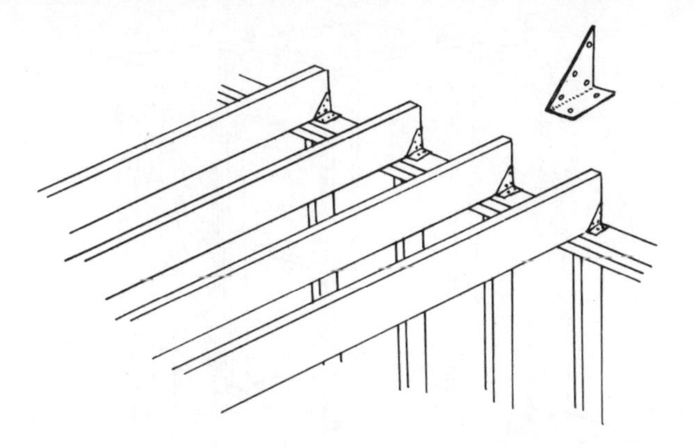

Fig. 33. Framing Anchor

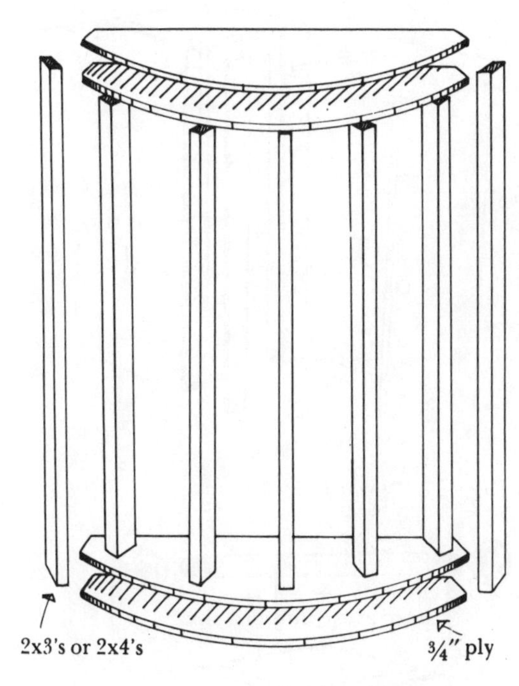

Fig. 34. Exploded View

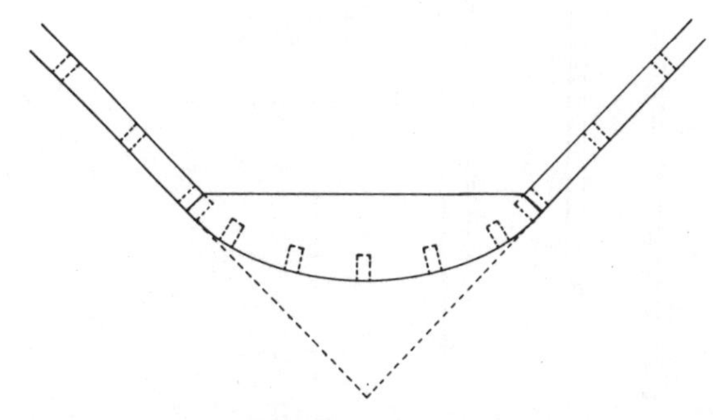

Fig. 35. Structure for Curved Wall

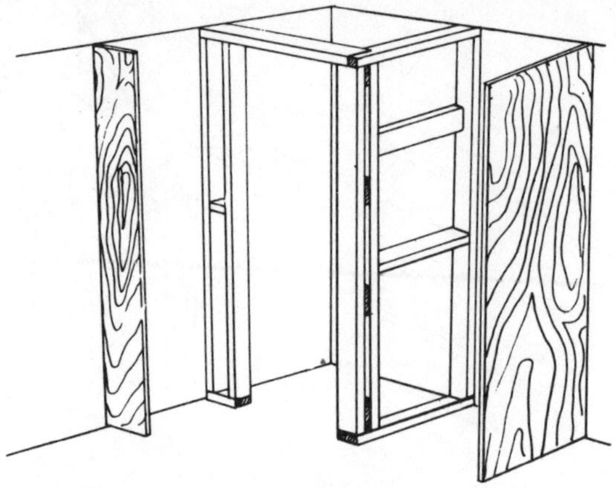

Plasterboard or hardboard can be curved and fastened over this surface; ⅜″ plasterboard can be bent lengthwise to a radius of 7½′, or 25′ widthwise. Obviously, you'll be better off applying the sheet horizontally in this situation. If you soak both surfaces of the sheet with a mop and a lot of water, lay it flat, and let the water soak in for an hour, you'll be able to bend it even more.

The curved plates at the top and bottom of the wall are just plywood cut with a saber saw. Plywood can also be bent to some extent by soaking it with water, letting it sit an hour and then bending. Scoring the back surface with shallow saw cuts running *across* the bend also helps.

Building a Closet with Walls

You can build a sturdy closet to any size you want by framing it with stud walls, leaving an opening for a door, and then covering the walls with plywood or even plasterboard. The minimum depth for a clothes closet is 24″, with the pole for clothes hangers located midway back at 12″ (*36* to *38*).

Fig. 36. Panels over 2x4 Structure

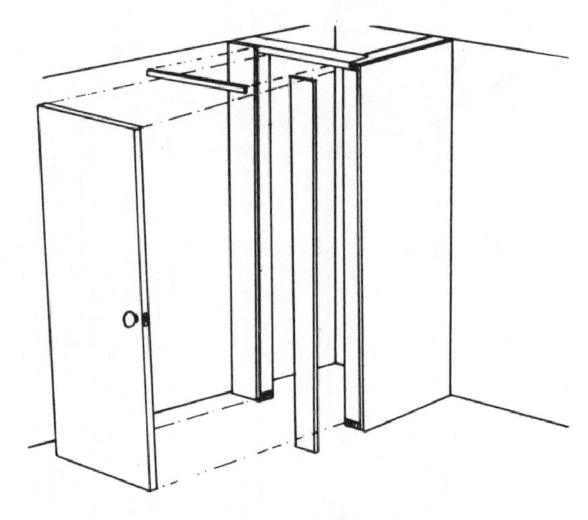

Fig. 37. Closet Parts

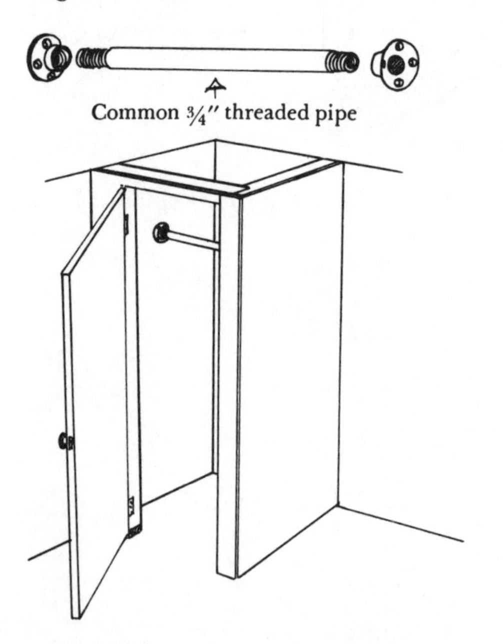

Fig. 38. Closet and Pole Assembly

PLASTERBOARD

Most new walls or partitions are covered with plasterboard. It's cheap, quick to put up, and easy to paint. Plasterboard, Sheetrock, wallboard, dry wall, or gypsum board are all the same thing—a core of plaster sandwiched between two layers of thin cardboard. See "Wood and Other Building Materials" for the different types and sizes available.

Taping is the process of sealing the seams between the sheets with *joint compound* and a special paper tape. If you do it neatly, all the pieces will look like one flat continuous surface.

Preparations

Measure the wall area to be covered, to figure how many 4′x8′ sheets you need. Decide what thickness you want—½″ is common for partitions and ceilings, ⅝″ for fire walls and certain sound-control walls. Check the plasterboard when it's delivered. Don't accept sheets with damaged edges or broken corners. The important front faces will probably be all right, since sheets are factory packed in pairs, face to face. Small dents in the backs are not important.

Don't *you* damage any edges when you carry the pieces in. You can separate the pairs by pulling off the paper tabs at the ends (*1*). Carry them carefully, watching out for the corners. It's best for two people to carry them, particularly if you're going around corners, upstairs. Store the pieces flat, out of the way of ongoing construction. Plan ahead so you don't have to move them again until you use them.

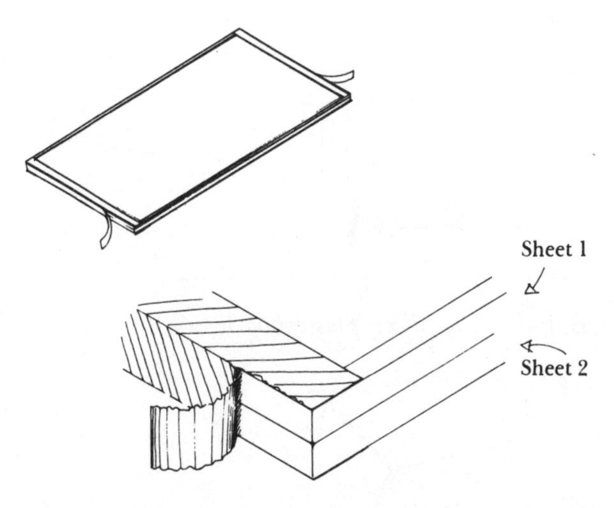

Fig. 1. Removing the end papers on plasterboard

For putting them up, you'll need a mat knife with extra blades, a tape measure, carpenter's square, a long straightedge, an inexpensive keyhole saw, goggles, and a nail apron. A chalk line, level, and small Surform rasp are helpful, but not necessary. For fastening to wood studs you need a hammer and plasterboard nails or a screw gun and special drywall screws. The nails come in two types—the annular ring (which we think holds best), and the galvanized. Use 1¼″ nails for single-layer plasterboard. The screw gun is faster, but not noticeably so, unless you're doing a 100′ length of wall. For metal studs, however, you must use the screw gun.

Measuring and Fitting

Plasterboard can be applied on walls with the long dimension vertical or horizontal. Just be sure that all vertical seams end up in the middle of a stud—except, of course, at the end studs. Ceiling pieces should be applied with the long dimension across the joists, end seams on the joists. Attach ceiling pieces before the wall pieces. The 16″ spacing of joists and studs means more full 4′x8′ sheets, and less cutting and taping. That's why good planning of your stud wall is important, with special care and foresight needed at the corners. When you must cut a full sheet to fit, take careful measurements of the space. If the fit will be tight—for instance, between

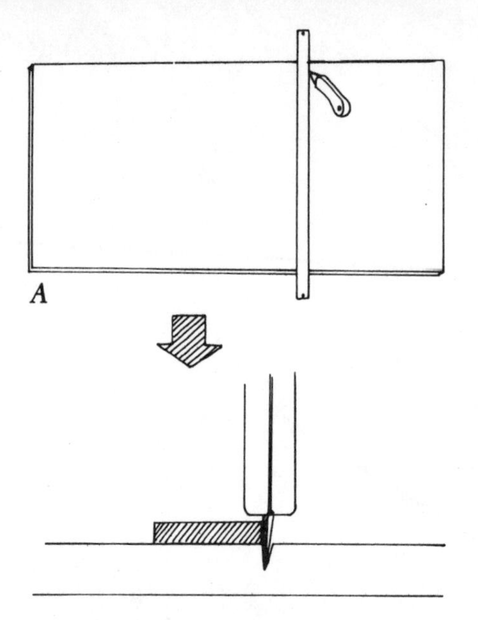

B. Pull knife along plasterboard

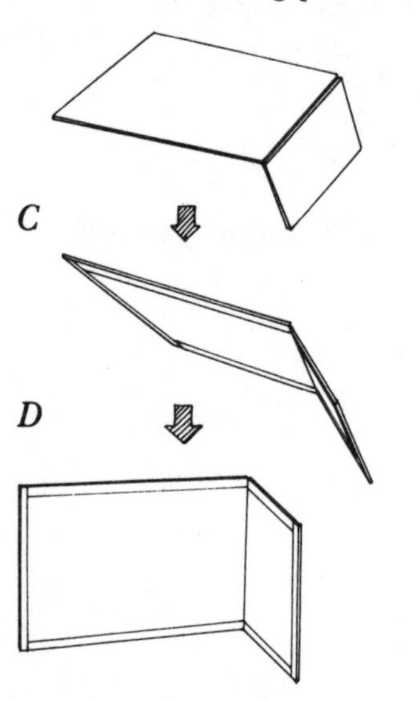

E. Bend board on score and stand up

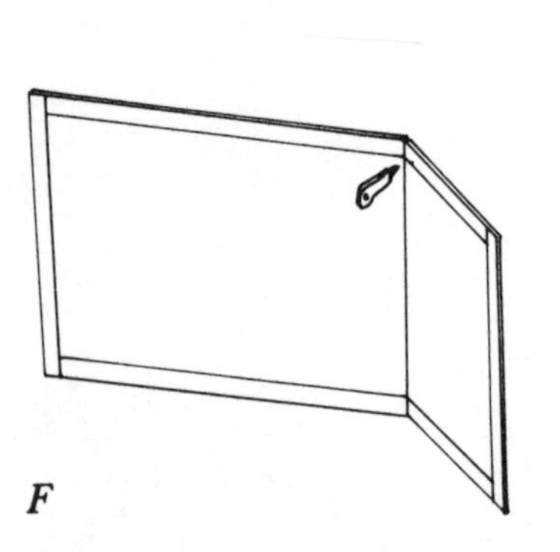

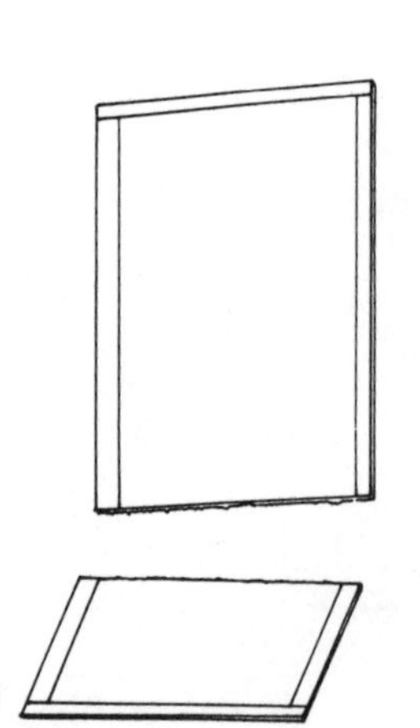

Fig. 2. Cut through from back

two other pieces—subtract $\frac{1}{16}''$ to $\frac{1}{8}''$ from your figures. Gaps can be filled easily during taping; it's unnecessary trouble to have to retrim a piece that's a shade too big.

Be safe when cutting—keep your other hand *behind* the knife. One efficient cutting technique is to leave the plasterboard on the pile or lay it on the floor, face up. Be sure no nails are under the piece. Lay your guide in place and hold it steady with one hand and by kneeling on it (*2A*). With the mat knife, score the line from edge to edge. The cut only has to go through the paper and slightly into the plaster (*2B*). Always make this cut on the front face. Stand the piece on edge. A sharp blow on the back of the cut with your knee or arm will neatly snap the plaster core; cut through the back of the back paper to completely separate the pieces (*2C* to *G*). Smooth rough edges with a small Surform.

When you have to make two cuts in one piece, such as cutting a 3′x6′ piece out of a 4′x8′ sheet, the first cut, no matter which direction, must still go from edge to edge. The plasterboard would not snap correctly if you made your first cut only three feet across. Completely detach the first section before making your second cut.

Cutting out L-shaped pieces is a special problem. You *can't* continue *either* cut from edge to edge, because you are cutting out an inside corner. You must cut all the way through the thickness of the board along the shorter cut, then make the second cut as usual. A keyhole saw will cut through the first part a lot easier than the mat knife; score the paper with the knife for a neat finished edge on the face, then saw.

Cut out holes for obstructions by scoring the outlines with the knife, then cutting through the surface with the keyhole saw. Measure carefully to locate the hole. Often the hole has to be enlarged further to be maneuvered into place. You can fill gaps later with compound or plaster.

Cutting a strip less than 2″ wide from a piece is difficult because it's hard to snap such a thin section evenly. The normal method is to score the face deeply, then to press your thumbs along the back of the cut and pull the strip back with your hands. On long cuts, only part of the strip will snap; repeat the process farther along the cut. You can save your thumbs by using a cheap pair of canvas pliers.

Application

Mark the location of the studs on the floor or ceiling so you'll know where they are when the plasterboard covers them. Otherwise you won't know where to place the nails. (Similarly, for ceilings, mark joist locations on the wall.) The first piece of plasterboard should go at the end where the 16″ spacing begins. Then you continue with full-width pieces till the other end. The first piece should be placed level, even if the floor is not level. Each succeeding piece is butted flush to the last one, so you don't have to worry anymore about the levels. If a level sheet makes a gap between it and part of the floor, you can trim the bottom to fit, or just leave it. Small gaps can be filled with compound or covered with molding. Plasterboard does not have to be perfectly cut or fitted. You can always fix things up later with plaster or joint compound.

To begin, put some nails in your apron, slip your hammer into your belt, and lift the first sheet into place. Try not to damage the corners when you lift the piece—if you have trouble, have someone help you lift it. Most people should be able to lift a full sheet of ½″ plasterboard.

Line the piece up and tack nail it near the two opposite corners—that is, nail only part way into the studs, in case you want to remove these nails to readjust the piece. This is all it takes to hold the piece while you put all the other nails in. Nails should be spaced 6″ to 7″ apart on ceilings, and 7″ to 8″ on center on walls. If a nail misses the wood behind it, pull the nail out. Indent each nail by dimpling the plasterboard surface with the hammer—hard enough so the nail goes below the original surface level, but not so hard that the paper surface breaks. This dimple is later filled with compound to cover the nailhead.

If you're using a screw gun, you can space the screws up to 12″ apart. The settings on the gun can be adjusted to automatically sink the screw heads and dimple the surface around them.

Try to use the factory edges around door and window openings, for cleaner lines.

Change mat knife blades often—a dull blade is slower and more dangerous.

Support pieces on the ceiling while nailing by wedging a T brace from floor to ceiling. A T brace could be just a 1′ 2x4 nailed to the end of a long 2x4. Have at least one person help.

Taping and Filling Dimples—First Coat

The goal is to produce a totally smooth surface, with no gaps or seams. Tape all seams between pieces, at ceiling, and at corners.

Use your 6″ putty (or taping) knife to transfer compound from a five-gallon can to the float and to apply the compound to the wall. Use your smaller knife to fit in a one-gallon can. The float is your carrier. Close the can immediately after you use it, so the stuff inside doesn't dry out.

On a wall, begin with the vertical seams. Coat them generously from top to bottom with compound in a strip slightly wider than the tape (*3*). If this is a tapered seam, fill almost to the face. If it's a nontapered seam, make it a thinner coating. In either case, err on the side of excess. Cut a length of tape (*4*) slightly longer than the seam and firmly imbed it into the compound with your fingers. Now pull your taping knife down the length of the seam at about a 30° angle to the wall (*5*). Press firmly; the idea is to squeeze excess compound out through the perforations and along the edges of the tape, at the same time flattening and smoothing out the tape.

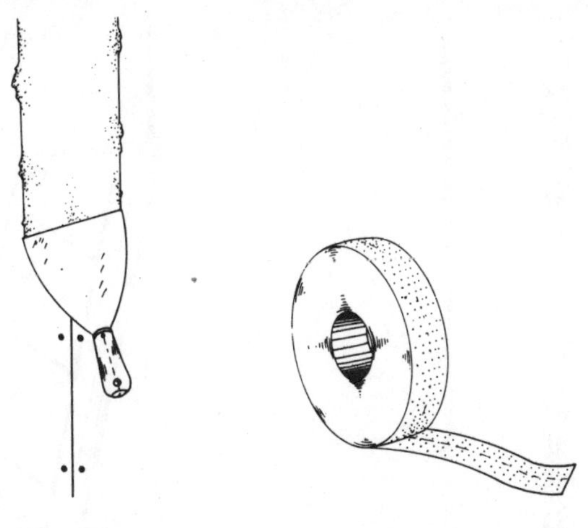

Fig. 3. First Layer of Compound

Fig. 4. Wallboard Tape

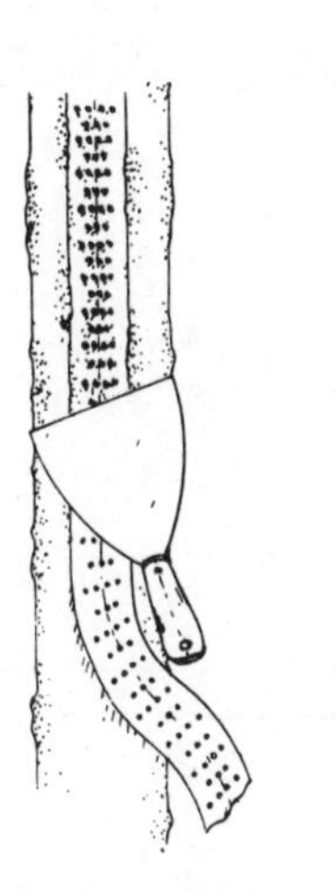

Fig. 5. Press tape into compound

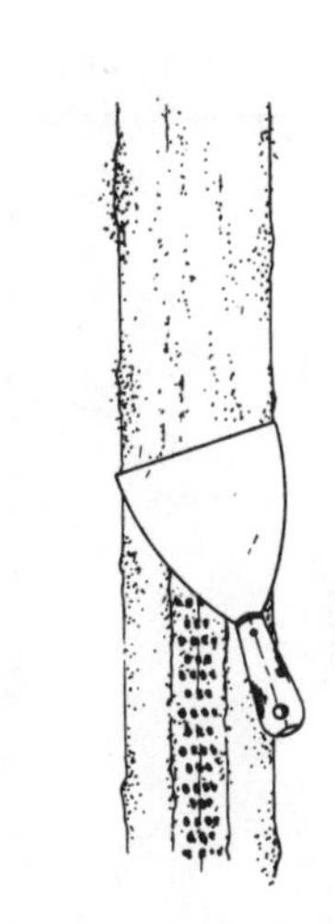

Fig. 6. Skim excess over tape

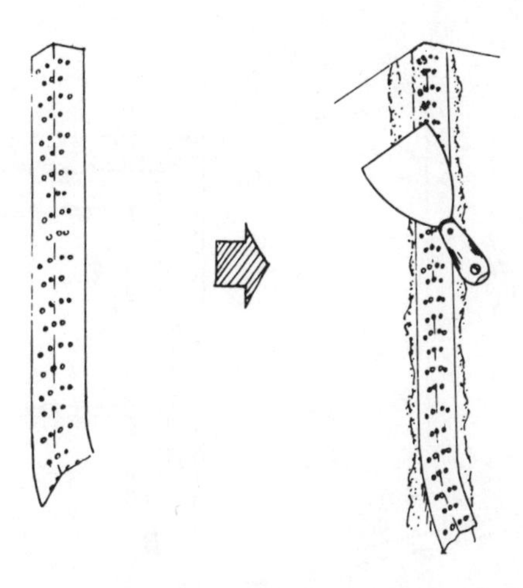

Fig. 7. Applying tape to inside corner

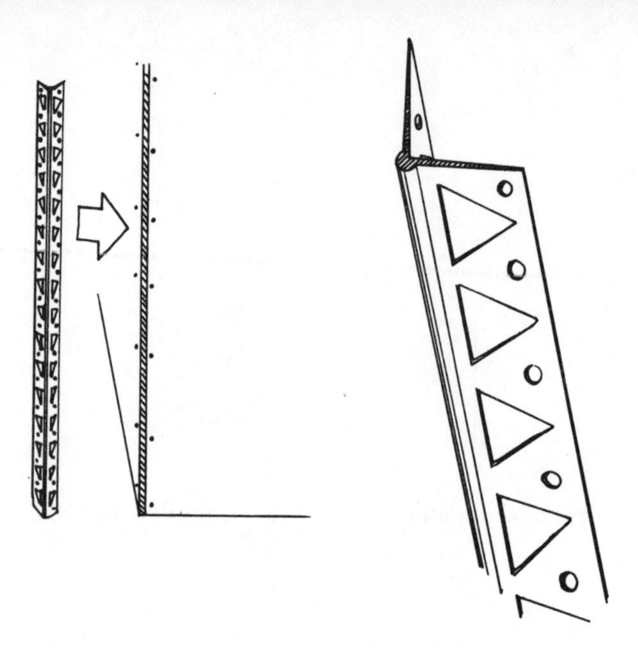

Fig. 8. Corner bead on outside corner

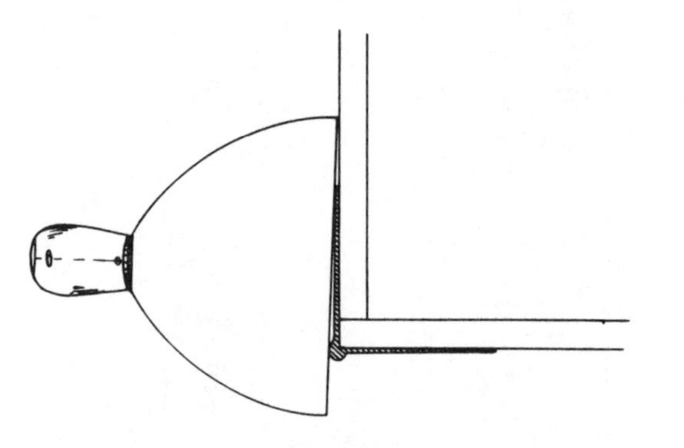

Fig. 9. Position of knife for applying compound over corner bead

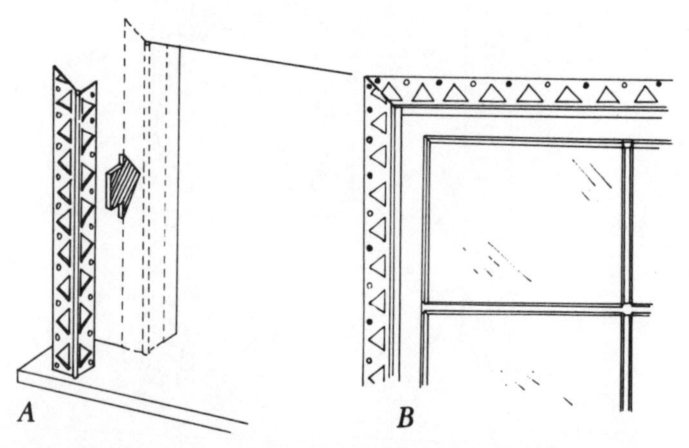

Fig. 10. Mitered corner bead around window

Straighten out any creases or bubbles in the tape; place the excess compound on your float. Now go over the tape once more (preferably with a larger knife, but the 6″ one will do), and apply a very thin skim coat over the tape (*6*). Make it as smooth as possible.

Do all the seams this way for the first coat. Where horizontal and vertical strips intersect, cut the second tape to butt against the first—don't overlap them. Be careful not to wrinkle the first piece when you smooth out the second.

The biggest problem is the nontapered seams. You need enough compound so the tape will stick everywhere, but as little as possible, to minimize the bulge.

As you do the seams, fill in the neighboring nail dimples approximately flush to the surface. Use the excess compound that you squeeze out from the tape. If any heads stick out, hammer them in first.

Taping inside corners

Fold the perforated tape in half along its precreased center. Put a first coat of compound on the two corner surfaces, and imbed the folded tape into the corner (*7*). Run your knife along one side, then the other, holding it about 1/16″ or less from the corner, to avoid cutting the adjacent side with the corner of the blade. There will be a slight bulge in the center that can be sponged away later. Continue with the second and third coat as usual.

Taping outside corners

Use *corner bead* under the compound (*8*). This is a metal strip, L-shaped in cross section, that helps protect the corners from being chipped or damaged. (See "Wood and Materials.") Cut the bead with shears, taking care not to bend it too much. Nail or screw it on through the precut holes. There is a flexible bead made for curved edges.

Now you can either tape over the corner bead or just compound without the tape. Taping assures the covering of the metal but may take a bit more time. Apply the compound and use a separate strip of tape for each side, just up to the metal bead at the corner. Rest the knife on the bead as you run it along (*9*). The bead should show when finished.

You can use corner bead around windows and doors instead of facing the surface with wood. Miter it to get a tight fit (*10A–10B*).

If you have a wall ending in the middle of a room, put corner bead on the edges of the open end and fill the area between with compound. Then run your knife or float along the two beads to make an even surface. You can also nail a facing piece of wood to the open end, which eliminates the need for corner beading.

Second Coat

Let the seams dry overnight. Next day apply a second coat of compound, using the float this time, feathering the edges out another couple of inches on either side (*11*). The feathering, or gentle tapering, gives a greater illusion of flatness.

Final Coats

Again let the seams dry overnight. Since you're probably not an experienced taper, you will probably find little ridges in the compound when it dries. Manufacturers recommend sanding these ridges down with fine sandpaper, but we find that method time-consuming and dusty, and it sometimes roughs up the paper face of the plasterboard. We prefer to go over the compound with a dense sponge, moist but not dripping, within a day or two after the second coat (*12*). The surface of the compound will form a thin pasty substance and you can smooth out the bumps and ridges. Be careful not to wet it so much that the tape starts to show through again. You'll notice that the compound shrinks a little as it dries. Therefore if a third coat is necessary, overfill the dimples a little. When dry, sponge again.

Hints and Tricks for Neat Taping

Always close the can of compound tightly when you're not actually digging into it. The excess compound from your float can be reused elsewhere on the wall (if it is free of lumps and dried pieces), but don't return it to the can. It will cause lumps. As you apply and smooth out the compound, watch out for dried-out chunks that cause tracks. Pick out the pieces and throw them away.

Keep your taping knife edges clean—dirt and lumps on them can also cause tracks. Use the edges of the waste box or one edge of the float to keep the knife clean as you work. Between coats, completely clean knives and float by scraping against each other. If you must wet them to clean them, dry them *immediately* to avoid rust.

Cracks more than ½" wide should be filled with plaster or smaller pieces of plasterboard before taping. Otherwise the compound will crack as it dries and shrinks.

Cracks in the seam after drying? Either your first coat was too heavy, the original seam was too wide, or possibly it was too hot in the room and the compound dried too quickly. Sponge the surface down as much as possible, then fill in with more compound.

Blisters under the dry tape? You didn't put enough compound under the tape. Cut the blister with a blade, dampen slightly, and insert compound. Press it flat and allow to dry. Then follow usual steps.

Keep the temperature between 55° F. and 70° F. throughout the job.

Don't forget to pound in *all* the nails.

Don't paint the wall until the compound is completely dry. Wait two to three days. Otherwise paint could discolor.

Above all take your time during each stage of plasterboard application and taping. Think while you're working, take lots of breaks. A good taping job is just a matter of care.

Fig. 11. Feather out further with float

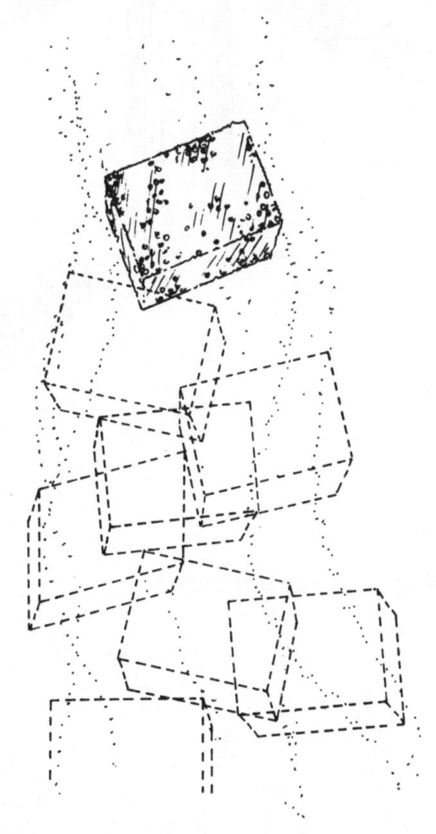

Fig. 12. Sponging down to smooth compound surface

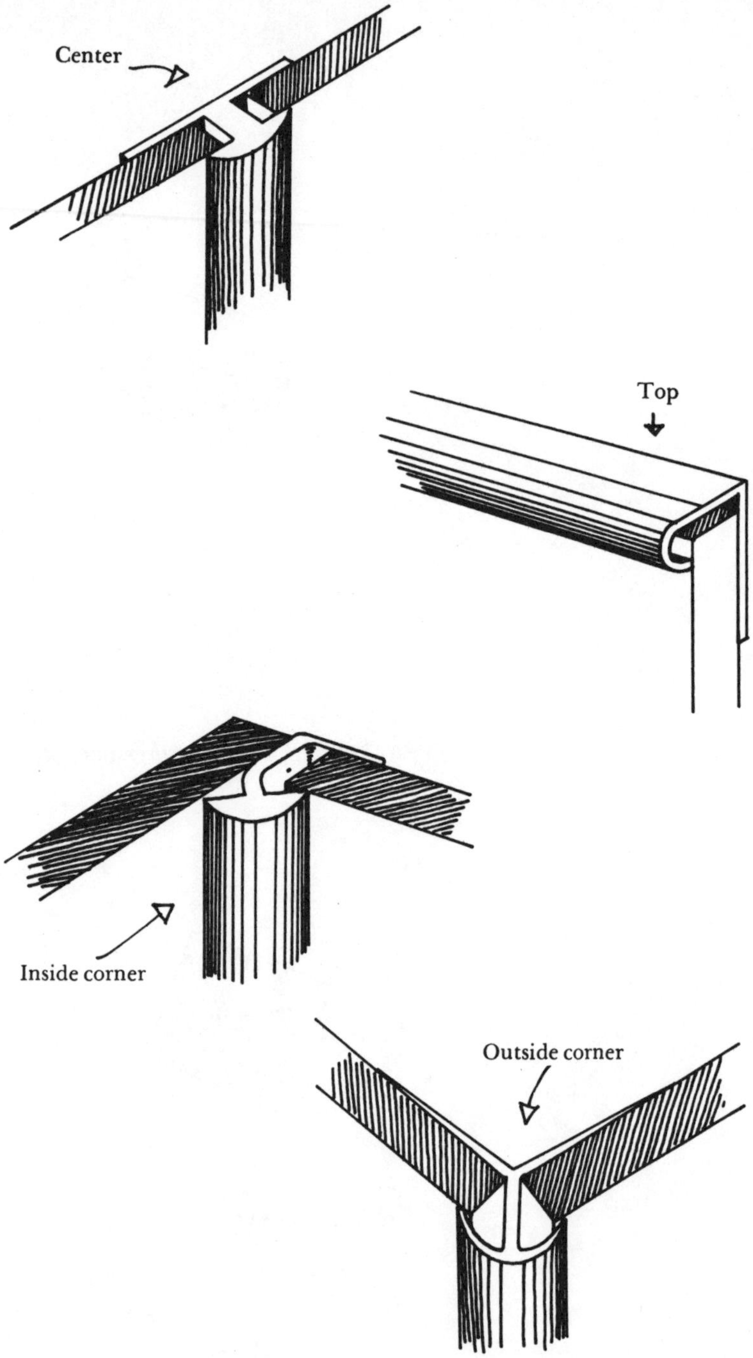

Center

Top

Inside corner

Outside corner

Fig. 1. Panel Moldings

NEW SURFACES FOR OLD—WALLS, CEILINGS, AND FLOORS

Sometimes a wall, ceiling, or floor is in such bad shape that it's necessary to make a whole new surface. "Wood and Other Building Materials" describes many of the available possibilities. Plasterboard application has been covered in the above section, and is most commonly used on new, rather than old, walls. Here, we'll go into the most common new surfaces that involve carpentry—wall paneling, ceiling tiles, and wood floors. Decide what you want, study this chapter, order, and go ahead.

You should review "Walls and Ceilings," too, since you have to know what the existing structure is like. And don't forget that cleaning and painting a wall—or even removing plaster to expose brick (described in "Finishing and Refinishing") can give you walls as good as new.

Old moldings

Often when a new surface is applied, it's possible to save the old moldings—baseboards, window copings, and the like—for reuse. Remove it carefully and take out the nails by reversing the piece and, with it supported so the nail has room to come out, pound the point backward till the head comes out far enough to be pulled. If you are putting molding up into brick or masonry, you should drill through the molding first or use the old holes. Then use a cut nail countersunk with your nail set.

Furring strips

When the old surface is too rough or broken for direct application of a new surface, and when the spacing of the existing studs or joists is wrong, it's usually necessary to apply furring strips. These are common-grade or construction-grade 1x2s, nailed securely to the existing structure (studs, joists, or masonry; see "Walls and Ceilings"). The furring strips are spaced at proper intervals for the new surface to be applied—16″ or 24″ on center for paneling, 12″ on center for ceiling tiles (shown in Figure 4), and so on. If the existing surface isn't level, the furring strips can be shimmed to level as described under Flooring Preparation later in this chapter.

Wall Paneling

Planning

Once you have decided you want paneling and have picked a type, you will need to know how much to get. Measure the ceiling height at its highest point and the wall lengths at the floor. Multiply these two to get the area, the total number of square feet. A wall 9′ high by 14′ long has 126 square feet.

A sheet 4′x8′ has 32 square feet, so you need four sheets. But if you do a little figuring, you will see that with the extra foot on top, it would be better to get five sheets. The reason for this is that you will put in three whole sheets leaving you half a sheet on the end. Once the half is in, you have half left to fill in the 1′x14′ space. You have enough material to do it, but if you are using paneling scored to give the effect of boards, you won't be able to match up the scores. Five sheets give more waste but a neater job.

If you can get 10′ panels, you would need only four. And if you are lucky enough to find 5′x10′ sheets, three will suffice. If you are using a solid unscored style, four 4′x8′s will do.

If the wall you are covering has a window or door in it, subtract the area from your total. Before you buy, however, make sure the scores will line up with the door and window area subtracted.

Moldings

Some types of paneling need divider moldings between the sheets. Some people prefer to put in dividers for all types; it's up to you. All paneling will need ceiling and floor moldings to add to the effect. You may also want corner moldings. Most of these are made of plastic and have tracks for the paneling to slide into, and are made to match the fake wood of the panel (*1*). These types are most necessary with the patterned panels. The top and bottom moldings can still be found in wood.

Figure how much molding you will need at the same time so you can purchase all of it together.

Molding usage

The plastic types that slip on the edges have to be put in place before the panel is up. To cut the plastic, score it and snap it off. Or cut it with a saw. When applying cove-shaped or quarter-round wood molding, you have to make a coped cut in order for the curves to match at the corners. This is done with a coping saw as shown in Figure 2.

Cutting and Fitting

There are right and wrong ways to cut a panel. With a handsaw, cut from the finished side. With a power saw, cut from the back side. Make sure it is well supported; it's pretty thin stuff. Always use a fine-toothed blade in your saw.

To cut the sheets to size, measure the ceiling height at each location. Subtract ¼″ or ½″ to allow for maneuvering in. Don't worry about the space; molding will cover up to a 2″ gap. If the last piece has to be cut, make sure you take the design into account. Scored panels have a groove on the edge to look like another score. Make sure you cut off the edge which is opposite the one that butts against the previous piece.

Room corners may be irregular. To get the exact shape of the wall, you can use a pair of dividers. With the panel in place against the corner, run the dividers along the wall and the panel edge touching it (*3*). The point of the divider will make a line exactly like the wall's contour. Cut along it.

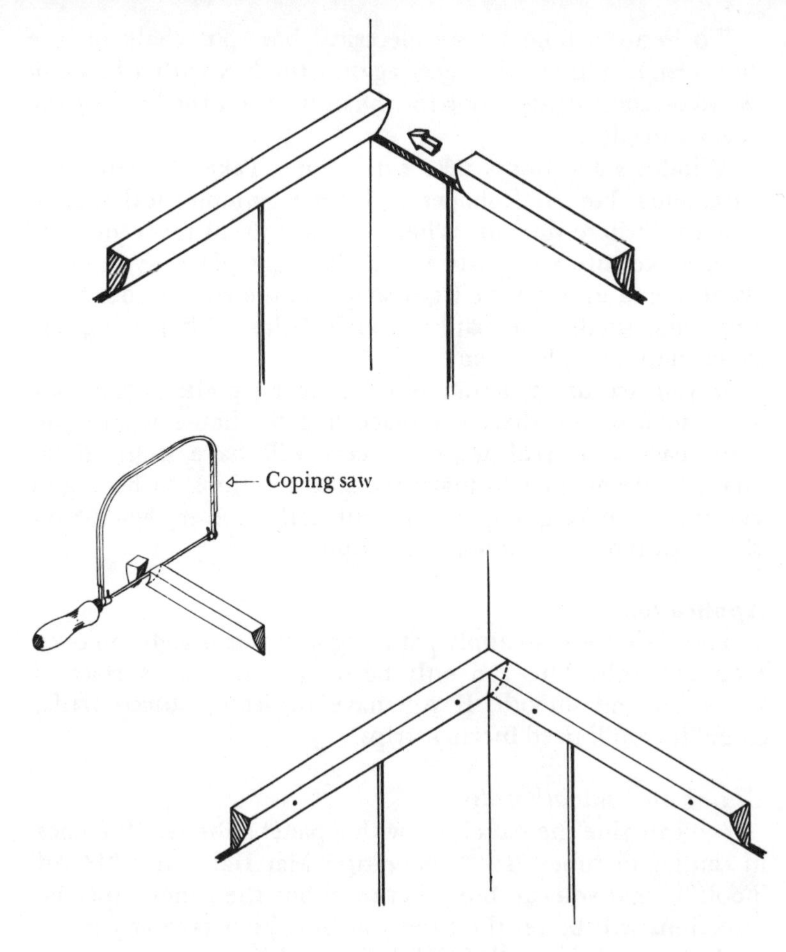

Fig. 2. Steps for coping a corner piece of molding

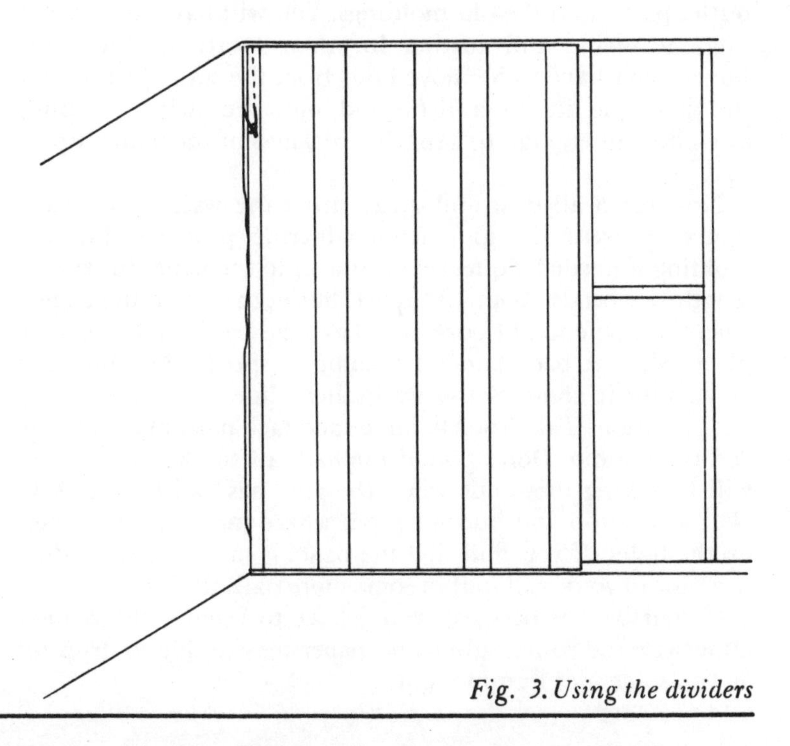

Fig. 3. Using the dividers

To locate a hole for an electrical box, put chalk on the box's edges and tap the sheet against the box with a block of wood on the outside. Look for the imprint on the back of the sheet and cut it out.

Windows and doors take extra time. Take accurate measurements. For all L-shapes and other asymmetrical shapes, take care where you cut. When you have to cut from the back side, make sure your cuts are in the right places—you don't want to end up with the right shape on the wrong side. Making some small-scale paper models helps with getting the right shape and placement.

If you are using actual wood veneer paneling, you may want to lean the sheets in place first to choose where you want each one—real wood veneers will have slight tonal changes. Remember to mark the stud locations on floor and ceiling, if you're going to nail. Put nails in every 8″–10″ on the vertical to assure adequate holding.

Application

The easiest way to apply paneling is to put it right on over your old wall. This can only be done if the old surface is fairly flat and smooth. If you have rusticated stucco walls, forget it; you'll need furring strips.

Gluing onto smooth walls

You can glue the panels on with a panel adhesive. It comes in caulk-gun tubes (see "Wood and Materials" and "Hand Tools"), that you can buy where you buy the panels. Specific directions will be on the tube you buy. First take any nails, hooks, etc., off the wall. Sand down or chip away any bumps you feel will be in the way. Remove fixtures and switch or outlet plates and the old moldings. You will have to do accurate measuring and cutting for their locations. Electrical boxes don't have to be moved out from the wall. Just loosen the retaining screws a little and pull the switch or plug assembly out enough to gain the thickness of the panel, up to ¼″.

Now clean all dust and grease from the wall. These substances prevent the glue from adhering properly. Put on molding if needed. Squeeze the glue onto the panel in several lengthwise beads about 12″ apart, but *not* right on the edges, where the glue might ooze out. Take the panel and put it in place. Slide it back and forth some to spread the glue and make sure it contacts the wall; then slide the panel to its final position. Take your hammer and tack nails in at the top and the bottom. Don't pound the nails all the way in, as you will be taking them out when the glue has had time to set. Nail at the top and bottom edges where molding can cover up the holes. If you find that the panel is not lying tight due to a bow in your wall, put in some more nails till it does.

If your wall is papered, you'll have to remove the paper. Otherwise the panels, glue, and paper may decide to drop in on you some nice humid night.

Nailing through to studs

If your wall is made of plasterboard and you know the location of the studs, nail on the panels. Special color-coordinated panel nails are made for this purpose and are sold at your panel dealer. You can of course glue to plasterboard.

Nailing to furring strips

If your old walls are badly damaged or uneven, you have to put up 1x2 furring strips, shimmed and leveled. You can either nail or glue to these. Make sure you have one strip straddling each panel seam just as with plasterboard.

To apply paneling to brick or masonry walls you must apply furring.

With all walls, if you are covering up damage, plug the holes first with plaster.

Finishing

When you have all the panels up, put in the moldings at the ceiling and floor, and around any doors and windows. You can replace the old moldings, or use new ones; for applying casing around wall openings, see "Doors and Locks."

Ceiling Tiles

Tiles not only enhance the ceiling appearance but muffle noise from above. If they're acoustical tiles, they'll absorb room noise, too.

The most common tiles are 12″ square. These are made by many manufacturers and come in numerous styles and colors. The other type comes in panels, usually 2′x4′, and is suspended from the ceiling.

Square Tiles

If your old ceiling is O.K., you can staple or glue the 12″-square tiles directly to it. Most ceilings, however, need to have furring strips nailed to the joists at 12″ on-center intervals.

First you must determine the square footage of the room. Add about 10 percent for waste and mistakes, and order what you need. Get furring, nails, glue, or a staple gun.

Begin by figuring out a pattern of tiles that will avoid very narrow strips along any wall. In a simple four-sided room, there should be a way of making sure no tile has to be cut to less than 6″ along the walls. This usually involves either *centering* the center row of tiles in the center of the room, or lining up the *edge* of the tiles along the center line of the room. Use graph paper to help you figure it out, then find and mark the joists, and put up the furring strips across them. These strips should of course be along where you've planned the tile seams to be.

You are ready to put the tiles up. You will have noticed by now the tiles have flanges on them and can only be joined a certain way. If you are stapling, the flanges are where the staples go (*4*). If you are gluing to furring strips, make sure the glue is over the part of the flange which will be resting on the strip. If you are gluing to a flat ceiling, put a blob in each corner—but not right up to the edge—and press in place.

When you have tiled up to the walls, you probably will have to cut the last tiles. Make sure you cut the correct side. Use your mat knife to score about a quarter of the way in, then snap. Trim edges with knife. If you happen to have an L-shaped piece to cut, either score the longer cut and cut all the way through the other, then snap the first, or use your saber saw.

For light fixtures, use this method: If the fixture falls anywhere within a tile, cut a hole only as large as necessary (*5*). Cut the tile in half through the center of the hole and slide each piece into place. This is done because you could never get the tile over the fixture. Also your edge pieces may have to have their male flanges cut off in order to get them in.

Always wear gloves and goggles. Many ceiling tiles are made partially of mineral fibers and can be harmful if the particles get in your eye or into an open cut.

Suspended Ceilings

These are used mainly for one of two reasons—either to lower very high ceilings or to hide the old ceiling. You can easily take part or all this type ceiling down when desired. You thus have easy access to the old ceiling.

There are many different kinds of suspended ceilings. The most common is made of material similar to the tiles. They are 5/8″ to 2″ thick, and come in sheets that measure 2′x4′.

Applying

To put this type of ceiling up, you need tracks or channels. The slabs of material lie in the channels, which are hung from the ceiling by wires.

First, do the same as you would for tiles—find the square footage to figure your materials, and make a layout plan to avoid narrow panels at the walls. Decide how low you want to hang the panels. Once this has been ascertained, use your chalk line to mark a level line all around the walls as a guide for the track supports. Next, mount the supports all around the room on this line (*6*).

When these are up, put in the tracks for the direction perpendicular to the ceiling joists. They are suspended by heavy-gauge braided wire from screw hooks or eyes mounted into the ceiling joists (*7*). The tracks of course are hung at 24″ on-center intervals to accept the slabs. Your ceiling will probably not be an exact multiple of 24″, so arrange the tracks according to your layout, with narrower intervals at both walls.

Once these are in, put in the cross tracks. They snap into the ones that are up and will need no other suspension. Before you put up the slabs, figure out what you will do about your overhead lighting. You can use the old box and install special recessed lighting fixtures that will sit right into the

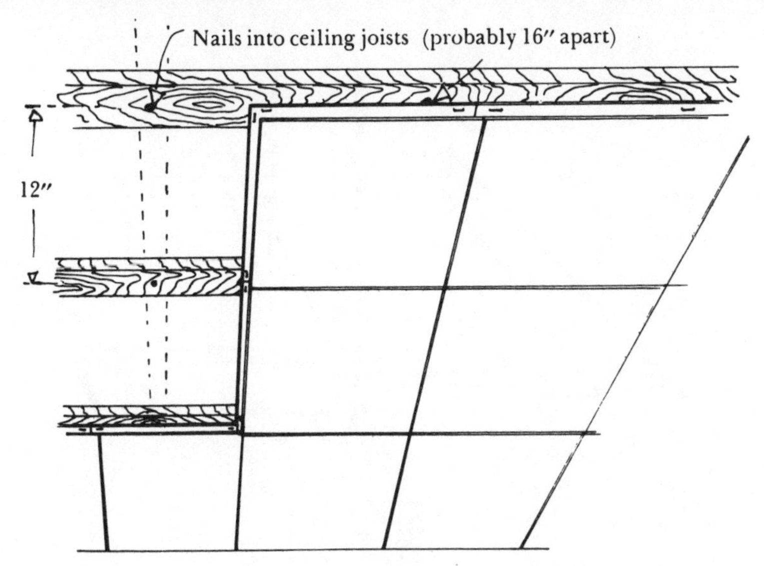

Fig. 4. Furring strips with tiles stapled to them

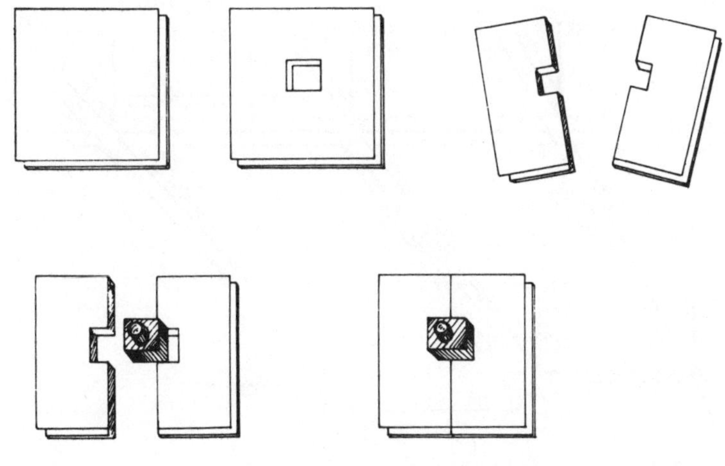
Fig. 5. Steps for cutting around an obstacle

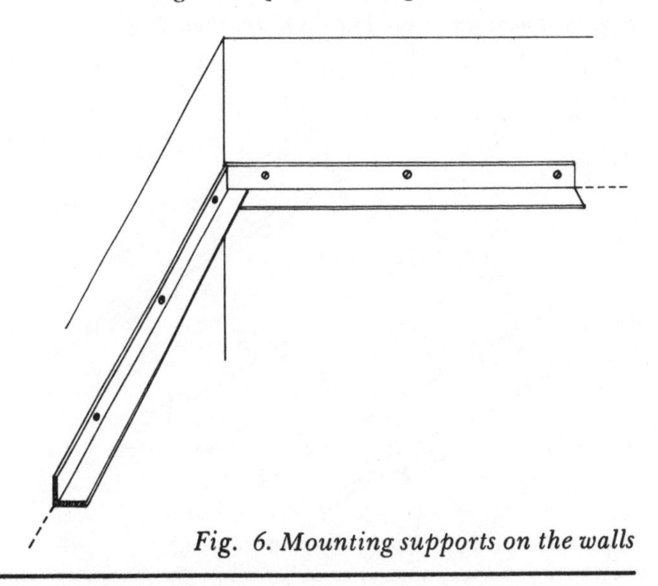
Fig. 6. Mounting supports on the walls

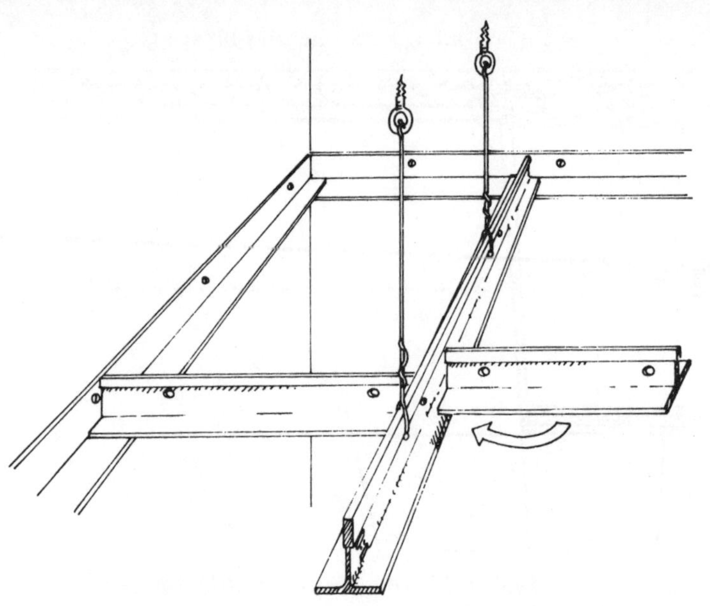

Fig. 7. Hanging the center supports and adding the cross tracks

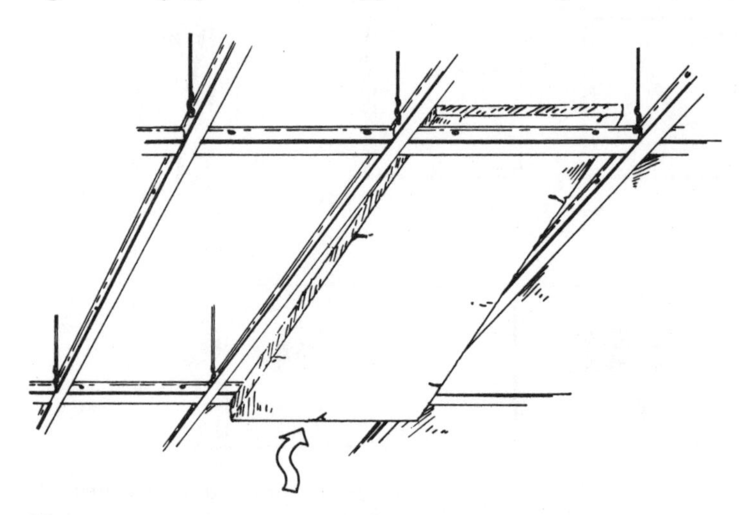

Fig. 8. Twist sideways and push up, then drop in

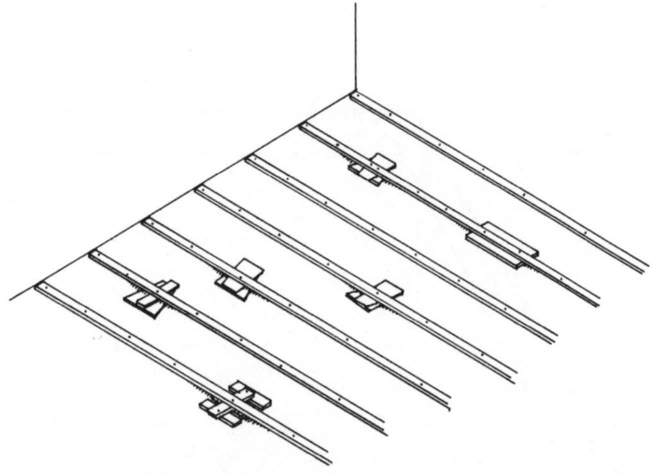

Fig. 9. Leveling overlay strips for flooring

new grid. Cut any holes you may need in the slabs with your saber saw.

To put up the slabs, slide them in sideways; then flatten them out and lower them into place (*8*).

Flooring

There are many types of finished wood flooring. The most common is tongue-and-groove strip flooring. The boards commonly come in widths of from 1½″ to 5³⁄₁₆″, and sometimes up to 12″, usually about 1″ thick. Lengths vary greatly, to permit staggered joints. You could also cover your floor with plywood or composition board. These are not as long-lasting, but can be attractive and much less expensive. *Parquet* floors come in blocks up to 12″ square, and assorted shapes and patterns, all very expensive. Whatever surface you use, it will need a coat of polyurethane or some other protective coating, unless it is adequately prefinished.

Preparation of Old Floors

The first step is to remove the baseboard moldings. Then nail down any loose and squeaking boards. Old floorboards will usually have been laid over a subflooring (either plywood or rough planks) supported on joists (see "Walls and Ceilings" for explanation of joist patterns). Use two flooring nails, nailed at opposing angles, wherever a board is loose (predrill holes slightly thinner than the nails).

You want a level, or at least a flat, surface for your new floor. Use a line level to determine the condition of your old floor. Strip flooring needs a flat surface to be applied to, since the strips will buckle without almost continuous support. Plywood sheets, however, can span large gaps, certainly up to 16″ with no problem.

If your floor is very flat, you can lay new flooring directly on top of it.

If it has irregularities, but minor ones, you can lay a subflooring of ½″ plywood directly on top of the old floor. If you're not sure how serious the imperfections are, lay a piece of ply across the dips and see how much it flexes—there should be practically no flexing. If the floor is bad in just a few spots, use shims in those spots. If the dips are large, make sure the ply is supported with shims at least every 16″.

If the floor is very bad, you will have to lay a whole new level subfloor, using 1x2s to support the ply. Run the 1-bys along the longest dimension of the floor at 16″ on-center intervals. When you come to a depression, level it out with shims (*9*). Shingles (No. 3 grade) are best for this. Start at the highest point. (If there is only one high point, you can sometimes chisel it down.) Always nail the 1-bys and shims into the old floor with flooring nails, which won't pull out. Work across, leveling each run to the next. Box around obstructions so the ply is supported on all sides (*10*).

When the level 1x2 strips are down, lay the subfloor. Use ¾″ ply or composition. Nail it to the 1-bys with flooring nails. Don't worry about small gaps between the sheets; the strip flooring can span them.

Strip Flooring

You order flooring by the square foot. It comes in bundles with the longest strips being about 4'. Figure in a large waste percentage over the actual square footage, due to bad pieces, cutting to fit, and so on.

Waste allowances for flooring of various sizes:

25/32"x1½"	add 55%
25/32"x2"	add 43%
25/32"x2¼"	add 39%
½"x2"	add 30%

For better grades, the waste may not be as great. Remember you can always get a little more, but most places won't

Lay the boards down in random lengths, making your own pattern of light and dark. Never line up any ends adjacent to each other; this damages the structural qualities. Keep them at least 5" apart (*11*). Knock each piece in place with a soft-headed mallet so you don't mar it.

Don't nail them down with a hammer unless you have only a small area. Rent a special machine for this, called a Porta-nailer. It is designed to fit right over the tongue before the next strip is slipped into place, and with a shot from the mallet it drives the staplelike nail right into place. No countersinking with a nail set is needed. It takes a little practice to master, but it's not hard. Make sure the rental man explains it thoroughly to you, and practice on some old wood with the real flooring.

When doing flooring, it's best to work with one other person. One chooses the pieces and does the cutting and the other knocks them in place and uses the machine.

The end pieces are set ¼" to ½" away from the wall to allow for expansion of the boards. The groove side of the first piece is laid facing the wall, with the tongue out, waiting for the groove of the next piece (*12*). There won't be enough room to fit the machine in till the third row. The same problem occurs at the last two rows. The normal procedure for these end rows is to drill into the face of the pieces, nail, countersink, and plug. Use at least two nails per piece, even for the real short ones. Otherwise you can space them up to a foot apart. Do not nail within 2" of the ends.

Cut pieces for the ends and around obstacles with your saber saw, from the back side. For long straight cuts use your circular saw with the combination or rip blade.

After all the strips are down, they have to be sanded to even out any irregularities and give you a good smooth surface to finish. You will need to rent a sander for this; again it's fairly inexpensive. Vacuum all the dust away and then, if the flooring is not prefinished, put one to three coats of polyurethane on to finish.

Finally, replace the baseboards; they will cover the expansion gap.

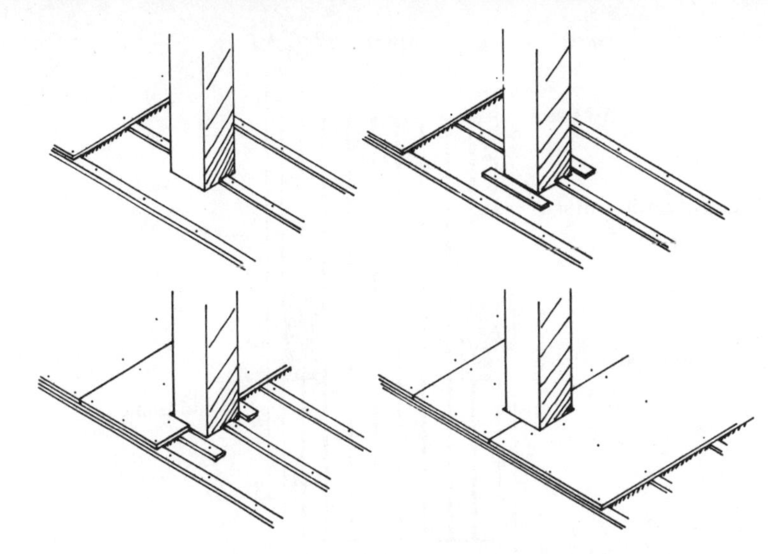

Fig. 10. Steps for working around an obstruction

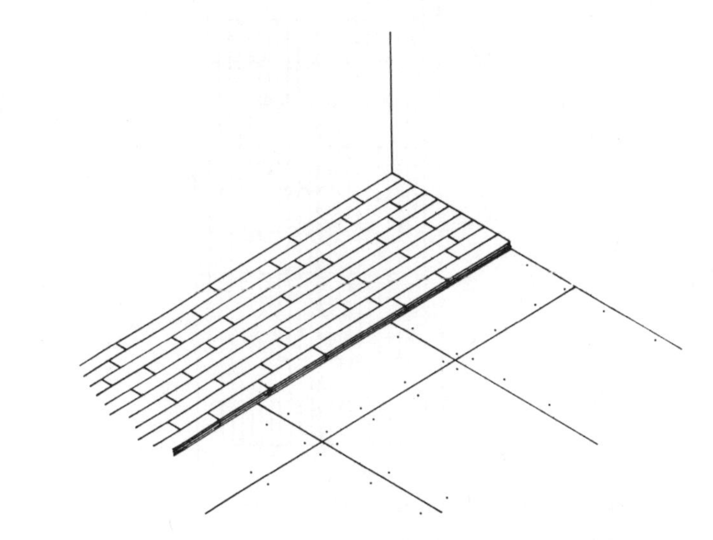

Fig. 11. Laying tongue-and-groove flooring

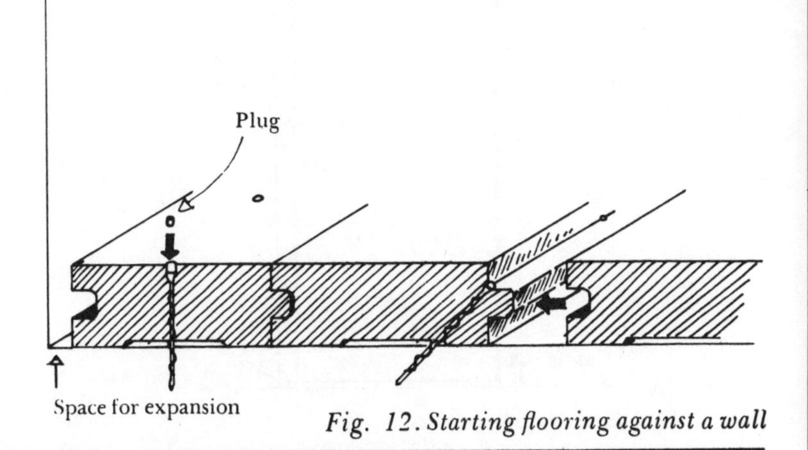

Fig. 12. Starting flooring against a wall

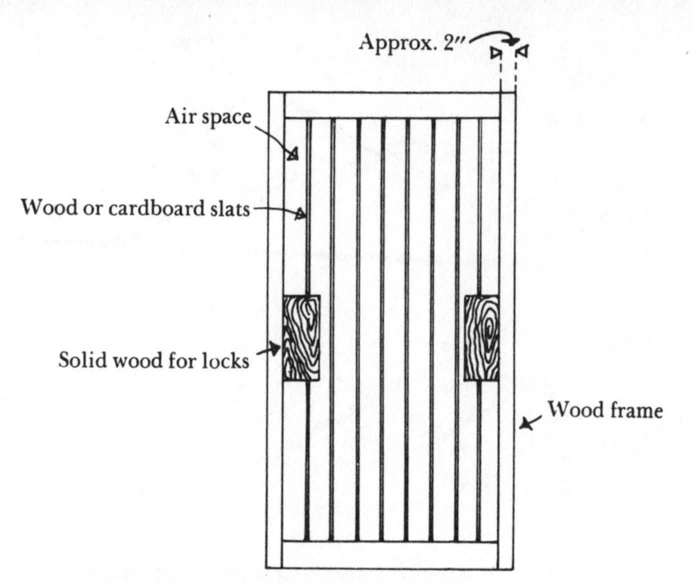

Fig. 1A Inside of Hollow Core Door

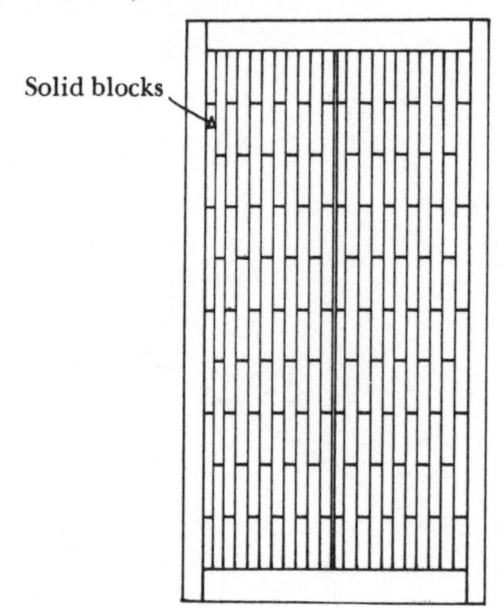

Fig. 1B Inside of Solid Core Door

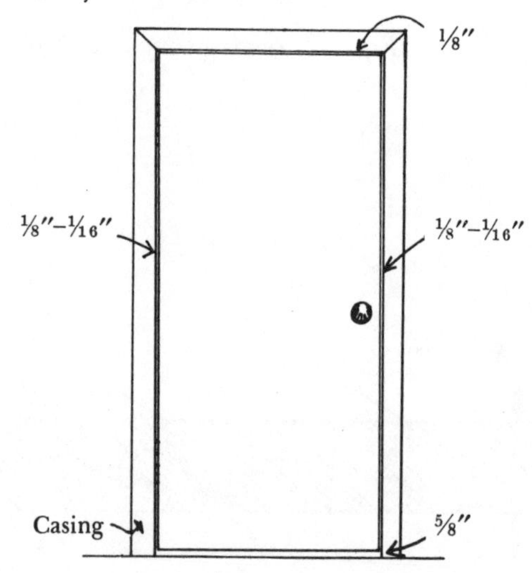

Fig. 2. Standard Measurements

DOORS AND LOCKS

Doors

The two main kinds of doors are panel and flush doors. The *panel door* has indented panels on its face, or perhaps louvers or glass windows. The *flush door* is more common these days. It may either be hollow or have a solid core, with flat sheets of ⅛″ plywood on both faces. The *hollow-core flush door* is filled with thin strips of wood or cardboard, with a solid wood frame about 2″ wide on the sides and about 3″ wide top and bottom (*1A*). The frame makes the door rigid and provides backing for the hinges and lock. Light and inexpensive but vulnerable to a good swift kick, the hollow-core door is used mainly for interior installations. The *solid-core flush door* has an interior of solid wood blocks (*1B*). It is heavier, stronger, more expensive, and a better thermal and sound insulator than a hollow-core door. Thus it's a good exterior door, though you can use it inside if you like to increase soundproofing between rooms.

Doors are available at your lumberyard in many sizes. The heights available are 6′, 6′6″, 6′8″, and 7′. The widths are 1′6″, 2′, 2′4″, 2′6″, 2′8″, and 3′. They come in thickness of 1¼″, 1⅜″, and 1¾″. The most common size is 2′8″x6′8″, with 1¾″ thickness for exterior and 1⅜″ for interior. Closet doors can be narrower.

Naturally, you should always try to get a door size that fits the doorway as exactly as possible. Doors are not meant to be cut down, though in old doorways it's sometimes necessary. When you build a new wall and doorjamb, always build to a standard door size. Leave ¹⁄₁₆″ to ⅛″ gap between door and jamb on the top and sides; leave ⅝″ at the bottom, or more if you plan a very thick rug there (*2*).

Doors are also sold *prehung,* hinged to their own preassembled jamb, usually of 1x4s (*3*). Some come with lock and knob installed, or with holes predrilled for them. They are most handy in new construction, where you can build the framing to exact size.

Fire doors are made either of metal or of wood with specially treated plaster and chemical salts inside. They must be installed in special steel jambs. These jambs come separately, or attached as a prehung unit. Fire doors should not be cut down at all. They are mainly for outer hall or stairway doors, as a deterrent to spreading fires. Check with your local building code for specifics, or ask at your lumberyard.

Plan which way your door will open, in or out, and on which side it will be hinged. Entrance doors always open inward, so there are no hinge pins on the outside, open to removal by unwanted visitors. Interior doors usually open *into* the room from the hallway for greater ease of passage, but it's up to you. Place the doorknob on the side with the light switch, so you can find the switch with a minimum of fumbling. Make sure it doesn't open on obstructions such as air-conditioners or against mirrors.

Installing a Door in an Old Jamb

If you are replacing an old door, check its fit in the jamb before removing it. See if it scrapes or sticks anywhere; mark those trouble spots on the jamb. Check the floor or the saddle, if there is one. The *saddle* is the hill-like piece of wood on the floor directly under the door (*4*). It seals the space below the door and often is needed as transition between carpet and bare floor.

Now you can remove the door. The easiest way is to knock the pins in the hinges up and out (bottom one first) with an old screwdriver and hammer (*5*). If you can, take the pins out while the door is closed. This separates the two leaves of each hinge so you can just lift the door off. On older doors the hinge pins may be painted in and need some chipping to get them out. Unscrew the hinge leaves from the door and jamb. Old hinges are often solid brass; if they are in good shape (unbent), you can reuse them on the new door. The screws are easily replaced if necessary. And you'll save time because these hinges are already mortised in the jamb.

Measure the jamb and the door dimensions. Check them both with a square. Consider any trouble spots you marked on the jamb. All this investigation will tell you what size door to get and how much trimming it may need. Decide what kind of door you need; get one the same thickness as the old one, if possible.

Check for warps on the new door with a straightedge along the face. Refuse delivery if it's warped.

Cutting doors to fit

If you're lucky, the new door will either fit or only need slight planing. Most literature on doors says they should *not* be cut down, that they might warp if too much structural frame is removed. This is a fine rule when you can build your jamb to exact size. Unfortunately, old jambs often leave you no alternative but to cut, unless you feel like completely rebuilding the jamb. Cuts of more than ½″ along the length should be made on the lock side of the door, since the other side receives the stress of the hinges. With a hollow-core door you *can* cut up to 1½″ off the lock side. Or take less off the lock side and cut the difference, up to ½″, on the hinge side. Use a good plywood blade for cutting and place your guide on the good side of the cutting line, so that if the saw strays it will mess up the waste part. Support the guide firmly at the middle so it doesn't bend with the force of the saw.

If you have to cut the height of a door, cut at the bottom. Top and bottom are usually designated on the edges of the doors. On a hollow-core door, the bottom piece of solid wood behind the veneer may only be a couple of inches high. Determine its height by sticking a pin through the face veneer to find where it stops. If you must cut more than this off the door, here's how to do it without leaving the bottom edge open and weakening the structure. Make your cut at the desired door height (*6A*). Remove the veneer from the piece

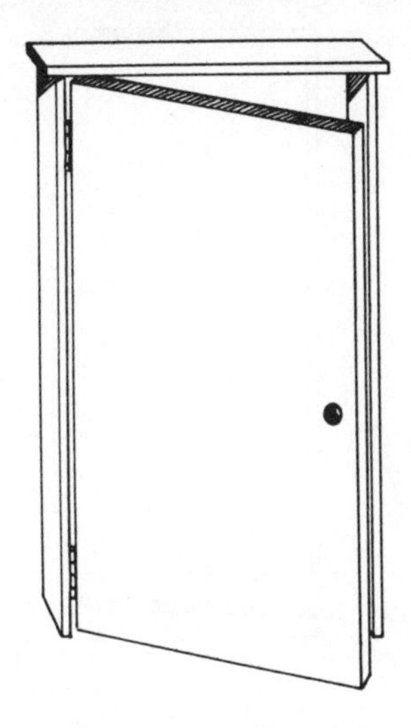

Fig. 3. Prehung Door

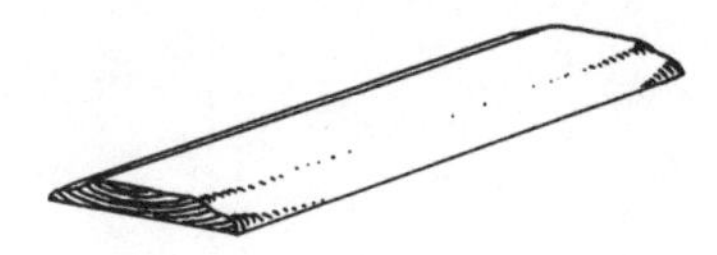

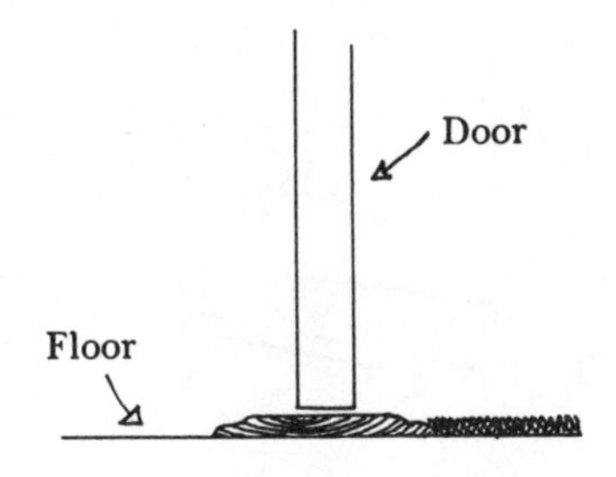

Fig. 4. A Saddle

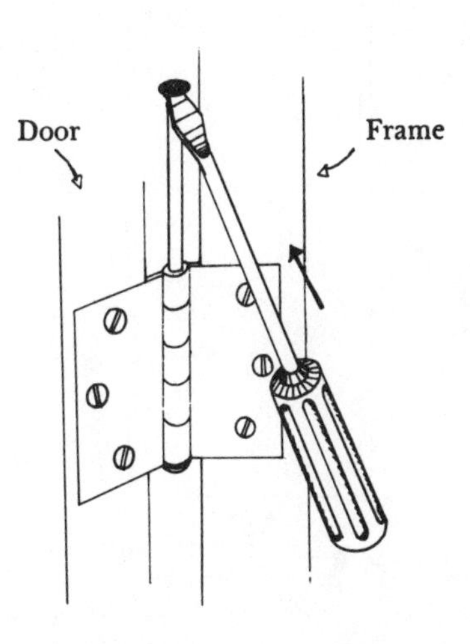

Fig. 5. Removing a hinge pin

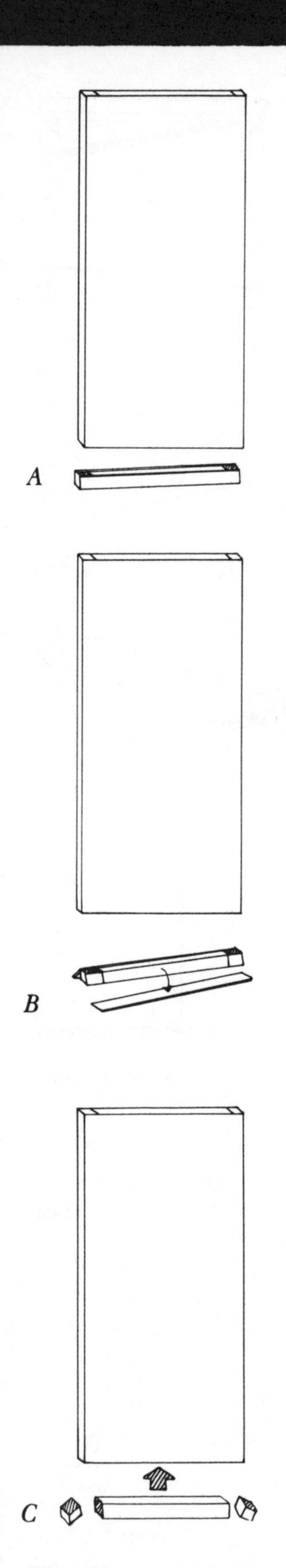

Fig. 6. Steps for shortening a hollow core door

you cut off, and then chisel the sections of side framing off the bottom piece (*6B*). Watch out for the corrugated nails, those things shaped like waffle potato chips. Now remove enough of the core sandwich from inside the bottom of the door (with a knife, and very carefully), so that you can slip the bottom block back inside (*6C*). Glue and screw it in.

Attaching the hinges

All you have to do to locate the hinges on the new door is transfer the hinge locations from the old door that was in the jamb. If the top hinge, for instance, was $10\frac{1}{2}''$ below the top of the old door, then it will be $10\frac{1}{2}''$ below the top of the new door. Be especially careful to maintain the same distance between the nonpin edge of the hinge and the door edge, usually about $\frac{1}{4}''$ (*7A*). And get the pin on the right side!

When you've outlined the hinge locations, mortise them (*7B*). This means you chisel out the wood within the hinge outlines to a depth equal to the thickness of one hinge leaf (*7C* to *7E*). Then the leaves can be set flush within the wood, leaving a minimum gap between jamb and door.

Separate the hinge leaves by pulling the pins out. Lay each leaf into its corresponding mortise on the door and trace the screw holes (*7F* and *7G*). Be sure to position them so the pin opening is on the *top* of the hinge—otherwise the pin can just fall out. We usually drill for and insert one screw per leaf, then check the door fit before drilling the rest. Insert the other halves of the hinge in the jamb mortise, putting one screw through each into the old holes. (Unless, of course, your new hinges need different holes.) Lift the door into place, standing on the hinge-pin side. Slip the top hinge leaves together first and insert the pin. Try to keep the bottom near its rightful position while doing this. Then slip the bottom leaves together. You may have to tap the jamb's bottom leaf a little with a hammer to get them to meet properly. Don't hit the door's leaf, or the door may swing out of your grip. If the bottom leaves don't meet, gauge the amount of error, take the door down, and adjust one leaf. With a slight error, try loosening the screw a little and setting a second screw in to hold the hinge in the correct position. For large errors you may have to remove the first screw, plug the hole with putty or doweling, and redrill.

Rehang the door. If everything fits, mark locations of the other screws, take the door down, drill, screw, and put up. If you've loosened that first screw in any leaf, leave it loose till the door is up, then tighten.

New Jambs

Figure 8 shows the proper elements in a professional doorjamb. The jamb itself is made of 1x4s or 1x6s ripped to fit the thickness of your finished wall. The double shims are very important. First, they make it easy to level the 1-bys the proper distance apart. Secondly, they isolate the jamb so the wall doesn't shake so much when the door is closed.

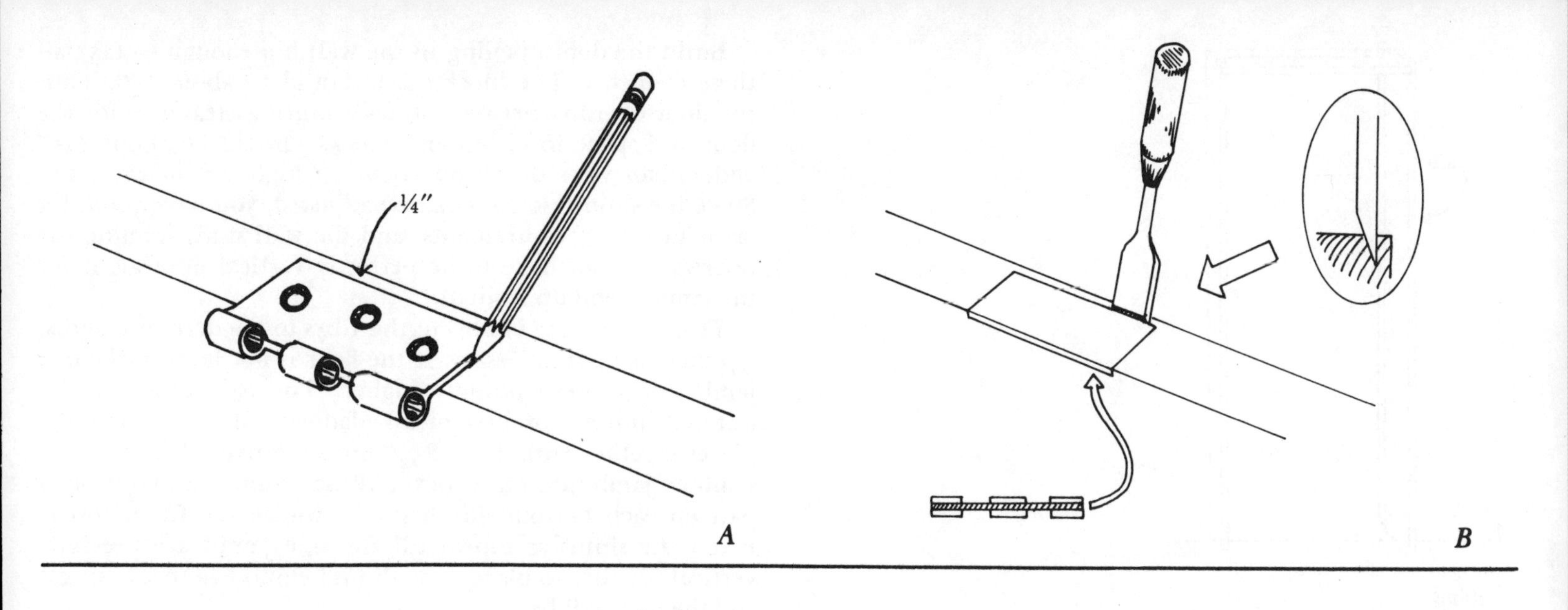

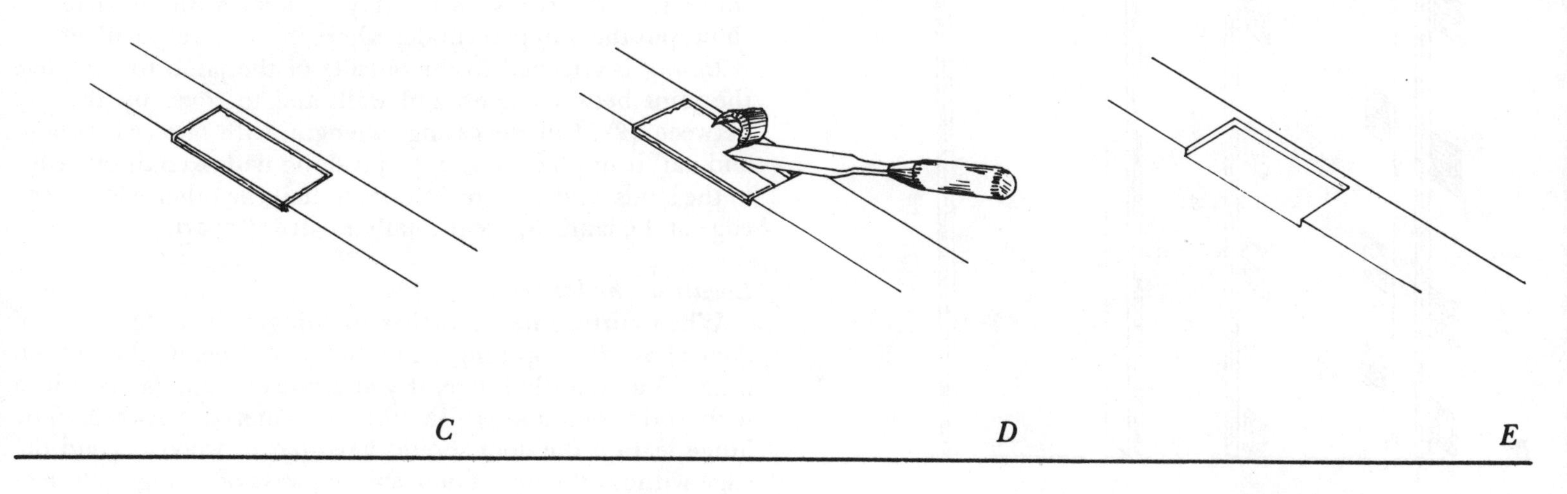

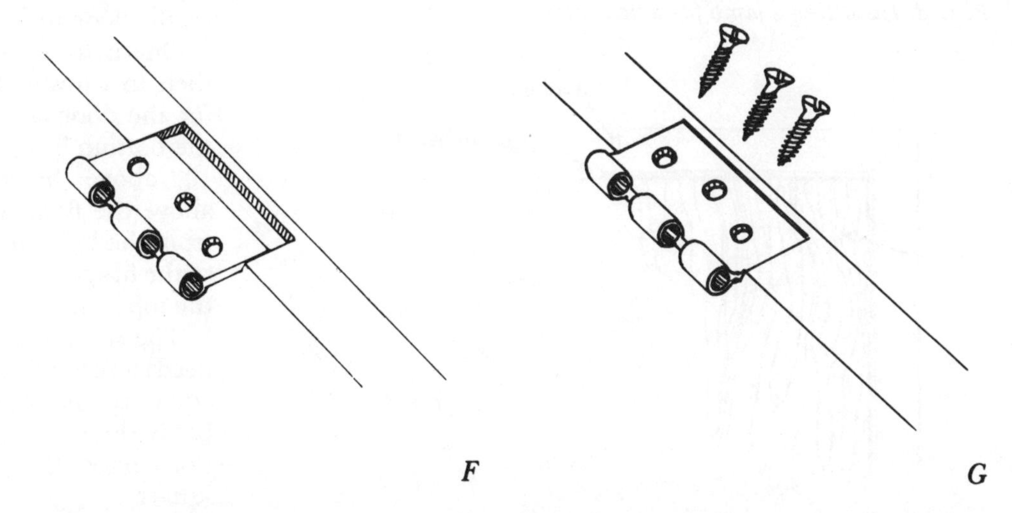

Fig. 7. Steps for installing a hinge

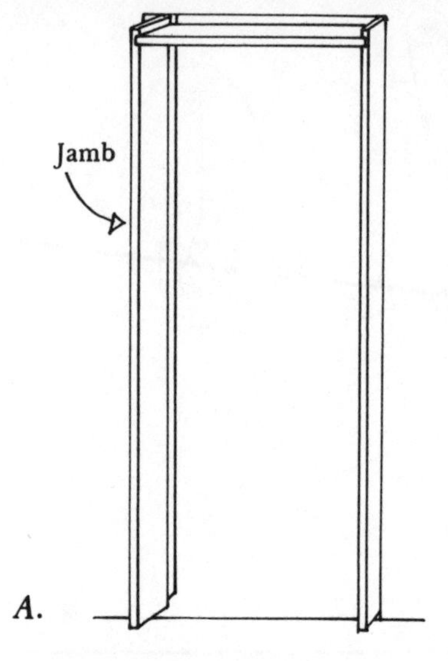

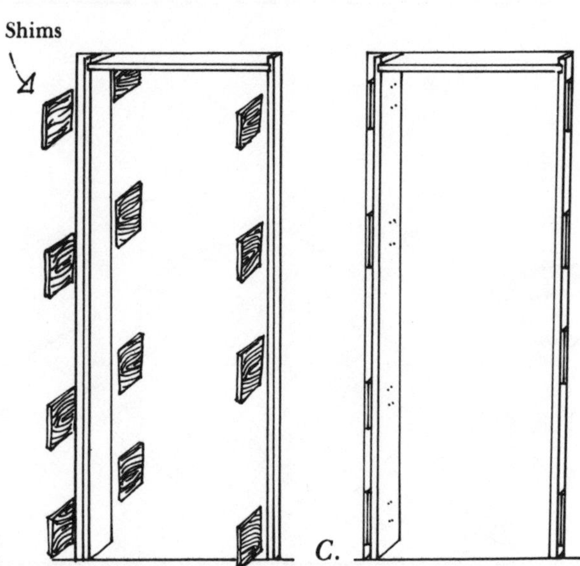

Fig. 8. Installing a jamb for a new door

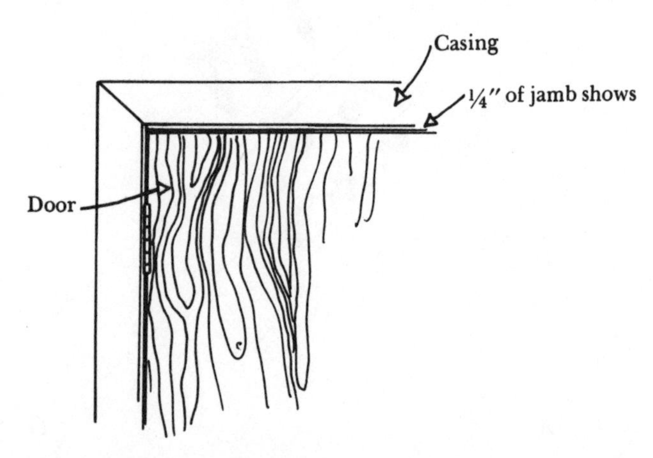

Fig. 9. Door Framing Detail

Build the door opening in the wall big enough to take all these elements. The block space should be about ½″. Thus the door opening between studs—counting clearance for the door and space for 1-bys and blocks—should be about 2¾″ wider than your door and about 2″ higher than the door. Since the shim thickness can be adjusted, you have room for error in your measurements; and the wall studs framing the doorway do not have to be perfectly vertical or straight for the jamb to end up straight.

To construct the jamb, cut the 1-bys to the desired lengths; rip them also, if necessary. If the floor is not level, make one jamb side correspondingly higher. The top joints can be mitered, butted, or best of all, dadoed. Glue and nail the pieces together with 2″ or 2½″ finishing nails. Insert the assembled jamb into place (*8A*). Wedge shims into position, a pair on each bottom side first to stabilize the frame (*8B*). Insert the shims as shown till the side jambs are precisely vertical. Be sure to place some shims behind where the hinges and the lock will be.

Use 2½″ finishing nails to fasten the jamb through the shims into the studs (*8C*). If you don't wish the nails to show, put them in pairs under where the door stop will go.

Casing is attached to the outside of the jamb to complete the joint between jamb and wall, and to cover up the gap between (*9*). Cut the casing to length, with mitered corners, and nail it in place. Use 2½″ finishing nails to nail one edge to the studs, and 1½″ or 2″ ones to nail the other edge to the edge of the jamb. Space the nails about 18″ apart.

Locating the hinges

When cutting new mortises for hinges, leave 5″ to 6″ of door above the top hinge, and 10″ to 11″ below the bottom hinge. The middle hinge, if you are using one (a good idea with solid-core doors), should be centered between. The hinge leaf on the door should have ¼″ of wood beyond the edge without the pin. There are two ways of lining up hinges on the door and jamb.

One is to mark the height of the hinges on the jamb and then to transfer those measurements to the door, adjusting for the door clearance at the top and bottom. For instance, the bottom hinge is placed on the jamb with its bottom edge 10″ above the floor. The door is to have a ⅝″ clearance above the floor, so the bottom of the hinge should be 9⅜″ from the bottom of the door. The door is *raised* off the floor, so the hinge is *lowered* on it to meet the jamb hinge. Measure the top hinge from the floor also, to minimize errors.

The second method of lining up is less prone to error, but needs a second person. Simply set the door in place on shims equal to the proper bottom clearance. While one person holds the door on one side, the other marks the hinge positions across the edge of the door and onto the jamb with a square.

Once the hinge locations are marked on door and jamb,

mortise and attach the hinges and then hang the door the same way as with a new door in an old jamb.

Door stop

A door stop (*10*) is nailed along the sides and top of the jamb, to stop the door when it closes. Cut the pieces to size, mitered or butted. Have someone close the door from the hinge-pin side, flush with the jamb, while you, on the other side of the door, line up the stop flush against the door on the edge where the lock will be. Open the door and mark the stop position. Tack it in place with 1½″ finishing nails, but don't drive the nails all the way yet. Close the door against it and locate the stop on the hinge edge about ¹⁄₁₆″ from the door. Tack it in place. Tack the top piece between the two sides pieces. Test the door. If everything is flush, hammer the nails in all the way. Congratulations, you have just saved yourself the cost of hiring someone to hang your door.

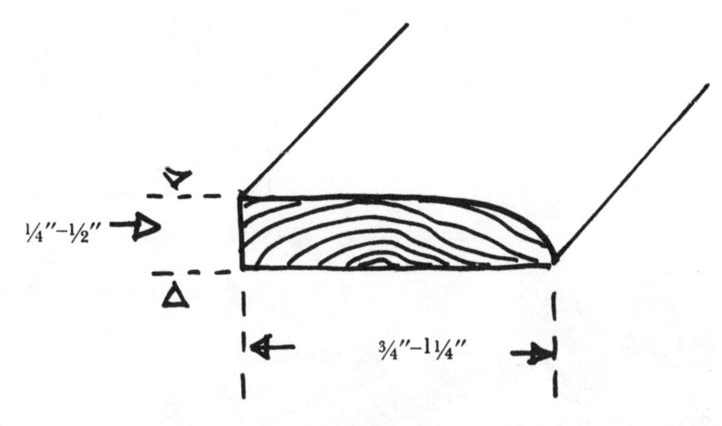

Fig. 10. Common Door Stop

Knobs and Locks

The most common knob-lock combination for interior doors is the passage lock. It has no key, just a simple bolt that opens when you turn the knob and catches automatically on closing. The privacy lock also has no key, but the bolt can be locked by pushing or turning a button on the knob or, in some cases, pushing in the knob itself. Mainly for bathrooms and bedrooms, it usually has a hole on the outside knob through which you can insert a pin or screwdriver to release the lock—in case a child gets stuck in the bathroom, for instance.

There are three main positions for key locks: in the knob, just above or below the knob, and independent of the knob. All three come with either the standard latch bolt (unfortunately, easy to force open with a credit card) or the more secure dead bolt. The locks that are independent of knobs are available with a special vertical-bolt mechanism. This is the strongest of the three kinds of bolt systems.

Most home locks have replaceable cylinders, that part of the lock which accepts the key. You don't have to change the entire lock if the cylinder breaks or if you want to change the key to your house or apartment.

The strongest lock of all is the police lock, a big-city phenomenon designed to prevent the front door from being battered in. One type has a double horizontal bar mechanism mounted on the center of the door. When the key turns, the bars slide in or out of slots in the jamb. The second kind has a heavy metal bar that, when locked, wedges between the lock case and a steel slot in the floor. (See Figure 17 in "Structure.") Unlocking the door slides the bar into a ring at the side, which keeps the bar steady but allows access.

Installation

The nice thing about knob locks (*11*) is ease of installation. Every new lock comes with detailed installation and

2⅝″
2 9/16″
2¼″
For door thickness 1⅜″ to 2¼″ regular; up to 2½″ special

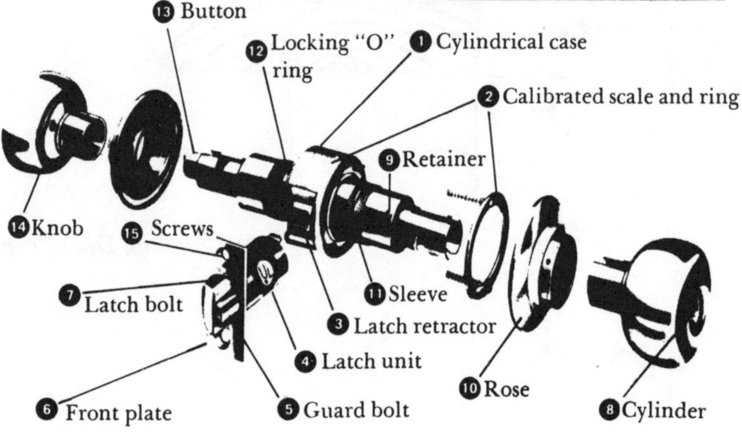

Fig. 11. Yale 5400 Series Locksets

mortising instructions, and a template to locate all the holes on the door exactly. The only special tools you will need will be a spade-type drill bit or hole-cutter for the lock cylinder and a bit for a latch unit. When you buy the lock, open it in the store, and read the directions to see what bits you need. Buy them at that time if you don't own the right ones.

The only installation steps you might have trouble with, even with the clear instructions, are shortening the cylinder screws and shortening the connecting pin in cylinder locks. The screws go through the door to hold the cylinder to the lock face. They are manufactured with two or three indentations along their lengths, allowing segments to be broken off neatly to fit your exact door thickness (*12*). Following the specific lock directions, try the cylinder with the screws at normal length. If they go all the way in and the lock is held flush and tight, fine. If the lock is loose, or the screws don't go all the way in, you have to shorten the screws.

You must be careful breaking a segment off. Hold the segment to be removed with a pair of pliers, just below the crease (*12*). Now try to bend the screw against a hard surface. Some screws will break off easily, others will give you trouble. The important thing to remember is not to damage the good threads with the pliers, and not to bend the good part of the screw. You can use a second pair of pliers on the screw head, but, again, be careful not to damage the head. If you mess up the screw, don't panic. You can usually find replacements at the hardware store.

The cylinder pin offers the same problem, only being flat and having no threads to worry about, it is easier to break off neatly. This is the part that actually turns as you turn the key. The main thing is not to twist it or bend it out of shape. If you do bend it accidentally, don't be afraid to bend it back as flat as you can.

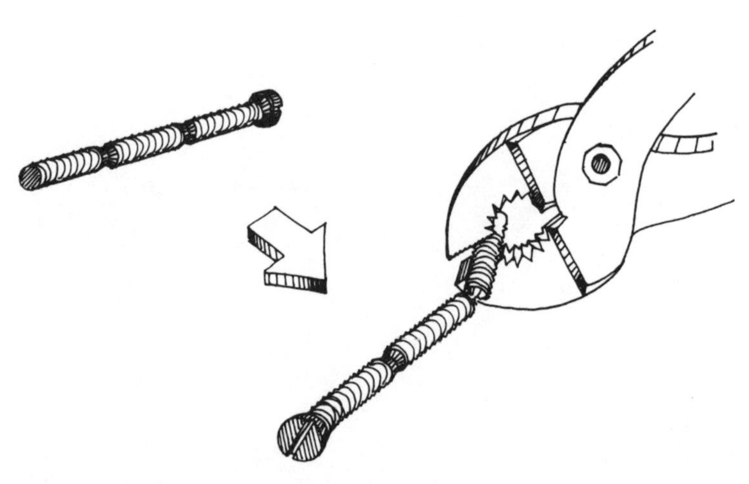

Fig. 12. Shortening a lock screw

SOUNDPROOFING

Soundproofing is an area misunderstood by many people. Putting egg cartons or rugs or acoustic tile on your walls will *not* soundproof your room completely; it will only absorb a good deal of the sound in the room. Recording studios and the like have similar material on the walls, but it is only one component of an entire system. The other component of soundproofing is stopping or at least minimizing the transmission of sound from one room to the next.

When building a wall or a whole room from scratch, it is easy to make it relatively soundproof. It is much more trouble to soundproof an old room. In some cases there is practically nothing you can do unless you're willing to spend a great deal of money.

To understand just what you can do, you first have to understand a few simple things about sound and how it travels. Then we can talk about what kinds of methods to use for specific sound problems.

Two Types of Sound

Airborne sound

Airborne sound travels through the air by setting up vibrations in the air's molecules. Each molecule excites its adjoining molecules and so on, right up to your ear, which picks up the vibrations and transforms them into the perception of sound. If a wall is between the sound source and your ear, the air molecules hit the wall and make it vibrate, which in turn makes the air vibrate on the other side and continue on to your ear. The more soundproof a wall is, the more it stops or dampens a lot of the movement; in effect, it "uses up" the vibration (*1*). A 2′-thick lead wall would stop vibration by sheer mass; more common sound-control walls substitute sneaky construction techniques to minimize the transmission of the vibration. And of course a soundproof wall will have no leaks, cracks, or holes for the airborne sound to leak through.

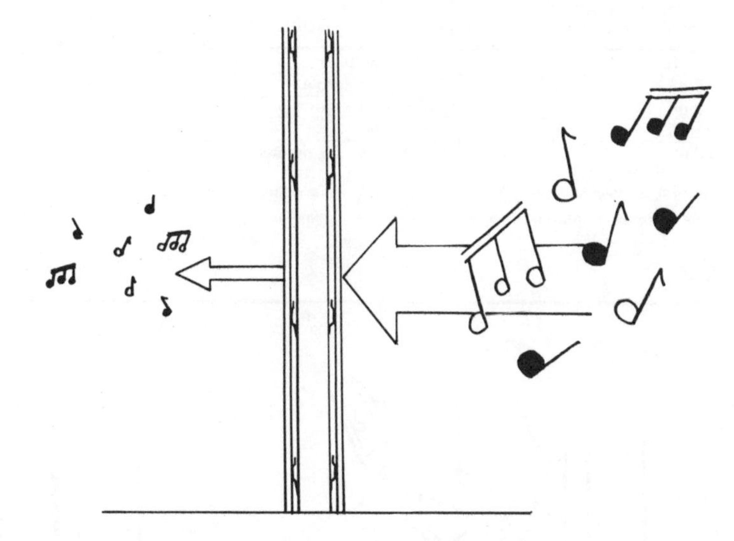

Fig. 1. How a sound-control wall cuts down what passes through it

Impact sound

Impact sound is generated by the impact of some object hitting part of the building structure. A sound like footsteps is transmitted by the structure itself along the joists. This is one of the hardest kinds of sound to stop. You can often hear people walking on the floor above you, even when you can't hear them talking. But if you were upstairs with them you would hear their voices much more than their footsteps. Machines such as washers, driers, and even vacuum cleaners cause an impact sound by vibrating against the floor and making it vibrate. If they could float in air, they wouldn't be as loud. They do radiate some sound through air, but the main problem is the impact sound. Hi-fi speakers transmit airborne sound, but when they sit on the floor they also create impact sound. Their strong vibrations are transmitted

RELATIONSHIP OF SOUND INTENSITY, LEVEL, AND LOUDNESS

INTENSITY (RELATIVE ENERGY - UNITS)	SOUND PRESSURE LEVEL (DECIBELS)	LOUDNESS
100,000,000,000,000	140	Jet aircraft and artillery fire
10,000,000,000,000	130	Threshold of pain
1,000,000,000,000	120	Near elevated train
100,000,000,000	110	Inside propeller plane
10,000,000,000	100	
1,000,000,000	90	Full symphony or band
100,000,000	80	Inside auto at high speed
10,000,000	70	Conversation, face to face
1,000,000	60	
100,000	50	Inside general office
10,000	40	Inside private office
1,000	30	Inside bedroom
100	20	Inside empty theater
10	10	
1	0	Threshold of hearing

NOTE:

The decibel number represents a ratio (actually 10 x the logarithm) of the Intensity measured to a reference intensity roughly equivalent to the threshold of hearing.

SUBJECTIVE EFFECT OF CHANGE IN SOUND PRESSURE LEVEL

CHANGE IN SOUND PRESSURE LEVEL	CHANGE IN APPARENT LOUDNESS
3 dB	Just perceptible
5 dB	Clearly noticeable
10 dB	Twice as loud (or 1/2)
15 dB	Big change
20 dB	Much louder (or quieter)

Glenn A. Kahley; Vincent G. Kling and Associates; Philadelphia, Pennsylvania
Lyle F. Yerges, Consulting Engineer; Downers Grove, Illinois

Fig. 2

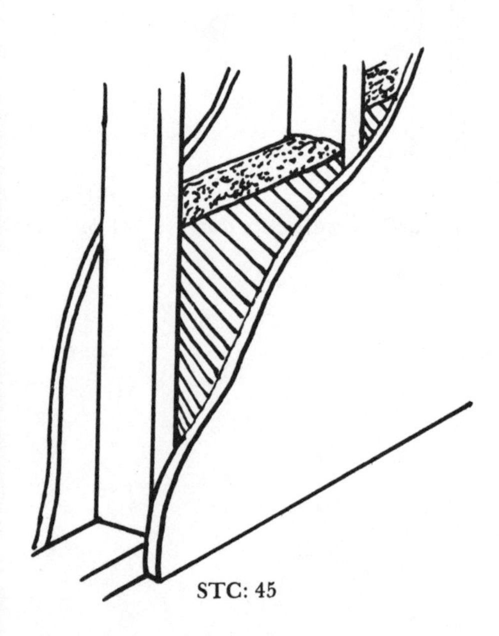

Normal wall with insulation — *A*

Double wall with insulation on one side — *B*

Wall with sound clips and insulation — *C*

Fig. 3. Different Walls and Their Ratings

through the speaker box to the floor. Impact sound is best stopped right at the source—a carpet under your neighbor's feet, some kind of floating platform or thick rubber pad for machines: something to soften the actual impact.

Sound Level

The relative *loudness* of sounds is shown by their decibel (dB) levels. A 3 dB change is about the smallest the human ear can detect. Any 10 dB rise is approximately equivalent to a doubling of the loudness, a 20 dB rise a quadrupling, and so on. Similarly, a 10 dB drop seems as if the sound is half as loud. Thus a sound of 50 dB is twice as loud as one of 40 dB. The table here will help you estimate the sound levels you're dealing with, and how much soundproofing you need (*2*).

A note to confuse you

The above information is all you really need to know about decibel levels for simple soundproofing systems. If you do further research, however, you may get confused by talk of *intensity* and *loudness* levels as related to decibels. The problem is that intensity and loudness are two different qualities, and not directly proportional. Many books incorrectly use the two terms interchangeably. The decibel level is really an objective measurement of the *intensity* of sound, which can be accurately found by instruments. *Loudness* is a subjective reaction to intensity and cannot be measured on any instrument. We *use* the decibel levels, however, to get a *relative* idea of loudness.

Sound Absorbers

Cardboard egg cartons, rugs, acoustical ceiling tile, sound board, fiberglass insulation, foam and pillows are all *sound absorbers*. They are porous, full of air, and sound tends to get "trapped" inside the air pores, like water in a sponge. The sound energy is actually absorbed and turned into heat. They also work because they do not vibrate very much.

Sound absorbers are indispensable to any sound-control system, but alone they do little to prevent sound transmission. They do prevent the sound in the source room from increasing in intensity by cutting down the reverberations, the sound reflections bouncing off the bare walls. This means less total sound to soundproof against. However, the *location* of the sound absorbers in the source room has no effect on the transmission to the next room; it's simply the overall absorption that matters, as far as the adjoining room is concerned.

Sound in a highly sound-absorbent room seems muffled. It leaves the source and dies. In a bare room, the reverberations may cause the sound to be indistinct or foggy. You may have trouble hearing over background noises. You can hear the difference in quality if you try talking in "soft" and "hard" rooms. Clap your hands, ring a bell. If you can't find such rooms, put a pillow in front of your face and speak into it; then stand in front of a bare wall and speak.

Thus you can control the *quality* of sound in a room by adding or removing sound-absorbent materials. Soften it with rugs on the floor or walls, drapes, easy chairs. If that's not enough then try acoustical ceiling tile or panels. Ceiling tile is enough for most living situations.

Sound absorbers are rated as to their efficiency at absorbing sound waves. This is called the *noise-reduction coefficient*. The highest ones are about 0.80 to 0.85, but the 0.50 to 0.60 range is fine. (An absorber with a 0.60 coefficient will absorb 60 percent of the sound that hits it.) The coefficient is a useful guide to tell which materials are better. It won't help you very much in telling how much sound you can eliminate.

Sound Control Walls and Their Construction

In many home situations, a sound-control partition is all you need to isolate the sound between two spaces—no extra floor or ceiling construction is needed. It must, of course, span from floor to ceiling and wall to wall. Such walls use special construction methods to break down the vibration paths from one side to the other.

Partitions are given an STC (Sound Transmission Class) rating by various acoustical associations. A wall of 51 STC stops approximately 51 dB of sound; 80 dB of sound on one side is heard as 29 dB on the other. It will seem about twice as good as a wall rated at 41 STC (10 dB less).

The STC rating is an average of a wall's protection against all sound frequencies under ideal conditions. Most walls aren't as effective against low-frequency sounds like an electric bass guitar as against middle- or high-frequency sounds; but for usual living conditions you needn't worry.

Background noise in one room serves to mask or cover noise from the source room. In a quiet home, just the background noise alone—street noise, radiators, all the sounds you never notice that are there all the time—probably have a level of at least 30 dB. This will mask 30 dB of noise from the next room. In such a home, a 50 STC wall would be adequate control for up to 80 dB of noise, since background noise will mask the remaining 30 dB. A room with relatively soundproof walls, floor, and ceiling will have little masking noise. Say the room is rated at 50 STC, and you play a radio in it at 70 dB. It probably can't be heard in the adjacent room above the usual background noise. But play the radio in the adjacent room at the same volume, and you'll probably hear it in the so-called soundproof room, where there is no masking noise.

A simple stud wall with ½" plasterboard on both sides has a 33 STC rating. This is a standard partition, with little sound-control value. The connection between the two sides is *too* complete—the structure transmits the sound easily. Also, the sound intensity is increased by reverberation *within* the narrow air cavity. By adding sound-deadening board or 3½"-thick mineral-wool blankets (*3A*), however, you can raise its STC way up to 45. The insulation cuts the reverberation between layers and absorbs a certain amount of the sound.

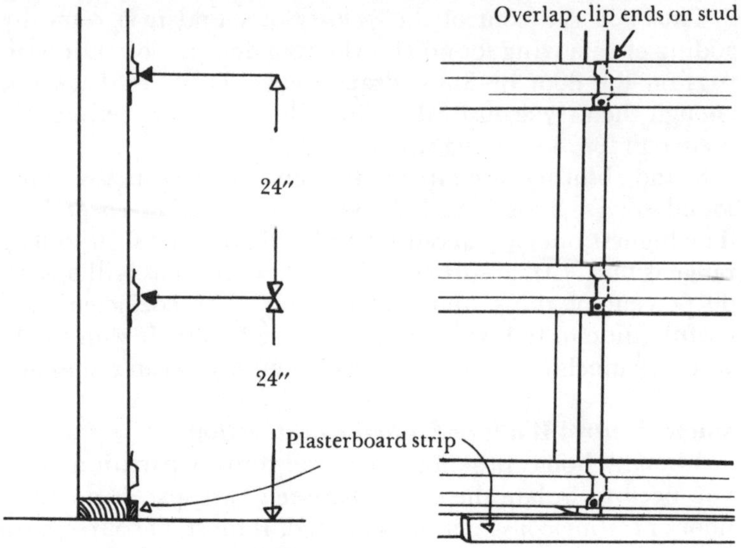

Overlap clip ends on stud
24″
24″
Plasterboard strip

Fig. 4. Set-up for Sound Clips

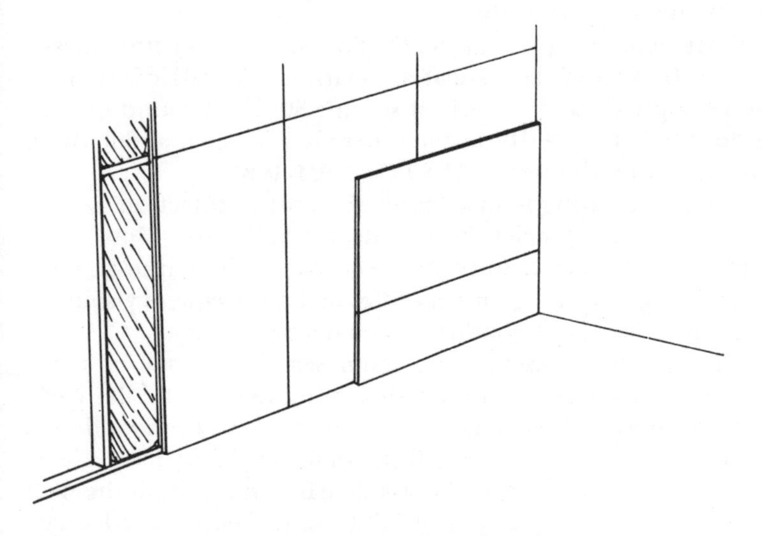

Fig. 5. Alternating two layers of plasterboard

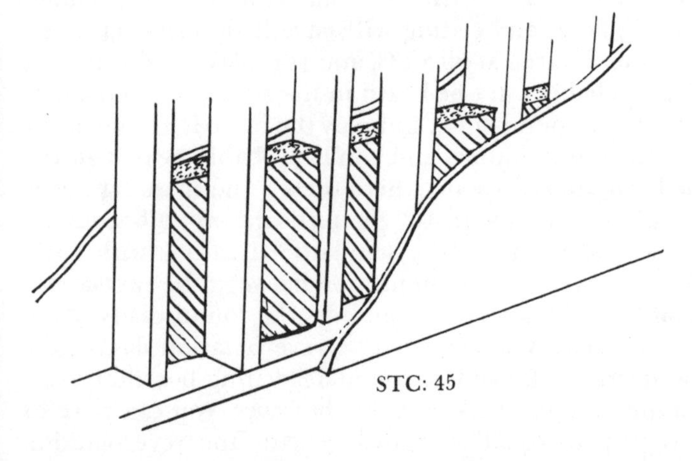

Fig. 6. Double Stud Wall

A lot more sound is cut out if the two sides are not directly connected. There are two very good ways to isolate the sides. One is simply to build two stud walls, at least 1″ apart, with 5/8″ plasterboard on the outer surfaces. Without any insulation, this wall has an STC of 45. Three-and-a-half-inch insulation in one of the walls increases the STC to about 51, good enough for any home (*3B*).

The second method is to build a regular stud wall, then to attach special metal channels, called RC-1 sound clips, horizontally to one side of the stud wall (*3C*). Place them 24″ on center from the floor (*4*). At the bottom, lay a 3″ to 4″ strip of plasterboard against the studs for backing (as protection against good swift kicks); don't fasten into it. Start the first clip just above it. Screw the 5/8″ plasterboard to the clips (use a screw gun, since you need dozens of screws); nail the plasterboard as usual to the other side of the wall. The clips hold it away from the wood, eliminating direct vibration transmission. The result is a 45 STC without insulation, and 52 STC with insulation—about the same as the double-wall construction. The advantage is that the sound clips are cheaper, easier, and quicker to put up than a second wall. The only disadvantage is that you need a screw gun to attach the plasterboard to the metal clips. Screw guns are expensive, but you can rent one cheaply.

You can considerably increase the STC of any of these walls by adding a second layer of plasterboard. Apply one layer horizontally and one vertically. Try to stagger the seams (*5*). Two layers of 5/8″ on each side of the double wall raise the STC about 5 dB, for instance. An RC-1 construction with two layers of 1/2″ plasterboard has a rating of 49 STC without insulation and 59 with.

In double-wall construction, the more space between the walls, the better the sound control, particularly in the low-frequency range. A 6″ space between plates might be as much as 5 dB better than a 1″ space.

Stuffing *extra* sound blankets in the cavity of either construction will not help all that much—at most a couple of decibels. For the cost, you'd be better off putting up an extra layer of plasterboard.

Another construction method you might come across is the staggered-stud system—essentially two staggered rows of studs sharing single wider plates at top and bottom (*6*). The problem, of course, is that the plates still provide a connection from one side to the other. With just a little more wood and labor you can have better sound control with two separate walls or with the sound clips.

Openings and leaks in walls

A little bit of manic compulsiveness is a great help in building a soundproof wall. A 1/2″ hole, a little crack, and the sound, like water, pours through the best wall. When you install the mineral wool, be sure it covers all the open space between the studs—side to side, and top to bottom. If you use sound-deadening board, be sure the seams meet tightly, and that it fits well against the floor, ceiling, and existing walls; fill any cracks with thin slices of soundboard, insulation, or caulk.

Take the same care with plasterboard. Butt the seams tightly. Cut closer to ceiling and floor contours than you would normally. The smaller the crack, the less chance of the joint compound cracking later. If you must leave large cracks or holes—around obstructions, for example—stuff small pieces of plasterboard in, then plaster over. Caulk the seam at the floor with a good silicon sealant. Put tape and joint compound on all other seams, including the wall/ceiling seam.

The big sound leaks are usually through electrical or plumbing lines, or through doorways. If at all possible, run any pipes through side walls; put a rubber strip around the pipe where it enters the wall to minimize vibration transmission; plaster or caulk the hole well. Unfortunately the pipes themselves transmit sound. Use flexible BX cable rather than hollow electrical conduit—the cable is a poor transmitter of sound.

Doorways are usually the main sound leak. If possible, place your entrance in a side wall rather than in the main "barrier" wall. Use solid-core doors, not hollow core. If the door must be in the main wall, one standard door will probably not be good enough, particularly for walls of 45 STC or more. Special soundproof doors and jambs, the kind that go "whoosh" when they close, are expensive. Check your lumberyard, or the acoustical-material manufacturers in your Yellow Pages.

The other alternative is to use two standard solid-core doors, one on each side of the wall (7). This is easy with the double-wall construction; with the RC-1 method, use 2x4 studs, not 2x3s; to insure enough space for the two doors and their stops. Fit each door as airtight as possible—only 1⁄16″ to 1⁄8″ gap all around. With luck, you'll locate special gaskets, or rubber door seals, for the door to close into—again check the acoustical-material manufacturers. Otherwise, fit the door stop tightly against the door, and weatherstrip all the cracks well.

Doorknobs and locks transmit a lot of sound. The manufactured soundproof doors have special hardware designed to thwart sound passage. A cheaper, easy solution for a standard door is simply not to use a knob or lock. Instead, put a handle on one or both sides; use two or three cabinet catches—the roller types are good—to hold the door shut.

You may want a window between rooms—perhaps you want to cut the noise, but still to keep an eye on the situation. A double window of regular plate glass with an air space between works fine. The glass should be well sealed, preferably with an acoustical or silicon sealant. Set the glass on thin blocks of rubber, isolating it from the wood frame. Special acoustical glass is prohibitively expensive at lumberyards, but not too bad direct from the manufacturers. Acoustic panes

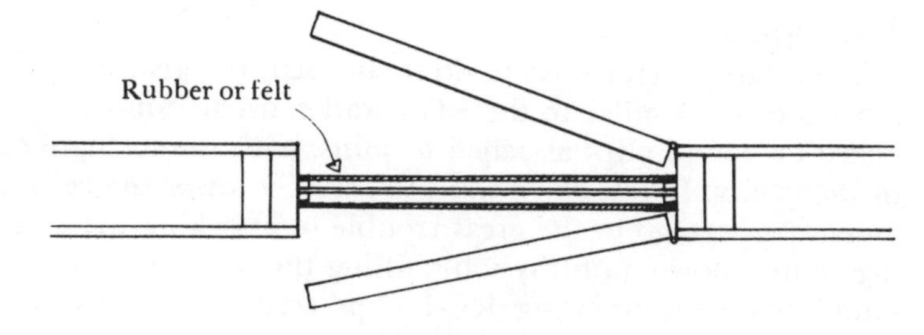

Fig. 7. Double Door in Sound Wall

are made of two sheets of glass laminated together with a thin air pocket between.

Special Sound Problems

Now that you know all about soundproof walls, know also they won't solve every problem. Airborne sounds can sometimes travel around the wall, particularly through the flanking walls, the ones that are common to both rooms. In general, when your new wall has an STC of more than 50, the flanking walls become important. The sound can simply travel along the flanking wall, right past the control wall, and out again into the "soundproof" room (*8*) ; or it can transmit its vibrations directly to the sound-control wall at the common joint. You should isolate your 50 STC wall from the old ones by using rubber strips between the end studs and the old walls. Also seal any holes and cracks in the old wall.

The only other hope is to build a sound-control wall in front of the flanking wall. Just put plasterboard on the front side—the flanking wall serves as the second half of the system.

Another weakness of sound-control walls is that they don't do well against impact sounds, such as someone stomping on your ceiling or even in the next room. Impact is best attacked at the source. The easiest thing is to put a rug on the floor where the noise originates. If you have a heavy-footed upstairs neighbor, the cheapest sound-control you can get is to buy him a carpet and tell him to walk barefoot. (Good luck.) Acoustical tiles on your ceiling won't help much. They only *absorb* sound within your room. The only real alternative is to add a new lowered ceiling. This is a big job, of course.

New ceiling

There are several ways to do it. In each, the goal is a ceiling somewhat similar to the RC-1 wall construction—plasterboard on sound clips attached to joists, with sound blankets in the cavity. If you don't want to actually lower the ceiling much, you can go to the great trouble of knocking the existing ceiling down (a filthy job), filling the joist cavities with sound blankets, screwing RC-1 clips across the joists, and screwing ½" plasterboard to the clips.

The easier way is to bolt 2x6 beams to opposite walls an inch or so below the existing ceiling; attach joists between them with joist hangers; staple fiberglass insulation between the joists; fasten the sound clips across; and screw the plasterboard to them. Run the joists across the shorter room dimension, of course. This method is easier, cleaner, and gives better sound control. It's still a lot of work.

New floor

In rooms designed for music rehearsal, you may want to prevent impact sound from getting out. A rug with a thick pad underneath is the easiest and cheapest method. It won't do a great job, however, for very loud music. The answer is a floating floor, constructed as shown here (*9*).

Lay the bottom row of 2x3 joists on edge. They needn't be nailed down—the weight of the finished floor will keep them in place. If the existing floor is bad—a lot of cracks, holes, etc.—you can first lay a subfloor under the new joists of ½" ply or composition board. Lay sound blankets between the joists. Lay a second layer of joists across the first layer, with strips of some sort of resilient material, like heavy rubber ⅛" or ¼" thick, between the pieces where they cross. You might glue these strips down with mastic, just to hold them till the upper floor is set, but *do not* tack or staple the strips down, since the metal would transmit vibrations. Also, *do not* nail the top layer of joists to the bottom layer. The whole idea is to break the connection.

Nail your flooring, ¾" ply or composition board, to the top joists. Stop the floor and joists just ⅛" or so short of the walls. Caulk or weatherstrip this gap around the perimeter of the top flooring layer. Again, the idea is to break the connection between the floor and the existing structure.

This floating floor may sound to you as if it will slide around in a slight breeze; it won't. The top layer is one massive raft of joists and plywood that won't budge one bit.

Summary of Noise Problems and Solutions

To control any sound, understand it. Find its source and try to locate its path. Is it airborne or impact sound? You can, of course, isolate two areas from each other with a new sound-control wall. With an existing wall, if a lot of sound is coming through, put up a sound-control wall in front of it. You lose a few inches of space, but gain quiet. Be sure, however, that the sound is not also traveling through a flanking wall, or along the floor or ceiling. The same wall will of course keep *your* noise out of neighbors' ears. If the noise seems to be coming from everywhere, there is not too much you can do, short of building a whole room within a room.

Deal with impact sound at its source. Put vibratory machines on pads, drummers on floating platforms, upstairs neighbors on carpets. If *you* are the noisy upstairs neighbor, buy yourself a carpet. If a room is too "hard," its sounds too muddy, soften it with rugs, upholstered furniture, acoustic tile. If sounds seem too dead, remove soft materials.

Soundproof floors and ceilings are really beyond the requirement of most home situations. They are more in the line of rehearsal studios. A good music studio is essentially a room within a room—four soundproof walls, a lowered ceiling on top, and a floating floor. Such a room is not all that difficult to build at home, but it's expensive. The materials for a good-sized room can run from one to two thousand dollars. If some of your walls are already in good shape, or your floor and ceiling pose no problems, you can get away with a lot less. Quiet is rare these days. If you have the resources and inclination, a relatively soundproof room is a gracious touch for music, study, relaxation, sleeping, or any concentrated endeavor.

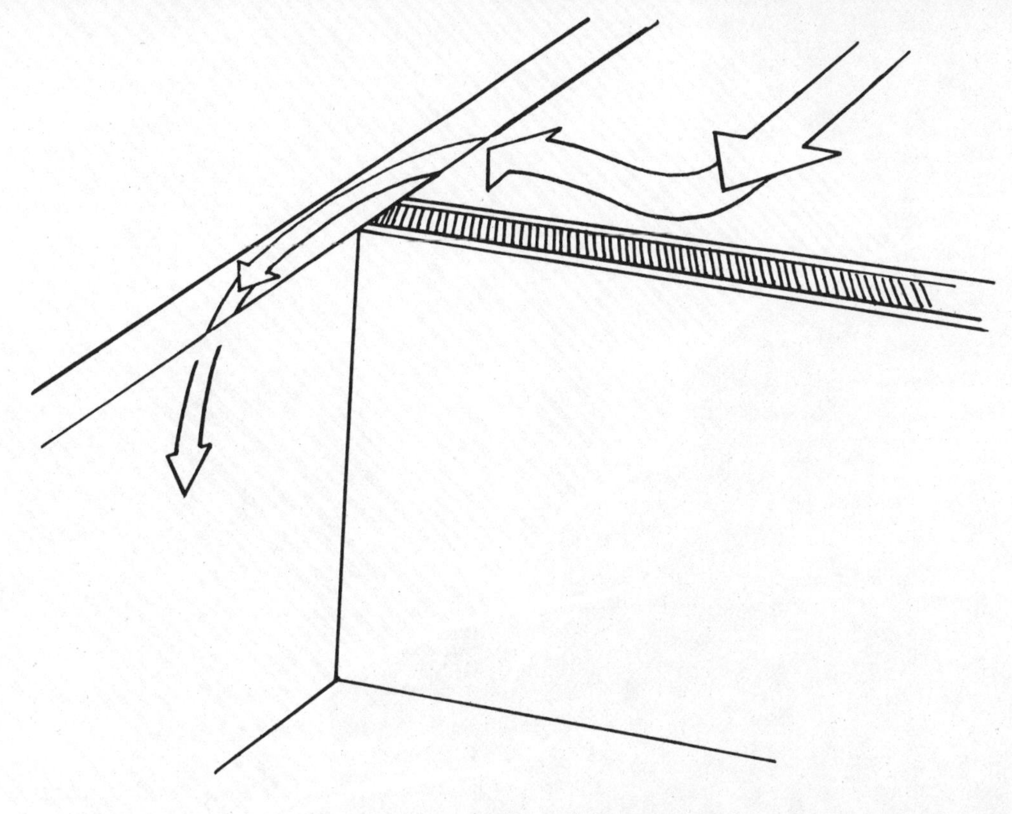

Fig. 8. How sound can pass around control wall

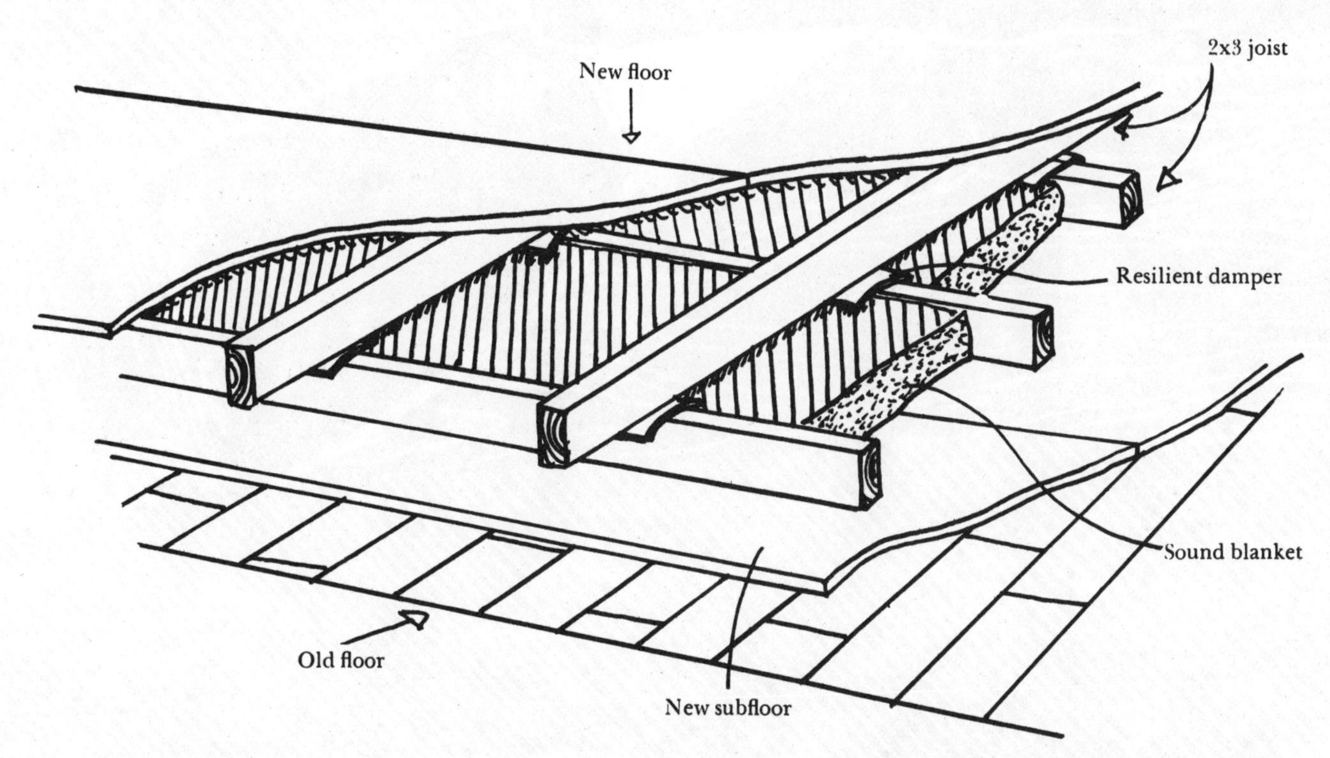

Fig. 9. Setup for Sound-Control Floor

FINISHING

12/FINISHING

How to Enhance and Protect Your Manifestations

This chapter deals with finishes on various surfaces. First we deal with the most important type of finishing for the carpenter, that of raw wood—how to protect and bring out the natural beauty of the wood in your new projects. We describe the different types of finishes, from stains, varnishes, and shellac, to paints; what they're good for and how to use them.

Wood finishing is largely a matter of taste; unfortunately, space does not permit us to present more than the basic techniques of finishing in this book.

A short section follows on refinishing furniture, concentrating on the essential process of removing old finishes. Finally, we deal with the refinishing and painting of the walls and surfaces in your home—how to remove old paint and wallpaper, and even how to expose a brick wall.

Furniture

A good finish, properly applied to a well-prepared surface, protects your projects from water, warping, nicks, and scratches. It can also enhance or change the color of the wood grain, and give the piece an overall luxurious glow. In this way you can closely match the wood to other wood pieces in your home.

Stains color the wood, bring out the grain. An oil-based stain also penetrates the wood and acts as a preservative against moisture. Wax, linseed oil, shellac, varnish, and polyurethane give stronger protection and can be applied either over or without the stain. And of course you can paint your wood.

Preparation

For an even and long-lasting professional finish, surfaces must first be sanded and cleaned. Check that you completed all the little construction details that might affect the sanding and finish—did you set all the nails, countersink the screws, putty or plug where you wanted? Did you rasp or sand down any large irregularities? You should now *remove,* if possible, any hardware like handles or hinges. The sanding will be much easier and the hardware won't get discolored by the finish you apply. Reattach them when the finishing is complete and the piece is dry.

The type and amount of finish sanding you do on a piece depends partly on what kind of wood it is and partly on just how fine you want it to be. Different types of wood are treated differently. For rough construction-grade wood, sand first with a coarse-grit paper and gradually work your way down to the finer ones. Don't expect to get the surface supersmooth. Much rough wood may even look better—at least more rustic—left unsanded. Clear lumber, hardwood, and veneer plys are fairly smooth when you get them. Start with a medium or fine paper and work down to the extra-fine. To find the correct grade of paper to begin with, test several grades on a scrap of the same wood the project is made of. Paper that is too rough will tear up the wood surface. Too fine a grade will take hours and mountains of paper to sand away rough spots. Take your time and experiment. Select that grade that makes the wood just a bit smoother; start with that and progress to the finer grades. Never skip more than one grade number at a time; it may seem like a lot of trouble, but it's really the quickest and most efficient way to get a smooth clean surface. (See Sandpaper in "Hand Tools.")

Always sand with the grain. Sanding across it causes tiny rips and burrs to appear. If you sand by hand, always use some type of sanding block. Your hand or finger may be fine for tiny areas but on larger ones you need a smooth backing for the paper. To sand in hard-to-reach places like interior corners, wrap the paper around a small piece of wood or cardboard. For very large jobs, use a power sander—if you don't own one, you can rent finishing or belt sanders. The latter will save you hours and hours on rough wood.

If you have exhausted all the grades and types of sandpaper and you still want to go smoother, use fine steel wool. Start with a low number grade and work down to 4/0, the finest. Wear gloves to protect your hands from the fibers, or cut an old rubber ball in two, stuff the steel wool in one half, and hold the ball as you rub.

When you are satisfied with the surface, you must dust and clean it. This is important, since any residue gets gunked up in the finish coatings. Use a tack cloth—a cheesecloth that has been impregnated with a "tacky" gluelike substance that picks up dust when rubbed lightly over the work. Any protruding burrs from raised grain will snag the rag, so get your surfaces smooth. Tack cloths cost less than a dollar. If you cannot find one, use plain cheesecloth, passing it over the work a few times to catch the resettled dust. You may also want to vacuum projects that have enclosed spaces.

Clear Finishes

Linseed Oil

For those who like the natural feel of wood, linseed oil, which darkens the wood slightly and brings out the grain, is the only acceptable finish. *Boiled* linseed oil is usually used. There are brand names of oil finishings that are made up mostly of boiled linseed oil. You can make your own by mixing two or three parts oil to one of turpentine or the equivalent. Oil will not keep the wood clean like wax or varnish. Marks will still show on the surface. Use it mainly on objects that you want to preserve but will not have much hand contact. There will be an oily odor that will last for a while.

Stain

Any stain looks different when applied to different wood. Your store may have charts for the good stains to show how they look on various woods, but even these are not totally reliable. Test your stain color on a scrap piece of the project wood, or on the project itself in a spot that doesn't show. A heavy coat or more than one coat will give a darker color. The color also lightens a bit as it dries.

To spread the stain, use a soft cloth, such as cheesecloth or an old undershirt. Paper towels tend to shred despite all those TV ads. A brush is fine but a little harder to control. Again, experiment with your application. Spread the stain evenly; you do not have to follow the grain. Better stain can be poured directly on the wood and spread without any blotting, assuming you spread it within five to ten minutes. Cheap stain blots very quickly, so pour it on the cloth first and then spread. Always work from the highest areas downward; then you can easily see and spread the drips before they blot. Wear rubber gloves; you want to stain the wood, not your hands.

Certain irregularities in the wood affect the evenness of staining—you may or may not feel they enhance the look. Knot areas may resist the stain due to the tightness of the grain. If so, go over these areas lightly with a rough sandpaper to open up the grain; restain carefully. End grain stains darker for the opposite reason—the grain is more open in these spots. You can try sanding these areas finer. Another problem may be glue that somehow snuck onto the finished surfaces. Though it may dry clear, glue is stain-resistant. The offending glue should be sanded off beforehand.

Read the directions on the can for drying time. Make sure the first coat is completely dry before you apply the second. One coat alone is adequate, but we usually put on at least two, or one with another kind of finish over it. For cleaning or thinning stain, use turpentine or any similar type of solvent.

Mixing Your Own Stains in Any Color

Buy a small tube of artists' oil paint at an art-supply store. Burnt umber and raw umber (browns) and burnt sienna and raw sienna (red-browns) are the most common earth colors, those closest to wood hues. Any color will work, however. If you want to stain your dresser turquoise, go right ahead.

Use one part boiled linseed oil and two to three parts turpentine. In a separate container, first thoroughly mix a very small amount of pigment into a little bit of turpentine (it dissolves more easily before the oil is added). Then pour in the proportion of turpentine you have chosen, add the oil, and mix some more. If you want it darker, mix more pigment with small amounts of turpentine separately, and *then* add it. This prevents lumps. Use as you would a store-bought oil stain.

It's a good idea to start out with a small measured amount of oil and turpentine, and an approximately measured tiny blob of pigment. Test out the result on a piece of wood.

Then make another mixture, adding more or less pigment according to your desire.

Paste Filler

For a completely smooth surface you will want to use a paste filler. This needs to be done only with the more open-grained varieties of hardwoods—oak, walnut, and mahogany for instance. The closer-grained woods such as birch or cherry need not be filled.

Filler is available in paste form and comes in many different colors to match the natural wood or a stain. You can even use a lighter color to accent the natural grain patterns.

Filler is applied after the stain is dry. Brush it on with a clean brush. Wait till it dries to a haze, about fifteen minutes or less. Wipe the surface with a rough rag *across* the grain (to help fill the pores) to remove the excess. Then take a clean rag such as cheesecloth and go over the surface thoroughly. Make sure you clean out any corners or edges; when the filler dries it will be very hard to remove. Let the piece sit for twenty-four hours to dry completely.

Finally, seal the surface with shellac thinned 50 percent with alcohol or turpentine; or, if you plan to varnish the piece, a mixture of one part varnish and two parts turpentine.

Varnish Stain

This is a stain mixed with a varnish, designed to save the trouble of applying separate coats of stain and varnish. It's messier, however, harder to control than either varnish or stain alone, and overlapping brush strokes may show when dry. We don't recommend this for really fine work.

Always use a brush to apply. Brush in one direction only, spreading evenly and working fast. Varnish stain gets tacky quickly, and hard to spread smoothly. Make sure you are working in a relatively dust-free area, otherwise the particles become imbedded in the finish. Let the piece dry in a warm room for at least twenty-four hours. Follow the specific directions on the can.

Varnish

Alone or over a stain (and filler if necessary), this is the finish on most fine woodwork. Make sure you use cabinetmaker's or furniture varnish, not marine varnish designed for use on boat hulls. (Marine varnish *is* useful for kitchen counters, and other work surfaces.) Varnish comes in high-gloss finish and in various semiglosses such as 80 percent, 50 percent, 25 percent gloss. It is clear and will darken wood slightly, enhancing the grain.

The surface of the wood should be clean and dry. The first coat can be thinned with turpentine. Brush on in one direction. Drying time should be at least twenty-four hours per coat, or whatever the can says. Fine cabinetmakers apply as many as thirty to forty very thin coats of varnish. The more you use, the harder and smoother the surface will be. You can be happy with two or three coats, however, or as many as six for a fine piece of furniture. After each coat except the last, sand the surface carefully with very fine steel wool to smooth out the tiny imperfections. Make sure the surface is completely dry. Dust the surface clean and apply the next coat. Repeat this process as desired. Rub the final coat with a soft lint-free cloth and a mixture of fine powdered pumice with water or paraffin oil, with the grain. Wash clean with sponge and water, and dry with smooth cotton or chamois.

Wax

Cake or butcher's wax is not a car wax, but a harder, more durable type designed for the wear and tear furniture undergoes. Two to four coats offer a protective and attractive finish over bare or stained wood. It tends to slightly darken wood. Wax offers extra protection over a varnished surface.

Clean unfinished surfaces thoroughly. Apply with a clean soft cloth, in a circular motion. Be generous but don't put it on so thick that you see ridges of wax. Let it dry to a haze—about five minutes (or as can says). Then buff it with a different cloth or a soft shoebrush. Don't wax the whole piece at once; work in sections to allow full time for drying, and to be able to keep track of your progress. Always buff between coats.

Polyurethane

Next to varnish this is the strongest finish. It is extremely hard and durable, a clear plastic originally designed for bowling alleys and now used widely for floors and furniture. It comes in gloss, semigloss, and satin. Satin is the dullest, but it still has a little gloss. Some paint stores have large drums of polyurethane and can sell you individual gallons much cheaper than the big brand names—look around.

Thoroughly clean the surface to be finished and apply with a brush or roller, smoothly and quickly. Polyurethane is probably the easiest finish of all to apply neatly. Use at least two coats for furniture and at least three for floors. The more coats, the more you can build up a thick hard surface that will fill small indentations in the wood.

With some polyurethanes, sanding the first coat afer drying is necessary for proper adhesion, unless you apply the second coat within twenty-four hours of the first. Check the can's directions.

Take adequate safety precautions. Polyurethane is extremely flammable and noxious to breathe. Ventilate the area well, extinguish any open flames, stoves, and heaters, and wear a breathing mask and gloves.

Shellac

Shellac is a very hard, but very brittle, quick-drying finish. It is also *not* waterproof. Thus it is not particularly reliable as a protective finish. It's good as a fast, easy finish for small projects that won't undergo much abuse; it is also effective as a sealer.

It comes in orange and white (which dries clear), gloss and flat. The orange is not like a paint, of course, but gives an orangish tint to the wood. It is best for darker woods. The common "four-pound cut" shellac should be thinned with an equal amount of denatured alcohol before application.

Rhoplex, the Wild Finish

This is an intriguing material whose decorative effects are just being discovered by artists.

It's a milky-white polymer emulsion that dries clear and very glossy. It's water-resistant, but nowhere near as *hard* as varnish. The unique thing about it is that it does not necessarily sit flat on the surface. If you swirl your brush the swirl marks will stay. The more you manipulate the stuff on the surface, the more thick and bubbly it gets.

Rhoplex was designed originally as a pigment extender for commercial paints. Thus you can add just a tiny bit of water-based pigment and get a whole lot of tinted Rhoplex. The result is not clear like a stain, but more opaque, like a jellyfish. The pigment can be any water-based paint or ink.

Experiment with the Rhoplex. Try it for decorative areas on door panels or your projects. Or just play with it on scrap pieces. You can get many odd effects—its many uses haven't even been scratched yet.

Buy Rhoplex at art-supply stores (it may be hard to find). The two most common types, both suitable for our purposes, are labeled AC-22 (for indoor use) and AC-234 (for outdoor use). Your brushes can be cleaned with just hot water, but clean them immediately.

Paint

Sealing the Wood Before Painting

If you want to paint unfinished new wood or refinished wood, you first have to seal it. Otherwise some of the paint will be absorbed into and expand the wood fibers, making the surface rougher and unattractive. Sealer fills the pores. It also stops water stains from showing or bleeding through the paint. It is usually white and semigloss in appearance. Apply it with a brush, lightly. When dry, sand it lightly with 6-0 paper or steel wool. Clean your brushes with paint thinner or turpentine.

Shellac can also be used as a sealer. It was used as such for many years before the development of paint sealers. Thin it out first with five parts of denatured alcohol. Shellac is a better sealer for things like large knots.

Paints

Paints are decorative, and can cover a multitude of flaws; they also protect the surface. There are two main groups of paints—oil-based, such as enamels and alkyds, and latex. The

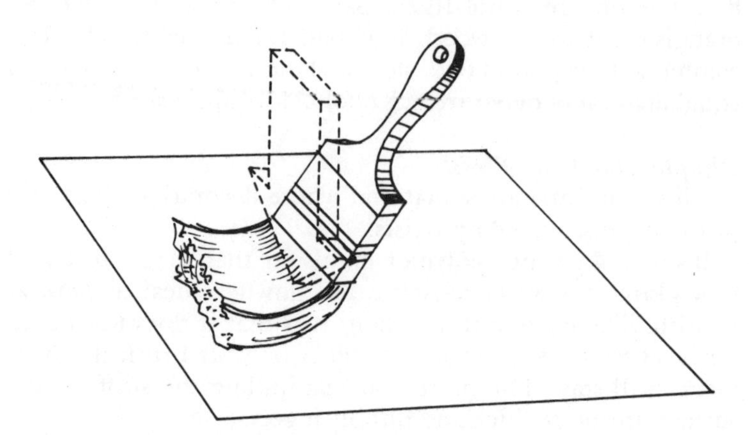

Fig. 1. By pressing the brush this way, you remove any material trapped at the top

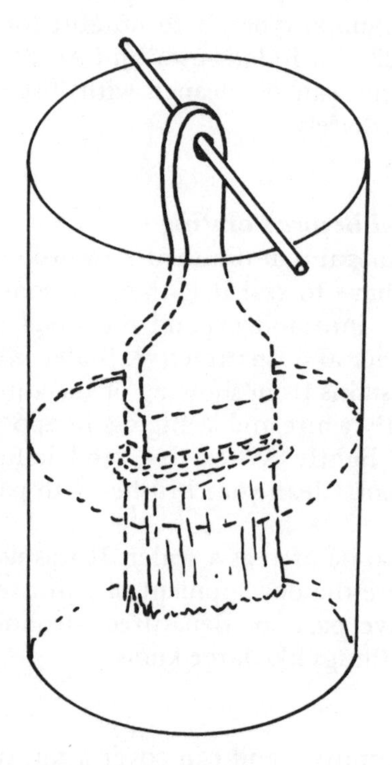

Fig. 2. How to store or soak a brush

oil-based paints are waterproof; enamels are usually high gloss, though semigloss and flat are now available. They are used mostly for bathrooms, kitchens, window sills, or on any furniture where you want protection and washability. Latex is a water-based paint, so it is not as waterproof. It is much easier to work with, though. It goes on more easily, dries in an hour, and drips can be cleaned up with water. Latex paints are mostly flat, but semigloss and high gloss are available; use latex where moisture is no problem.

Have the store mix any paint for you on their machine. Mix it again by hand, if you don't use it right away.

For the enamel and other oil-based paints, a natural-bristle brush should be used. For the latex or water-based paints, use an artificial-bristle brush such as nylon.

Try to get good brushes and good paint. It's worth it in the long run. Cheaper paint takes more coats to cover and will not hold up as long. When you are buying white paint, read the label to check the pigment-to-vehicle percentages. Most good paints will tell you this. The better ones will have a higher percent of pigment. Try to get titanium as opposed to zinc. Zinc has a tendency to yellow.

Place the piece to be painted on blocks, if possible, so the bottom edges will not stick to the newspaper or dropcloth underneath when the paint dries. In general, always work from the top down. If possible, set the piece so that most of the surface to be painted is horizontal. This minimizes dripping. With a piece like a bookcase, paint the under surfaces, let it dry, turn it over, and paint the top surfaces last.

Cleaning Your Brushes

Wipe the brush on scrap wood or newspaper to get off the excess paint (*1*). Clean your latex brushes with hot water and soap—under a faucet is fine. Squeeze the excess paint out with your fingers or paper towels. Use turpentine on enamel brushes. Soak an enamel brush in turpentine a short time, then press it on paper at the spot where the bristles enter the handle. This is where paint tends to accumulate—which is why we try not to dip the brush into the paint can that far. Pressing from both sides loosens it up.

After cleaning, soak your brushes in the proper cleaner overnight. Suspend the brush so the bristles do not touch bottom and get pushed out of shape (*2*). Suspending also prevents pigment from building up and clogging the bristles. The next day, reclean the brushes if necessary. Store your cleaned brushes hanging free—that's what the holes in the handles are for. Otherwise the bristles might get compressed into some unfortunate shape. Follow all these steps and your good brushes will last indefinitely.

Refinishing Furniture

Paint and Varnish Removers

Refinishing old furniture requires the removal of the old finish or paint first. Does your piece have a lot of intricate scrollwork, undercuts, interior corners? The finish on such a piece is so difficult and time-consuming to remove by hand that you should consider taking it to a professional refinisher's shop. These shops can dip the piece into a large vat of chemical remover and do a quick, good job. Ask a few places for an estimate and decide if the time and aggravation saved are worth it to you.

To do it yourself, buy a can of paint or varnish remover. Unfortunately the flammable, more dangerously caustic type, seems to work best. The thicker kind—called paste remover—is better when you have to work on vertical surfaces, since it is less likely to run or walk on you. Check the directions on the various cans or consult with the hardware store people before buying one. Some types also remove stain, which you may not want.

Wear rubber gloves, goggles, a breathing mask, and long sleeves when using paint removers. Keep the room well ventilated. Place the piece to be refinished on blocks on top of thick layers of newspaper (*not* plastic dropcloths). Follow the instructions on the can exactly. Usually you apply the remover with a brush, then let it bubble and eat away at the old finish. Scrape it off with a putty knife at the time suggested by the directions, but be careful not to mar the wood by scraping too hard. Use steel wool for the finer areas. Dump all the residue in a glass container—the chemicals eat up metal or plastic. Remember that the old finish should come off easily. If it doesn't, don't keep scraping—apply some more of the remover, and let it set a little while longer. When done, wash the surface thoroughly with whatever solvent is recommended on the can—usually turpentine or alcohol, sometimes water.

Removing Stain

If the varnish remover does not remove the stain, you can use a commercial furniture bleach, after *all* the varnish, or other finish is off. Most bleaches come in two solutions which you mix just before applying. Wear rubber gloves, protective clothing, and goggles. Apply it with a brush, let it dry. Clean it off with water or whatever the directions suggest.

One problem with bleach is that it may also harm or remove the patina on very old wood—that mellow surface that comes only with age.

When all the finish is removed, clean and sand the surface very carefully, so as not to go through the patina. Then proceed as you would with new wood.

Walls and Ceilings

We all have areas in our homes that need some kind of refinishing. Perhaps it's crumbling paint in the bathroom, or some wallpaper you'd like to get rid of. Or maybe you just want to paint a room a new color. Maybe you just want to live with a new color.

Removing Old Paint

Flaking, bubbling, or peeling areas of paint must be removed and the surface fixed up or your new layer of paint will just peel off the same way. There's no easy way to do this. Professionals use blowtorches for large areas, but this is no technique to try on your own. You can try a paint remover, like the kind used for furniture, but it's a bit dangerous to use on a ceiling where the chemical might fall on your head. It's also expensive. We've done it for small ceilings, but we don't advise it. If you must, buy the thickest kind possible, so it has less tendency to drip.

The usual solution is to simply scrape the paint off by hand with a paint scraper. It's a lot of work. If you have a very large space, and a lot of bad areas, consider hiring someone with a blowtorch. Or plan on nothing else for a few days.

Wear goggles, and something to keep the flakes out of your hair. A breathing mask is a good idea—you may not notice the need till you wake up the next day with a sore throat. Use a sharp scraper; resharpen it if you need to. Start with the very loose areas and work outward. Tap bubbles and flakes lightly to start, then scrape out from them. Try not to chip into the wall itself. Always chip or scrape as far as you can go with the loose paint. Better now so it won't come down later along with your new paint. There's no need, usually, to take off the paint that adheres tightly. If there are any hairline cracks in a plaster surface, trace along them with the edge of your scraper. This will spread them out a bit and give you a space to be filled. The result with most walls will look like shallow lakes within a slightly raised layer of old paint. To prevent further peeling at the edges of old paint areas, taper the edges (of the lakes) with steel wool. You can also fill the lakes with joint compound or spackle. The latter, of course, gives a much flatter surface to be painted over. Fill any deep holes with plaster (or wood putty for wood surfaces) and then joint compound over it for a smoother surface. Wait several days, at least, before applying paint over new plaster or compound; otherwise the paint may crack as the plaster dries out.

Removing Wallpaper

Peeling wallpaper must be removed before you paint. Even if the wallpaper seems to be fine, you should remove it anyway before painting. Paint *may* cause it to loosen or peel later—you never know. That's especially likely with water-base paints.

There are several methods of attack. First try to pull off by hand as much paper as possible where it is loose, and try to scrape off the rest. This probably won't get you too far, but it will give you an idea of how thick the paper is and whether there's more than one layer. Next try soaking the paper with a sponge dipped in hot soapy water. This solution may dissolve some pastes enough so the paper will scrape off easily. Don't use boiling water unless you enjoy pain.

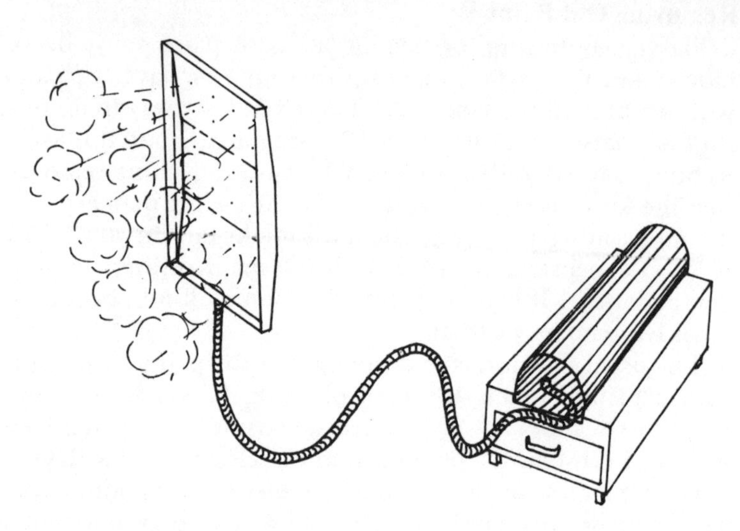

Fig. 3. Wallpaper Steamer

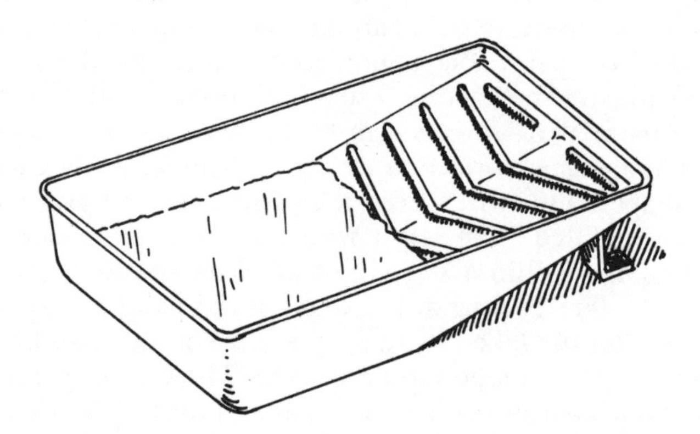

Fig. 4. Roller Pan

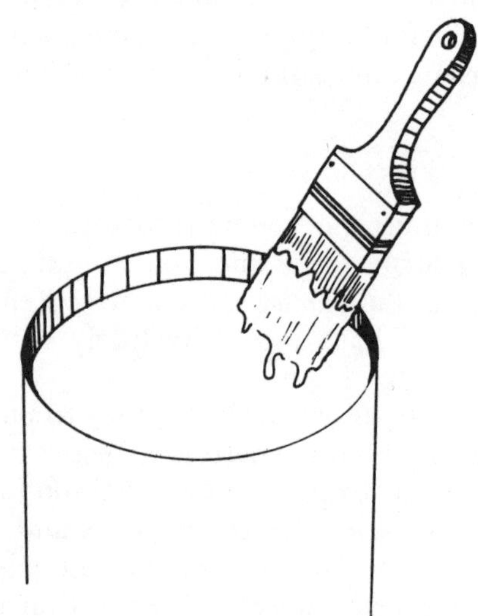

Fig. 5. Dip brush in about halfway

You can also buy commercial wallpaper removers, most of which are powders you mix with hot water and apply to the paper. It's worth a try, but in many cases this method seems no more efficient then soap and hot water. It's the *hot* and the *wet* that seem to do the most work.

The most efficient method is to steam the paper off. You can rent a commercial electric wallpaper steamer for under ten dollars a day (*3*). This is a large flat rectangular pan with a trigger to control the steam flow. The pan is attached to a steam-maker you fill with water. There is a long hose for distance work. You hold the pan over the wall, let the steam soak through and dissolve the glue, then scrape off the wallpaper. Try not to chip the wall. Work over a small area at a time, about twenty square feet. When you're done, fill any holes or cracks.

Painting

Your surfaces are ready to be painted. Clear out the area, cover the floor and what can't be evacuated with dropcloths or newspapers. Use masking tape on edges where you want the paint to stop—windowpanes, moldings, and so on. It takes time to tape, but you can save more time later by not having to be so careful with the brush. Remove the tape as soon as the paint is dry.

If you're going to enamel some areas and use latex on others, do the enamel first. Latex—if it drips—can be easily washed off enamel without smearing the enamel. Wear a hat and goggles for ceilings.

Rollers are much quicker than brushes for large flat areas to be covered with latex. You can finish the neat edges with a brush. Make sure you buy a roller-holder that allows for a pole extension to be threaded into the handle. Your own broom or mop handle may fit just fine. Rollers come in 6″ and 9″ sizes, for all different purposes—rough surfaces, smooth ones, different paints. Rollers are hard to clean for reuse, so it's usually smarter to buy two cheap ones. The roller pans also come wide enough for the two sizes of rollers. Don't buy a 6″ pan for a 9″ roller.

Fill your roller pan up to the washboard ribs (*4*). Dip the roller in, then roll it on the ribs a few times to take the excess off. Roll the paint on slowly so you don't splatter. Don't be afraid to go over an area once or twice to even out the paint. The outline of latex roller strokes should disappear when dry, so don't worry about them. (Don't leave thick ridges, of course.) When using a brush, dip it into the pan or can so that only half the bristles are covered, then wipe the excess off on inside edge of can (*5*).

Enamel should always be applied with a brush because of its thickness and tendency to get tacky very quickly. When applying, let it flow off the brush onto the surface lightly; brush in one direction only, and do not go over it. Too much brushing causes brush strokes which may show when the paint dries.

Exposing Brick Walls

The walls of many old buildings are brick with an inside layer of plaster (with or without lath). Exposed brick walls are dramatically beautiful. But the dust, mess, and work of the removal can be traumatic. The best time to do this job is *before* you move into a place—it's difficult to live with. Also consider hiring a professional to sandblast the wall clean.

Brick walls are usually exterior walls or those neighboring the next building. If you're not sure if you have a brick wall, chip away a small section of plaster and see. (In an apartment building, the super should be able to tell you.) If you rent, you might inquire of your landlord for permission to expose the wall. Some consider this to be property damage; others are more enlightened.

There are six steps to expose and finish a brick wall.

1. Remove everything you can from the room, cover well what's left. Plaster dust will try to seep in everywhere. You can cut the dust flow somewhat by hanging plastic dropcloths completely around the work area. Tape them to the ceiling. Try to leave an open window within the "tent" for ventilation. Collect lots and lots of small cardboard boxes to dump the plaster in; small because a large box of plaster is too heavy for most people. Remove all wooden baseboards, moldings, hanging objects, hooks, electrical face and switch plates, etc. When the plaster is gone, any electrical lines will protrude from the wall. This may be illegal in your area—check. If so, they will have to be moved or replaced. Wear goggles and a breathing mask for this work. Cover your hair with a scarf or hat. Take frequent rests to get some fresh air.

2. With a hammer and a large flat brick chisel, start chipping away the plaster. Always work from the top down, so debris won't fall on your head. Chisel an even line across the top right through to the brick. Do the same at each end of the wall. This prevents plaster on the ceiling or other walls from being pulled off accidentally. Now you remove the plaster off the brick any way you like. Beat it with a hammer (don't use a good nail hammer—this ruins the head) and the plaster should start to fall off in large chunks. Slip the chisel or a crowbar behind the plaster and you can pry off large areas. Remove any lath when it gets free, being careful not to loosen plaster at the ceiling seam. Throw large plaster chunks into the boxes as you go.

At the end of this step there will still be thin bunches of plaster sticking in a few places. Also a white residue, called *efflorescence,* which is the result of moisture seeping in through the bricks over the years.

3. Wire-brushing removes the remaining plaster and some of the efflorescence. On a very small wall you can use a hand wire brush. No trick to it—just a lot of elbow grease. Scrape away till the surface looks clean.

Much quicker and easier is an electric drill, preferably high-speed, with a wire-brush attachment. It still takes a lot of work, though. The cheapest drill you can get will be fine, perhaps best; you certainly don't need accuracy, and whichever drill you choose is likely to get a little beat up. Get a number of cheap wire-brush attachments; they wear out very easily on the hard brick surface. Keep shifting the brush angle to the wall to get maximum use out of each one. Get some extra carbon brushes for the motor. When done clean the room of all the dust.

4. Some efflorescence will still remain. Most of it can be removed with a solution of one part muriatic acid to five parts water, mixed in a plastic or rubber pail. Buy the acid at a hardware store. It can burn you badly, so wear rubber gloves, goggles, and long sleeves, and try not to splash it around. The diluted solution won't burn immediately, but wash your skin off quickly if you do get any on you. (Putting some baking soda in your washing water helps neutralize the acid.)

Apply with a hard-bristle scrubbing brush on a pole. There's no need to scrub; let the solution settle into the brick till the efflorescence disappears. Add more if necessary. It should bubble and fizz if it's working. Realize that you may not be able to get rid of *all* the efflorescence.

Dispose of the remaining acid solution in a proper way (not down your sink). Call your sanitation department for advice.

5. Wash the wall down with lots of water, containing a little ammonia to neutralize the acid.

6. When the wall is dry, you can seal it with shellac, varnish, or polyurethane, either gloss or flat. Or you can leave the brick uncoated, if you don't mind the increased possibility of further efflorescence. If there are any holes or cracks in the mortar which seem to be letting air in, patch them with a premixed powdered mortar before sealing.

REFERENCE

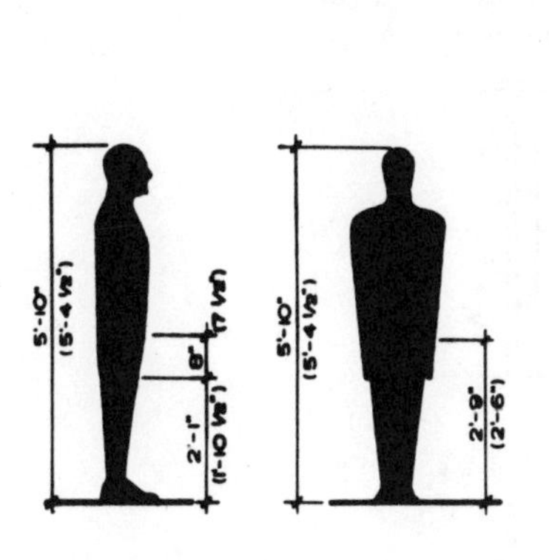

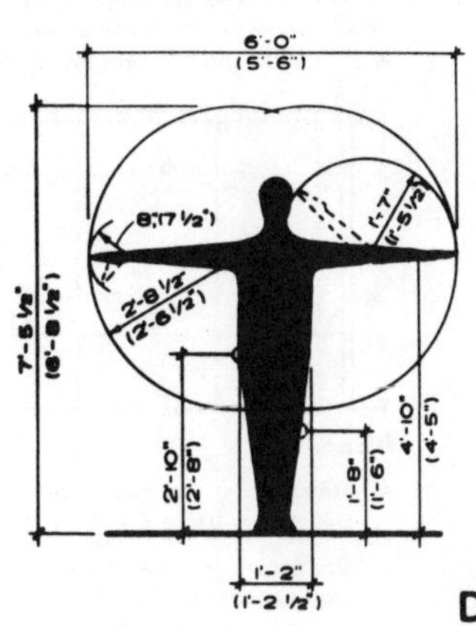

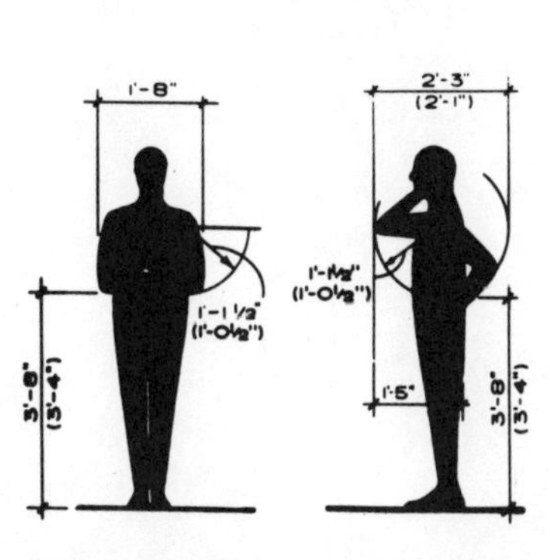

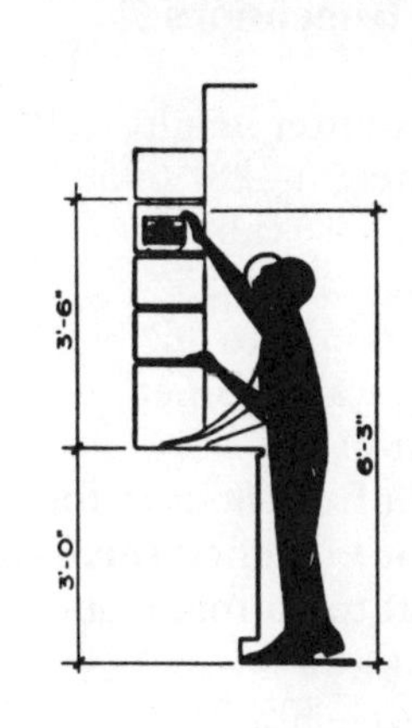

Dimensions of the Human Figure

KEY TO DIMENSIONS

2'-3" MEN
(2'-1") WOMEN

GENERAL NOTES

1. Dimensions shown are based on the average or normal adult.
2. Clearances are generally minimum and should be increased when conditions allow.
3. Seating heights and table top heights shown on this page may be varied slightly; refer to furniture pages.

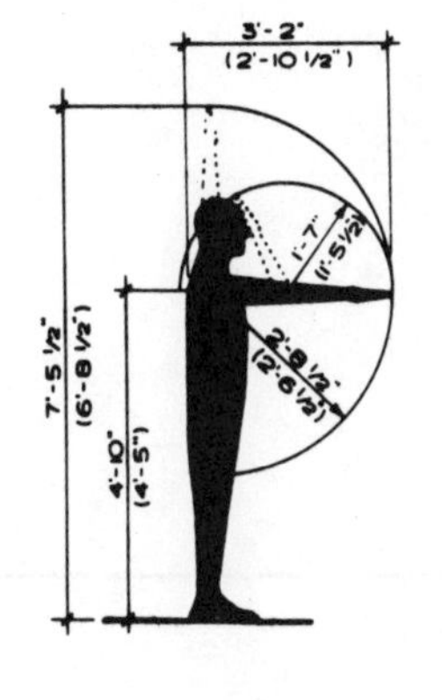

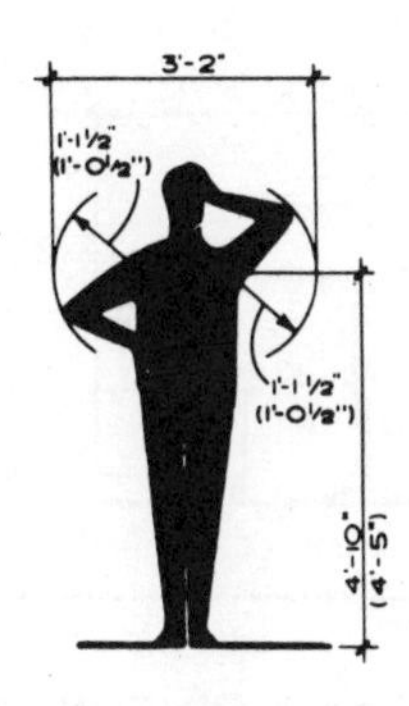

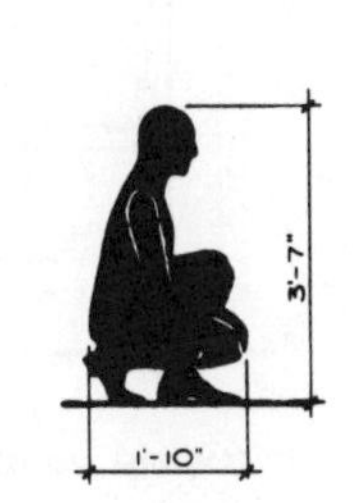

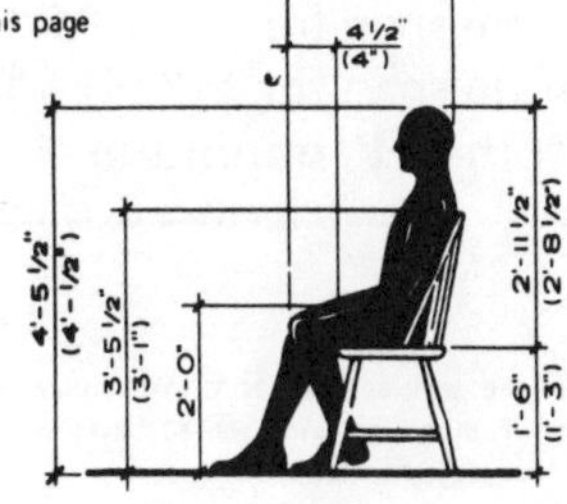

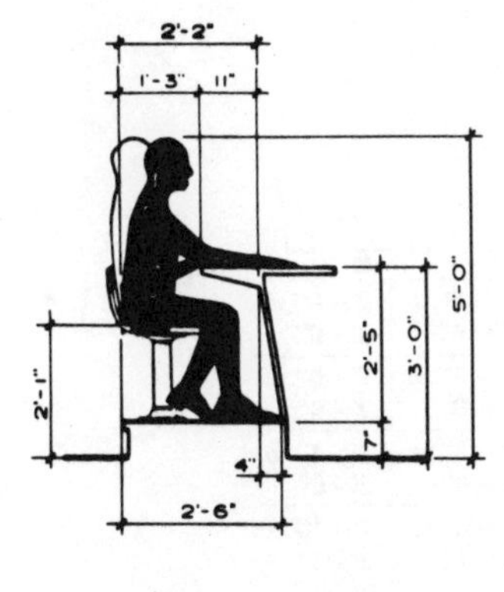

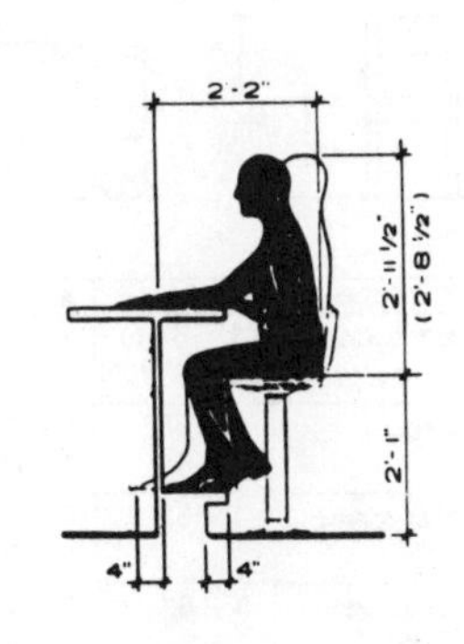

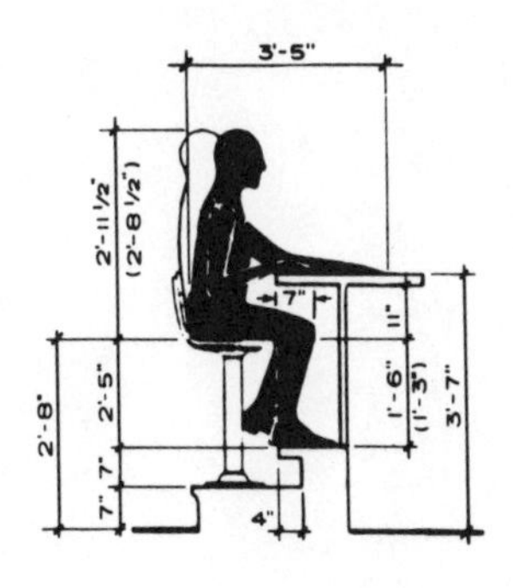

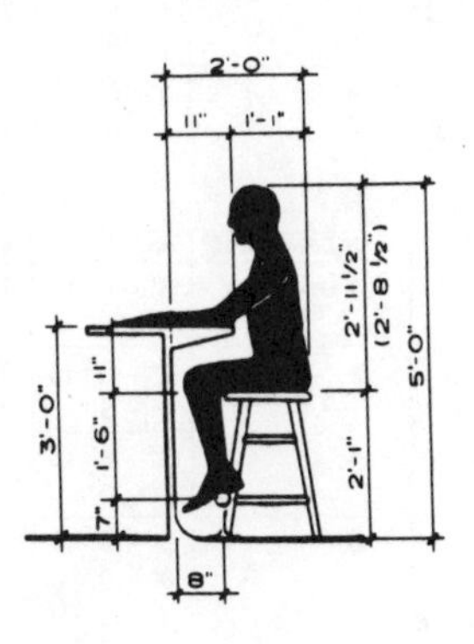

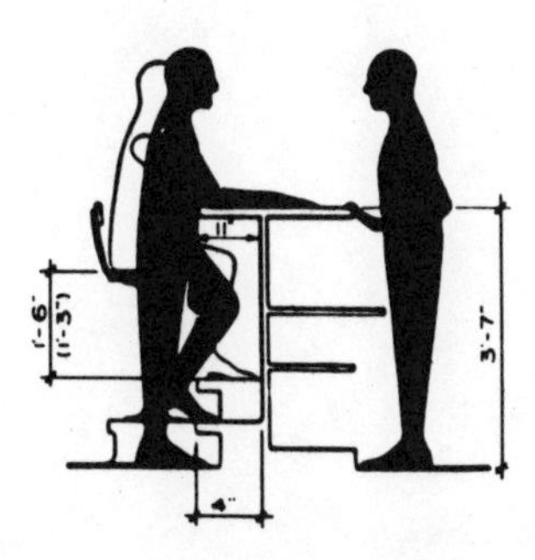

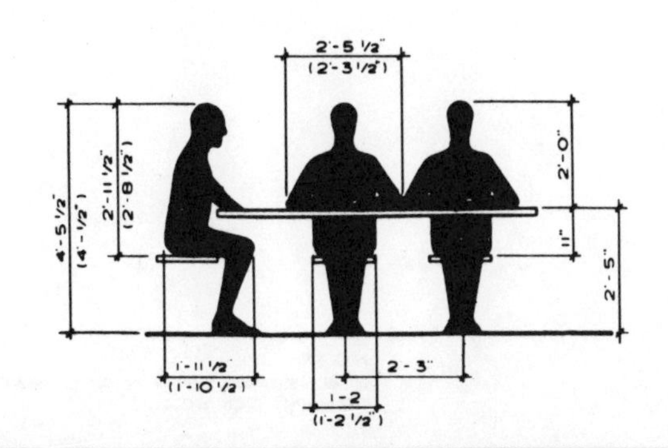

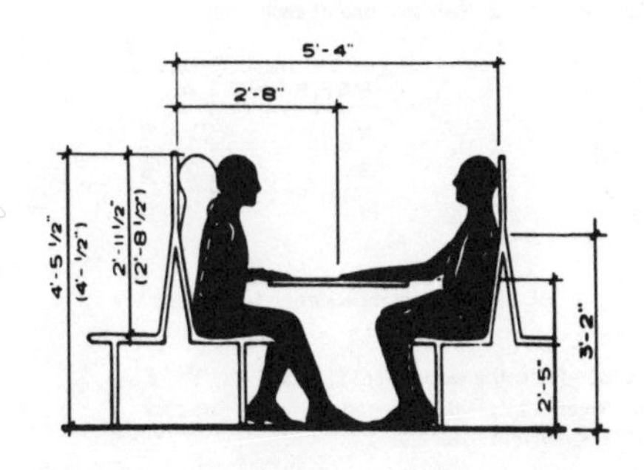

Useful Dimensions

Work counter height—36″
Table height—28″ to 30″
Kitchen base cabinet
  depth—22″ to 25″
  height—33″ to 36″
Kitchen wall cabinet
  depth—13″
  height from floor to top—82″ to 84″
  height to highest shelf—72″
  height to bottom from
    sink—22″
    stove—30″
    work counter—15″ to 18″
Toe space at bottom of cabinets—3″ to 4″ high by 3″ to 4″ deep
Finger space (between top of books, records, etc., and bottom of shelf above) —1½″ minimum
Shelf
  maximum span for 1-by—34″
  maximum span for ¾″ ply—32″
Closet depth—24″ minimum

TABLES

Minimum sizes are satisfactory for drink service; larger sizes for food. Tables with wide spread bases are more practical than four legged tables.

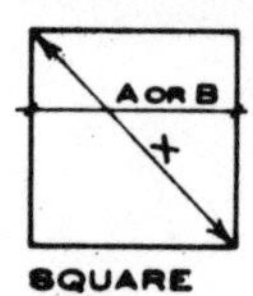

SQUARE

PERSONS	A OR B	X
2	1'– 8" to 2'– 6"	2'– 5" to 3'– 6"
4	2'– 6" to 3'– 0"	3'– 6" to 4'– 3"

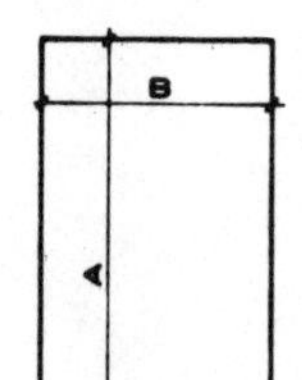

RECTANGLE

PERSONS	A	B
2 (ON ONE SIDE)	3' – 6" to 4' – 0"	2' – 0"
2	2' – 0" to 2' – 6"	2' – 0" to 2' – 6"
4	3' – 6" to 4' – 0"	2' – 0" to 3' – 0"
6	5' – 0" to 8' – 0"	

Tables wider than 2'– 6" will seat one at each end.

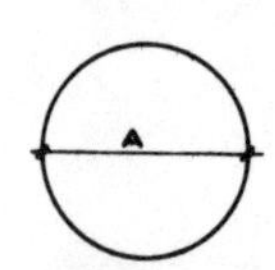

PERSONS	A
2	2'– 0"
3	2'– 6"
4	3'– 0"

CIRCLE

Round tables are usually recommended only for seating 5 persons or more.

"A" dim. depends on the perimeter, (1'– 10" – 1'– 2" per person), necessary to seat required number. For cocktails, 1'– 6" is sufficient.

Anthony J. Amendola, AIA, Forest Hills, New York

BOOTHS

Local regulations determine actual booth sizes. Tables are often two inches shorter than seats, and may have rounded ends. Circular booths have overall diameter of 6'– 4" ± .

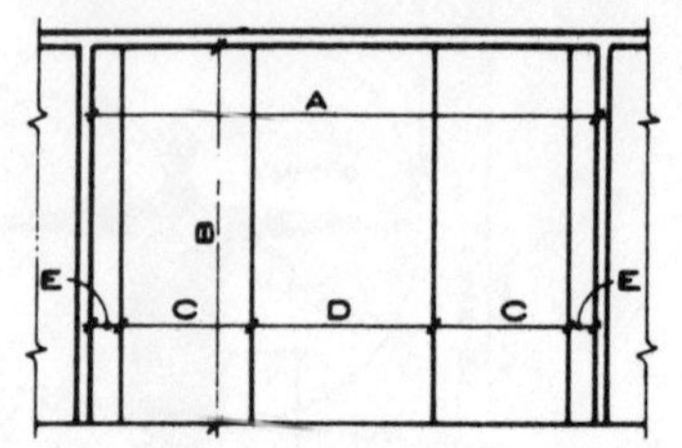

PLAN

A With sloping seat back: 5'– 4" to 6'– 2". Without sloping seat back: 5'– 4" to 6'– 2".

B One person per side: 2'– 0" to 2'– 6".
Two persons per side: 3'– 6" to 4'– 6".
Recommended max. for serving and cleaning 4'– 0".

C 1'– 6" ±

D 2'– 0" to 2'– 6".

E 2" to 6"

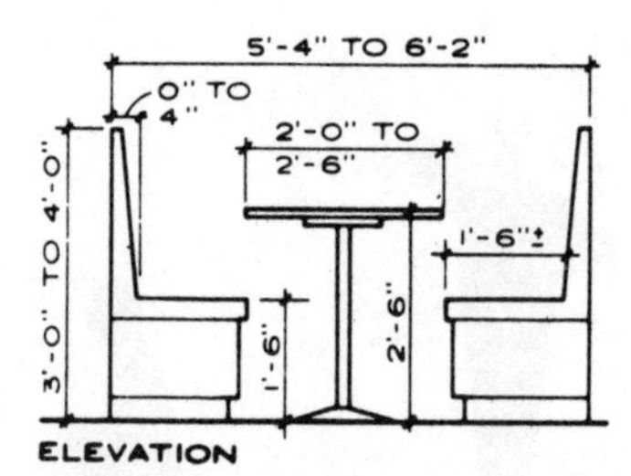

ELEVATION

CHAIR AND CHAIR DIMENSIONS

Chair rail heights are determined by dimension D.

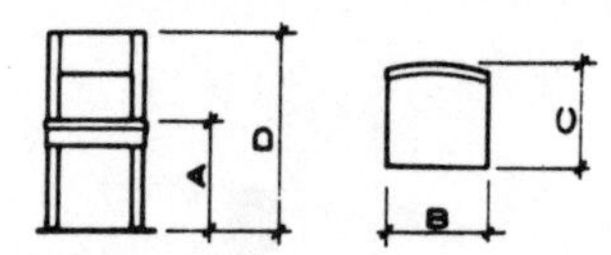

TYPES	A	B	C	D
STRAIGHT	1'– 5" to 1'– 6 ½"	1'– 2" to 1'– 4"	1'– 2" to 1'– 4"	2'– 8" to 3'– 0"
ARM	1'– 5" 1'– 6"	1'– 7" to 2'– 0"	1'– 3" to 2'– 0"	2'– 0" to 3'– 6"
TAVERN	1'– 5"	1'– 5" to 1'– 8"	1'– 3" to 1'– 6"	2'– 4" to 2'– 6"
DINING ROOM	1'– 6"	1'– 6" to 1'– 9"	1'– 6" to 1'– 10"	2'– 10" to 3'– 3"

NOTE:
DIMENSIONS SHOWN ARE NOT NECESSARILY DRAWN TO SCALE

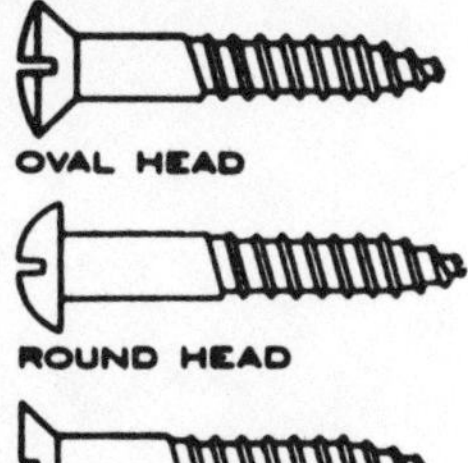

WOOD SCREWS (IN INS)

DIAM.	DECI. EQUIV.	LENGTH
0	.060	1/4 – 3/8
1	.073	1/4 – 1/2
2	.086	1/4 – 3/4
3	.099	1/4 – 1
4	.112	1/4 – 1 1/2
5	.125	3/8 – 1 1/2
6	.138	3/8 – 2 1/2
7	.151	3/8 – 2 1/2
8	.164	3/8 – 3
9	.177	1/2 – 3
10	.190	1/2 – 3 1/2
11	.203	5/8 – 3 1/2
12	.216	5/8 – 4
14	.242	3/4 – 5
16	.268	1 – 5
18	.294	1 1/4 – 5
20	.320	1 1/2 – 5
24	.372	3 – 5

Length intervals = 1/8" increments up to 1", 1/4" increments from 1 1/4" to 3", 1/2" increments from 3 1/2" to 5".

LAG BOLT (IN INCHES)

DIAM. (IN INS)	DECI. EQUIV.	LENGTH
1/4	.250	1 – 6
5/16	.313	1 – 10
3/8	.375	1 – 12
7/16	.438	1 – 12
1/2	.500	1 – 12
5/8	.625	1 1/2 – 16
3/4	.750	1 1/2 – 16
7/8	.875	2 – 16
1	1.00	2 – 16

Length intervals = 1/2" increments up to 8", 1" increments over 8".

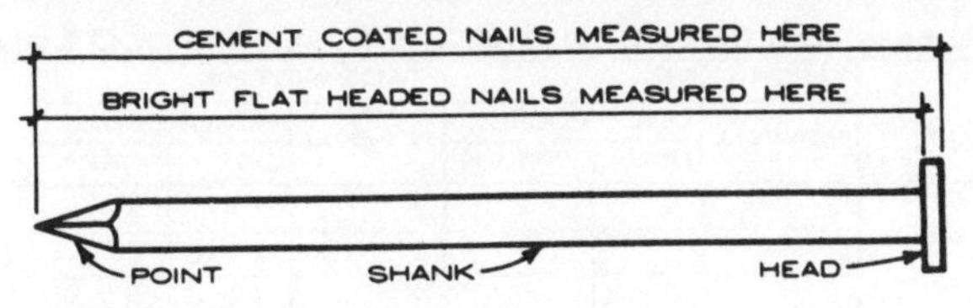

COMMON NAIL

LENGTH (IN INS.)	PENNY	GAUAGE	DIAM. OF HEAD (IN INS)	NO. OF NAILS PER LB.	SAFE WORKING RESISTANCE TO LATERAL SHEAR-LB.
1	2	15	11/64	847	
1 1/4	3	14	13/64	543	
1 1/2	4	12 1/2	1/4	296	
1 3/4	5	12 1/2	1/4	254	
2	6	11 1/2	17/64	167	48
2 1/4	7	11 1/2	17/64	150	
2 1/2	8	10 1/4	9/32	101	64
2 3/4	9	10 1/4	9/32	92.1	
3	10	9	5/16	66	80
3 1/4	12	9	5/16	66.1	96
3 1/2	16	8	11/32	47.4	128
4	20	6	13/32	29.7	160
4 1/2	30	5	7/16	22.7	
5	40	4	15/32	17.3	
5 1/2	50	3	1/2	13.5	
6	60	2	17/32	10.7	

SHEET METAL & THREADING SCREWS

SHEET METAL GIMLET POINT.

Hardened, self-tapping. Used in 28 to 6 gauge sheet metal; aluminum, plastic, slate, etc. Usual head types.

SHEET METAL BLUNT POINT.

Hardened, self-tapping. Used in 28 to 18 gauge sheet metal. Made in 4 to 14 sizes and usual heads.

THREAD CUTTING-CUTTING SLOT.

Hardened. Used in metals up to 1/4" thick. Sizes: 4 to 5/16", in usual head types. (Flat, oval, round, etc.).

NOMINAL AND MINIMUM DRESSED SIZES OF SOFTWOOD LUMBER PRODUCTS

LUMBER PRODUCT	THICKNESS			FACE WIDTHS		
	NOMINAL	MIN. DRESSED		NOMINAL	MIN. DRESSED	
		DRY	GREEN		DRY	GREEN
BOARDS				2	1 1/2	1 9/16
				3	2 9/16	2 5/8
				4	3 9/16	3 5/8
				5	4 1/2	4 5/8
				6	5 1/2	5 5/8
	1	3/4	25/32	7	6 1/2	6 5/8
	1 1/4	1	1 1/32	8	7 1/2	7 5/8
	1 1/2	1 1/4	1 9/32	9	8 1/2	8 3/4
				10	9 1/2	9 3/4
				11	10 1/2	10 3/4
				12	11 1/2	11 3/4
				14	13 1/2	13 3/4
				16	15 1/2	15 3/4
2 INCH DIMENSION				2	1 1/2	1 9/16
				3	2 9/16	2 5/8
				4	3 9/16	3 5/8
				6	5 1/2	5 5/8
	2	1 1/2	1 9/16	8	7 1/2	7 5/8
				10	9 1/2	9 3/4
				12	11 1/2	11 3/4
				14	13 1/2	13 3/4
				16	15 1/2	15 3/4
DIMENSION OVER 2 INCH THICK DRY OR GREEN				3	2 5/8	
				4	3 5/8	
	2 1/2	2 1/8		6	5 1/2	
	3	2 5/8		8	7 1/2	
	3 1/2	3				
	3 1/2	3 1/8		10	9 1/2	
	4	3 5/8		12	11 1/2	
				14	13 1/2	
				16	15 1/2	

LUMBER PRODUCT	THICKNESS		FACE WIDTHS	
	NOMINAL	MAX. M.C. 19%	NOMINAL	MAX. M.C. 19%
FINISH	3/8	5/16	2	1 1/2
	1/2	7/16	3	2 9/16
	5/8	9/16	4	3 9/16
	3/4	5/8	5	4 1/2
	1	3/4	6	5 1/2
	1 1/4	1	7	6 1/2
	1 1/2	1 1/4	8	7 1/4
	1 3/4	1 3/8	9	8 1/4
	2	1 1/2	10	9 1/4
	2 1/2	2	11	10 1/4
	3	2 9/16	12	11 1/4
	3 1/2	3 1/16	14	13 1/4
	4	3 9/16	16	15 1/4

NOTE: Max. M.C. 19% = maximum moisture content 19%

Span Charts

These span charts may seem complicated at first, but they're not. Use them as guidelines when you're not sure what size lumber to use as joists across a structure—whether it's a ceiling, a loft bed, or a counter. Use the Ceiling Joists table for any structure that has only a dead load, such as a dropped plasterboard ceiling or a plywood countertop. If the structure also is meant to carry a live load (namely you), a structure such as a loft bed that has an actual floor on top of it, use the Attic Floor Joists table.

These are standard charts for architects, who are building houses and dealing with all sorts of conditions. If you use the span distances in the first column of either table, those under the heading "E=1,000,000," you'll have no problems. (If you're interested, the "E" stands for the *modulus of elasticity*, the allowable amount of which is determined by local building code.) The heading "Span 'L' Limited by Deflection" simply means that if you follow the suggested guidelines, the joist will not bend more than 1/360 of its span.

Example: You want to span a distance of 6′4″ with joists to support a loft bed. If you space them 16″ on center, 2x4s will not be strong enough; they are suggested only up to a span of 6′0″. You would have to use 2x6s, which can span up to 9′2″. Or you could use 2x4s if you placed them 12″ on center, since under these conditions they can span up to 6′6″.

CEILING JOISTS

MAXIMUM ALLOWABLE LENGTHS "L" BETWEEN SUPPORTS

SIZE (NOMINAL) IN INCHES	SPACING (C TO C) IN INCHES	SPAN "L" LIMITED BY DEFLECTION				
		E=	1,000,000 Ft. In.	1,200,000 Ft. In.	1,400,000 Ft. In.	1,600,000 Ft. In.
2 X 4	12	L=	9– 4	10– 0	10– 6	11– 0
	16	L=	8– 7	9– 2	9– 8	10– 1
	24	L=	7– 7	8– 1	8– 6	8–11
2 X 6	12	L=	14– 2	15– 1	15–10	16– 7
	16	L=	13– 1	13–11	14– 8	15– 4
	24	L=	11– 8	12– 5	13– 1	13– 8
2 X 8	12	L=	18– 6	19– 8	20– 8	21– 7
	16	L=	17– 2	18– 3	19– 3	20– 1
	24	L=	15– 4	16– 4	17– 2	17–11
2 X 10	12	L=	22–11	24– 4	25– 7	26– 9
	16	L=	21– 5	22– 9	23–11	25– 0
	24	L=	19– 2	20– 5	21– 6	22– 5
2 X 12	12	L=	27– 2	28–11	30– 5	
	16	L=	25– 5	27– 1	28– 6	29– 9
	24	L=	23– 0	24– 5	25– 8	26–10

NOTES:

(The ceiling joist span lengths are based on the following):

1. Maximum allowable deflection = 1/360 of the span length.
2. Dead load:
   A. Weight of joist (40 lbs. per cu. ft.).
   B. Lath and plaster ceiling (10 lbs. per cu. ft.).
3. Live load: none.

ATTIC FLOOR JOISTS

MAXIMUM ALLOWABLE LENGTHS "L" BETWEEN SUPPORTS

SIZE (NOMINAL) IN INCHES	SPACING (C TO C) IN INCHES	SPAN "L" LIMITED BY DEFLECTION				
		E=	1,000,000 Ft. In.	1,200,000 Ft. In.	1,400,000 Ft. In.	1,600,000 Ft. In.
2 X 4	12	L=	6– 6	6–11	7– 4	7– 8
	16	L=	6– 0	6– 4	6– 8	7– 0
	24	L=	5– 3	5– 7	5–10	6– 1
2 X 6	12	L=	10– 1	10– 8	11– 3	11– 9
	16	L=	9– 2	(– 9	10– 4	10– 9
	24	L=	8– 1	8– 7	9– 1	9– 6
2 X 8	12	L=	13– 4	14– 2	14–11	15– 7
	16	L=	12– 2	13– 0	13– 8	14– 3
	24	L=	10– 9	11– 5	12– 0	12– 7
2 X 10	12	L=	16– 9	17– 9	18– 8	19– 7
	16	L=	15– 4	16– 4	17– 2	17–11
	24	L=	13– 6	14– 5	15– 2	15–10
2 X 12	12	L=	20– 1	21– 4	22– 5	23– 6
	16	L=	18– 6	19– 7	20– 8	21– 7
	24	L=	16– 4	17– 4	18– 3	19– 1

NOTES:

(The attic floor joist span lengths are based on the following):

1. Maximum allowable deflection = 1/360 of the span length.
2. Dead load:
   A. Weight of joist (40 lbs. per cu. ft.).
   B. Lath and plaster ceiling (10 lbs. per cu. ft.).
   C. Single thickness of flooring (2.5 to 3.0 lbs. per sq. ft.).
3. Live load: 20 lbs. per sq. ft. of floor area.

GUIDE TO APPEARANCE GRADES OF PLYWOOD (1)

	GRADE SYMBOLS (2)	DESCRIPTION AND MOST COMMON USES	VENEER GRADES			COMMON THICKNESS (3)					
			FACE	BACK	INNER PLIES	1/4"	3/8"	1/2"	5/8"	3/4"	1"
INTERIOR TYPE	N-N, N-A N-B INT-DFPA	Natural fin. cab. quality. lor both sides select all heartw'd or all sapw'd veneer. For furniture of natural fin., cab. doors, built-ins, special items.	N	N A B	C					•	
	N-D INT-DFPA	Natural finish paneling. Special order.	N	D	D	•					
	A-A INT-DFPA	Interior applications where both sides will be on view. Built ins, cabinets, furniture, partitions. Face smooth and suitable for painting.	A	A	D	•	•	•	•	•	•
	A-B INT-DFPA	Uses similar to Int. A-A but appearance of 1 side is less important and 2 smooth solid surfaces are required.	A	B	D	•	•	•	•	•	•
	A-D INT-DFPA	Interior use where appearance of 1 side only is important. Paneling, built-ins, shelves, partitions.	A	D	D	•	•	•	•	•	•
	B-B INT-DFPA	Int. utility panel for 2 smooth sides. Permits circular plugs. Paintable.	B	B	D	•	•	•	•	•	•
	B-D INT-DFPA	Int. utility panel for 1 smooth side. For backing, sides of built-ins, shelving (industry).	B	D	D	•	•	•	•	•	•
EXTERIOR TYPE	A-A EXT-DFPA (4)	Use where appearance of both sides is important. Fences, built-ins, signs, boats, cabinets, commercial refrigerators, tote boxes, ducts.	A	A	C	•	•	•	•	•	•
	A-B EXT-DFPA (4)	Use similar to A-A EXT panels, but where appearance of 1 side is less important.	A	B	C	•	•	•	•	•	•
	A-C EXT-DFPA (4)	Exterior use where appearance of only 1 side is important. Sidings, soffits, fences, structural uses, farm bldgs., commercial refrigerators.	A	C	C	•	•	•	•	•	•
	B-C EXT-DFPA (4)	Outdoor utility pan. for farm serv. & work bldgs.	B	C	C	•	•	•	•	•	•
	B-B EXT-DFPA	Outdoor utility pan. with solid paintable faces.	B	B	C	•	•	•	•	•	•

NOTES: (1) Sanded both sides.
(2) Available in Group 1, 2, 3, or 4.
(3) Standard 4x8 panel sizes. Others available.
(4) Also available in Structural 1

GUIDE TO CONSTRUCTION GRADES OF PLYWOOD

	GRADE SYMBOLS (1) (2)	DESCRIPTION AND MOST COMMON USES	VENEER GRADES			COMMON THICKNESS, INCHES (3)							
			FRONT	BACK	INNER PLIES	1/4	5/16	3/8	1/2	5/8	3/4	7/8	1 1/8
INTERIOR TYPE	STANDARD INT-DFPA (4)	Unsanded interior sheathing grade for floors, walls, and roofs	C	D	D		•	•	•	•	•	•	
	STRUC-TURAL I and STRUC-TURAL II INT-DFPA	Unsanded structural grades where strength is most important. Made only with exterior glue. Structural I limited to Group I species for all plies. Structural II permits Group 1, 2, or 3 species.	C	D	D		•	•	•	•	•	•	
	UNDERLAY-MENT INT-DFPA (4)	Underlayment or combination subfloor-under layment for resilient floor coverings, carpeting. Ply beneath face is C or better. Sanded or touch-sanded as specified.	C plug'd	D	C & D	•		•	•	•	•		
	C-D PLUGGED INT-DFPA (4)	Utility built-ins, backing for wall & ceiling tile. Not under layment substitute. Unsanded or touch-sanded.	C plug'd	D	D		•	•	•	•	•		
	Z-4-1 INT-DFPA (5)	Comb. subfloor-under layment. For resil. Floor cov., carpet, wd. strip flooring. Exterior glue for moist areas. Unsanded or touch-sanded.	C plg	D	C & D								•
EXTERIOR TYPE	C-C EXT-DFPA (4)	Waterproof bond for subflooring & roof deck, siding for service bldgs. Unsanded.	C	C	C		•	•	•	•	•	•	
	C-C PLUGGED EXT-DFPA (4)	Base for resilient floors & tile backing in moist areas. Refrig. or controlled atmosphere rooms. Sanded or touch-sanded.	C plg	C	C	•		•	•	•	•	•	
	STRUC-TURAL I C-C EXT-DFPA	Engineered applications requiring full exterior – type panels of all group 1 woods. Unsanded.	C	C	C		•	•	•	•	•	•	
	PLYFORM CLASS I & II, B-B EXT-DFPA	Concr. forms, high re-use factor. Sanded 2 sides. Edge-sealed, mill-oiled. Special restr. on species. Also in HDO.	B	B	C					•	•		

NOTES: (1) Interior grades available with exterior glue.
(2) Avail. + & g 1/2, 5/8, 3/4, 1 1/8, except Plyform.
(3) Standard 4x8 panel size. Others available.
(4) Available in Group 1, 2, 3, or 4.
(5) Available in Group 1, 2, 3 only.

GUIDE TO SPECIALTY GRADES OF PLYWOOD (1)(2)– SEE NOTES AT BOTTOM LEFT.

	GRADES SYMBOLS	DESCRIPTION AND MOST COMMON USES	VENEER GRADES: FRONT	VENEER GRADES: BACK	VENEER GRADES: INNER PLIES	COMMON THICKNESS (") (3): 5/16	3/8	1/2	5/8	3/4	1
INTERIOR	DECORATIVE PANELS	Rough textured faces. Accent walls, etc.	B+	D	D	•	•	•			
INTERIOR	PLYRON INT-DFPA	Hardboard faces standard; standard, tempered, screened, smooth. Counter tops, shelves, cab. doors, floors.			C D			•	•	•	
EXTERIOR TYPE	HDO (6) EXT-DFPA (4)	High density overlay, hard resin fiber Paint not req. Conc. forms, signs, acid tanks, cabs., ctr. tops.	A B	A B	C pl.	•	•	•	•	•	•
EXTERIOR TYPE	MDO (4) EXT-DFPA (6)	Medium density overlay, smooth opaque resin fibe: heat fused. Paint base. Siding etc.	B	B C	C Cpl.	•	•	•	•	•	•
EXTERIOR TYPE	303 SPEC-SID'G EXT-DFPA (5)(6)	Special surface treatm't. See following pages.	B+	C	C		•	•	•		
EXTERIOR TYPE	T I-II EXT-DFPA (5)	1/4x3/8 parallel grooves. Sid'g.	C+	C	C				•		

NOTES FOR SPECIALTY GRADES OF PLYWOOD (See Table)

(1) Sanded both sides except for decorative or other surfaces.
(2) Available in group 1, 2, 3, or 4 unless otherwise noted.
(3) Standard 4'x 8' panel size. Others available.
(4) Also available in Structural I.
(5) Panel sizes 4'x7', 8', 9', 10'.
(6) Also horiz. lap siding 3/8 x 12, 16 to 16' long.

Foster C. Parriott; James M. Hunter & Associates; Boulder, Colorado

VENEER GRADES

N	Special order "natural finish" veneer. Select all heartwood or all sapwood. No open defects. Some repairs allowed.
A	Smooth & paintable. Neat repairs permissible. Also used for natural finish in less demanding applications.
B	Solid surface veneer. Circular repair plugs & tight knots permitted.
C	Knotholes to 1". Occasional knotholes 1/2" larger within certain limits in specified section. Limited splits permitted. Mini-m veneer for Exterior-type plywood.
C	Improved C veneer with splits limited to 1/8" wide & knotholes & borer holes limited to 1/4" by 1/2"
D	Permits knots & knotholes to 2 1/2" wide & 1/2" larger under certain specified limits. Limited splits permitted.

KEY DEFINITIONS

GROUP
Refers to species used in manufacturing plywood. Species are classified into groups based on stiffness. Group number in grade-trademarks refers to species of face and back veneers.

STANDARD SHEATHING
Interior sheathing grade replacing former C-D grade. Subflooring, wall sheathing, roof decks, pallets, crates, some engineered applications.

STRUCTURAL I, STRUCTURAL II
Unsanded grades for use where strength is most important. Examples: box beams, gusset plates, stressed skin panels, shipping containers, bins. Both grades made only with exterior glue. Structural I is limited to Group I species and available in all exterior grades. Structural II permits Groups 1, 2, or 3 species.

IDENTIFICATION INDEX NUMBER
2 numbers separated by a slash appearing in grade-trademarks of all sheathing grades. Numbers indicate spacing in inches for supports, no. on left for roof decking with 35 p.s.f. minimum loading. No. on right for subflooring with 100 p.s.f. loading.

CLASS I, CLASS II
Applies only to Plyform grades. Indicates species mix permitted under Product Standard PS-166. Class I is stronger than Class II.

CLASSIFICATION OF SPECIES

GROUP 1
Douglas fir 1
Larch, Western
Pine, Southern
- Loblolly
- Longleaf
- Shortleaf
- Slash

Tanoak

GROUP 2
Cedar, Port Orford
Douglas fir 2
Fir
- California red
- Grand
- Noble
- Pacific silver
- White

Hemlock, Western
Lauan
- Red
- White

Pine, Western white
Spruce, Sitka

GROUP 3
Alder, red
Cedar, Alaska yellow
Pine
- Lodgepole
- Ponderosa

Redwood

GROUP 4
Cedar
- Incense
- Western red

Fir, subalpine
Pine, sugar
Poplar, Western
Spruce, Engelman

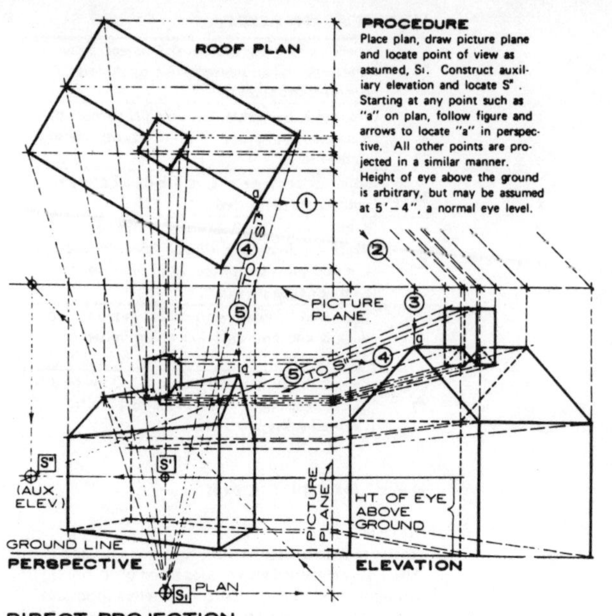

PROCEDURE
Place plan, draw picture plane and locate point of view as assumed, S_1. Construct auxiliary elevation and locate S". Starting at any point such as "a" on plan, follow figure and arrows to locate "a" in perspective. All other points are projected in a similar manner. Height of eye above the ground is arbitrary, but may be assumed at 5'–4", a normal eye level.

DIRECT PROJECTION

PROCEDURE WHEN VANISHING POINTS ARE OFF BOARD

1. Draw any arc from VP_1.
2. Place a cardboard cutout against curve of circle.
3. Place head of t-square against cutout, making sure both ends of head are always touching cutout, and then draw lines.

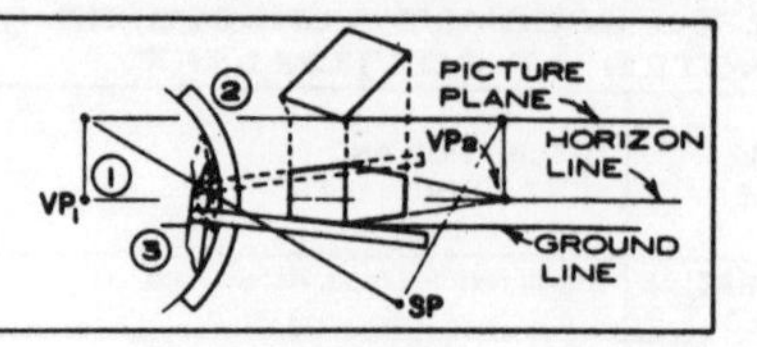

PROCEDURE (ONE POINT)
Draw A.B.C.D., section which is cut by P.P., at any desired scale, and locate S' (point of view in elevation) on line of sight from S_1. Locate the 45° vanishing points V_L and V_R on either side of S' and as distant as S_1 is from the picture plane. All lines parallel to P.P. will remain parallel and all plane figures parallel to P.P. will show their true shape. Vertical lines will be vertical in perspective. Horizontal lines parallel to P.P. will be horizontal. Horizontal lines perpendicular to P.P. will vanish at S'. Horizontal lines at 45° to F.P. (used to measure distances ⊥ to P.P.) will vanish at 45°VPs.

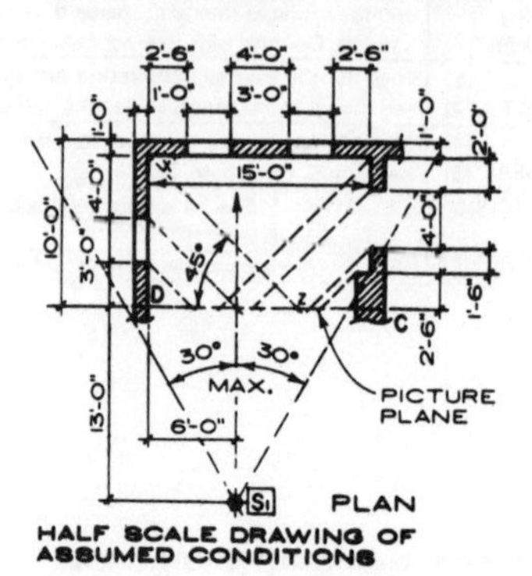

HALF SCALE DRAWING OF ASSUMED CONDITIONS

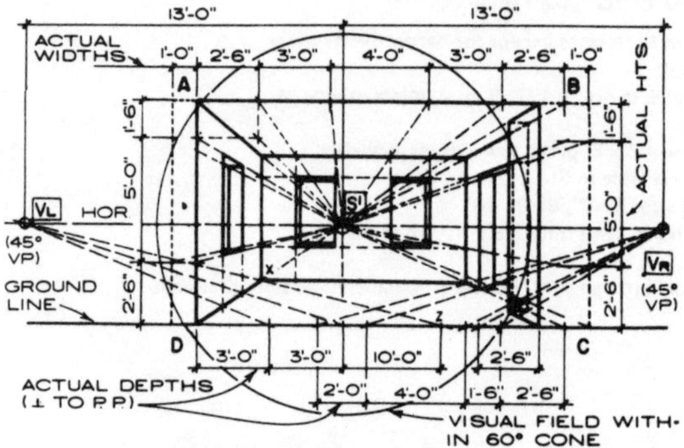

"ONE-POINT" OR PARALLEL PERSPECTIVE

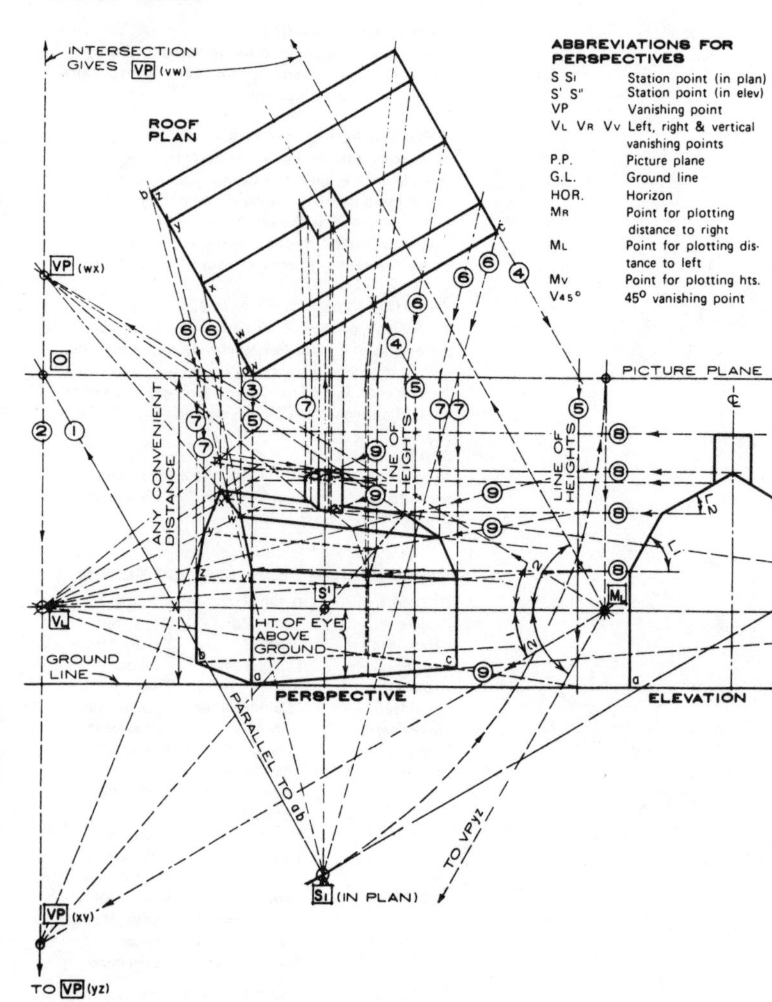

TWO POINT OR ANGULAR PERSPECTIVE ALSO KNOWN AS "OFFICE METHOD"

ABBREVIATIONS FOR PERSPECTIVES

S S_1	Station point (in plan)
S' S"	Station point (in elev)
VP	Vanishing point
V_L V_R V_V	Left, right & vertical vanishing points
P.P.	Picture plane
G.L.	Ground line
HOR.	Horizon
M_R	Point for plotting distance to right
M_L	Point for plotting distance to left
M_V	Point for plotting hts.
$V_{45°}$	45° vanishing point

PROCEDURE FOR TWO POINT PERSPECTIVE

PLAN: Assume picture plane (P.P.) and locate plan of object as desired. Assume point of view, or station point, S_1. To minimize apparent distortion, this point is commonly taken about opposite the center of the drawing, and far enough away to keep the field of view within about 60° latitude.

ELEVATION: Locate ground line where convenient. Place elevation as indicated, or measure heights directly on any vertical "Line of Heights". Locate S' on vertical through S_1 and at assumed height above ground line.

PERSPECTIVE: Through S' draw horizon. Draw parallel to principal horizontal lines of object through S_1 (in plan), and project intersections with P.P. down to the horizon, giving principal vanishing points V_L and V_R.

NOTE: To find VPs for inclined lines, swing S_1 about 0 into P.P. and project to horizon at M_L. Draw through M_L parallel to actual slopes (angles 1 and 2) to intersection with vertically projected line through V_L. Vanishing points for inclined lines are not absolutely essential, but are frequently found very useful as is shown in the determination of the inclined lines of the gambrel roof in this perspective. Follow arrows and numbered lines. See figures 4 and 1 on following page.

Recommended Books

The magazines *Popular Science* and *Popular Mechanics* are the handiest sources we know for keeping up with new developments of tools and products. They also have articles on tool use, comparisons and use of materials like adhesives, plus suggestions and plans for projects.

How to Work with Tools and Wood, edited by Robert Campbell and N. H. Mager (Pocket Books), is a great buy if you want a handy reference book on tool and tool use, with emphasis on hand tools. It's a small paperback for under $2. The book was originally published by Stanley Tools, who know what they are doing, but be aware that the discussion and illustrations are *only* about Stanley tools.

Modern Carpentry and *Modern Woodworking,* both by Willis Wagner (Goodheart-Willcox), are valuable reference books. The first one is a good source of information about general carpentry, particularly house construction; we consult it a lot on our own jobs. The second volume deals more with cabinet work, and includes more information on table tools. Both books are really textbooks, and assume you either know quite a lot already or have a teacher to show you things; but you should find them very useful as you gain experience.

Architectural Graphics Standards, by Charles G. Ramsey and Harold Sleeper (John Wiley and Sons, Inc.), is the bible for architects and builders. It's a reference book for any information you could possibly need, from hardware and materials descriptions, to building details and specifications for everything from a wastepaper basket to a gymnasium. If the price is too much for you (around $45), look for used copies or check it out in a library.

Other sources of information are published by the Government Printing Office, which include a number of fine books and pamphlets on all aspects of carpentry; and the manufacturers of building materials, such as U.S. Plywood and U.S. Gypsum, who are more than happy to send you free pamphlets on the use and installation of their products.

If you would like to know more about structure, particularly how it relates to architecture, the clearest and best book we know is *Structure in Architecture,* by Mario Salvadori in collaboration with Robert Heller (Prentice-Hall). Another very interesting book, one written for the layman, is *The New Science of Strong Materials,* by J. E. Gordon (Walker and Co.). Both books helped clarify our own ideas on structure.

Two excellent books on soundproofing, from which we learned a lot, are *Handbook of Noise Control,* by Cyril Harris (McGraw-Hill), and *Sound, Voice and Vibration Control,* by Lyle F. Yerges (Van Nostrand Reinhold).

There are many good volumes on wood finishing, yet few of them totally agree on much beyond the basics. You should take a look at several to see what *you* like. One book we like, which should give you a solid foundation, is *How to Paint Anything,* by Hubbard Cobb (Macmillan). It's about all types of wood finishes, not just paint. Also, some manufacturers of finishing materials have helpful pamphlets, though oriented toward their own products. One of the best is *Constantine's Wood Finishing Manual,* put out by Albert Constantine and Son, Inc. (see Mail-Order Sources).

Mail-Order Sources

If you have trouble finding what you need in your home area, there are a number of very good mail-order houses. In fact, their prices, even with shipping costs, are often better than you can find at home. Some of these places also have showrooms and stores. Catalogues are available from all of them, either free or for a nominal cost; these catalogues will provide you with many happy hours, plus suggest a lot of exciting ideas and solutions to problems. For addresses, see below.

Try Silvo and U.S. General for practically anything, but particularly for good buys on discontinued tools. Their prices are about as low as you can find anywhere. For particularly fine tools, especially imported ones, try Garrett Wade, Leichtung, or Woodcraft. Brookstone's has a dazzling variety of hard-to-find, specialized tools. If you need fine hardwoods, veneers of any kind, or patterned inlays, try Constantine or Craftsman; they also carry fine tools, finishing materials, and hard-to-find hardware for cabinet work.

Albert Constantine and Son, Inc.
2050 Eastchester Road
Bronx, N.Y. 10461

Brookstone Company
124 Vose Farm Road
Peterborough, N.H. 03458

Craftsman Wood Service
2727 South Mary Street
Chicago, Ill. 60608

Garrett Wade
302 Fifth Avenue
New York, N.Y. 10001

Leichtung
701 Beta Drive #17
Cleveland, Ohio 44143

Silvo Hardware Company
107–109 Walnut Street
Philadelphia, Pa. 19106

U.S. General
100 General Place
Jericho, N.Y. 11753

Woodcraft Supply Corp.
313 Montvale Avenue
Woburn, Mass. 01801

Glossary

Airborne sound: Sound transmitted through the air by the vibrations of adjoining air molecules.

Arbor: A shaft or axle, as on a circular saw.

Base molding: Wood molding that runs along the bottom of a wall, covering any gap between floor and wall.

Beam: A main horizontal supporting member.

Bevel: A slant; any angle other than 90°.

Bind: Be constrained, get stuck; as a saw binding in the kerf.

Blind dado (also stopped or closed dado): A dado, or groove, that does not continue to the end of the piece.

Blind hole: A hole that does not go all the way through a piece.

Brace: An added support for stability.

Brad: A small wire nail (1″ or less in length) with a head similar to that of a finishing nail.

Brushes: (in a motor): Little carbon-graphite blocks that brush against and carry electrical current to the armature.

Built-in: A piece of furniture that uses as part of its structure the floor, walls, or ceiling—or some combination of the three.

Butt joint: A joint where the end of one piece is placed flush to a second piece at a 90° angle.

Cantilever: The part of a horizontal member or structure that overhangs its support.

Casing: The pieces attached around the outside of a doorjamb to complete the joint between jamb and wall.

Cat: A crosspiece placed between studs for bracing and stability.

Chuck: A mechanism for holding a tool or work; such as a drill chuck, which holds the drill bit.

Cleat: A block or piece of wood fastened to a surface to support another piece such as a shelf or joist.

Collet: A collarlike clamping device in a tool; as the collet on the router shaft that tightens onto and holds the bit.

Compression stress: A pushing force.

Conduit: A hollow metal pipe used for carrying electrical wires between junction boxes.

Counterbore: To enlarge the top of a screwhole so the screw head can be sunk *below* the surface.

Countersink: To enlarge the top of a screwhole so the screw head can be sunk *flush* to the surface; also to drive a nail flush or below the surface.

Cripple stud: A stud shortened to leave room for an opening in a wall (such as a door or window).

Cross grain: At a right angle to the direction of the grain.

Cutting plan: A sketch showing how one or more large pieces (such as plywood sheets) can be cut up into several smaller pieces.

Dado: A groove cut into a board, usually to accept the edge or end of another piece.

Decibel: A unit used to measure the intensity of sound; abbreviated dB.

Door stop: Molding pieces fastened to the inside of a doorjamb that stop the door when it closes.

Dowel: A cylindrical piece of wood.

Dynamic load: A load that moves or vibrates, such as a person or a motor.

Edge: The thin dimension of a piece of wood; e.g., the 2″ side of a 2x4.

End grain: The end of a piece of wood, where the grain fibers are severed.

Face: The wide dimension of a piece of wood; e.g., the 4″ side of a 2x4.

Facing: A frame on the front of a cabinet for a lipped or overlay door to close against.

Factory edge: Any edge of a manufactured panel, as in plywood. The factory edge is usually straight enough to use as a straightedge guide.

Flange: A projecting rim or edge.

Flush: Even; level; two adjacent surfaces on the same plane.

Frame: A skeleton structure; the name sometimes given to a wall structure before the sheathing is applied.

Furring strips: Pieces (usually 1x2s) attached to a rough surface (like masonry) to give a level nailing surface for a covering (like plasterboard). They are also used to create an air pocket (for insulation) between structure and covering.

General construction: Construction other than furniture—such as stud walls, floors, ceilings.

Green lumber: Wood not kiln-dried; it has a high moisture content and will thus shrink a lot.

Hardwood: Wood from deciduous trees (they shed

their leaves every autumn). Most, but not all, hardwoods are harder than softwoods. Mahogany, birch, maple, oak, walnut are common hardwoods. Poplar is a hardwood that is actually softer than many softwoods.

HEADER: In a wall, the horizontal structural member over an opening.

HEX: Hexagonal, or six-sided; as a hex nut.

HOLLOW WALL: Wall with a cavity behind the outer surface, such as a stud wall covered with plasterboard. A masonry wall is *not* a hollow wall.

IMPACT SOUND: Sound generated by the impact of some object hitting part of the building structure; as footsteps on the floor.

JAMB: The frame around an opening, such as a doorjamb or window jamb.

JOINT: The junction of two pieces held together by various means.

JOIST: A horizontal structural member used to support a floor or ceiling.

KERF: The slot made by a saw blade.

KILN-DRIED: Lumber carefully dried and seasoned in a large oven. Such wood should not shrink too much.

KNOT: The intersection of a branch with the trunk of the tree.

LAG BOLT: A large screw with a hex or square bolt head.

LATERAL STRESS: Stress applied at an angle to the horizontal.

LATH: Thin wood or metal strips fastened to studs or joists as a base for plaster.

LAYOUT: The measuring and locating of work lines and shapes.

LEDGER STRIPS: Cleats attached to a frame to support joists or other horizontal pieces.

LEVEL: Parallel to the ground.

LOAD: Any weight.

LOAD-BEARING: Something that supports a load other than itself.

MASONRY: Brick, stone, concrete, or similar materials.

MITER JOINT: A joint where each piece is cut to half of the total angle of the joint; most common is a 90° joint with each piece cut to 45°.

MORTISE: A hollow or indentation made in a surface to accept a corresponding projection from a second surface.

NOMINAL SIZE: The dimensions by which a piece of lumber is described, though its actual size is somewhat smaller.

NOTCH: A cutout made in the surface of a piece.

ON CENTER: Measured from the center of one piece to the center of the next, in a series of pieces; e.g. in 16″ on center studs, the initial measurement is from the *end* of the first piece to the center of the second; abbreviated o.c.

PARTITION: An interior wall.

PATINA: The fine mellow surface on wood that comes only with age.

PLATE: The top or bottom horizontal of a stud wall.

PLUMB: Exactly vertical.

PLUNGE CUT: A cut started within a piece of wood, as opposed to at an edge.

POST: A vertical member used as a structural support.

PREFABRICATED: Built or assembled before being installed in place; as a prefabricated cabinet. As opposed to a built-in.

PRENAILING: Starting nails in a piece before putting it into position to be fastened; usually with the nail points just protruding through the back of the first piece.

RABBET: A groove or dado along an edge.

RECIPROCATING MOTION: To and fro.

RIP: To cut along the grain.

RUNNERS: Guides for something (particularly a drawer) to slide on.

SCORE: A thin, shallow cut in some material, usually so the material can then be snapped in two.

SCREW DRILL: A bit that drills for the thread, unthreaded shank, and head of a screw all at once.

SET SCREW: In a tool, a small screw that is tightened to clamp another part tight.

SHAFT: A cylindrical part of a tool, such as the shaft of a screwdriver.

SHEATHING: In a wall, any covering attached to the studs and joining them together like a skin; often plasterboard, plywood, or paneling.

SHIM: A small piece of wood used to help plumb or level.

SOFTWOOD: Wood from evergreen trees; generally, but not always, softer than hardwoods. Fir, pine, redwood, cedar are all softwoods.

Sound absorbers: Soft, porous materials that trap sound inside them like a sponge, and thereby stop reverberations. They do not stop sound transmission, however. Acoustic tile, rugs, foam, egg cartons are common sound absorbers.

Spline: A thin strip of wood, metal, etc., that fits into slots in adjoining pieces to form a joint.

Staggered: Arranged in an overlapping pattern.

Static load: A load that does not move on its own; e.g., books, boxes. Also called stationary or dead load.

Straightedge: Anything with one edge in a straight line, such as a metal rule or the factory edge of a piece of plywood. Usually used as a guide for tools or layout.

Stress: The force or strain exerted by a load upon an object.

Strip floor: A floor with a surface of tongue-and-groove wood planks.

Stud: A vertical structural member in a wall.

Stud wall: A wall whose basic structure is wood or metal studs between floor and ceiling plates.

Tacking: Nailing one piece to another, but leaving the nail heads sticking out for easy removal; useful for aligning pieces temporarily.

Tang: In a tool, the thin shank or tongue that inserts into the handle.

Template: A guide for a definite cutting pattern or shape—like a cookie cutter.

Tension stress: A pulling force.

Toenail: To nail at an angle to the surface.

Toe space: The space left at the bottom of an object, such as a cabinet, to allow room for your toes as you stand directly in front of it.

Toggle switch: A switch with a projecting knob or lever.

Torque: Turning power.

Trimmer stud: The shorter studs to which the ends of the header are fastened.

Veneer: A thin layer of wood, such as veneer tape; also the layers in plywood.

E

F

G

R

S